Encyclopedia of Ion Implantation

Encyclopedia of Ion Implantation

Edited by **Jared Jones**

NY RESEARCH
PRESS

New York

Published by NY Research Press,
23 West, 55th Street, Suite 816,
New York, NY 10019, USA
www.nyresearchpress.com

Encyclopedia of Ion Implantation
Edited by Jared Jones

International Standard Book Number: 978-1-63238-140-8 (Hardback)

Printed in China.

Contents

	Preface	**VII**
Section 1	**Physics of Implantation**	**1**
Chapter 1	**High-Resolution Ion Implantation from keV to MeV** Sébastien Pezzagna and Jan Meijer	**3**
Chapter 2	**Coulomb Heating Behaviour of Fast Light Diclusters in Si<100> Direction** R. C. Fadanelli, M. Behar and J. F. Dias	**25**
Chapter 3	**Radio Frequency Quadrupole Accelerator: A High Energy and High Current Implanter** Yuancun Nie, Yuanrong Lu, Xueqing Yan and Jiaer Chen	**41**
Section 2	**Characterization**	**63**
Chapter 4	**Spectroscopic Ellipsometry of Ion-Implantation-Induced Damage** Denis Shamiryan and Dmitriy V. Likhachev	**65**
Chapter 5	**Lattice Strain Measurements in Hydrogen Implanted Materials for Layer Transfer Processes** Caroline Moulet and Mark. S. Goorsky	**81**
Chapter 6	**Determination of Near-Neighbour Bonding in the Mn-Implanted GaSb Crystals** Anna Wolska	**105**
Section 3	**Materials Applications**	**127**
Chapter 7	**Ion Implantation in Phase Change $Ge_2Sb_2Te_5$ Thin Films for Non Volatile Memory Applications** Stefania Maria Serena Privitera	**129**

Chapter 8 **Si Nanocrystal Arrays Created in SiO_2 Matrix by High-Energy Ion Bombardment** 153
Irina Antonova

Chapter 9 **Neon and Manganese Ion Implantation into AlInN** 183
Abdul Majid

Chapter 10 **Nitridation of GaAs Surface by Low Energy Ion Implantation with *In Situ* Control of Chemical Composition** 217
Valery Mikoushkin

Chapter 11 **Implantation Damage Formation in GaN and ZnO** 237
Katharina Lorenz and Elke Wendler

Section 4 Advances in Processing **265**

Chapter 12 **Optical Waveguides Fabricated by Ion Implantation/Irradiation: A Review** 267
Ovidio Peña-Rodríguez, José Olivares, Mercedes Carrascosa, Ángel García-Cabañes, Antonio Rivera and Fernando Agulló-López

Chapter 13 **Ga^{3+} Focused Ion Beam for Piezo Electric Nano Structuration Fabrication** 315
D. Rémiens, D. Deresmes, D. Troadec and J. Costecalde

Chapter 14 **Annealing Effects on the Particle Formation and the Optical Response** 327
Akira Ueda, Richard R. Mu and Warren E. Collins

Chapter 15 **Ion Implantation for the Fabrication of Plasmonic Nanocomposites: A Brief Review** 355
Umapada Pal and Ovidio Peña Rodríguez

Chapter 16 **Surface Modification by Ion Implantation to Improve the Oxidation Resistance of Materials for High Temperature Technology** 389
Hans-Eberhard Zschau and Michael Schütze

Chapter 17 **Ion Implantation-Induced Layer Splitting of Semiconductors** 417
U. Dadwal, M. Reiche and R. Singh

Permissions

List of Contributors

Preface

This book examines the constantly transforming technology of ion implantation. Ion implantation is an advancing field, owing to the recognition of its several commercial undertakings. Enhancements in equipment, comprehension of beam-solid interplays, applications to novel materials, optimized characterization strategies and more current advancements to employ implantation of nanostructure formation point to novel directions for ion implantation are described in this all-inclusive book.

This book is a comprehensive compilation of works of different researchers from varied parts of the world. It includes valuable experiences of the researchers with the sole objective of providing the readers (learners) with a proper knowledge of the concerned field. This book will be beneficial in evoking inspiration and enhancing the knowledge of the interested readers.

In the end, I would like to extend my heartiest thanks to the authors who worked with great determination on their chapters. I also appreciate the publisher's support in the course of the book. I would also like to deeply acknowledge my family who stood by me as a source of inspiration during the project.

Editor

Section 1

Physics of Implantation

1

High-Resolution Ion Implantation from keV to MeV

Sébastien Pezzagna and Jan Meijer
RUBION, Ruhr-Universität Bochum
Germany

1. Introduction

With the trend in miniaturisation of electronic, optoelectronic or photonic devices, it becomes necessary to control, at the highest level, the number and positioning of dopant atoms which are introduced in the devices by ion implantation. As well, new promising fields of research like quantum information processing in the solid-state also require an exact placement of implanted atoms which could be operated as the quantum bits of a future quantum computer (Ladd, 2010). In these purposes, an ideal ion implantation setup would provide deterministic single ions and nanometer spatial resolution.

In this chapter we will mainly focus on the resolution of ion implantation in the keV to MeV energy range. We will review the different ion implantation setups available at Bochum university and show the precision at which ions can be implanted along the whole energy range, that is from surface to bulk implant. Particular attention will be given to the "nanoimplanter" system with which a lateral resolution of 15 nm has been demonstrated. The nanoimplanter combines a low energy (5 kV) ion beam with the tip of an atomic force microscope (AFM) in which a nano-hole is drilled and used as a movable mask. Furthermore, a technique able to deterministically implant single ions will be presented. A section will also be dedicated to our work oriented towards the creation of defect centres in diamond which can find applications for many purposes: quantum computing, single photon sources, magnetometry, bio-labelling, drug delivery or sub-diffraction optical imaging. This latter application is directly beneficial for the ion implantation community: single defects related to nitrogen in diamond and produced by ion implantation can be imaged with a resolution of 6 nm. Sub-diffraction optical imaging offers a unique way to visualize single implanted ions at a precision approaching the one of the lattice parameter. We will use this technique along the chapter in order to precisely determine the resolution which can be achieved by the different ion implantation techniques. We have therefore implanted nitrogen ions in diamond with each of these techniques at the highest focus possible and used optical nanoscopy to directly image the implantation spots. Although we present here results obtained with nitrogen, the ion accelerators are versatile and the implantation can be extended to any kind of ions at the same resolution.

2. Spatial resolution of ion implantation and its limiting factors

In the process of ion implantation, the spatial resolution is defined as the precision in the final position of an implanted ion in the target material. It can be seen as a volume distribution in which the implanted ions will end up. To some extend, in the case of a single ion, the spatial resolution can be assimilated as a position uncertainty around a most probable position. In this chapter, we will discuss mainly the lateral resolution. The depth resolution (implantation depth and its uncertainty) is directly related to the kinetic energy of the ion and it will be considered to have only intrinsic physical limitations. We assume that for all the implantation techniques presented here, the ion kinetic energy E is well known and that $\Delta E/E$ is negligeable.

The lateral resolution is also intrinsically-limited due to ion-matter interactions. However, the technical limitations related to the focusing capabilities of the implantation setup are in most cases predominant. In this section, the intrinsic limitations will be first briefly reviewed while the techniques to obtain the best focus will be presented in detail and discussed in section 4.

2.1 Ion straggle

Whatever the focusing capabilities of an ion implantation setup, ion straggle represents a physical intrinsic limit of the spatial resolution which can be achieved. For a kinetic ion propagating into a target material, straggling occurs due to the multiple collisions that the ion experiences at the atoms of the material. It results in a broadening of the implantation volume and can be associated to an uncertainty in the final position of each single implanted ion. The straggle is defined as the square-root of the variance of the ion distribution. It is a complex process which depends on several parameters such as the ion kinetic energy, the nature of the target material, the temperature, the crystal orientation, etc.

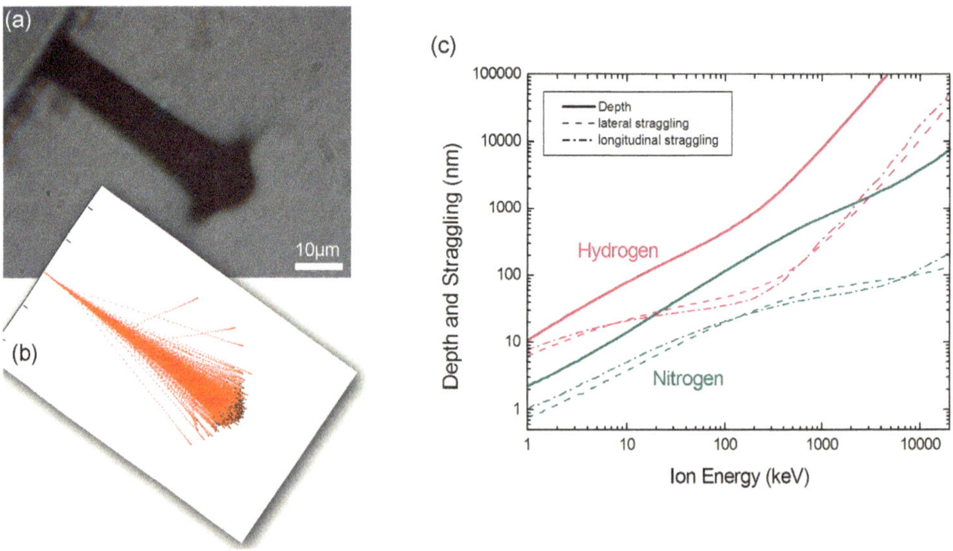

Fig. 1. (a) Optical image (side view) of a diamond sample implanted with a 3 MeV focused proton beam at high fluence. (b) SRIM simulation of 3 MeV protons implanted in diamond. (c) Plot of the ion range and straggle (lateral and longitudinal) simulated with SRIM for hydrogen and nitrogen implanted into diamond.

Figure 1a is an experimental illustration of ion straggle. It shows the side-view of a diamond implanted at high fluence with 3 MeV protons. Defects are created along the whole ion trajectory, mainly vacancies and interstitials. Above a certain ion fluence, the diamond structure gets amorphised, allowing to "see" the implantation-induced defect region. At this energy, most of the protons stop at the average depth of 49 µm, where a lateral broadening associated to straggle can be observed. In this experiment, the ion beam size is about 8 µm and the lateral resolution of the implantation is broadened up to around 10 µm due to straggle. This experiment can be simulated using SRIM (Ziegler, 2008) as shown in figure 1b. The obtained ion range is 48 µm with a lateral straggle of 1.3 µm. Figure 1c is a plot of ion range (depth) and straggle as a function of ion energy in the case of nitrogen or hydrogen ions implanted in diamond, simulated with SRIM. This graph shows that for applications requiring very high precision in the implantation of impurity atoms and aiming at nanometer resolution, a low ion energy has to be used in order to ensure a low ion straggle.

2.2 Ion channeling

Ion channeling is another effect which may be important while implanting ions into crystal substrates. Ions propagating along low-index crystallographic axes and planes penetrate deeper and produce less defects. Channeling leads therefore to a loss of the resolution in the depth direction. For a given crystal axis of the target material, channeling occurs within an acceptance angle around this axis depending on several parameters including ion nature, ion energy and sample temperature (Hobler, 1996). The part of the implanted ions undergoing channeling is hardly predictable. An experimental characterisation of channeling in the case of nitrogen ions implanted in diamond has been shown recently using secondary ion mass spectrometry (Toyli, 2010). The measured depth profiles show that a small part of the implanted ions can penetrate twice deeper as the main implantation depth.

In order to avoid channeling during implantation, the sample has to be tilted away from any channeling direction. In the case of a substrate cut along one of the principal crystallographic planes, an angle of 7° with respect to the ion beam is generally the best compromise. This is used in semiconductor technology with cubic semiconductors like silicon and germanium.

2.3 Other factors

Due to the damage induced by ion implantation, the samples generally need to be annealed in order to restore the crystal lattice. The heating process may lead to diffusion of the implanted species and therefore a broadening of the implanted volume and a loss in the resolution. In the case of diamond, the diffusion of impurity atoms is extremely small (Craven, 2009; Harte, 2009) and, under the annealing conditions that we use (2 hours at 800°C), the implanted nitrogen atoms do not diffuse to a measurable extent. This is an important point since the determination of the ion implantation resolution will be measured in diamond samples that are annealed after being implantated.

In the case of insulating samples, the surface can charge up during ion implantation. This phenomenum may be important especially in the case of low energy ions which can be electrostatically deviated from their initial trajectory. This possible loss of resolution can be avoided by depositing a thin conductive layer on the sample before implantation or by exposing the sample to an electron shower.

3. Optical nanoscopy to measure the spatial resolution

In this section, we describe the optical methods that have been chosen to characterise the resolution of the different ion implantation techniques. We introduce the test "object" of the whole chapter, the nitrogen-vacancy (NV) centre in diamond, a very common defect in natural diamonds, responsible for pinkish colouration. Nowadays, synthetic diamonds with extremely low impurity levels can be grown by chemical vapour deposition (CVD). The concentration of nitrogen and boron (the most common impurity atoms in diamond) can be reduced to a few parts per billion. The availability of such pure substrates enables therefore an unambiguous optical characterisation of artificially-created defects.

Recently, new optical imaging techniques have broken the diffraction limit. Especially, high resolution imaging has been demonstrated using NV centres as point objects to be imaged. One of these techniques called stimulated emission depletion (STED) microscopy will be introduced in this section (Hell, 1994). Using this method, NV centres have been revealed with a lateral resolution of ~ 6 nm (Rittweger, 2009). We propose therefore to use STED microscopy in order to characterise the focusing capabilities of our ion implantation techniques. Indeed, NV centres can be easily and reliably fabricated in diamond by implantation of nitrogen followed by annealing. Typically, annealing 2 hours at 800°C in vacuum allows to restore the crystal lattice and to form the N-V bonding by diffusion of vacancies to the nitrogen atoms.

3.1 Imaging single defects in diamond

Far-field optical microscopy using focused light is an efficient way to spatially map optical centres in diamond. There are hundreds of optical centres known in diamond. Amongst these defect centres, our interest for imaging the focus of ion beams goes to the nitrogen-vacancy centre, for several reasons. The NV centre has strong optical absorption and emission, single centres can therefore be imaged using standard confocal microscopy (Gruber, 1997), it can be easily fabricated, it has a high temperature stability (up to 1500°C) and does not bleach under laser excitation. Figure 2a shows the structure of an NV centre. It consists of a substitutional nitrogen and a carbon vacancy at an adjacent lattice site. It can exist in different charge states. A room-temperature photoluminescence spectrum of a diamond containing NV centres is shown in figure 2b. This spectrum shows the zero phonon line of the negatively charged NV^- centre at 637 nm and a broad vibronic band up to 800 nm. The zero phonon line of the neutral NV^0 is also visible at 575 nm. The inset in figure 2b depicts the energy diagram of the negatively charged NV^-.

To image NV centres in diamond we use a homemade confocal microscope. An excitation laser beam is focused on the sample through a microscope objective down to an effective confocal volume of almost 1 μm^3. The fluorescence light emitted from the excited volume is collected by the same objective and sent to the detection part of the microscope passing through a dichroic mirror (reflecting the excitation wavelength). The fluorescence is detected using an avalanche photodiode after passing through a confocal pinhole to increase the signal-to-noise ratio. A typical saturation curve for a single NV centre (fluorescence intensity vs laser power) is plotted in figure 2d. To produce an image, the sample is scanned using a (x,y,z) piezo-element and the fluorescence is collected point after point (figure 2c). The optical resolution of a standard confocal microscope is limited by diffraction: single defects appear as a characteristic "confocal spot" of ~ 300 nm diameter. The resolution d is given by $d = \lambda / 2$ NA where λ is the optical wavelength and NA is the numerical aperture of the objective.

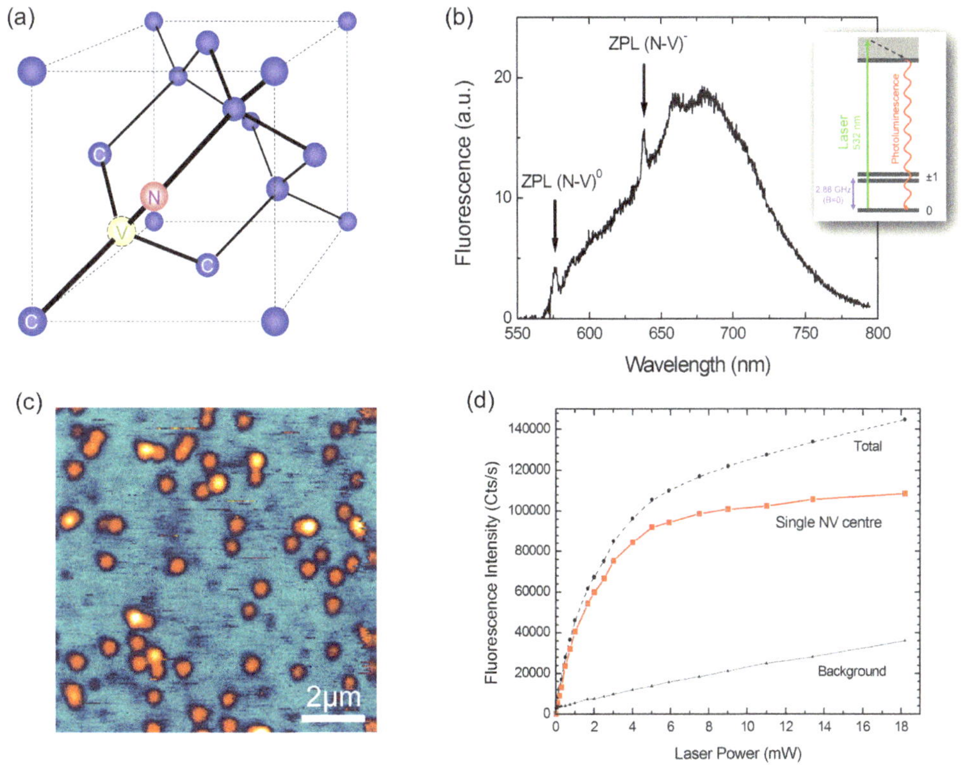

Fig. 2. (a) Structure of a nitrogen-vacancy (NV) centre in diamond, made of a substitutional nitrogen atom and a neighbouring carbon vacancy. (b) Photoluminescence spectrum of an ensemble of NV centres at room temperature with laser excitation at 532 nm (courtesy of B. Naydenov). The inset is a simplified energy level diagram of the negatively charged NV centre (Manson, 2006). (c) Confocal scan of a diamond containing single NV centres. (d) Saturation curve of the fluorescence intensity of a single NV centre as a function of the laser power.

If two or more emitting centres are separated by a distance smaller than the resolution of the microscope, they will not be optically resolved. Some of the ion implantation techniques presented in this chapter promise a beam focus in the range of tens of nanometers or even less. Confocal microscopy will therefore not provide enough optical resolution and sub-diffraction imaging techniques will be required.

3.2 High-resolution optical microscopy

A few new optical techniques have proven to break the diffraction barrier. One of these techniques is particularly adapted to image NV centres in diamond: stimulated emission depletion (STED) microscopy (Hell, 1994; Rittweger, 2009). STED microscopy is a far-field fluorescence microscopy technique which switches off fluorescent objects outside a nano-sized region using the process of stimulated emission. The simplified scheme of a STED microscope is shown in figure 3. Two superimposed laser beams are required. A green excitation beam (532 nm) having a Gaussian intensity distribution and a second beam (775 nm) having a torus intensity profile with a zero in the centre. The second beam is used in order to deplete by stimulated emission the excited level of the fluorescent objects excited by the first beam. Since the depletion beam has a zero of intensity in the centre, depletion occurs outside the centre, above a certain intensity threshold. A fluorescent object at the very centre is therefore not affected by the STED beam and shows normal fluorescence. The resolution d of a STED microscope scales with the square root of the intensity I of the depletion beam and is sensitive to the steepness of the intensity profile around the centre. It is described by:

$$d \approx \frac{\lambda}{2NA\sqrt{1+I/I_S}}$$

where λ is the optical wavelength, NA is the numerical aperture of the objective, and I_S is the characteristic saturation intensity of the NV centres. The setup that has been used for the measurements presented in this chapter is equipped with pulsed excitation and depletion beams (~ 100 ps pulse duration).

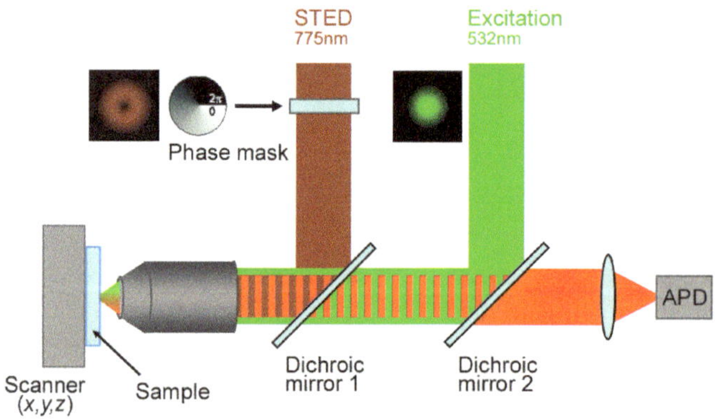

Fig. 3. Simplified scheme of the stimulated emission depletion (STED) microscope used at the Max Planck Institute in Göttingen, Germany (group of S. W. Hell).

They are synchronized to arrive simultaneously on the sample. The fluorescence signal incoming from the reduced nano-sized area is collected after application of the pulses. The images are obtained by scanning the sample mounted on a piezo-stage.

STED microscopy is possible with NV centres because they possess a strong electron-phonon coupling and therefore a large PL sideband (excitation and depletion beam can be spectrally well separated). Moreover, their high photo-stability allows to apply high laser intensities for the depletion beam and therefore to reach high resolutions.

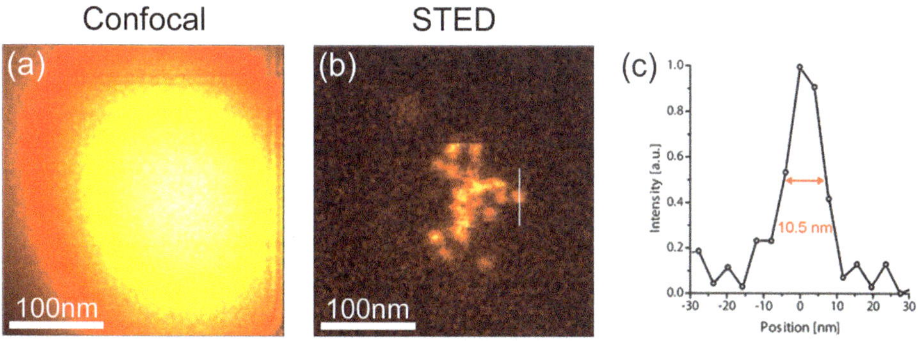

Fig. 4. Optical imaging by far field microscopy of NV centres created in diamond by implantation of 5 keV nitrogen ions through a 120 nm aperture (Pezzagna, 2010b). Optical measurements have been done at the MPI Göttingen, Germany (group of S. W. Hell). Both images are 300x300 nm^2 scans of the diamond surface. (a) Confocal imaging is diffraction-limited to ~ 300 nm. (b) STED breaks the diffraction limit and allows to visualise each optical centre individually. (c) Intensity profile (line in (b)) of a single NV centre imaged by STED microscopy.

An example of the imaging capabilities of STED microscopy in comparison with confocal microscopy can be seen in figure 4. In this purpose, NV centres have been created in diamond by implantation of nitrogen (5 keV, fluence $3x10^{13}$ cm^{-2}) through a nano-aperture of diameter ~ 120 nm. Figure 4a shows the confocal fluorescence image of the diamond surface where the nitrogen ions have been implanted. From the observed bright spot it is impossible to retrieve the size of the aperture through which the NV centres have been created. The number of NV centres present within the spot may be estimated by integrating the intensity and comparing it to the intensity of a single NV centre. Note that for a small number of NV centres, the study of the photon statistics using a "Hanbury Brown Twiss interferometer" can also give access to the number of emitting centres (Jelezko, 2006). However, in both cases, no spatial information can be found. In contrast, the STED microscopy image is shown in figure 4b. Due to the strong increase in optical resolution, almost all NV centres can be imaged separately. This allows to determine precisely the size of the aperture used during the ion implantation. Note that implanting a higher nitrogen ion fluence and therefore obtaining a higher density of NV centre would give an even better image of the 120 nm aperture. Figure 4c shows the intensity profile of a single NV centre corresponding to the line in figure 4b. The optical resolution of the STED microscope is here about 10 nm.

4. Ion implantation setups from keV to MeV

In this section, the different ion implantation techniques that are in use at the university of Bochum are presented. They cover a wide ion energy range from keV to MeV and therefore implantation depths from the very first nanometers to several micrometers or tens of micrometers. For each of these techniques we will show and discuss their capabilities in terms of resolution.

4.1 4 MV Tandem accelerator

Different types of accelerators provide ions with high kinetic energy. Ion implantations with kinetic energies from a few keV up to a few MeV can be produced by electrostatic accelerators. The main parts of an accelerator are the ion source, the generator which provides the acceleration voltage and the mass analyzing magnet. The setup of the ion source depends on the desired ion species and the requested ion current. There are a large number of sources available. The generation of the high voltage can be provided electrically (e.g. using a Crockoff-Walton circuit) or by mechanical charge transport as realized in a van de Graff or in a Pelletron accelerator. In the MeV regime, the electrical solutions provide a higher stability but are of higher complexity. The mass separation is easily achieved by magnetic fields. For low energy beams it is possible to install the mass separation before the acceleration takes place in order to reduce the necessary field strength. Beside these compounds, ion focusing and steering systems are necessary. The properties of an accelerator are defined by its energy resolution, beam size and current for a given ion species. The energy resolution $\Delta E/E$ depends on the stability of the ion source and the high voltage generation. Using modern electronics, $\Delta E/E < 10^5$ are typically available.

The 4 MV Dynamitron Tandem accelerator available at the university of Bochum can provide ions in the energy range 300 keV to around 50 MeV, depending on the ion charge state. Ion currents up to 100 μA can be reached. Elements are available as kinetic ions provided they can be negatively charged. This accelerator is suitable for deep implantation of ions in so-called bulk conditions, generally far from the surface influence. For the implantation of large structures in a short time, the system is equipped with an ion projection setup (not presented here). For the implantation of fine structures or of single ions we use two different techniques: either directly focusing the beam to the sample using a strong magnetic field or implanting through a mica mask containing nano-channels with high aspect ratio.

4.1.1 Focusing with single lens superconductive solenoid

Microbeam systems are under development since the seventies. Most systems are based on quadrupole lenses to focus MeV proton beams for analytical purposes. For a maskless ion implantation, the focusing of heavy ion beams is necessary. Additionally, a fast change of the focusing power is useful. Electrostatic lenses are too weak for this purpose and only magnetic fields meet all the requirements.

A maskless and focused ion implantation below 1 μm can be achieved using a single-lens system based on a superconductive 14 T solenoid without any ferromagnetic core. Such a system has been developed in Bochum. A focusing power of 100 MeV a.m.u. can be achieved which is enough to focus any ion in the energy range from 500 keV to more than 20 MeV. Moreover, a fast and easy change of ion species and/or ion energy can be done within 30 minutes. The chromatic aberration is comparable to the one of quadrupole systems. However, the spherical aberration is about one order of magnitude smaller. The smallest spot size achieved is about 400 nm, which corresponds to the expected theoretical value (as considering the diaphragm size and demagnification factor). However the alignment and the focusing are extremely sensitive and the beam focus can hardly be reduced below 1 μm. An electrostatic octupole placed inside the lens is used in order to scan the beam and to correct the lens aberrations. A scheme of the system can be seen in figure 5a. A standard beam focus can be estimated from figure 5b. It shows an array of spots made of a few NV centres which has been created by the focused implantation of ~ 30 nitrogen ions (energy 11 MeV) per spot. In these conditions, the NV centres are 4.2 μm deep and the expected ion straggle is about 120 nm. It can be seen that the beam focus is in the range of 1-2 μm. Better focuses have nevertheless been achieved for the creation of pairs of close-by NV centres (Meijer, 2005; Neumann, 2010) or of intrinsic TR12 centres (Naydenov, 2009).

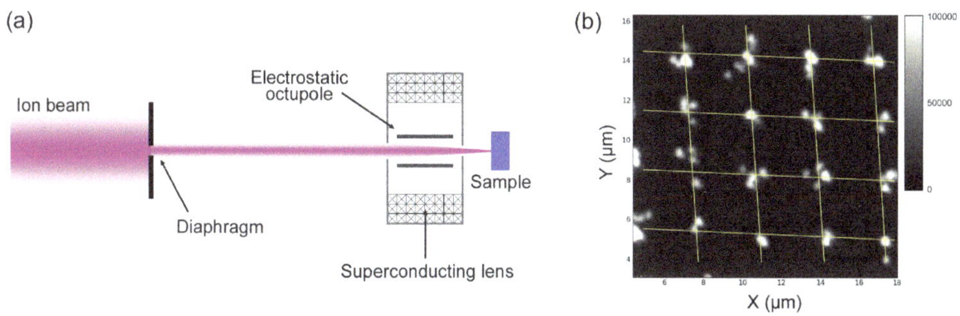

Fig. 5. (a) Scheme of the ion microbeam setup. The ion beam delivered by the accelerator is formed by an aperture and focused with the superconductive solenoid lens on the target. An in-lens electrostatic octupole is used for correction of the lens aberration and for scanning the ion beam (Pezzagna, 2011a). (b) Confocal fluorescence microscopy scan of an ultrapure diamond implanted with 11 MeV nitrogen ions. The focused beam has been scanned in order to produce a pattern of NV centres with a distance of 3 μm. The average number of implanted N per spot is 30. Each bright spot corresponds to a single NV centre.

4.1.2 Collimating through nano-channels in mica

Another possibility to obtain high-resolution implantation using MeV ions is to use a mask. However, fabricating a mask able to collimate an MeV ion beam to a few tens of nanometers is not trivial. It requires small channels inside a thick material to stop energetic ions, and therefore with a high aspect ratio. In what follows, we show how mica can be used in such a way, how nano-channels with an aspect ratio higher than 100 can be fabricated and how ions can succesfully be implanted through the nano-channels. The nano-channels are fabricated in a two step process. First, mica (muscovite) layers of 5 to 30 μm thickness are irradiated by swift heavy ions, samarium with an energy of 1,6 GeV. The energy is high enough so that the ions go through the mica layer while they produce an ion track of amorphous material (a few nm in diameter) across the mica sheet. The amorphous material in the tracks presents a higher etcheability and can therefore be removed chemically in a second step (Khan, 1981). We use an aqueous solution of hydrofluoric acid at 10%. The result of such a process can be seen in figure 6a. Nano-channels with very smooth walls are formed in the mica, all having the same rhombus-shaped opening. The size of the channels depends on the time the layer is left in the acid solution. Channel sizes below 30 nm have been demonstrated in 5 μm thick mica (Pezzagna, 2011b). Figure 6b shows the result of the implantation of nitrogen ions in diamond through such a mica layer with nano-channels. The fluorescence scan of the diamond after annealing reveals the creation of NV centres. Each bright spot corresponds to the fluorescence of one or several NV centres. We have unfortunately not yet the possibility to show high-resolution STED microscopy pictures of the single spots, however, previous measurements indicate that average spot size is approaching the limit of the ion straggle. This result shows an increase in resolution of one order of magnitude in comparison with the focusing using a superconductive lens. Regarding the high aspect ratio of the nano-channels, it is however important to characterise the transmission of ions through the mica layer. We first can indirectly estimate the transmission by measuring the average number of NV centres created by the implantation of N through a single nano-channel. We assume a creation yield of NV centres of 15% (Pezzagna, 2010a). By measuring the second order autocorrelation function of the photon emission statistics coming from each fluorescence spot (Hanbury Brown, 1956), we found out that approximately 4% of the ions have been transmitted. To verify this estimation, we performed a direct measurement of the transmission by shooting 1 MeV nitrogen ions on a mica layer with nano-channels and using an ion detector placed directly after it. The layer has been mounted on an 100 μm diameter aperture which can be tilted with respect to the ion beam. In this experiment, the rate as well as the energy of the transmitted ions have been measured as a function of the tilt angle. At maximum, the transmission is about 1% and an energy broadening of the transmitted ions of about 175 keV is observed. If the mica is slightly tilted (1.7°), the transmission strongly drops down and the energy maximum is shifted down to 650 keV with a FWHM of about 500 keV.

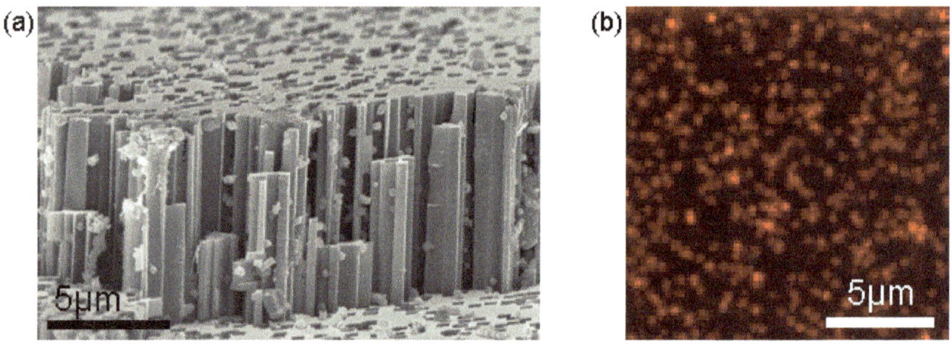

Fig. 6. (a) Mica layer with nano-channels (Pezzagna, 2011b). (b) Confocal fluorescence scan of an ultrapure synthetic diamond implanted with 1 MeV nitrogen ions through the mica layer. Each spot corresponds to one or more nitrogen related optical centres.

Details of the experiment can be found in reference (Pezzagna, 2011b). The energy broadening and the decrease of the energy maximum of the transmitted ions indicate that ions are scattered at the walls of the nano-channels where they lose energy. We attribute this to a misalignment of the ion beam and the nano-channels. Charging up of the mica layers may also contribute to limit the transmission. Further investigations have to be done in order to improve the transmission, however this technique offers an implantation resolution for MeV ions approaching the straggling limit. In the future, single nano-channels will be used to collimate high energy ion beams in order to achieve patterned implantations with high resolution.

4.2 100 kV negative ion accelerator

In the energy range of a few tens of keV or less, it is possible to use resist and electron beam lithography to achieve high-resolution ion implantation. Typically, using PMMA photoresist, one deposits a few hundreds of nm on top of the substrate. This means that the energy of the ion has to be low enough to preclude any transmission through the resist mask.

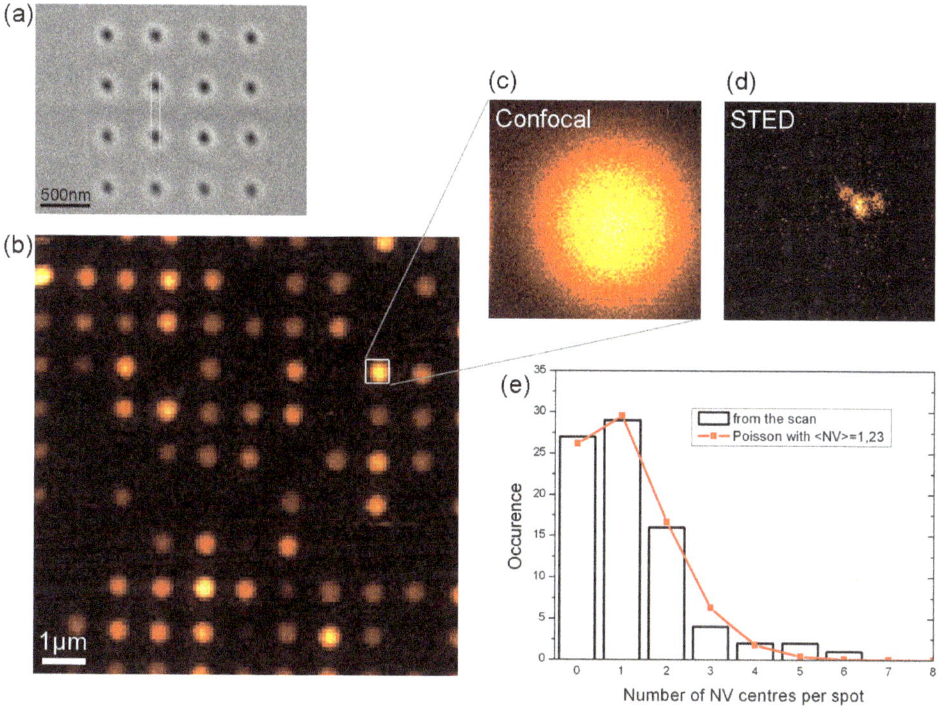

Fig. 7. (a) SEM image of a PMMA photoresist mask patterned by electron beam lithography. The estimated hole diameter is 65 nm and the resist thickness is 230 nm. (b) Confocal fluorescence image of the surface of a diamond implanted with 55 keV CN^- molecules through a PMMA mask. (c) Zoom on the inset in (b). (d) STED image of the same confocal spot, the single NV centres can be resolved. (e) Distribution of the number of NV centres per spot, fitted by a Poisson distribution with an average value of 1,23.

The creation of NV centres in diamond with this technique has already been demonstrated with 80 nm resolution (Spinicelli, 2011; Toyli, 2010). A compromise has to be found between the resist thickness and the smallest details that one wants to achieve into the photoresist. The use of a PMMA mask suits the 100 kV negative ion accelerator that is operated in Bochum. A SEM picture of a PMMA with ~ 65 nm holes made by e-beam lithography is shown in figure 7a. The creation of NV centres by implantation of CN^- molecules is shown in figure 7b. Here also, confocal microscopy cannot provide informations on the spatial resolution. We have therefore used STED microscopy on the brightest spot containing the highest number of NV centres (figure 7c and 7d). The dimension of the spot (although not circular) is about 60-80 nm which is in good agreement with the size of the holes observed in the PMMA layer.

Note that this accelerator enables the implantation of CN^- molecules which can be a unique method to create a pair made of an NV centre and a close by ^{13}C. As it will be seen in the last section, the electron spin associated to an NV^- centre can be coupled with the nuclear spin of a ^{13}C atom (provided both are separated by only a few nm) to build a quantum register operating at room temperature.

4.3 5 kV nanoimplanter

In order to aim at the highest positioning precision of implanted ions, one has to reduce the ion energy to a few keV or less. The 5 kV "nanoimplanter" is dedicated to ion implantation with the smallest straggle possible and therefore targeting nanometer resolution. It combines a low energy ion gun with an atomic force microscope (AFM), the tip of which is pierced with a nano-hole (Meijer, 2008; Pezzagna, 2010b). Focusing such a low energy (a few keV or less) ion beam to nanometer spot size is almost impossible due to chromatic aberrations. The principle of the setup is therefore to use the nano-hole in the AFM tip as a collimator of the ion beam which can moreover be addressed at any place of a substrate. This system offers the possibility for example to previously image with the AFM tip a structure in which ions have to be implanted (a photonic crystal, a planar micro-LED, a quantum dot, etc.) and then to run the implantation process. The setup built in Bochum is equipped with a gas-source ion gun (5 kV, SPECS IQE 12/38) combined with a Wien mass filter (E×B). This provides positive species of any inert gas (He, Ar, Ne, Xe, Kr, N_2) as well as reactive ions by use of O_2 or H_2. However, this source can be replaced by a liquid metal ion source to extend the implantation possibilities to other atoms. The ion currents can be in the range 10 μA to less than 1 pA and the acceleration voltage can be tuned from 5 kV down to 0,2 kV. A supplementary single electrostatic lens is used in order to focus the beam to the sample. The minimum beam diameter obtained at the target sample position can be reduced by using an aperture a few μm large placed before the single lens. The sample holder is mounted on the piezo-electric scanner of the AFM which allows nanometer placement in a range of 60 μm. The AFM scanner itself is mounted on a long-range step-motor table providing μm placement over several centimeters. The ion beam current is measured using a Faraday cup fabricated directly on the sample holder and using a picoamperemeter. In the case of very low currents, an electron multipler (detection efficiency > 90%) allows to detect single ions. In order to align the AFM tip to the ion beam, an in-situ microscope objective (10x) gives a live view of the sample. The AFM system (PI Instrument) works in contact mode and uses piezoresistive cantilevers (Persaud, 2006). The AFM cantilever and tip are made of silicon nitride and the tips are hollow. A view of the system is shown in figure 8a. It is mounted on a vibration-free optical table and an ion pump is used during the ion implantation with the tip in order to avoid vibrations. Figure 8b is a view inside the chamber and figure 8c is a scheme of the setup showing the Faraday cup and the electron multiplier (both not visible in figure 8b) which are used to measure the ion beam current.

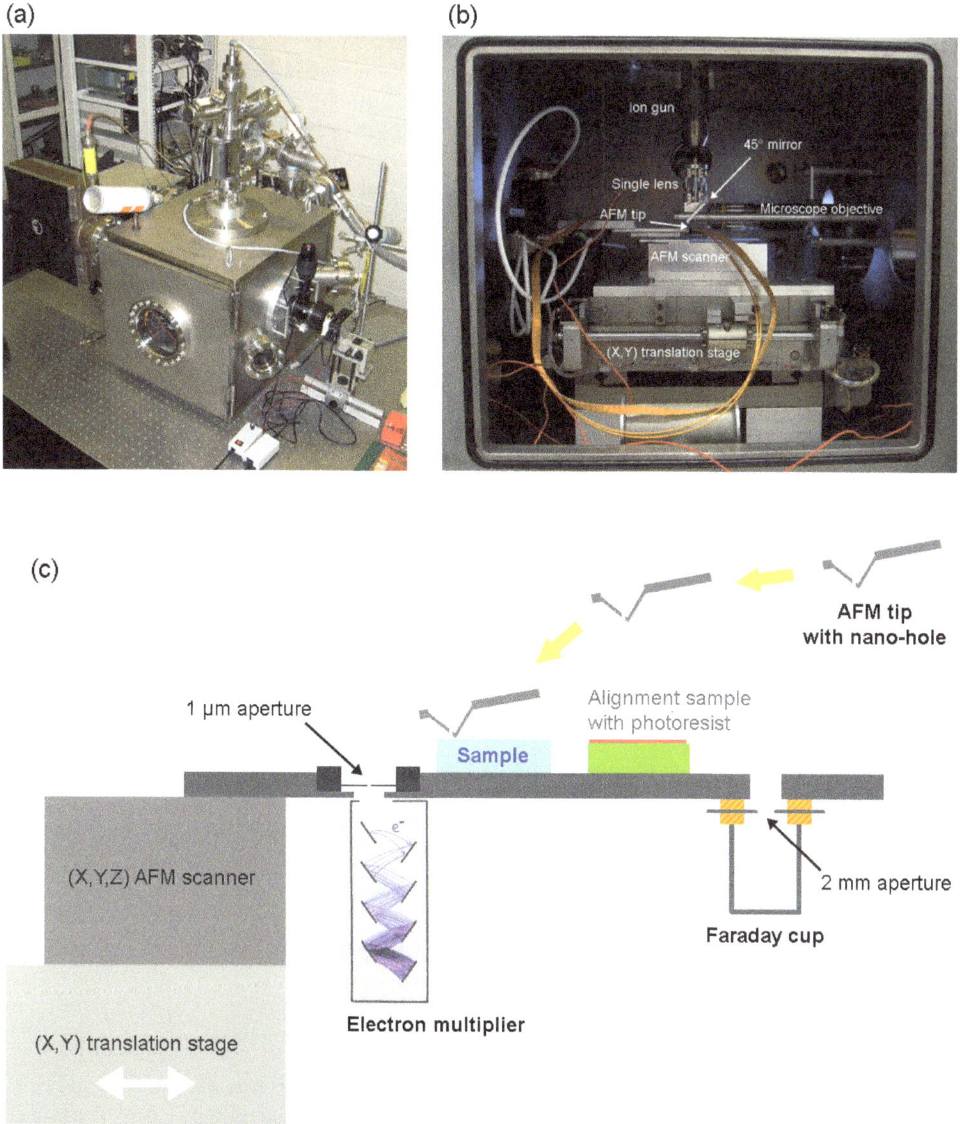

Fig. 8. 5 kV ion implanter combined with an AFM. (a) Picture of the chamber placed on a vibration isolated optical table. (b) View inside the chamber. (c) Scheme of the sample holder. The ion beam is fixed, therefore the sample holder can be moved using the combination of a long range translation stage and the scanner of the AFM. The AFM tip is controlled by another (X,Y,Z) arm.

4.3.1 Drilling the nano-hole by focused ion beam

The nano-holes used to collimate the ion beam are drilled by focused ion beam (FIB) from the backside of the AFM tip (figure 9a and 9b). A thin (a few nm) conductive layer is generally deposited before the FIB process in order to avoid any charging up of the tip which is based on silicon nitride. A Faraday cup may be placed below the AFM tip which has to be drilled so that as soon as a hole is pierced an ion current can be measured and the ion beam stopped. It is generally preferred not to drill the hole directly at the summit of the tip (figure 9c) because it may be damaged or closed when in contact with the substrate to be implanted. The hole shown in figure 9c has a diameter of about 80 nm. We choose the facet presenting the highest incidence angle with respect to the ion beam in order to increase the hole lifetime under ion bombardment. Due to the fact that the pyramidal tip is hollow and thin, the holes can be drilled in a few seconds using 30 keV Ga^+ ions and a 10 pA current. A typical example of collimated implantation through such a hole can be found in figure 4. It will be described in section 4.3.3 how the size of the holes drilled by FIB can be reduced.

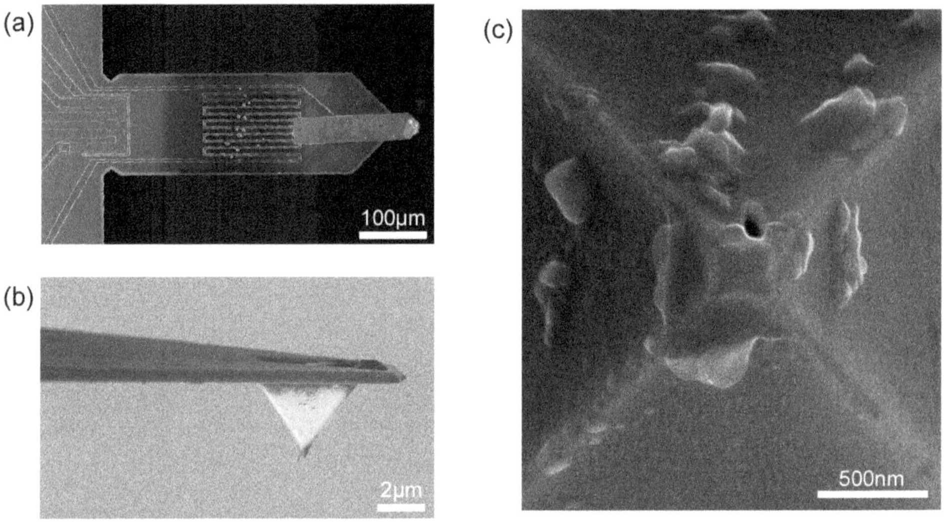

Fig. 9. (a) SEM image of the piezo-resistive cantilever at the end of which the tip is glued. (b) SEM image of the hollow pyramidal AFM tip. (c) SEM image of the summit of the tip where a nano-hole has been drilled by focused ion beam milling.

4.3.2 Alignment between AFM tip and ion beam

We use a piece of silicon covered with PMMA photoresist which is sensitive to the ion beam. The position of the ion beam, as well as its diameter (figure 10a), can be precisely measured in situ using a microscope objective (10x). The AFM tip is then aligned to the imprint of the ion beam (figure 10b) which is confirmed by the shadow of the tip eclipsing a part of the ion beam. In most of the applications, we use a large beam diameter (~ 20 μm) which allows an easy alignment.

If the hole is placed close to the edge of the tip, there will be an implanted halo around the tip. This should however be avoided for some specific applications. In such a case, one can either use a smaller beam focus (however with a more delicate alignment step), or use another type of tip in which the hole is located in the middle of a much larger cantilever thus allowing to use large beams and still ensure excellent alignment.

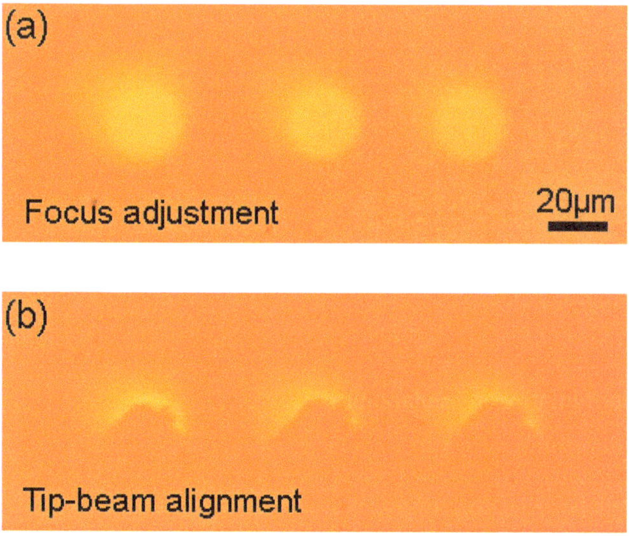

Fig. 10. In situ beam characterisation using PMMA photoresist. (a) Impact of the ion beam on PMMA at different focusing voltages of the single lens for focus adjustment. (b) Alignment of the mobile AFM tip to the fixed ion beam.

4.3.3 Reducing the size of the nano-holes

It has been shown that ion beam irradiation could lead to a reduction in the size of nano-apertures. Li (Li, 2001) named this effect ion beam sculpting. We observe the same phenomenon for the nano-holes in the AFM tips (Pezzagna, 2010b). Therefore, there exists a way to control the closing or even opening of the nano-holes in a more or less limited way. This allows to reach extremely small collimation of the ion beam and to perform ion implantion at very high spatial resolution. Figure 11a shows the creation of NV centres by implantation through a hole, the size of which has been previously reduced from 100 nm to below 20 nm by ion beam sculpting. The STED image reveals an implanted spot (FWHM of 21 nm) in which the single NV centres can no more be resolved (figure 11b). The optical resolution of the STED microscope is here of 10 nm. A critical point for this technique to ensure the collimating of the ion beam is the distance between the nano-hole and the substrate. The use of the AFM tip guaranties an excellent control of this distance. In case the hole is far away from the surface (a few µm), undesired scattered ions may broaden the implantation resolution. The technique using the pierced AFM tip provided to date the highest precision in the creation of NV centres in diamond. It is important to note that using the nano-hole as collimators, any kind of ions can be implanted and not only nitrogen.

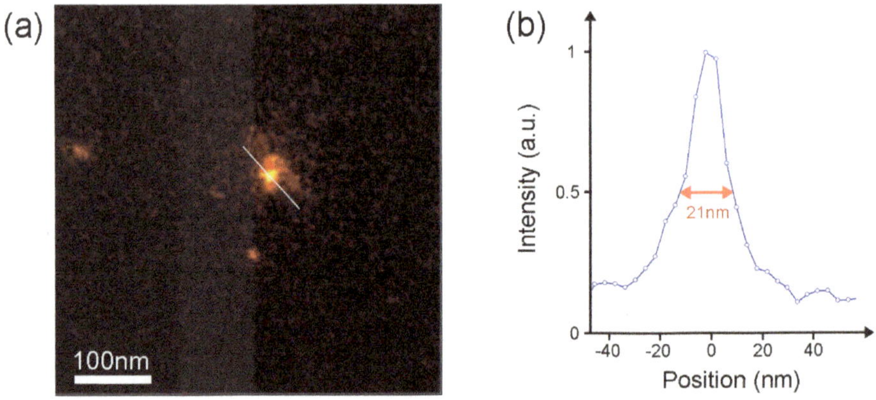

Fig. 11. (a) STED microscopy scan of the spot of NV centres created by implantation of 5 keV nitrogen ions through a reduced nano-hole. (b) Intensity profile corresponding to the line seen in (a). The optical resolution of the STED microscope is here 10 nm.

4.4 Towards deterministic ion implantation

An ideal ion implantation setup would provide nanometer spatial resolution of deterministic single ions. We have just seen that resolution below 20 nm is possible and that furthermore it can be even improved. However only a statistic number of ions can be implanted. Although single implanted ions can be detected in some specific cases where the secondary electrons produced by the impacts in the target are recorded, it would be much more advantageous to directly shoot deterministic single ions from the accelerator. In this view, the most promising method relies on an ion Paul trap out of which single ions can be extracted and further accelerated and focused. Schnitzler et al. show that deterministic single Ca^{+} ions can be shot out of such a trap and detected by an electron multiplier (Schnitzler, 2009). In terms of focusing, nanometer regime can theoretically be reached provided the ions in the trap can be cooled down to their motional ground state. Energies in the keV range should be reacheable as well as extraction rates in the kHz regime.

5. Applications of optical centres in diamond

There are several hundreds of optical centres known in diamond (Zaitsev, 2001) which cover a large spectral range from ultraviolet to near infrared. The most prominent of them are the nitrogen-vacancy NV centre, the silicon-vacancy SiV centre, the nickel-related NE8 centre, or the chromium-related centres. All of these defects in diamond have been observed as single centres and have shown single photon emission under optical excitation. One of the greatest advantages of optical defects in diamond is their high stability and the possibility to create some of them artificially, especially by high-resolution ion implantation, in bulk diamonds but also in nano-diamonds. Diamond-based robust single-photon-sources using NV centres (Kurtsiefer, 2000) are nowadays finding remarkable applications in many fields of physics (applied and fundamental). They have been used for quantum cryptography (Beveratos, 2002; Alléaume, 2004) or single photon interference (Jacques, 2007). Electrically triggered single-photon-sources are however still missing.

In biology and medicine, NV centres imbedded in nanodiamonds are widely used as fluorescent labels (Faklaris, 2008; Fu, 2007). They combine the advantages of high brightness, excellent photostability and non-toxic nature of the nano-diamonds. The continuous imaging of the digestive system and the cellular development of living worms feeded with nano-diamonds has for example been demonstrated (Mohan, 2010).

Another promising application of the NV centre is quantum registers. The electron spin associated to one single NV^- centre can be polarised and read out optically and manipulated using microwave pulses (Jelezko, 2004). Moreover, the coherence time of the electron spin can be as long as 2 ms (Balasubramanian, 2009) in isotopically engineered diamonds (with reduced [^{13}C] concentrations) which has no equivalent in the solid state at room temperature. The achievement of basic quantum information protocols necessitates multipartite entangled states. Several qubits have to be coupled and controlled independently in order to create such quantum states. This can be achieved in diamond at room temperature; quantum registers using the interaction between the electron spin of an NV defect and other neighbouring spins in the diamond matrix (nuclear spins of ^{13}C isotopes or of the nitrogen atom itself) have been demonstrated at 300K (Gurudev Dutt, 2007). A scalable quantum information architecture would however require the coupling between distant NV centres. In the case of ultrapure CVD diamond substrates, the distance should not exceed 20 nm (to keep coupling prevailing over decoherence). A quantum register has also been demonstrated using the dipolar magnetic coupling between the electron spins of two NV centres separated by 9 nm (Neumann, 2010). These NV centres have been produced by the focused implantation of 2 MeV nitrogen ions. To create scalable NV centres further efforts are however necessary: increase the creation yield from N to NV during annealing (Pezzagna, 2010a) and achieve deterministic implantation of single nitrogen ions.

High sensitivity magnetic sensors can also be realized using the spin properties of the NV centres (Balasubramanian, 2008; Degen, 2008; Maze, 2008). In the case that a magnetic field is applied in the vicinity of an NV centre, the ground state sublevels (m_s=±1) are splitted by the Zeeman effect. This splitting can be measured by optically-detected magnetic resonance technique. The frequency splitting between the two resonance lines is directly related to the amplitude of the magnetic field. More sophisticated and sensitive techniques are based on the determination of the magnetic field through the dephasing that it induces in Ramsey-type experiments (Taylor, 2008). Two magnetometry approaches are possible: using a single NV spin attached at a scanning probe or using ensemble of NV centres. The first approach provides high resolution imaging while the second reaches high detection sensitivity.

For future quantum communication, the NV centre is however not the ideal single-photon-source (Aharonovitch, 2011; Pezzagna, 2011a). Both in terms of spectral width and of emission rate, other defect centres have been found (especially the "Chromium-related centres") which provide higher photon rates with narrower emission bandwith. Unfortunately, it has not yet been possible to produce these centres in a reproducible way neither by ion implantation nor during diamond growth. As a result, there is still a large field to explore in the search of an ideal single-photon-source in diamond and the versatility of ion implantation makes it be the technique of choice.

6. Conclusion

Ion implantation is a unique tool which makes it possible to artificially place impurity atoms in any host substrate. New implantation techniques allow for more precisely controlling the position of implanted ions. In this chapter, we have reviewed some of the techniques providing the highest focus for ion beams on a large energy range. To give a relevant comparison, we have implanted nitrogen ions in diamond with each of these techniques in order to produce the so-called nitrogen-vacancy (NV) defect centres. An extremely precise determination of the lateral resolution was made possible by optically imaging the single NV defects produced by implantation and taking advantage of the exceptional imaging capabilities of new sub-diffraction optical microscopy. There is a physical limit to the ion implantation resolution, mainly due to straggle and channeling. For a given ion and target material, this limit depends only on the ion's kinetic energy. Therefore, the implantation resolutions obtained with techniques providing ions in different energy ranges should also be compared with the intrinsic limit at the corresponding energies.

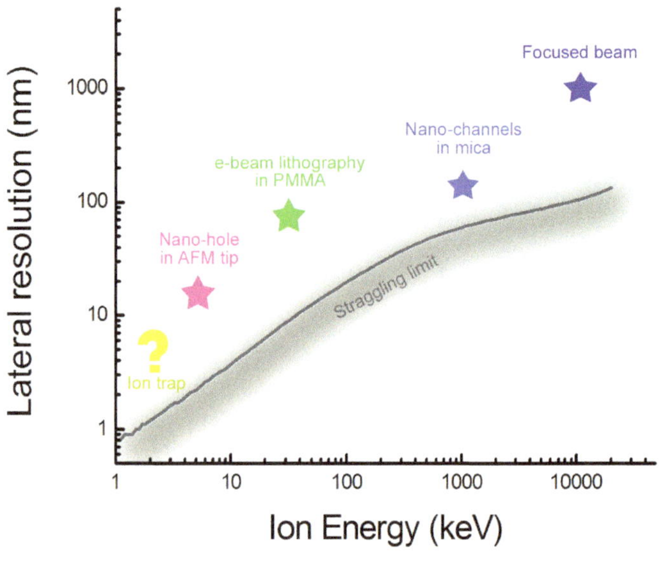

Fig. 12. Lateral resolution vs ion energy plot of the state of the art ion implantation of nitrogen in diamond. The grey line represents the lateral straggle for nitrogen in diamond.

Figure 12 is a plot of the state of the art lateral resolution which has been demonstrated at different ion energies in the case of nitrogen ions implanted in diamond. The lateral straggling is also plotted. The highest implantation precision (15 nm) has been obtained by collimating the ion beam using an AFM tip pierced with a nano-hole (Pezzagna, 2010b). It is possible to even improve this resolution as it has been shown that nano-hole diameters can be reduced using the ion sculpting technique (Li, 2001). Note that the AFM-implanter system provides ions possessing the lowest energy of all the techniques presented here and therefore the lowest ion straggle. In the case of MeV ion beams, the collimation by nano-channels in mica offers a precision approaching the straggle limit (Pezzagna, 2011b). Since this is a very new technique, further research will be necessary, especially to improve the transmission of ions through the nano-channels.

Apart from the placement precision, the control of the number of implanted ions becomes crucial in smaller implanted devices. Promisingly, the method based on an ultracold ion trap (Schnitzler, 2009) presented in section 4.4 has already made possible the deterministic implantation of countable single ions for the first time.

7. Acknowledgment

The authors acknowledge the financial support of the Volkswagen Foundation. The authors acknowledge fruitful collaborations with the groups of Prof. Fedor Jelezko (Ulm University, Germany), Prof. Jörg Wrachtrup (Stuttgart University, Germany), Prof. Stefan W. Hell (Max Planck Institute Göttingen, Germany), Prof. Jean-François Roch (Ecole Normale Supérieure de Cachan, France) and Prof. Alexander Zaitsev (Staten Island University, USA).

8. References

Aharonovich, I.; Castelletto, S.; Simpson, D. A.; Su, C.-H.; Greentree, A.-D. & Prawer, S. (2011). Diamond-based single-photon emitters. *Reports on progress in Physics*, Vol. 74, pp. 076501

Alléaume, R.; Treussart, F.; Messin, G.; Dumeige, Y.; Roch, J.-F.; Beveratos, A.; Brouri-Tualle, R.; Poizat, J.-P. & Grangier, P. (2004). Experimental open-air quantum key distribution with a single-photon source. *New J. Phys.*, Vol. 6, p. 92

Balasubramanian, G.; Chan, I. Y.; Kolesov, R.; Al-Hmoud, M.; Tisler, J.; Shin, C.; Kim, C.; Wojcik, A.; Hemmer, P. R.; Krueger, A.; Hanke, T.; Leitenstorfer, A.; Bratschitsch, R.; Jelezko, F. & Wrachtrup, J. (2008). Nanoscale imaging magnetometry with diamond spins under ambient conditions. *Nature*, Vol. 455, pp. 648-652

Balasubramanian, G.; Neumann, P.; Twitchen, D.; Markham, M.; Kolesov, R.; Mizuochi, N.; Isoya, J.; Achard, J.; Beck, J.; Tissler, J.; Jacques, V.; Hemmer, P. R.; Jelezko, F. & Wrachtrup, J. (2009). Ultralong spin coherence time in isotopically engineered diamond. *Nature Materials*, Vol. 8, pp. 383-387

Beveratos, A.; Brouri, R.; Gacoin, T.; Villing, A.; Poizat, J.-P. & Grangier, P. (2002). Single photon quantum cryptography. *Phys. Rev. Lett.*, Vol. 89, p. 187901

Craven, J. A.; Harte, B.; Fisher, D. & Schulze, D. J. (2009). Diffusion in diamond. I. Carbon isotope mapping of natural diamond. *Mineralogical magazine*, Vol. 73, pp. 193-200

Degen, C. L. (2008). Scanning magnetic field microscope with a diamond single-spin sensor. *Appl. Phys. Lett.*, Vol. 92, p.243111

Faklaris, O.; garrot, D.; Joshi, V.; Druon, F.; Boudou, J.-P., Sauvage, T.; Georges, P.; Curmi, P. A. & Treussart, F. (2008). Detection of single photoluminescent diamond nanoparticles in cells and study of the internalization pathway. *Small*, Vol. 4, pp. 2236-2239

Fu, C.-C.; Lee, H.-Y.; Chen, K.; Lim, T.-S.; Wu, H.-Y.; Lin, P.-K.; Wei, P.-K.; Tsao, P.-H.; Chang, H.-C. & Fann, W. (2007). Characterization and application of single fluorescent nanodiamonds as cellular biomarkers. *Proc. Natl. Acad. Sci. USA*, Vol. 104, pp. 727-732

Gruber, A.; Dräbenstedt, A.; Tietz, C.; Fleury, L.; Wrachtrup, J. & von Borczyskowski, C. (1997). Scanning confocal optical microscopy and magnetic resonance on single defect centres. *Science*, Vol. 276, pp. 2012-2015

Gurudev Dutt, M. V.; Childress, L.; Jiang, L.; Togan, E.; Maze, J.; Jelezko, F.; Zibrov, A. S.; Hemmer, P. R. & Lukin, M. D. (2007). Quantum register based on individual electronic and nuclear spin qubits in diamond. *Science*, Vol. 316, pp. 1312-1314

Hanbury Brown, R. & Twiss, R. Q. (1956). Correlation between photons in 2 coherent beams of light. *Nature*, Vol. 177, pp. 27-29

Harte, B.; Taniguchi, T. & Chakraborty, S. (2009). Diffusion in diamond. II. High-pressure-temperature experiments. *Mineralogical magazine*, Vol. 73, pp. 201-204

Hell, S. W. & Wichmann, J. (1994). Breaking the diffraction resolution limit by stimulated emission: stimulated-emission-depletion fluorescence microscopy. *Optics Letters*, Vol. 19, No.11, (June 1994), pp. 780-782

Hobler, G. (1996). Critical angles and low-energy limits to ion channeling in silicon. *Radiation effects and defects in solids*, Vol. 139, pp. 21-85

Jacques, V.; Wu, E.; Grosshans, F.; Treussart, F.; Grangier, P.; Aspect, A. & Roch, J.-F. (2007). Experimental realization of Wheeler delayed-choice gedanken experiment. *Science*, Vol. 315, pp. 966-968

Jelezko, F.; Gaebel, T.; Popa, I.; Gruber, A. & Wrachtrup, J. (2004). Observation of coherent oscillations in a single electron spin. *Phys. Rev. Lett.*, Vol. 92, p. 076401

Jelezko, F. & Wrachtrup, J. (2006). Single defect centres in diamond: a review. *Phys. Stat. Sol. (a)*, Vol. 203, pp. 3207-3225

Khan, H. A.; Khan, N. A. & Spohr, R. (1981). Scanning electron microscope analysis of etch pits obtained in a muscovite mica track detector by etching in hydrofluoric acid and aqueous solutions of NaOH and KOH. *Nucl. Instr. and Methods*, Vol. 189, pp. 577-581

Kurtsiefer, C.; Mayer, S.; Zarda, P. & Weinfurter, H. (2000). Stable solid-state source of single photons. *Phys. Rev. Lett.*, Vol. 85, pp. 290-293

Ladd, T. D.; Jelezko, F.; Laflamme, R.; Nakamura, Y.; Monroe, C. & O'Brien, J. L. (2010). Quantum computers. *Nature*, Vol. 464, pp. 45-53

Li, J.; Stein, D.; McMullan, C.; Branton, D.; Aziz, M. J. & Golovchenko J. A. (2001). Ion-beam sculpting at nanometre length scales. *Nature*, Vol. 412, pp. 166-169

Manson, N. B.; Harrison, J. P. & Sellars, M. J. (2006). Nitrogen-vacancy center in diamond : model of the electronic structure and associated dynamics. *Phys. Rev. B*, Vol. 74, p. 104303

Maze, J.; Stanwix, P. L.; Hodges, J. S.; Hong, S.; Taylor, J. M.; Cappelaro, P.; Jiang, L.; Gurudev Dutt, M. V.; Togan, E.; Zibrov, A. S.; Yacoby, A.; Walsworth, R. L. & Lukin, M. D. (2008). Nanoscale magnetic sensing with an individual electronic spin in diamond. *Nature*, Vol. 455, pp. 644-647

Meijer, J; Burchard, B.; Domhan, M.; Wittmann, C.; Gaebel, T.; Popa, I.; Jelezko, F. & Wrachtrup, J. (2005). Generation of single color centers by focused nitrogen implantation. *Appl. Phys. Lett.*, Vol.87, pp. 261909

Meijer, J.; Pezzagna, S.; Vogel, T.; Burchard, B.; Bukow, H. H.; Rangelow, I. W.; Sarov, Y.; Wiggers, H.; Plümel, I, Jelezko, F.; Wrachtrup, J.; Schmidt-Kaler, F.; Schnitzler, W.

& Singer, K. (2008). Towards the implanting of ions and positioning of nanoparticles with nm spatial resolution. *Appl. Phys. A*, Vol.91, pp. 567-571

Mohan, N.; Chen, C.-S.; Hsieh, H.-H.; Wu, Y.-C. & Chang, H.-C. (2010). In vivo imaging and toxicity assessments of fluorescent nanodiamonds in Caenorhabditis elegans. *Nano Letters*, Vol. 10, pp. 3692-3699

Naydenov, B.; Kolesov, R.; Batalov, A.; Meijer, J.; Pezzagna, S.; Rogalla, D.; Jelezko, F. & Wrachtrup, J. (2009). Engineering single photon emitters by ion implantation in diamond. *Appl. Phys. Lett.*, Vol. 95, p. 181109

Neumann, P.; Kolesov, R.; Naydenov, B.; Beck, J.; Rempp, F.; Steiner, M.; Jacques, V.; Balasubramanian, G.; Markham, M. L.; Twitchen, D. J.; Pezzagna, S.; Meijer, J.; Twamley, J.; Jelezko, F. & Wrachtrup, J. (2010). Quantum register based on coupled electron spins in a room-temperature solid. *Nature Physics*, Vol.6 , pp. 249-253

Persaud, A.; Ivanova, K.; Sarov, Y.; Ivanov, T.; Volland, B. E.; Rangelow, I. W.; Nikolov, N.; Schenkel, T.; Djakov, V.; Jenkins, D. W. K.; Meijer, J. & Vogel, T. (2006). Micromachined piezoresistive proximal probe with integrated bimorph actuator for aligned single ion implantation. *J. Vac. Sci. Technol. B*, Vol. 24, pp. 3148-3151

Pezzagna, S.; Naydenov, B.; Jelezko, F.; Wrachtrup, J. & Meijer, J. (2010). Creation efficiency of nitrogen-vacancy centres in diamond. *New J. Phys.*, Vol.12, pp. 065017

Pezzagna, S.; Wildanger, D.; Mazarov, P.; Wieck, A. D.; Sarov, Y.; Rangelow, I.; Naydenov, B.; Jelezko, F.; Hell, S. W. & Meijer, J. (2010). Nanoscale engineering and optical addressing of single spins in diamond. *Small*, Vol.6, pp. 2117-2121

Pezzagna, S.; Rogalla, D.; Wildanger, D.; Meijer, J. & Zaitsev, A. (2011). Creation and nature of optical centres in diamond for single-photon emission - overview and critical remarks. *New J. Phys.*, Vol.13, pp. 035024

Pezzagna, S.; Rogalla, D.; Becker, H.-W.; Jakobi, I.; Dolde, F.; Naydenov, B.; Wrachtrup, J.; Jelezko, F.; Trautmann, C. & Meijer, J. (2011). Creation of colour centres in diamond by collimated ion-implantation through nano-channels in mica. *Phys. Stat. Sol. (a)*, Vol. 208, pp. 2017-2022

Rittweger, E.; Han, K. Y.; Irvine S. E.; Eggeling, C & Hell, S. W. (2009). STED microscopy reveals colour centres with nanometric resolution. *Nature Photonics*, Vol.3, pp. 144-147

Schnitzler, W.; Linke, N. M.; Fickler, R.; Meijer, J.; Schmidt-Kaler, F. & Singer, K. (2009). Deterministic ultracold ion source targeting the Heisenberg limit. *Phys. Rev. Lett.*, Vol. 102, p. 070501

Spinicelli, P.; Dréau, A.; Rondin, L.; Silva, F.; Achard, J.; Xavier, S.; Bansropun, S.; Debuisschert, T.; Pezzagna, S.; Meijer, J.; Jacques, V. & Roch, J.-F. (2011). Engineered arrays of nitrogen-vacancy color centers in diamond based on implantation of CN- molecules through nanoapertures. *New J. Phys.*, Vol.13, pp. 025014

Taylor, J. M.; Cappellaro, P.; Childress, L.; Jiang, L.; Budker, D.; Hemmer, P. R.; Yacoby, A.; Walsworth, R. & Lukin, M. D. (2008). High-sensitivity diamond magnetometer with nanoscale resolution. *Nature Physics*, Vol. 4, p. 482

Toyli, D. M.; Weiss, C. D.; Fuchs, G. D.; Schenkel, T. & Awschalom, D. (2010). Chip-scale nanofabrication of single spins and spin arrays in diamond. *Nano Letters*, Vol.10, pp. 3168-3172

Zaitsev, A. M. (2001). Optical properties of diamond. A data handbook. (Springer) ISBN 3-540-66582-X
Ziegler, J. (2008). The stopping range of ions in matter, SRIM-2008, online at http://srim.org

2

Coulomb Heating Behaviour of Fast Light Diclusters in Si<100> Direction

R. C. Fadanelli, M. Behar and J. F. Dias
Universidade Federal do Rio Grande do Sul
Brazil

1. Introduction

Beams of molecules and cluster ions are useful tools in fundamental research with promising applications in materials science and plasma physics. In particular, significant coherence effects (vicinage effects) have been predicted theoretically and in some cases experimental molecular stopping forces show clear deviations from simple additivity rules concerning the projectile constituents (Arista, 2000; Brandt et al., 1974). Other effects related to the correlated motion of cluster atoms, such as Coulomb explosion (Gemmell et al., 1975), enhanced electron emission (Rothard et al., 1990), desorption and sputtering (Matsuo et al., 1997; Yamagushi et al., 1995) have also been reported and reviewed in recent publications (Yamada et al., 2003).

In the case of crystalline materials, ions entering nearly parallel to a particular crystal axis or plane become channelled as their motion is guided by correlated collisions with target atoms. The average transversal momentum of channelled atomic particles increases due to the inelastic scattering with the target electrons and scattering at displaced target atoms. This effect enhances the number of close encounters with the atomic rows and is named transverse heating. It has been observed that charge-changing processes of fast heavy ions may even lead to transversal cooling, namely a reduction of the transversal energy (Assmann et al., 1999).

In addition to the channelling motion, molecular ions undergo a break-up process, since they lose their bonding electrons due to ionisation in the first monolayers of the material. The combination of these two correlated motions, namely the ion channelling and the break-up of the cluster under quasi-Coulomb forces leads to a transverse *Coulomb heating*. Pioneering investigations of this Coulomb heating have been performed using the transmission technique. (Poizat et al., 1972). On the other hand by measuring the energy spectra of backscattered protons from H^+, H_2^+, and H_3^+ beams (Caywood et al., 1971; Tombrello et al., 1973) they were able to deduce the dynamics of the molecular break-up process during channelling effects in Si. But no attempt was done to quantify the Coulomb heating. Only much later the Coulomb heating was at last estimated from dechannelling profiles of hydrogen molecules. However, the computer simulations employed in such study strongly overestimated the experimental results. (Khodyrev et al., 2002).

In this work we review on a quantitative determination of the transverse Coulomb heating energy (abbreviated *Coulomb heating*) for H and H_2, B and B_2, and C and C_2 beams irradiated on the Si<100> direction. To this end, we used the Si-Kα x-ray production and also the backscattering yield (RBS-C) of the mentioned single and molecular beams channelling along the aforementioned direction. Since vicinage effects do not affect the K-shell x-ray and backscattering yields, because the characteristic impact parameters for both processes are much smaller than the typical distance for vicinage effects, differences between experimental results for single and molecular beams are directly linked to the Coulomb explosion during the channelling motion. The experimental results for the Coulomb heating were compared to results of advanced computer simulations being the agreement quite good. These simulations were done in order to understand the physics lying behind the Coulomb heating effects.

In the present chapter we will show the experimental results obtained for H and H_2, B and B_2, and C and C_2 beams. In the case of the last two beams the energy /amu was changed in a wide interval and the obtained Coulomb heating results as a function of the potential energy stored in each ion of the molecule seems to follow on a straight line indicating some kind of "universal behaviour".

2. Experimental procedures

The experiments were carried out at the Ion-Implantation Laboratory of the Federal University of Rio Grande do Sul, Porto Alegre, Brazil. In all cases, monoatomic and cluster beams were irradiated on crystalline silicon targets after the alignment procedure and both backscattered ions and x-ray emissions as a function of the tilt angle were simultaneously detected and recorded.

In what follows, the experimental procedures are described according to the ion/cluster combination used.

2.1 H and H_2 beams

The experiments were carried out at the Ion-Implantation Laboratory of the Federal University of o Grande do Sul. The 500 kV electrostatic accelerator has delivered beams of H^+ and H_2^+, at 150 keV/amu with an average current reduced down to about 100 nA for H^+ and 50 nA for H_2^+ (these currents are typically smaller than the ones used for ion implantation). In order to ensure identical conditions, we kept the same proton flux hitting the target for all particles and clusters under study. Furthermore, the currents were reduced by an additional factor of 10 in the case of backscattering measurements. The Si (100) crystal was mounted on a goniometer for the alignment procedure (Azevedo et al., 2002). We have fixed the azimuthal angle Φ = 22.5° (relative to the plane [100]) in order to avoid a full azimuthal averaging procedure around the <100> channel, which would be time consuming and could lead to unacceptable damage of the sample.

The backscattered particles were detected by a surface-barrier detector located at 165° with respect to the incident beam. The overall resolution of the detector plus the electronic system was for protons better than 7 keV. The 1.74 keV Si Kα x-rays emitted from the target were detected by Si(Li) detector with an energy resolution of 180 keV at 5.9 keV. This detector was mounted at 135° with respect to the incident beam. This constrained the measurements only to directions around the Si <100> one. Once the center of the channel was determined, a sweep from -5° to +5° was performed in steps of 0.2° around the center of the channel. At each stage the corresponding RBS of the incident beam and the yield of the x-ray produced by the Si target were recorded.

The vacuum system of the analysing chamber consists of an oil-free turbomolecular drag pump with a liquid nitrogen trap, capable of reaching a final pressure below 10^{-7} mbar. It is important to point out that the samples were cleaned and etched using 10% HF before each measurement to remove the native surface oxide, in order to ensure a better channelling procedure.

2.2 C, C_2 and B, B_2 beams

In these cases the incident beams were provided by the 3 MV Tandetron of the Ion-Implantation Laboratory of the Universidade Federal do Rio Grande do Sul. The experimental procedure was basically the one that was described in section 2.1 with two major differences: first, the samples were mounted on a four axis goniometer and the channelling procedure was performed with He beams of 1.2 MeV following the same procedure as described above; second, subsequently the position of the sample was changed by using the vertical axis of the goniometer in order to get a fresh point not previously damaged by the alpha beam. Following we have changed to a C^+ (or B^+) beam and performed the sweep around the channel <100> and recording at each 0.2° the corresponding RBS and x-ray yield following the procedure described above. Once this procedure was finished we have changed to a C_2^+ (or B_2^+) beam. On this occasion we moved again the sample to another fresh point of the sample free of damage and we have repeated the above-described procedure.

As can be observed from the mentioned description an extreme care has been taken in order to avoid or minimize the damage induce by the incoming beam, during the scanning procedure. These procedures have to be taken because the x-ray emission is very sensitive to the damage or amorphization produced in the channel.

In the present case the overall resolution of the detector and electronic system was around 30 keV for C and 25 keV for B ions (FWHM) respectively.

It should be mentioned that for C, C_2 and B, B_2 the energy interval spanned in the present experiment went from 900 keV/atom up to 2200 keV/atom, for B and from 800 to 2200 keV/atom for C respectively.

For each beam energy, the x-ray yields as well the RBS-C spectra were obtained as an average from five up to six (depending on the RBS-C cross section) independent measurements performed under the same condition of ion flux. The uncertainties were calculated taking into account the independent measurements as well as the fitting to the spectra (to be described in Section 3).

3. Experimental results

3.1 H and H_2 data treatment and results

Measurements of close encounter events giving raise to x-ray emission and backscattering under channelling conditions provide a sensitive method to investigate variations and redistribution of the ion flux close to a row of atoms and can be used to quantify the heating process inside crystal channels. The results for the x-ray emission induced by H and H_2 beams at 150 keV /atom are depicted in Fig. 1 (Fadanelli et al., 2004) as a function of the projectile entrance angle. Here, large tilt angles correspond to nearly random direction. As can be observed from the figure, atomic and molecular beams at the same energy, flux and fluence per atom act differently in the Si induced x-ray production.

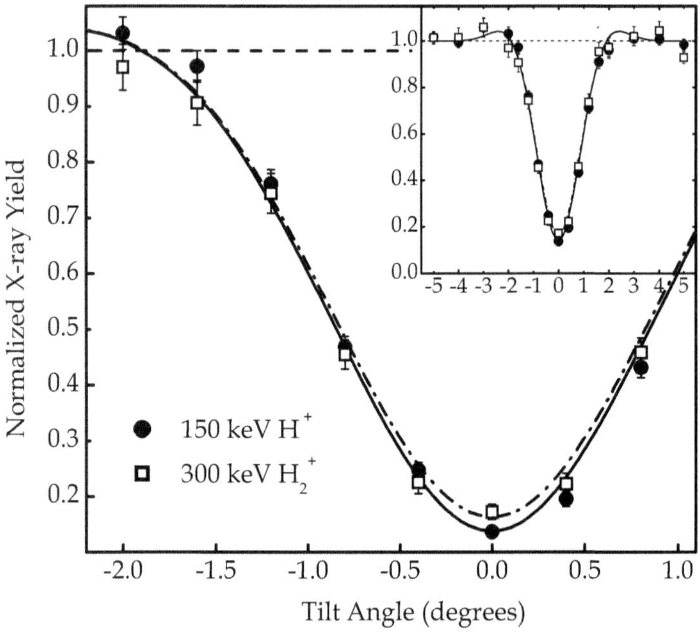

Fig. 1. Si x-ray emission as a function of the tilt angle induced by H and H_2 beams. In the inset is shown the results of the whole sweep obtained between -5° and +5° (Fadanelli et al., 2004).

First it should be emphasized that we do not observe any molecular effects at random direction (within 2% of difference), feature that indicates that no interference molecular effects, like the vicinage one, affect the results of the present experiments. On the other hand there are significant differences at small polar angles. In comparison to H, the yield for H_2 is 26% larger at the central channel direction.

In order to obtain quantitative results from the experiments for the Coulomb heating, it is necessary to find the dechanelling rate as a function of the tilt angle $\alpha(\psi)$. Since the Coulomb explosion time is about one order of magnitude smaller than the oscillation period time of the ion inside the channel, the Coulomb heating can act as an "astigmatic lens" for the channelled ions (Sigmund, 1992). In order to extract the experimental values for the Coulomb heating we have used the two-beam model (Götz et al., 1988; Lindhard, 1965), which have been successfully used in previous works (dos Santos et al., 1997; Fadanelli et al., 2004). The two-beam model is, in fact, a particular case of the discontinuous model used to solve the master equation for the transverse-energy distribution (Götz et al., 1988). In this model, the beam is considered as a sum of two beam fractions distinguished by their transverse energy range: the fraction of ions below a critical value is called "channelled". The remained fraction is called "dechannelled". Under the restriction of a nearly perfect crystal we have that the dechannelled fraction $\chi(z)$ is given by:

$$\frac{d\chi}{dz} = \alpha\left[1 - \chi(z)\right] \tag{1}$$

where z is the depth inside the channel and $[1- \chi(z)]$ is the channelled fraction which as well as $\chi(z)$ are both function of the penetration depth z. The parameter α is a weak function of the energy, so in agreement with previous works we have taking as independent of the energy in the studied depth range of 400 nm. We have calculated that in this zero approximation the final error in the Coulomb heating results is less than 5%. The α value was determinate for each incident angle ψ from the fitting of the RBS-C spectra. For this analysis we have used the electronic stopping to convert the energy loss in depth z as well as the Rutherford cross section to determine the backscattering yield (Chu et al., 1978) and each spectrum was convoluted with the experimental resolution and energy loss straggling. To describe the surface peak, we have considered a thin random layer in the surface.

In order to determine the H_2 corresponding α parameters the RBS-C fittings were obtained following the same procedure as described above. In Figure 2 are shown two well

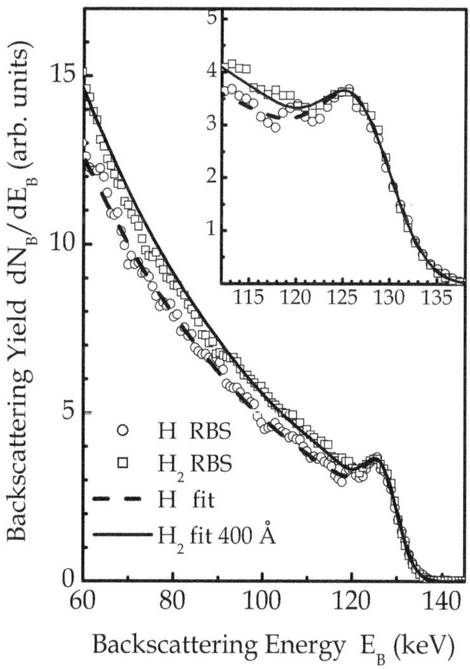

Fig. 2. RBS yield as a function of the energy. Curves stand for the fitting procedure according to the two-beam model assuming that the Coulomb explosion effectively takes place at a critical depth of 400 Å (Fadanelli et al., 2004).

channelled H and H_2 RBS-C spectra (Fadanelli et al., 2004). It can be clearly seen that the molecular RBS-C spectrum shows an increasingly larger dechannelling (for lower backscattering energies) than the corresponding to the H one. In this kind of experiments the dechannelling due to the Coulomb explosion seen in RBS spectra is more striking than the one observed in the x-ray experiments. This happens because as will be described below there is a delay in the Coulomb explosion effect. Consequently, the dechannelling effect is also a clear signature of the Coulomb heating effect.

At this point it should be mentioned that the Coulomb heating effect does not occur immediately after the molecule enters into the sample. In fact the molecular break-up begins soon after the molecular ions penetrate a few monolayers in the target. However the Coulomb explosion usually takes several hundreds of angstroms of ion penetration before changing the ion flux and thus the transverse energy distribution. The fitting of the RBS-C spectrum of the H_2 molecule, after few angstroms does not reproduce the experimental results. Although the flux change is a continuous process, for simplicity we will assume in the fittings that before a certain distance Δz the molecular beam flux is the same as the monoatomic one and that all changes in flux due to the explosion are complete after a certain Δz. Consequently, the obtained Δz values are a first-order estimation of the heating delay.

In this sense in order to obtain a good fit we have assumed that Δz is different from zero. However, we have obtained several possible ($\alpha(\psi)$, Δz) yielding a reasonable fit to the molecular spectra. Then, we fitted the x-rays integrated results as well (not shown in Fig. 1) to ensure that a correct pair is chosen. The two-beam model function was modified to take into account the K-shell ionisation cross sections. Finally a unique pair ($\alpha(\psi)$, Δz) that fits simultaneously the x-ray yields and the RBS-C spectra was found. In the present case the obtained Δz value was 400 angstroms.

When the $\alpha(\psi)$ is known, it could be found the Coulomb heating energy through the Hamiltonian of the transverse ion motion which, for small ψ angles, reads

$$E_T = E\psi^2 + U(x,y) \tag{2}$$

where E_T is the transverse ion energy, E is the ion initial energy and $U(x,y)$ is the channelling potential. Starting from Eq. (2) and taking into account the mentioned features of the cluster ion fragmentation, the Coulomb heating energy is finally given by

$$Y_{molecular}(\psi) = Y_{atomic}\left(\sqrt{\psi^2 + \Delta E_c / E}\right) \tag{3}$$

where $Y_{molecular}$ is the x-ray or RBS cluster ion yield, Y_{atomic} is the monoatomic ion corresponding one, and ΔE_c is the Coulomb heating energy. Therefore, in order to determine ΔE_c, it is necessary to determine $\alpha_{molecular}$ in the center of the channel and the value $\Delta\psi$ that leads to $\alpha_{atomic}(\Delta\psi) = \alpha_{molecular}(0)$, as shown in Fig. 3. Therefore, from Eq. (3), we have $\Delta\psi = \sqrt{\Delta E_c / E}$.

Then the experimental Coulomb heating of H_2^+ at 150 keV/amu obtained from the present experiment is 2.6 ± 0.6 eV. This value is significantly smaller than the estimated free

Coulomb heating (4.5 eV), indicating either a target screening or a molecule alignment inside the target.

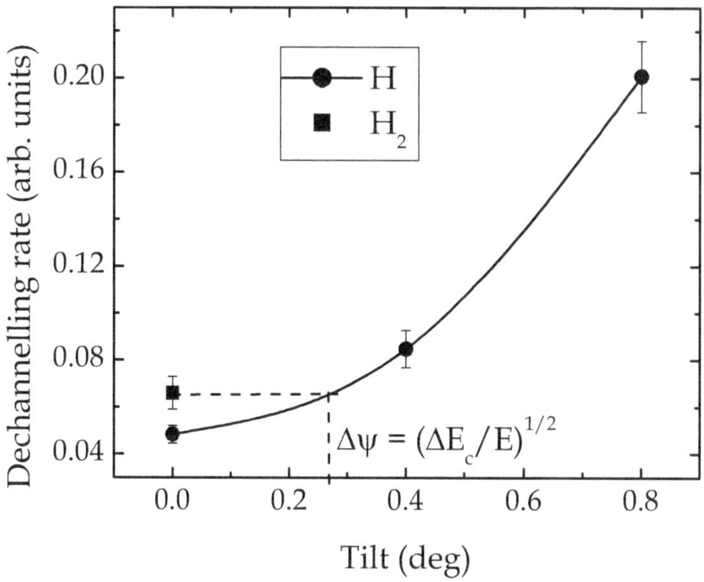

Fig. 3. Dechannelling rate as a function of the tilt angle obtained via TBM fittings of RBS spectra. The depth delay Δz used in this example is 400 Å.

3.2 Heavier ions data treatment and results

In what follows, only a brief summary of B_2^+ and C_2^+ results are shown. The data treatment is essentially identical to the one used for H_2^+. The main difference, besides the ion type, is the different range of energies used, instead of one single ion energy as done for H_2^+. Again, we do not observe any molecular effects at random direction (within 2% of difference) in RBS or x ray measurements.

In the aligned direction, however, it is possible to observe an increase of the x ray, as shown in Figure 4, and RBS yields, as depicted in Figure 5, (Fadanelli et al., 2008) for the diatomic ions. This is the signature of Coulomb heating for the heavier diatomic projectiles. For each dicluster it is possible to see an increase of the Coulomb heating effect as a function of energy. This behaviour indicates an important role played by the screening inside the target.

In Figure 5 it is shown both the raw experimental results for RBS yields (symbols) and the fitting procedures not taking into account the depth delay for the explosion (dashed line) and considering the delay (full line). The beam is aligned with the channel <100>. As in case of H^+ and H_2^+, to describe the surface peak, we have considered a thin random layer in the surface (about 20 Å). The importance of the delay is emphasized by comparing the dashed and the full lines.

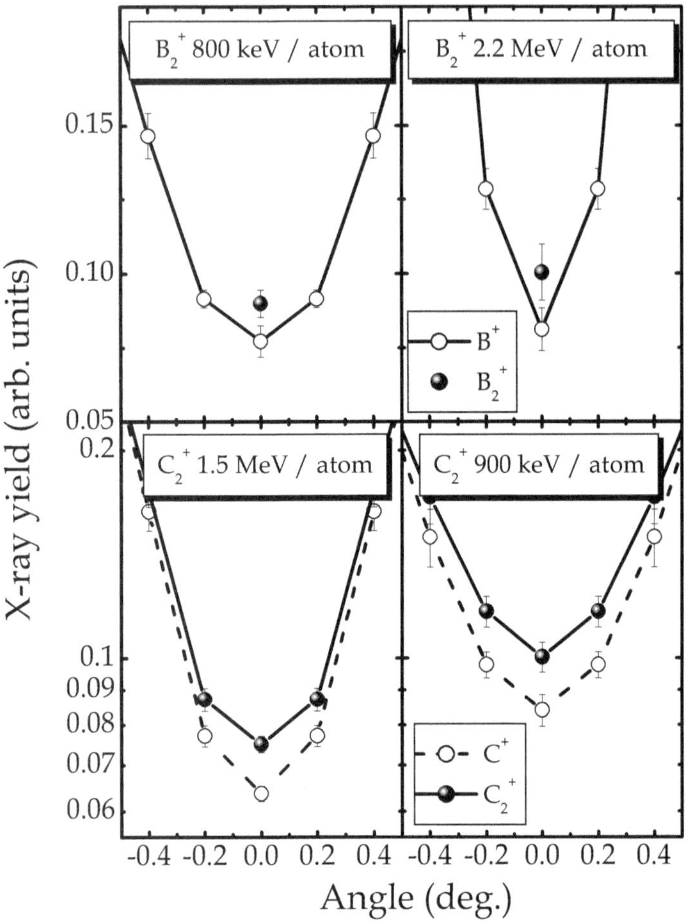

Fig. 4. Si x-ray yields near the center of the Si(100) direction. It is clear that the Coulomb heating signature for B_2^+ and C_2^+ increases with the energy and the atomic number of the molecule constituints (Fadanelli et al., 2008).

After performing essentially the same data treatment procedure as used for H_2^+, we show in Figure 6 the Coulomb heating results for B_2^+ and C_2^+ (Fadanelli et al., 2008; Fadanelli et al., 2010). As can be observed, in both cases the Coulomb heating is a function of the corresponding beam energy. For B_2 the Coulomb heating raises from 9 eV up to 17 eV. Instead for C_2 it increases from 14 up to 30 eV. At the same time the Coulomb delay explosion is also is a function of the beam energy and ranges from 1200 Å up to 2500 Å. Details about the simulation results will be given in Section 4.

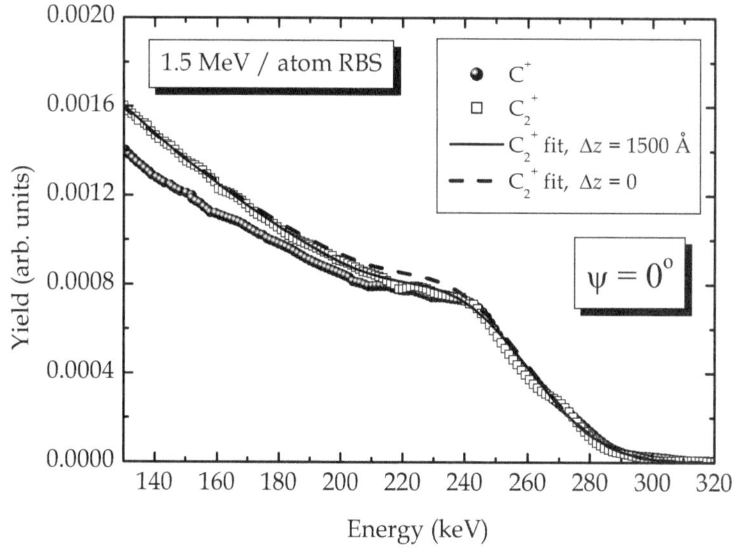

Fig. 5. C^+ and C_2^+ RBS spectra at 1500 keV / atom, taken under alignment conditions (Fadanelli et al., 2008).

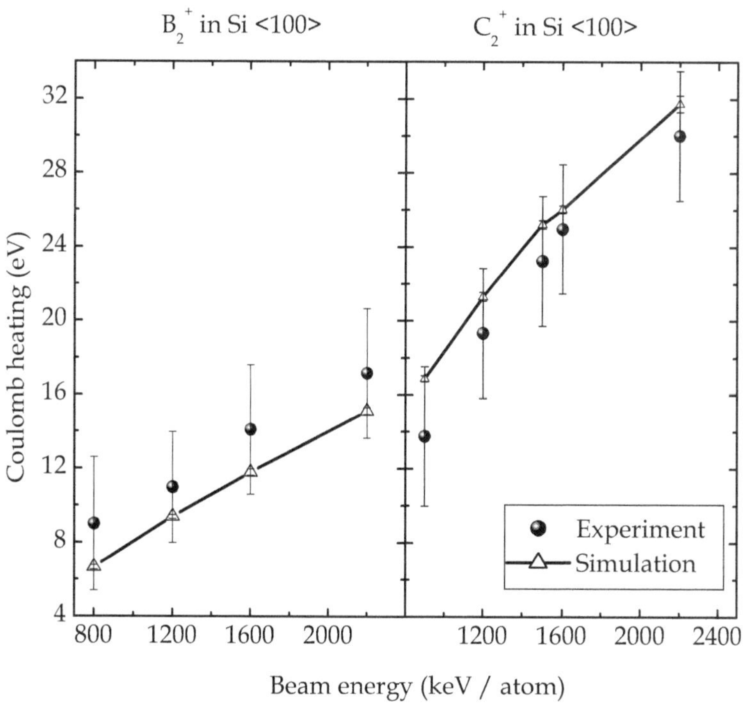

Fig. 6. Coulomb heating results for B_2^+ and C_2^+ (Fadanelli et al., 2008, 2010).

4. Simulations

Classical Monte Carlo simulations were performed for all corresponding experimental cases. To this end, we have modelled the interaction between the ions and the Si<100> target atoms using Molière continuum potential (Lindhard, 1965 & Gemmell, 1974) generated by sixteen strings, including a correction to take into account the thermal vibrations (Eckstein, 1991). The ions were allowed to move inside the channel defined by the four central strings. The interactions among the molecule fragments were modelled using either Yukawa potential (for H_2^+) or, if the number of electrons bound to the ion is large, static Dirac-Hartree-Fock-Slater potentials (Salvat et al., 1987).

In order to include a non-uniform energy loss, it was used the unitary convolution approximation (UCA) (Azevedo et al., 2000; Schiwietz et al., 1999). UCA results consist in energy loss values as a function of the distance between a given ion and the channel strings. The angle between the molecular axis and the motion direction, as well as the initial position of the projectile impinging the channel are chosen randomly. The ion charge state is chosen randomly from a gaussian distribution centred in the equilibrium charge value Z_{eq} with known width Δq (Schiwietz et al., 2001). Due to the relatively small amount of experimental information about B and C equilibrium charge states under channelling conditions in our energy range, we have selected the interpolation formulae of Ref. (Schiwietz et al, 2001), which corresponds to the ones obtained for amorphous Si. This choice seems reasonable, since in Ref. (Jiang et al., 1999a, 1999b) have investigated equilibrium charge states for different ions impinging on Si films under channelling and random conditions. In particular Jiang group found that for N at energies similar to the present ones or higher, the equilibrium charge states were achieved and are independent on the channelling or random conditions of the target film. The charge state is, finally, kept constant along all considered ion path and the Molière potential screening parameter for a given ion path is evaluated accordingly. It is important to point out that, in this way, charge changing effects along each ion motion were not taken into account in our simulations.

In this procedure we assume that some of the electrons from the impinging B_2^+ and C_2^+ molecules are suddenly removed along the first monolayers and, consequently, the molecule undergoes a break-up process. We assume that each ion reaches its own equilibrium charge state as soon as the break-up process begins. In this way, our simplified simulation does not take into account a more detailed molecular fragmentation description. In fact, orientation, velocity and position-dependent effects on the molecule break-up (as, for instance, wake effects (Arista, 2000)) not included here are expected to decrease the obtained theoretical values for the Coulomb heating.

In order to study the explosion effects on the ion flux, the simulations were carried out only along the first 4000 Å. There are two reasons for this depth choice. First, at 4000 Å, it is safe to assume that the potential energy stored in the molecule is nearly completely transferred to the molecular fragments motion. Thus, no further increase on the transverse energy can be expected from the Coulomb explosion after that depth. Second, a significant amount of ions is still channelled. Therefore, the ion flux inside the channel is still concentrated near the channel center, originating the so-called flux peaking.

Fig. 7 shows, as an example, the ion flux for both C and C_2 beams (Fadanelli et al., 2008) averaged along the depth (4000 Å). The ion flux for the C^+ beam displays the well-known flux peaking, i.e. a strong enhancement of the flux near the center of the channel. The C_2^+ beam still shows the flux peaking effect. However, as presented in Figure 7, this effect is considerably damped for the molecular beam due to the Coulomb explosion, which enhances the transverse energy of the molecule fragments and, consequently, produces higher ion fluxes far from the channel center.

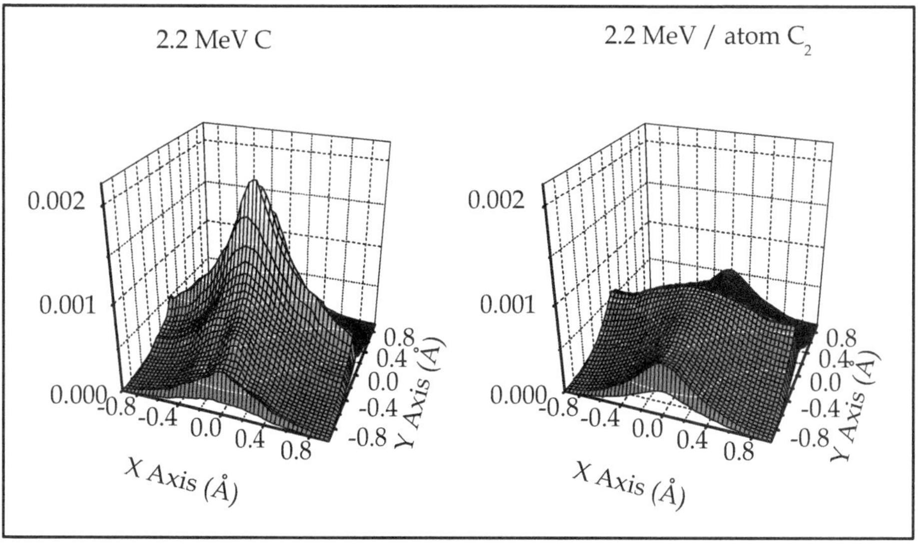

Fig. 7. C and C_2 fluxes inside the <100> Si channel obtained via simulation of aligned 2.2 MeV / atom C and C_2 beams (Fadanelli et al., 2008).

In order to evaluate the average transverse explosion energy transfer to the molecule fragments, we considered, for each beam energy, the distributions of transversal energies at 4000 Å depth in the target, far after the Coulomb explosion process. Two of these distributions, as obtained from our simulations (Fadanelli et al., 2004, 2008), are shown in figure 8 which corresponds to 150 keV / atom H^+ and H_2^+ (left panel) and 2.2 MeV/atom C^+ and C_2^+ beams (right panel). As can be observed for the same yield, the transversal energy distributions for the molecular ions are shifted to higher energies, a consequence of the Coulomb-heating process. The theoretical values of the Coulomb heating were then obtained by shifting (dashed curve) the atomic distribution in Fig. 8 to agree with the molecular one at higher transversal energies. This is because only higher transverse energies ions have high probability of undergo close encounter events that yields x-ray emission and backscattering. The same procedure was carried on for B^+ and B_2^+ (not shown).

By using the described simulation procedure, we have found the following values for Coulomb heating: 2.5 ± 0.2 eV for H_2^+ and, for the heavier molecules, the results are shown in Figure 6, where it is possible to compare with the experimental results. An observation of figure 6 shows that the theoretical curve follows the tendency of the experimental results, being the agreement, within the experimental errors, quite reasonable.

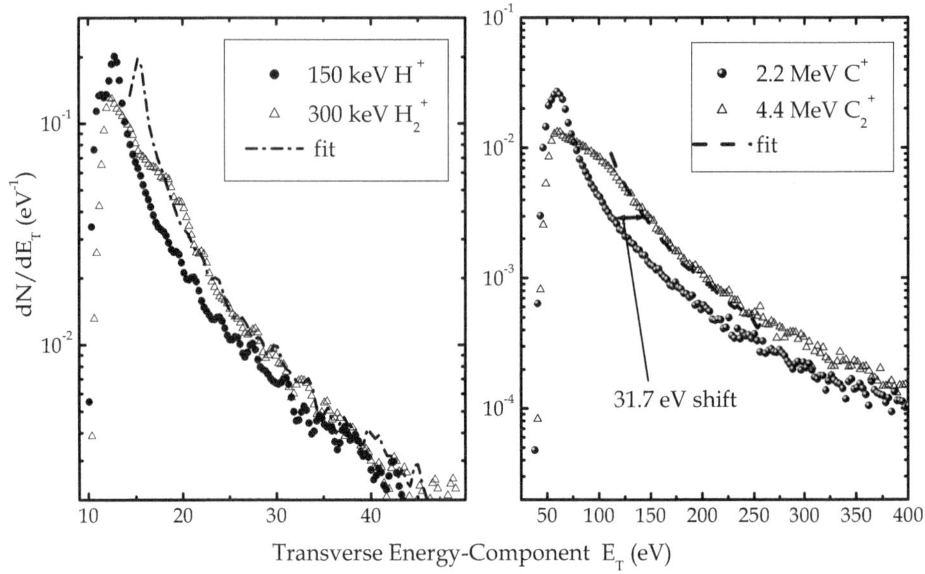

Fig. 8. Simulated transversal-energy distribution for 150 keV/atom H and H_2 (left) and 2.2 MeV/atom C and C_2 beams (right). Energy shifts are necessary in order to match the molecular and atomic beams distributions at transverse energies above 25 eV (H_2) and 150 eV (C_2) (Fadanelli et al., 2004, 2008).

5. Discussions

Since both experimental and simulated results indicate an increase of the Coulomb heating values as a function of the energy, an attempt was made in order to check whether a more general behaviour could be drawn from these data. Then, an approach was devised: instead of plotting the Coulomb heating results as a function of the ion kind and energy, we calculated the values of the potential energy per ion of the molecule in order to compare to the Coulomb heating results. To this end, we have used the DHFS calculations interpolated to the mean charge value of the ions. In Fig. 9, (Fadanelli et al., 2010) the Coulomb heating values, obtained from all the described experiments are plotted together as a function of the stored potential energy. All results lay in a single straight line, indicating some kind of "universal behaviour" where the Coulomb heating produced by the break-up of the molecules is a linear function of their potential energy per ion. In addition, the average Coulomb heating is given by:

$$\frac{1}{4\pi}\int_{\Omega}\left(U_0 \operatorname{sen}^2\theta\right)\operatorname{sen}\theta\, d\theta\, d\phi = \frac{2U_0}{3} \qquad (4)$$

where U_0 is the stored potential energy per ion and θ is the angle between the molecule axis and the initial molecule velocity.

Thus, assuming an isotropic Coulomb explosion, a plot of the Coulomb heating as a function of U_0 should be well fitted by a straight line bearing a slope of approximately 0.67. The fitted line in Fig. 9 has a slope of 0.60 ± 0.02, which is compatible to this prediction, namely 0.67. The difference could be attributed to second-order effects, e.g. the wake potential (Arista, 2000), not taken into account in our simulations. Therefore, the slope of the straight line of Fig. 9 indicate that two third of the Coulomb explosion energy produces the Coulomb heating. The remaining one-third gives place to the longitudinal explosion of the molecules and therefore do not contribute to the Coulomb heating, contributing to the beam straggling instead (Fadanelli et al., 2006).

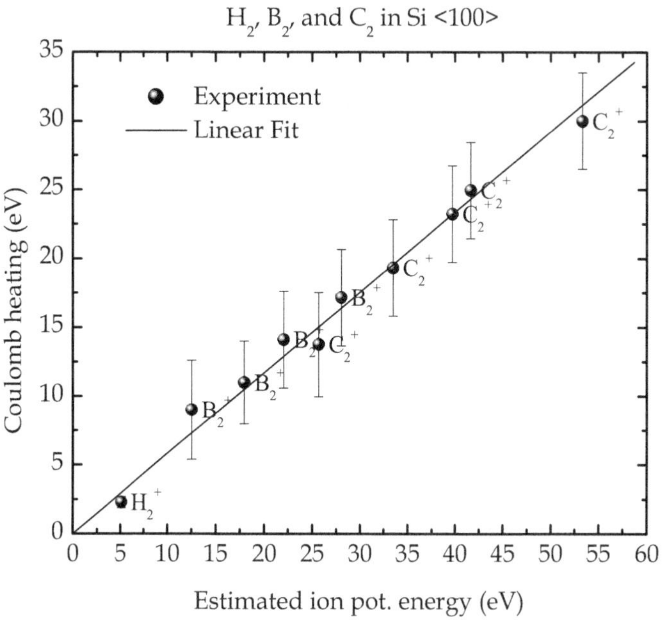

Fig. 9. Experimental Coulomb heating of H_2^+, B_2^+ and C_2^+ projectiles as a function of the estimated DHFS screened potential energy between the projectile components. All data seems to lay in a single straight line, with a slope about 0.6. The slope seems compatible with the expected 2/3 value (Fadanelli et al., 2010).

6. Conclusion

In summary, the present work shows that by combining monoatomic and molecular channelling RBS measurements with x-ray yields obtained by a scan around the Si <100> direction we were able to obtain the diatomic beams Coulomb heating values. Theoretical calculations are in good agreement with the experimental values. When the present results are plotted as a function of the potential energy per ion of the molecule, we have found that all the points fall on a straight line. This feature could indicate a "universal behaviour" for the Coulomb heating in the Si <100> channel. In addition, the line slope (0.6) shows compelling evidence that, at least for light molecules, the Coulomb heating corresponds to two-thirds of the stored potential energy per ion. Although this appears to be a general rule, it remains to be checked for heavier dicluster molecules travelling in other channels of Si (for instance, <110> and <111>) and other crystals.

7. Acknowledgment

The authors are indebted to the Brazilian agencies CNPq (Conselho Nacional de Desenvolvimento Científico e Tecnológico) and CAPES (Coordenação de Aperfeiçoamento de Pessoal de Nível Superior) for the support of this work.

8. References

Arista, N. R. (2000). Stopping of molecules and clusters. *Nuclear Instruments and Methods B*, Vol. 164-165, (April 2000), pp. 108-138, ISSN 0168-583X

Assmann, W.; Huber, H.; Karamian, S. A.; Gruner, F.; Mieskes, H. D.; Andersen, J.U.; Posselt, M. & Schmidt, B. (1999). Transverse cooling or heating of channeled ions by electron capture and loss, *Physical Review Letters*, Vol. 83, No. 9, (August 1999), pp. 1759-1762, ISSN 0031-9007

Azevedo, G. de M.; Grande, P. L. & Schiwietz, G. (2000) Impact-parameter dependent energy loss of screened ions, *Nuclear Instruments and Methods B*, Vol. 164 - 165, (April 2000), pp. 203 – 211, ISSN 0168-583X

Azevedo, G. de M.; Behar, M.; Dias, J. F.; Grande, P. L.; da Silva, D. L. & Schiwietz, G. (2002). Random and channeling stopping powers of He and Li ions in Si, *Physical Review B*, Vol. 65, No. 7, (January 2002), pp. 075203-1 - 075203-9, ISSN 1098-0121

Brandt, W. ; Ratkowski, A. & Ritchie, R. H. (1974). Energy-loss of swift proton clusters in solids. *Physical Review Letters*, Vol. 33, No. 22, (November 1974), pp. 1325-1328, ISSN 0031-9007

Caywood, J. M.; Tombrello, T.A. & Weaver, T. A. (1971). Backscattering from crystals using molecular hydrogen ions, *Physics Letters A*, Vol. A37, No. 4, (December 1971), pp. 350 - 352, ISSN 0375-9601

Chu, W. K.; Mayer, J. W. & Nicolet, M.-A. (1978). *Backscattering Spectrometry*, Academic Press, ISBN 0121738507, New York

Dos Santos, J. H. R.; Grande, P. L.; Behar, M.; Boudinov, H.; & Schiwietz, G. (1997). Angular dependence of the electronic energy loss of 800-keV He ions along the Si[100] direction, *Physical Review B*, Vol. 55, No. 7, (February 1997), pp. 4332 - 4342, ISSN 0163-1829

Eckstein, W. (1991) *Computer Simulation of Ion-Solid interactions*, Springer-Verlag, ISBN 0387190570, Berlin

Fadanelli, R. C.; Grande, P. L.; Behar, M.; Dias, J. F.; Schiwietz, G. & Denton, C. D. (2004). Coulomb heating of channeled H_2^+ and H_3^+ molecules in Si, *Physical Review B*, Vol. 69, No. 21, (June 2004), pp. 212104-1 - 212104-4, ISSN 1098-0121

Fadanelli, R. C.; Grande, P. L.; Behar, M.; Dias, J. F.; Czerski, K. & Schiwietz, G. (2006). Interplay between the Coulomb explosion and vicinage effects studied using H_2^+ molecules under channeling conditions, *Physical Review B*, Vol. 73, No. 24, (June 2006), pp. 245336-1 - 245336-7, ISSN 1098-0121

Fadanelli, R. C.; Behar, M. & Dias, J. F. (2008). Coulomb heating of channeled C^+ and C_2^+ molecules in Si, *Physical Review A*, Vol. 77, No. 5, (May 2008), pp. 052901-1 – 052901-6, ISSN 1050-2947

Fadanelli, R. C.; Behar, M. & Dias, J. F. (2010). Coulomb heating behavior of fast light diclusters, *Physical Review B*, Vol. 81, No. 13, (April 2010), pp. 132101-1 - 132101-4, ISSN 1098-0121

Gemmell, D. S. (1974). Channeling and related effects in motion of charged-particles through crystals, *Reviews of Modern Physics*, Vol. 46, No. 1, (January 1974), pp. 129 - 227, ISSN 0034-6861

Gemmell, D. S.; Remillieux, J.; Poizat, J.-C. ; Gaillard, M., J.; Holland, R. E. & Vager, Z. (1975). Evidence for an alignment effect in motion of swift ion clusters through solids, *Physical Review Letters*, Vol. 34, No. 23, (June 1975), pp. 1420-1424, ISSN 0031-9007

Götz, G. & Gärtner, K. (1988). *High Energy Ion Beam Analysis of Solids*, Akademie-Verlag, ISBN 3055004973, Berlin.

Jiang, W.; Grotzschel, R.; Pilz, W.; Schmidt, B. & Moller, W. (1999a). Random and channeling stopping powers and charge-state distributions in silicon for 0.2 - 1.2 MeV/u positive heavy ions, *Physical Review B*, Vol. 59, No. 1, (January 1999), pp. 226 - 234, ISSN 0163-1829

Jiang, W.; Grotzschel, R.; Pilz, W.; Schmidt, B. & Moller, W. (1999b). Random and channeling stopping powers and charge-state distributions in silicon for 0.2 - 1.2 MeV/u positive heavy ions, *Physical Review B*, Vol. 60, No. 1, (July 1999), pp. 714 - 716, ISSN 0163-1829

Khodyrev, V. A.; Kulikauskas, V. S. & Yang, C. (2002). The Coulomb explosion of molecular ions in channeling conditions, *Nuclear Instruments and Methods B*, Vol. 195, No. 3-4, (October 2002), pp. 259 - 268, ISSN 0168-583X

Lindhard, J. (1965). Influence of crystal lattice on motion of energetic charged particles, *Matematisk-Fysiske Meddelelser Udgivet af det Kongelige Danske Videnskabernes Selskab*, Vol. 34, No. 14, (November 1965), pp. 1 - 64.

Matsuo, J.; Toyoda, N.; Akizuki, M. & Yamada, I. (1997). Sputtering of elemental metals by Ar cluster ions, *Nuclear Instruments and Methods B*, Vol. 121, No. 1-4, (January 1997), pp. 459-463, ISSN 0168-583X

Poizat, J. C. & Remillieux. J. (1972). Channeling effects on transmission of MeV H_2^+ ions through a thin gold single-crystal, *Journal of Physics Part B: Atomic and Molecular Physics*, Vol. 5, No. 5, (1972), pp. L94.

Rothard, H.; Kroneberger, K.; Veje, E.; Clouvas, A.; Kemmler, J. ; Koschar, P. ; Keller, N. ; Lencinas, S. ; Lorenzen, P. ; Heil, O. ; Hoffmann, D. & Groeneveld, K. O. (1990). Experimental-study of molecular and cluster effects in secondary-electron emission, *Physical Review B*, Vol. 41, No. 7, (March 1990), pp. 3959-3967, ISSN 0163-1829

Salvat, F.; Martinez, J .D.; Mayol, R. & Parellada, J. (1987). Analytical Dirac-Hartree-Fock-Slater screening function for atoms (z=1-92), *Physical Review A*, Vol. 36, No. 2, (July 1987), pp. 467 - 476, ISSN 1050-2947

Schiwietz, G. & Grande, P. L. (1999). A unitary convolution approximation for the impact-parameter dependent electronic energy loss, *Nuclear Instruments and Methods B*, Vol. 153, No. 1 - 4, (June 1999), pp. 1 - 9, ISSN 0168-583X

Schiwietz, G. & Grande, P. L. (2001). Improved charge-state formulas, *Nuclear Instruments and Methods B*, Vol. 175 - 177, (April 2001), pp. 125 - 131, ISSN 0168-583X

Sigmund, P. (1992). Scattering and stopping of swift diatomic molecules under Coulomb explosion, *Nuclear Instruments and Methods B*, Vol. 67, No. 1 - 4, (April 1992), pp. 11 - 16, ISSN 0168-583X

Tombrello, T.A. & Caywood, J. M. (1973). Molecular-ion effects in backscattering from aligned silicon crystals, *Physical Review B*, Vol. 8, No. 7, (October 1973), pp. 3065 - 3070, ISSN 0163-1829

Yamada, I.; Matsuo, J. & Toyoda, N. (2003). Cluster ion beam process technology, *Nuclear Instruments and Methods B*, Vol. 206, (May 2003), pp. 820-829, ISSN 0168-583X

Yamagushi, T.; Matsuo, J.; Akizuki, M.; Ascheron, C. E. ; Takaoka, G. H. & Yamada, I. (1995). Sputtering effect of gas cluster ion-beams, *Nuclear Instruments and Methods B*, Vol. 99, No. 1-4, (May 1995), pp. 237-239, ISSN 0168-583X

3

Radio Frequency Quadrupole Accelerator: A High Energy and High Current Implanter

Yuancun Nie[1,2], Yuanrong Lu[2,*], Xueqing Yan[2] and Jiaer Chen[2]
[1]*State Nuclear Power Technology R&D Center, Beijing*
[2]*State Key Lab of Nuclear Physics and Technology (Peking University), Beijing*
China

1. Introduction

In this chapter, issues of applying the radio frequency quadrupole (RFQ) accelerator to ion implantation for material modification and microelectronics will be discussed at first. Subsequently, several examples of the RFQ-based implanters will be introduced, with an emphasis on the ladder IH-RFQ accelerator at Peking University, of which the output beam energy spread is optimized to be much lower than the conventional value by means of a novel beam dynamics design strategy. The chapter is divided into 6 sections, which are the introduction, advantages of RFQ as implanter, problems and solutions of RFQ-based implanters, several RFQs aiming at ion implantation, ladder IH-RFQ at Peking University, and summary.

RFQ accelerator, proposed by I. M. Kapchinsky and V. A. Teplyakov in 1970, can simultaneously focus, bunch and accelerate low energy beam extracted from ion source directly, based on the rf electrical field of a modulated quadrupole transport channel[1]. The details of its principle and development could be found in many literatures[2,3]. Taking advantages of its high current, compact size and good ion selectivity, etc, RFQ has been used widely in the field of particle accelerator, after 40 years' development both in the beam dynamics design and in the rf cavity structure [4]. At present, RFQs can provide high current (up to hundreds mA) ion beams at energy level of MeV over the mass range from Hydrogen to Uranium and the 5 MHz to 500 MHz frequency range[5].

2. Advantages of RFQ as implanter

Many efforts have been made to apply the RFQ accelerator to ion implantation. Generally speaking, the beam line of the RFQ-based implanter consists of the ion source, low energy beam transport (LEBT), RFQ accelerator and the target, equipped with power system, controlling system, vacuum and cooling systems, etc. The advantages of RFQ-based implanter are concluded as follows:

- The beam extracted from the ion source is accelerated further by the following RFQ, so the beam energy can reach the order of several MeV, which fulfills the requirement of deep deposition.

---

* Corresponding Author

- The current of heavy ion beam accelerated by the RFQ could be in the mA-range. For an implant dose of 10^{18} ions/cm^2, the implantation time needed is less than 3 minutes/cm^2 using an ion beam with average current density of 1 mA/cm^2.
- RFQ can simultaneously accelerate positive and negative ions with the same charge to mass ratio[6]. It makes full use of the rf power, and more importantly it avoids the charge accumulation while doubling the implant flux density.
- For a given RFQ, it is available not only for the designing ion beam, but also for other ion species, as long as the injection energy and the inter-electrode voltage are adjusted appropriately. For example, the 36 MHz IH-RFQ in the UNILAC accelerator of GSI can accelerate all the heavy ions with charge to mass ratio greater than 1/65 from 2.2 keV/u to 120 keV/u[7].
- Compared to the electrostatic accelerator, high voltage breakdown and high energy X-ray radiations are reduced greatly for this case, because high voltage terminal is avoided and the rf inter-electrode voltage in RFQ is normally below 100 kV.
- The equipment is very compact. For instance, the 26 MHz ISR-1000 RFQ at Peking University can accelerate $^{16}O^+$ beam from 22 keV to 1 MeV within 2.6 m, and the cavity diameter is 0.75 m[8].

3. Problems and solutions of RFQ-based implanters

The RFQ-based implanter has some drawbacks, i.e. the output beam energy is normally fixed for a given RFQ, and the output beam energy spread is quite big (usually the full width at half magnitude, FWHM energy spread is around 3%).

To make the output beam energy variable, two methods have been used, and they are:

- Making the resonant frequency of the RFQ cavity variable. It is mostly applied to the 4-rod RFQs. For instance, at Frankfurt University several VE (Variable energy) RFQs have been developed, and the output energy range could be doubled and even more while the ground plate connected to the stems is moved to vary the frequency (VF)[5]. Hitachi Ltd. also invented a variable energy RFQ linac with an external tunable one-turn coil, and the variable resonant frequency range is from 9.6 to 39 MHz, which corresponds to an energy variation of 16 times[9].
- Changing the output energy by using a high voltage platform (HVP) or a Spiral Loaded Cavity (SLC). The variation in the beam energy is continuous in the range determined by the HVP, for example in Ref.[10], the 60 MHz RFQ implanter with the HVP at voltage higher than 250 kV could provide a beam energy variable from 500 keV to 1.5 MeV for ions with charge to mass ratio 2/16. At Frankfurt University, a SLC was coupled to the RFQ designed for the acceleration of N^+ from 50 keV to 1.5 MeV to vary the beam energy in a range of ±200 keV[11].

To reduce the output energy spread, non-adiabatic bunching method could be used and there are also two ways:

- The first one is the external bunching method. A beam dynamics design with energy spread 0.6% has been realized for $^{14}C^+$ beam using this method[12]. However, the pre-buncher used to produce a pulsed beam from a DC beam before the RFQ makes the system complex, and the total beam transmission efficiency is constrained by it.
- The second one is the internal discrete bunching method. Most recently, the low energy spread beam dynamics design for ions with charge to mass ratio greater than 1/14 was

accomplished[13]. Optimization procedure of the design as well as its preliminary beam experiment will be discussed in detail.

4. Several RFQs aiming at ion implantation

In the following Table 1, parameters of several typical RFQs for ion implantation in the world are summarized.

Affiliation	RFQ structure	Ions or charge-to-mass ratio	Energy (MeV/u)	Current (mA)	Frequency (MHz)	Length (m)	Note
GSI	Split coaxial	Ar^+,Kr^+	0.045	5	13.5	9.4	Operation
Hitachi	Inductance-Capacitance	P^+,N^+	0.043	5	17.3	2.3	Operation
Hitachi[9]	4-rod	1/31	0.036	1	21	2.3	VE(VF)
Peking University	Integral Split Ring	O^+,O^-, N^+	0.0625	0.5	26	2.6	Operation
Nissin	4-rod	B^+,N^+,O^+	0.0835	0.2	33.3	2.22	Operation
Seoul[10]	Modified 4-vane	2/16	0.0625±0.0313	2	60	1.57	VE(HVP)
Simazu	4-vane	B^+,N^+,P^+	0.09	0.1-0.4	70	2.1	Operation
Frankfurt	4-rod	B^+	0.09	15	81	1.2	Operation
Frankfurt[11]	4-rod	1/14	0.107±0.014	10	108.5	2.0	VE(SLC)

Table 1. Parameters of some RFQs for ion implantation (after Ref.[14])

5. Ladder IH-RFQ at Peking University

Since the year 2008, a 104 MHz ladder type IH-RFQ has been developed at Peking University, for the acceleration of the ion species with mass-to-charge ratio smaller than 14 from 2.86 keV/u to 35.71 keV/u. As a specific feature, the output beam energy spread is as low as 0.6% achieved through the internal discrete bunching method, which makes potential applications of the RFQ feasible, such as accelerator mass spectrometry (AMS) and ion implantation. Design procedures and the most recent experimental results of the ladder IH-RFQ will be described in detail.

5.1 Beam dynamics design of the ladder IH-RFQ

In recent years, RFQ based ^{14}C AMS application has been studied extensively at the Institute of Heavy Ion Physics (IHIP), Peking University. Great efforts have been made to reduce the output FWHM energy spread of the RFQ. Non-adiabatic bunching method could be used owing to the lack of space charge effect since the current is usually weak. A beam dynamics design with energy spread 0.6% has been realized by means of the external bunching method. However, the total beam transmission efficiency is constrained by the pre-buncher used to produce a pulsed beam before it is injected into the RFQ. For this reason, we tried another possibility, the internal discrete bunching method proposed by J. W. Staples at LBNL to reduce the output energy spread without the pre-buncher[15]. The final low energy spread beam dynamics design for $^{14}C^+$ RFQ was accomplished. Optimization procedure of the design along with its performances will be reported.

The operating frequency 104 MHz was chosen because of the limitation of the existing rf power transmitter, and our interest of investigating the ladder type IH-RFQ structure [13], in which the four electrodes are supported alternately by erect stems mounted to the external cavity from top to bottom equidistantly, different from the traditional 4-rod RFQ and IH-RFQ structures. Properties of the ladder IH-RFQ will be studied in the next section.

According to the strategy of the internal discrete bunching, the whole RFQ beam dynamics design is divided into five sections, different from the conventional four-step method developed at LANL [16], consisting of the radial matcher, the buncher section, the drift section, the transition section and the acceleration section. With the buncher and drift sections serving as an internal non-adiabatic buncher, the new method reduces the output longitudinal emittance significantly and transverse emittance slightly compared with the traditional one. However, neither this strategy nor the conventional method takes into account the match between the beam size and the acceleration channel carefully, leading to beam envelope mismatch and even particle loss. According to the beam matching equation for weak current

$$\varepsilon_{\mathrm{tn}}^2 = \frac{a_0^4 \gamma^2 \sigma_{\mathrm{t}}^2}{\lambda^2} \tag{1}$$

where $\varepsilon_{\mathrm{tn}}$ is the normalized rms (root mean square) emittance, a_0 is the transverse rms beam radius, γ is the relativistic gamma, λ is the rf wavelength, and σ_{t} is the transverse phase advance without current. σ_{t} can be obtained by solving the transverse motion equation, i.e. the Mathieu Equation, and we have

$$\sigma_{\mathrm{t}}^2 \propto \frac{B^2}{8\pi^2} + \Delta \tag{2}$$

where B is the focusing parameter for the RFQ accelerator, while Δ is the defocusing factor, which are defined as follows

$$\begin{aligned} \Delta &= \frac{qe\pi^2 AV \sin\varphi_{\mathrm{s}}}{2m_0 c^2 \beta_v^2 \gamma} \\ B &= \frac{qe\lambda^2 FV}{m_0 c^2 a^2 \gamma} \end{aligned} \tag{3}$$

where A, V, φ_{s}, F are the acceleration coefficient, inter-electrodes voltage, synchronous phase and focusing factor, respectively, β_v is the relativistic velocity, a is the minimum aperture radius between two opposite electrodes of one acceleration cell. Based on the matching design method [17], we enhance the focusing parameter B gradually as the defocusing factor $|\Delta|$ increasing due to the growing acceleration coefficient A, during the first stage of the transition section that serves as the shaper section in the four-step method. Nevertheless, in the acceleration section, B is decreased slowly since $|\Delta|$ is reduced because of the rapidly increasing beam energy. The above improvements make the transverse phase advance σ_{t} constant, thereby the beam matched, accordingly suppress the emittance growth and minimize the particle loss.

In addition, there is an approach to reduce the RFQ energy spread furthermore, which can be shown by

$$\Delta w_{\max} = \left[2\xi eAVW_s \left(\varphi_s \cos\varphi_s - \sin\varphi_s \right) \right]^{1/2} \tag{4}$$

where $\Delta w_{\max}$ is the separatrix height that indicates the maximum beam energy spread, ξ is the charge-to-mass ratio q/A_0 of the ion, e is the charge of an electron, and W_s is the kinetic energy of the synchronous particles. It can be seen that the energy spread can be reduced by decreasing A and the inter-electrode voltage V while increasing the synchronous phase φ_s, where lower A means lower electrode modulation m. It means that φ_s should be as large as possible even close to 0° at the last acceleration cell in order to achieve low energy spread, not necessarily be -30° as usual. For our case, the frequency 104 MHz is relatively high for ^{14}C ions with $q/A_0=1/14$, which destines a small aperture radius a in order to obtain sufficient focusing strength under a certain V. Small aperture radius means small electrode-tip radius r_0 to avoid excessive Kilpatrick coefficient[2]. To relax the manufacture pressure of the electrodes, we have to enhance V to increase a, unfortunately, it conflicts with reducing not only the energy spread but also the power consumption. As a result of the compromise, an inter-electrode voltage 60 kV is used.

The main parameters were generated by the matched and equipartitioned design code MATCHDESIGN developed at IHIP[17], while the non-adiabatic buncher section was designed by the code of Staples[15]. Beam transport simulation was performed by the program RFQDYN derived from PARMTEQ[16]. In order to make a as large as possible, we tried to reduce B during the transition section and the acceleration section while ensuring the balance between defocusing and focusing of the beam. It finally induced 2% particle loss at the transverse direction. The final value of the synchronous phase φ_s was set to be -6° at the end of the acceleration section, which decreased the output beam energy spread and additionally quickened the acceleration process. The modulation parameter m was also evolved carefully, in view of that large modulation is advantageous to particle acceleration, but disadvantageous to reducing the energy spread.

Figure 1 shows us the evolutions of the focusing parameter B as well as the acceleration coefficient A along with the cell number for the final beam dynamics design. The other

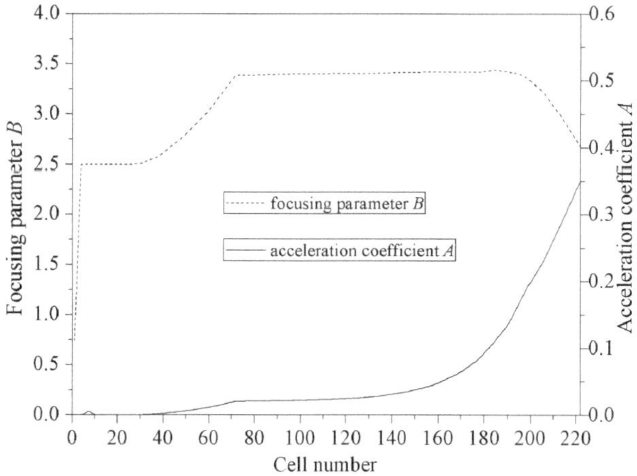

Fig. 1. Variations of the focusing parameter and the acceleration coefficient with the cell number of the ladder IH-RFQ.

dynamics parameters of the RFQ are plotted in Fig. 2. The rapidly advance of φ_s leads to a small portion of particles lose their balances of the longitudinal motion and can't be accelerated effectively. This explains why a tail appears in either the output energy spectrum or the output phase spectrum. The resultant transmission is 97.6%, while 80% particles are in the phase range of ±15° around φ_s, and 90% for ±30°. Meanwhile, more than 90% particles are limited by the energy range ±15keV, with the FWHM energy spread 0.6%. The total length of the electrode is about 1.1 m to accelerate the $^{14}C^{+}$ beam from 40 keV to 500 keV, which is supposed to be slightly longer than that of the external bunching method due to the implanted non-adiabatic buncher. In view of the overall simplified system and much higher transmission efficiency in total, the internal non-adiabatic bunching strategy is a competent candidate to design the RFQ for AMS application.

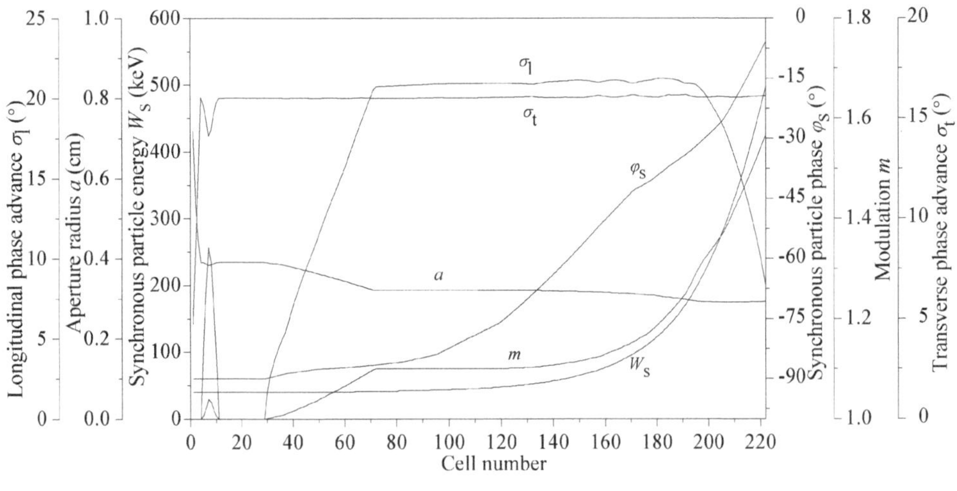

Fig. 2. The main beam dynamics parameters varying with the cell number of the ladder IH-RFQ.

Responses of the beam transmission efficiency and the output beam energy spread to input Twiss parameters (α, β), current and emittance (normalized, rms) are presented in Fig. 3, with the design values (0.31, 2.19 cm/rad), 0 mA, 0.042 mm-mrad, respectively. Figures 3(a) and 3(b) illustrate that neither the beam transmission efficiency nor the output beam energy spread is sensitive to the Twiss parameters, since all the fluctuations are acceptable within the plotted ranges. The influences of the input emittance on the transmission efficiency and output energy spread are observed but tolerable considering that even when it is twice the design value, the transmission efficiency and energy spread are still higher than 90% and below 1.5%. A remarkable feature is that the transmission efficiency and energy spread are better than 93% and 0.8% at the current 1 mA, despite the beam dynamics was optimized at 0 current.

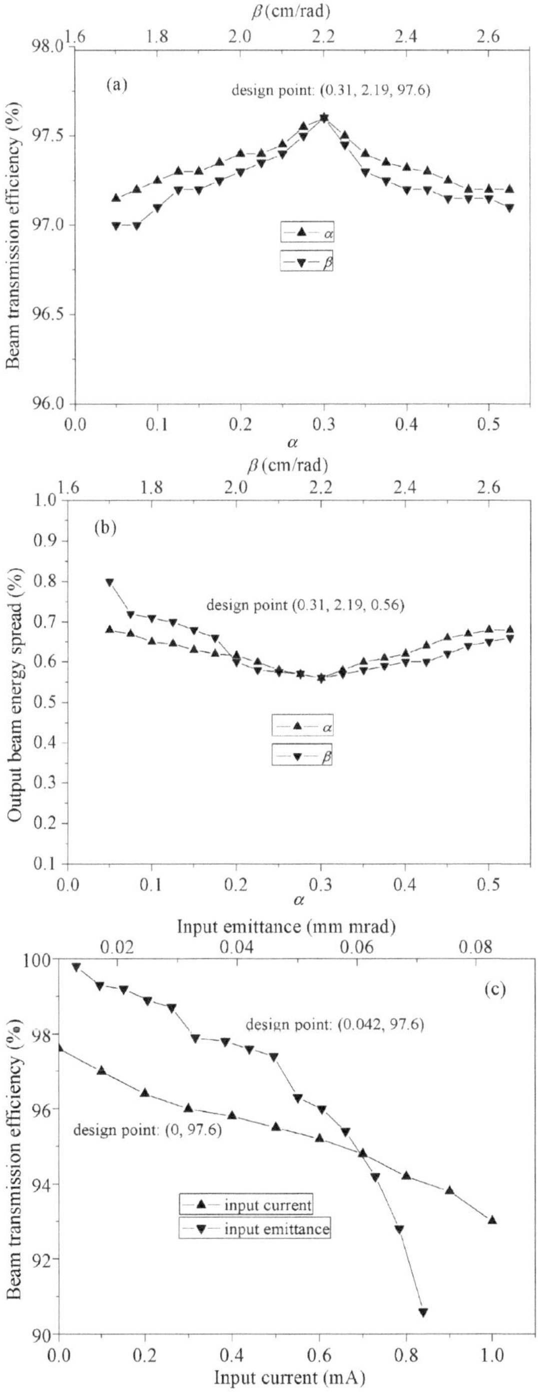
β (cm/rad)
1.6
1.8
2.0
2.2
2.4
2.6
98.0
97.5
97.0
96.5
96.0
Beam transmission efficiency (%)
(a)
design point: (0.31, 2.19, 97.6)
α
β
0.0
0.1
0.2
0.3
0.4
0.5
α
β (cm/rad)
1.6
1.8
2.0
2.2
2.4
2.6
1.0
0.9
0.8
0.7
0.6
0.5
0.4
0.3
0.2
0.1
Output beam energy spread (%)
(b)
design point (0.31, 2.19, 0.56)
α
β
0.0
0.1
0.2
0.3
0.4
0.5
α
Input emittance (mm mrad)
0.02
0.04
0.06
0.08
100
98
96
94
92
90
Beam transmission efficiency (%)
(c)
design point: (0.042, 97.6)
design point: (0, 97.6)
input current
input emittance
0.0
0.2
0.4
0.6
0.8
1.0
Input current (mA)

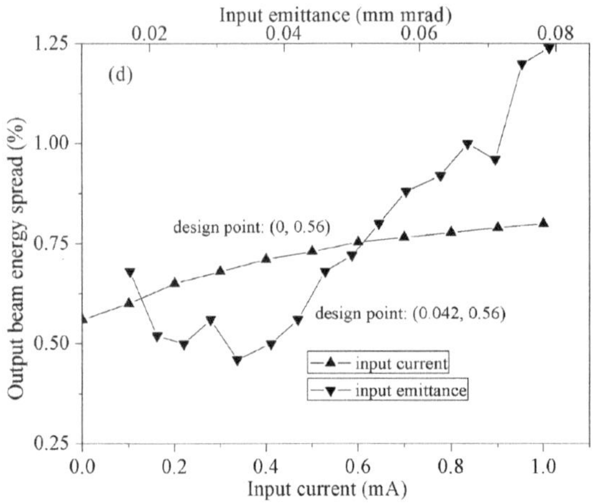

Fig. 3. Stability studies of the beam dynamics design of the ladder IH-RFQ: beam transmission efficiency versus Twiss parameters in (a), output beam energy spread versus Twiss parameters in (b), beam transmission efficiency versus input current and emittance in (c), output beam energy spread versus input current and emittance in (d).

The design has been proved to be suitable for various ion species with different charge-to-mass ratio q/m_0 and even different frequencies, as long as the operating parameters are chosen appropriately. According to (3), it can be seen that the focusing strength B is proportional to q/m_0, λ^2 and V. Meanwhile, the length of one acceleration cell is $\beta_v\lambda/2$. For given electrodes structure, when q/m_0 or λ changes, V and β_v i.e. the injection energy $W_i=1/2m_0v^2$ should be adjusted to keep B and the injection cell length unaltered, hence achieve the matching between the acceleration channel and the input beam. The beam dynamics parameters are listed in Table 2.

Ion species	$^{14}C^+$	$^{13}C^+$	$^{12}C^+$	$^4He^+$	$^4He^+$	$^4He^+$
Frequency f (MHz)	104	104	104	104	100	106.4
Inter-electrode voltage V (kV)	60	55.7	51.4	17.1	15.9	17.9
Injection energy W_i (keV)	40	37.1	34.3	11.4	10.6	12.0
Output energy W_{out} (keV)	500	464	428	143	132	150
Transmission T (%)	97	97	97	97	97	97
Energy spread (%)	0.6	0.6	0.7	1.0	1.1	1.0

Table 2. Beam dynamics parameters of the ladder IH-RFQ

5.2 RF design of the ladder IH-RFQ

According to the above beam dynamics design, the average electrode-tip radius is 3.1 mm corresponding to the average aperture radius 3.5 mm. Figure 4 shows the resonant model of CST Microwave Studio (MWS)[18] simulations for the ladder IH-RFQ. Simulation results of the electromagnetic field validated that the ladder IH-RFQ operates at $H_{21(0)}$ mode. As a consequence, the resonant frequency of it is unsurprisingly more than two times that of IH-

RFQ and about 1.3 times that of 4-Rod RFQ under the same transverse dimension, making it very attractive at frequencies above 100 MHz. Moreover, in respect that the electrodes are supported by completely symmetric stem structures, the quadrupole field of the ladder IH-RFQ is ideal unlike the 4-Rod RFQ and IH-RFQ, the dipole components of which are typically 2-3% and 0.5%, respectively[19].

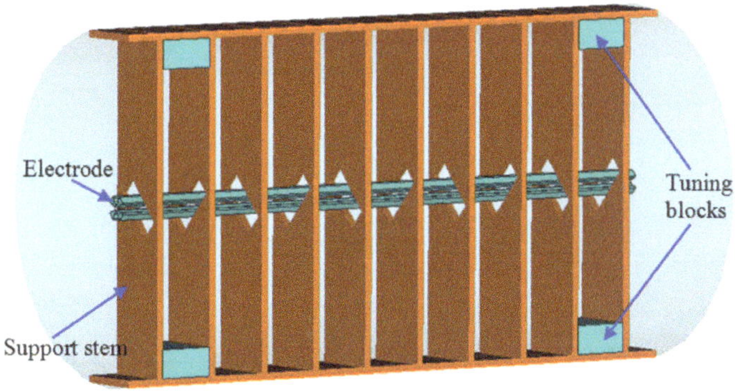

Fig. 4. MWS simulation model of the ladder IH-RFQ.

Many authors have reported their studies on the RFQ cavity modeling[20,21]. The conventional electrical model are shown in Fig.5, consisting of the equivalent resistance, inductance, capacitance and parallel shunt impedance, denoted by R, L, C and R_p, respectively. According to the definitions of intrinsic quality factor $Q_0=\omega W/P$ and shunt impedance $R_p=V^2/P$, with the resonant frequency $\omega=2\pi f=1/(LC)^{1/2}$, average energy stored in one period $W=1/2I^2L=1/2CV^2$ (where I is peak current and V is peak voltage), and power dissipation $P=1/2I^2$R, we have

$$Q_0 = \frac{1}{R}\sqrt{\frac{L}{C}} = \frac{R_P}{2}\sqrt{\frac{C}{L}} \tag{5}$$

For RFQ cavity having N basic resonant cells, the relation between the total R, L, C, I and the corresponding values of one cell R_{cell}, L_{cell}, C_{cell}, I_{cell} will be

$$I = NI_{cell}\,,\ C = NC_{cell}\,,\ L = L_{cell} / N\,,\ R = R_{cell} / N \tag{6}$$

when ignoring the end effect. Combining (5) with (6) we find

$$Q_0 = Q_{0,cell}\,,\ R_p = R_{p,cell} / N \tag{7}$$

where cell denote the values of single cell. For ladder IH-RFQ, Fig.6 shows the simplified circuit of one cell with

$$C_{cell} = 2C_{stem} + C_{rod}\ ;\ L_{cell} = L_{stem} / 2\ ;\ R_{cell} = R_{stem} / 2 \tag{8}$$

In (8) it is assumed that the resistance and inductance are mainly attributed to the stems, while the capacitance arises from the two pairs of opposite electrodes and adjacent stems with C_{rod} several times greater than C_{stem} generally.

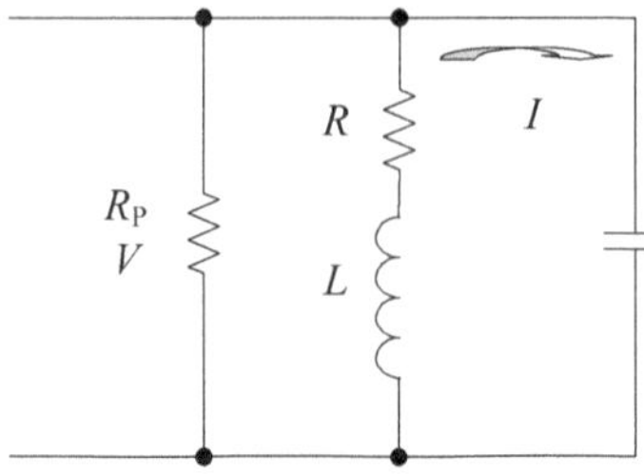

Fig. 5. Equivalent circuit of RFQ cavity.

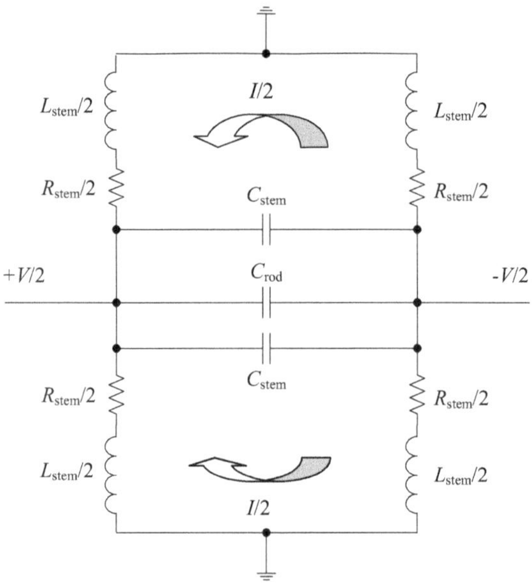

Fig. 6. Equivalent circuit of single cell for the ladder IH-RFQ.

Based on the equivalent circuit analysis, table 3 lists the MWS-simulated rf properties of different RFQ structures at the same transverse size, which is in agreement with the theoretical predictions.

RFQ Structures	Frequency (MHz)	Shunt impedance (kΩ)	Quality factor	The first High-mode Frequency (MHz)
IH-RFQ	46.5	273	5934	137.2
4-rod RFQ	77.9	155	5077	143.9
Ladder IH-RFQ	103.6	100	5192	156.3

Table 3. Comparison of different RFQ structures at the same transverse size

Mode analysis of the RFQ was performed using the eigenmode solver of MWS. Fig. 7 shows the electric field, magnetic field and surface current distribution. It can be seen that the two parts of inductance and resistance of one stem are in parallel. These results agree with the electromagnetic field distribution of a H_{210} mode, and the expectation from Fig.6.

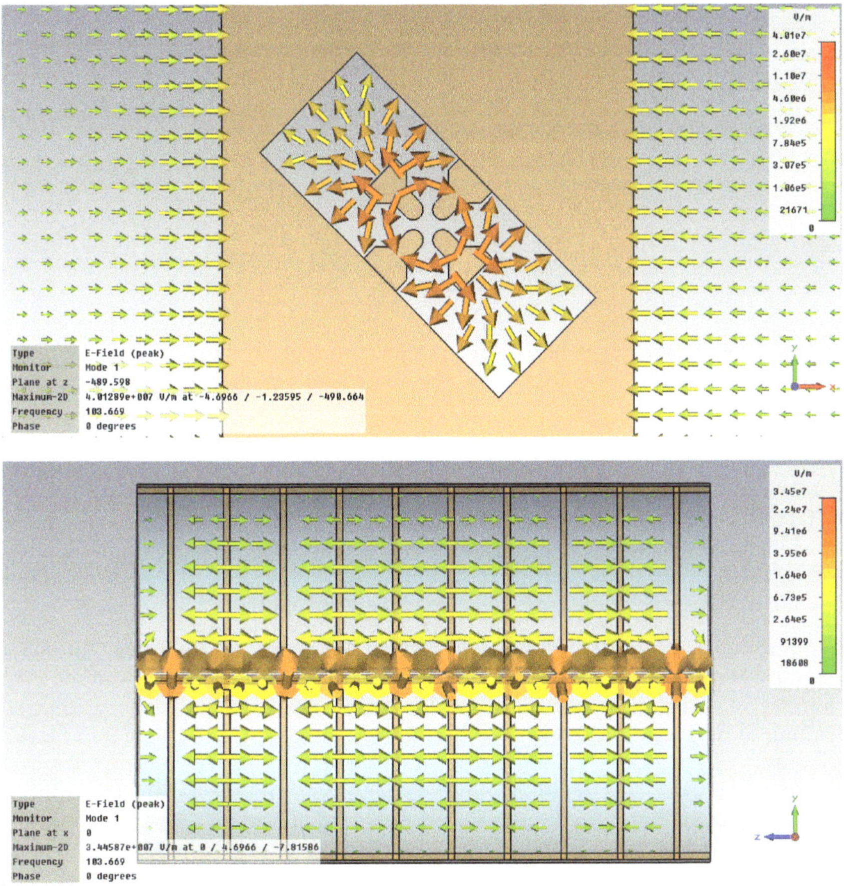

(a) Electric field among electrodes (up) and stems (bottom) of the ladder IH-RFQ

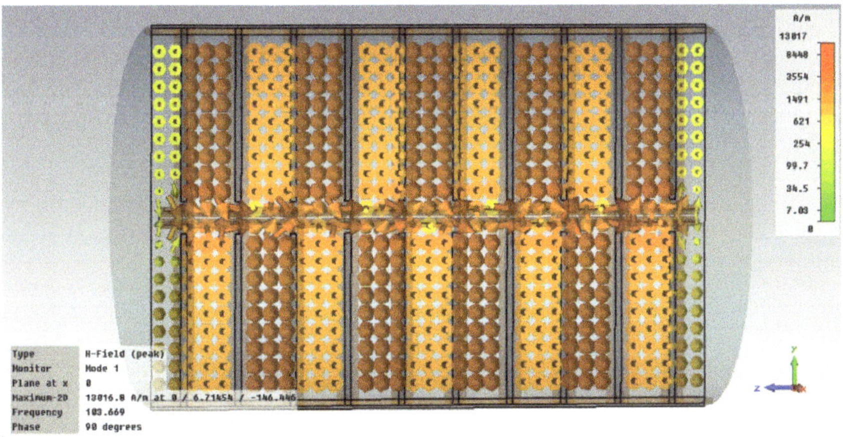

(b) Magnetic field distribution of the ladder IH-RFQ

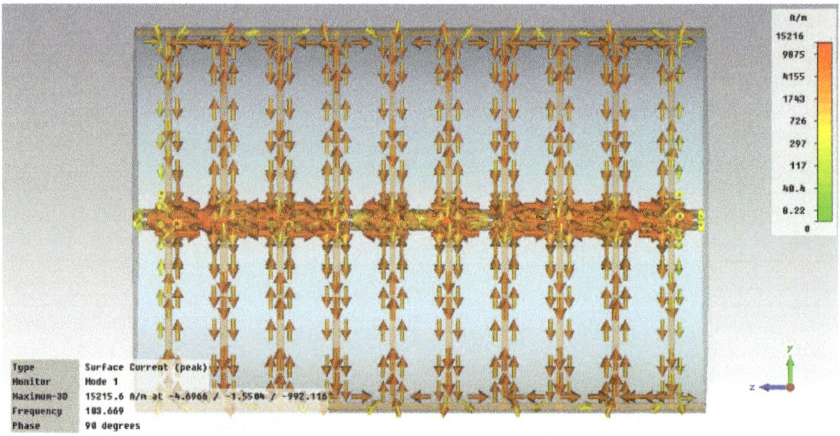

(c) Surface current flow of the ladder IH-RFQ

Fig. 7. MWS-simulated results of the electromagnetic field and surface current flow of the ladder IH-RFQ.

The RFQ structure was optimized by parameter sweeping to find the maximum Q_0, R_p, and minimum P at 104 MHz. Some conclusive results are as follows.

1. Stem number N_{stem}: The beam dynamics design established the inter-electrodes aperture and the electrode length, therefore the first parameter to be studied is the stem number N_{stem} used. Bigger N_{stem} means shorter distance between adjacent stems d, leading to decreasing L_{cell} and C_{cell}, but slightly increasing R_{cell} in view of more serious proximity effect. Taking (6) into account, R, L will decrease while C increase because of the increment in C_{stem}, finally resulting in bigger f as illustrated by MWS simulations. The cavity diameter D has to be scaled to keep f 104 MHz. Fig.8 shows the relevant simulation results, according to which 10 stems was employed.
2. Cavity diameter D: As a variable D is always used to make f unaltered when sweeping the other parameters. When f needs to be lowered, D should be extended to make L_{stem}, C_{stem} as well as R_{stem} bigger.
3. Stem spacing d: Over the 95 mm to 115 mm d range in 5 mm increments with N_{stem}=10, f decreases while R_p and Q_0 increase due to smaller C_{stem}, R_{stem} and bigger L_{stem} with increasing d. After adjusting D, it was found that R_p and Q_0 are approximately proportional to d, taking into account the end effects and the mechanical feasibility d=110 mm was eventually chosen.
4. Stem width W_{stem}: The increase in C_{stem}, and decrease in L_{stem} as well as R_{stem} with W_{stem} over the simulation range result in increasing f and Q_0 but decreasing R_p. By adjusting D, it could be seen that there is a peak value of R_p, this is why W_{stem}=130 mm was adopted to make power loss least.

The inter-electrodes voltage is usually supposed to be constant longitudinally when carrying out the beam dynamics design. So it has long been an important topic that how to tune the voltage unflatness of the RFQ owing to the end effect and modulated electrodes to ensure a good beam quality. For the ladder IH-RFQ, adjustable tuning plates as shown in Fig.4 can be inserted into two neighboring stems symmetrically up and down, which is similar to the case of the 4-Rod RFQ[22]. The tuning plates serve as short pieces of stems, changing their inductances thus the frequency and field strength. To verify their tuning capability, corresponding simulations were performed for different tuning plate heights 0, 50 and 100 mm, of which the results are plotted in Fig.9. Here electric field is equal to the voltage between relevant electrodes, since the electrodes are unmodulated with uniform aperture radius. It can be seen that the field unflatness achieves its optimum value 2% when the tuning height is 100 mm. Tuning plates cause increment in the eigenfrequency of RFQ, so f is generally set to be several percents lower than the desired value to leave enough tuning space.

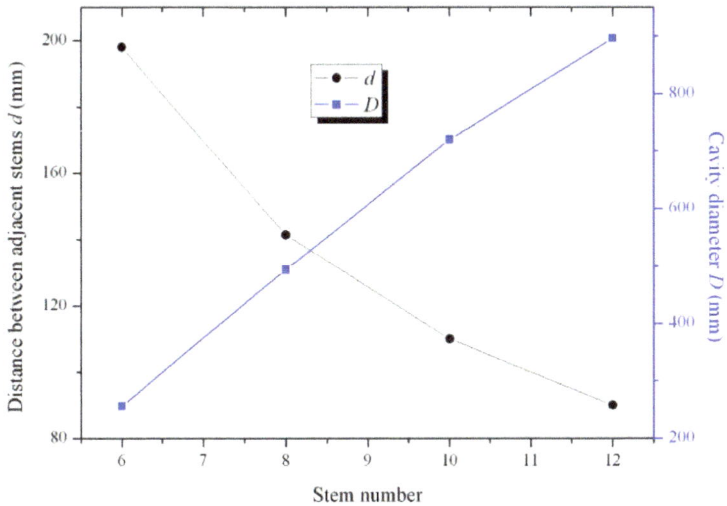

(a) structure parameters

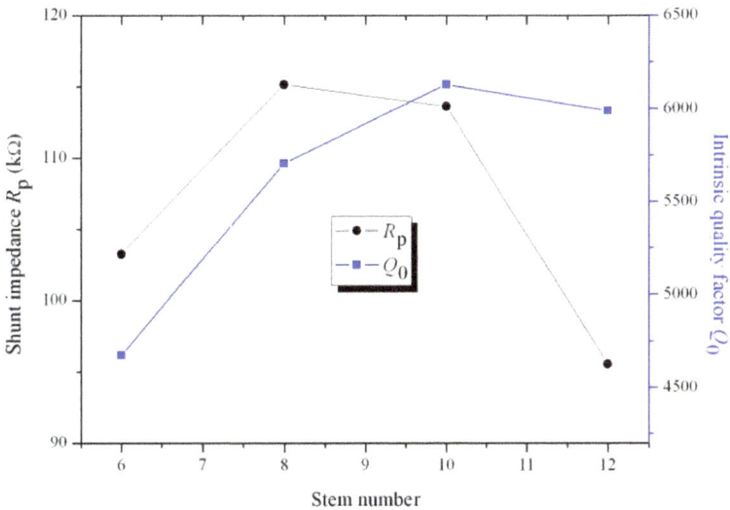

(b) rf properties

Fig. 8. Variations of the ladder IH-RFQ parameters with stem number.

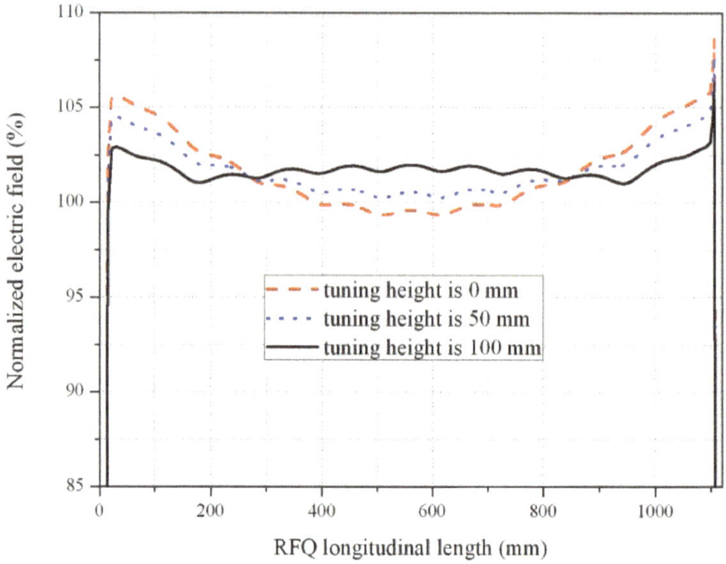

Fig. 9. Variations of the field unflatness with tuning plates height of the ladder IH-RFQ.

5.3 Experimental studies on the ladder IH-RFQ

The constructed ladder IH-RFQ cavity is shown in Fig.10. The geometrical parameters are as follows: the cavity is 720 mm in diameter and 1123 mm in length; 10 supporting stems with width and spacing of 130 mm and 110 mm, respectively, are used. The stems are mounted on two fixing plates up and down to improve the mechanical performance and in the

Fig. 10. The constructed ladder IH-RFQ cavity.

interest of installation. The stems and plates are made of oxygen free copper, and there are water-cooling channels in them. The inner wall of the cavity tank is well plated with copper of more than 0.1 mm in thickness to improve the rf performances. In order to fix the stems and plates on the tank, four copper blocks with tuning thickness 30 mm are used, which will make the resonant frequency higher and improve the field unflatness at the same time.

On the basis of transient simulations with MWS, a magnetic coupling loop with size of about 90 mm×50 mm in between two neighboring stems was established for rf power feeding of the ladder IH-RFQ. The critical coupling condition was met after proper adjustment of the loop orientation, and the result is shown in Fig.11(a). The cold measurement was performed by means of a vector network analyzer and the capacitance perturbation method. S-parameters were measured using an Angilent 8753E network analyzer, and the loaded quality factor Q_L was obtained by 3 dB method. Consequently, the unloaded or intrinsic quality factor Q_0 could be computed using $Q_0=Q_L(1+\beta_c)$, where the coupling factor β_c is defined as Q_0/Q_e with the external quality factor Q_e. During the experiment, a very small pick-up loop was used to weaken the influence of it on the cavity properties, as a result, $\beta_c \approx 1$ for critical coupling, i.e. $Q_0=Q_e$. Referring to the shunt impedance, the perturbation capacitor method was adopted as described in Ref.[23,24]. According to the definitions of intrinsic quality factor $Q_0=\omega W/P$ and shunt impedance $R_p=V^2/P$, we have

$$R_p = \frac{2Q_0}{\omega C} \tag{9}$$

When using a perturbation capacitance ΔC, the perturbed resonant frequency of the cavity f_{per} becomes

$$f_{per} = \frac{1}{2\pi\sqrt{L(C+\Delta C)}} \tag{10}$$

So the effective capacitance of the cavity can be thus obtained by

$$C = \frac{\Delta C}{\left(f / f_{per}\right)^2 - 1} \tag{11}$$

Finally R_p is calculated using (9), and the square root of it is treated as the relative value of the inter-electrode voltage. The locations of the perturbation capacitor are moved from stem to stem as shown in Fig.11(b), in the interest of the average R_p and voltage unflatness. Figure 12 is the adjustment result of the feeding point, and Fig.13 shows the S_{21} parameter with the loaded quality factor 1861 at 106.7 MHz. The cold measurement result of the shunt impedance is 102 kΩ compared with the simulation value 110 kΩ. The measured inter-electrode voltage unflatness and the simulated result are compared in Fig.14.

Finally, the high-power test and beam test of the ladder IH-RFQ were performed[25]. The rf and measurement system of the experiment is plotted in Fig.15. We used one pulse generator to control the ECR source[26] and the rf transmitter at the same time. The rf power is provided by the FM Transmitter, 816R-4, operating in the 88~108 MHz frequency range with the maximum rf output of 27.5 kW, the excitation of which is from a CE 802A solid-state FM wideband exciter. The input signal of the exciter is given by a signal modulation system consisting of a CW signal source, the pulse generator, a high frequency (HF) switch and a 30 dB solid-state power amplifier. The output of the transmitter is applied to the RFQ cavity via a low-pass filter and directional coupler. The beam current was measured using Faraday cups with the help of standard pick-up resistances.

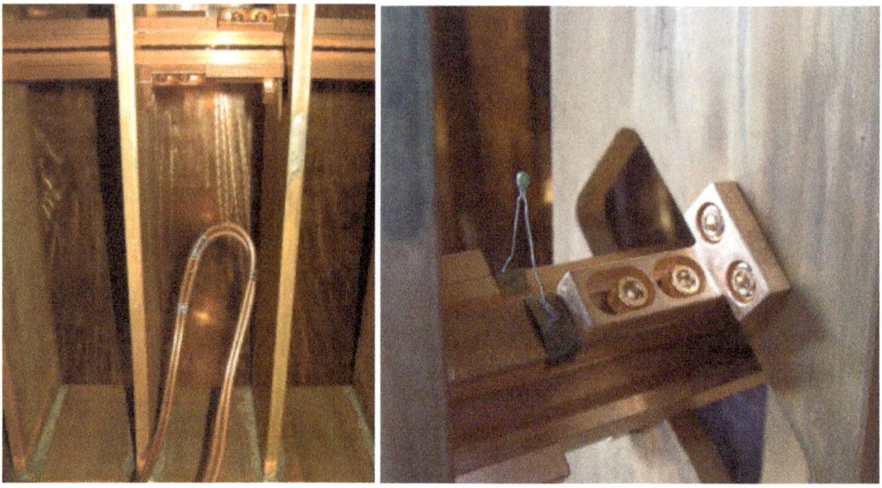

(a) (b)

Fig. 11. The critical coupling loop (a) and the perturbation capacitor (b) of the ladder IH-RFQ.

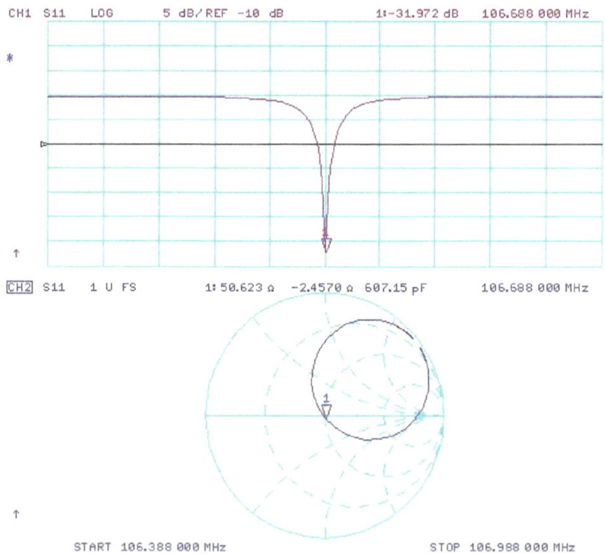

Fig. 12. The critical coupling point of the ladder IH-RFQ: the S_{11} parameter (up) and Smith Chart (down).

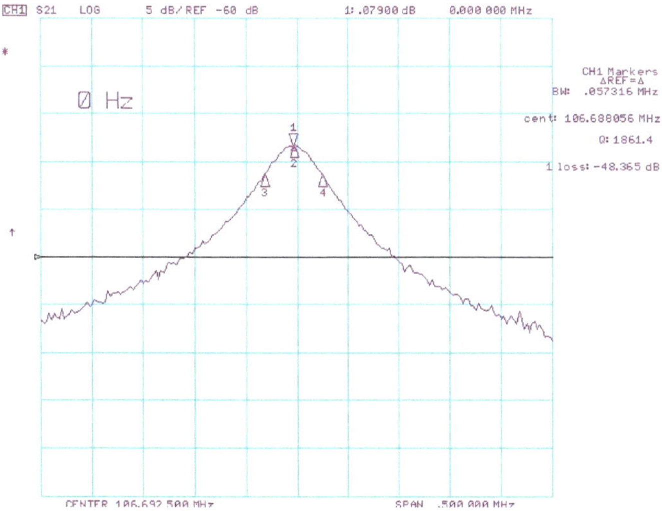

Fig. 13. The measured S_{21} parameter of the ladder IH-RFQ.

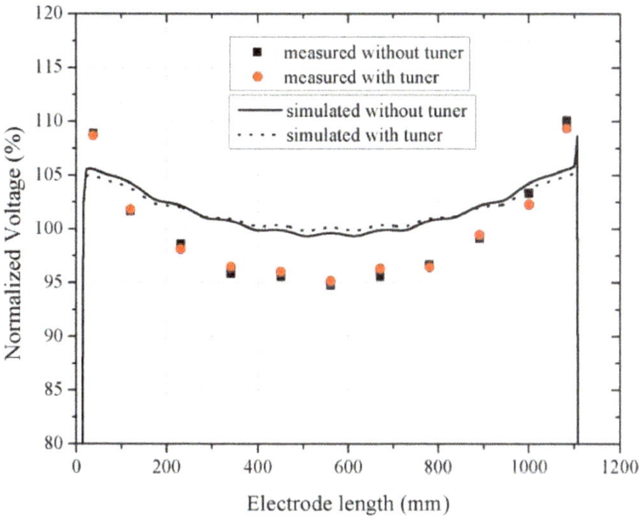

Fig. 14. Simulated and measured voltage unflatness of the ladder IH-RFQ.

Commissioning of the transmitter was carried out at first, because it had been more than 10 years since the last operation. During the experiment, a water load with 50 Ω impedance was used to consume the rf power from the transmitter. The inter-electrode voltage was started to be measured when the high vacuum in the RFQ cavity reached 10^{-5} Pa with both cooling water and high rf power. To obtain the inter-electrode voltage, a high purity Ge detector cooled by liquid nitrogen was used to measure the energy spectrum of X-ray [8]. An ORTEC computer multi-channel system was employed to deal with the signal, which consists of a preamplifier, a master amplifier, a PCI computer multi-channel card and a notebook PC. The maximum electron energy in the measured Roentgen ray spectrum corresponds to the inter-electrode voltage. The system was calibrated using two standard radiation source ^{241}Am and ^{152}Eu. After adjusting the amplification factor properly, γ ray of ^{241}Am 59.54 keV locates at channel 316 while ^{152}Eu 344.31 keV at channel 1889, then two peaks appear at channel 661 and channel 1340 corresponding to ^{152}Eu 121.78 keV and ^{152}Eu 244.66 keV, respectively. The above 4 points fit a well linearity, which means the calibration is successful.

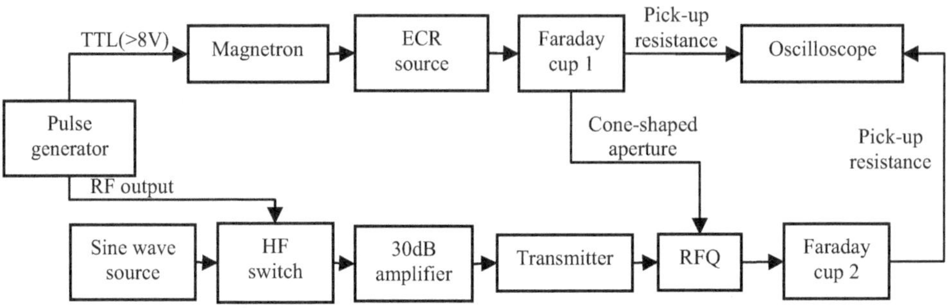

Fig. 15. The rf and measurement system of the experiment for the ladder IH-RFQ.

From the measured Roentgen spectrum at rf power 26.4 kW as shown in Fig.16, it can be found that the inter-electrode voltage reaches 72.1 kV, which is higher than the required value 60 kV. What should be paid attention to is that the spectrum was measured at the glass window locating near the entrance of the RFQ. According to Fig.14, the inter-electrode voltage unflatness is about ±7%, as a result, the average value of it along the cavity longitudinally should be 72.1/1.07 kV, i.e. 67.3 kV in this power case. All the measured voltages are normalized in this way. Figure 17 plots the variation of the square of the inter-electrode voltage with rf power, from which it can be seen that the linearity is well. The specific shunt impedance is defined as $R_s = (V^2/P) \bullet L = R_p \bullet L$, where L=1123 mm is the length of the RFQ. The slope of the fitting line that means the shunt impedance is 162 kΩ, so the corresponding specific shunt impedance is 178 kΩ•m. The result is better than that of simulation and cold measurement, which is because that the conductivity was set to be 5.0×10^7 S/m not 5.9×10^7 S/m in MWS simulations and the rf properties were improved via advance power test. The temperature of cooling water increased from 15.7 °C to 23.4 °C while the rf power increasing from 0 to 29.7 kW, and no sparking phenomenon appeared during the experiment. The results above indicate that the RFQ cavity possesses well mechanical strength and is water-cooled effectively.

The beam test has been performed using $^4He^+$ beam. The output current and transmission efficiency of the ladder IH-RFQ varying with the rf power fed into it are shown in Fig. 18 when the extraction current from the ECR source is 30 μA. The maximum transmission efficiency reaches 70%. During the experiment, the highest output beam is 210 μA, and the transmission efficiency is better than 70% at current of several tens μA. It should be figured out that the transmission efficiency is mainly limited by the unmatched beam injected to the RFQ, especially for high current beam. Further efforts will be made to improve it. The following works will also include the beam test of heavier ions, measurement of the energy spectrum, and some application studies on the RFQ as an implanter.

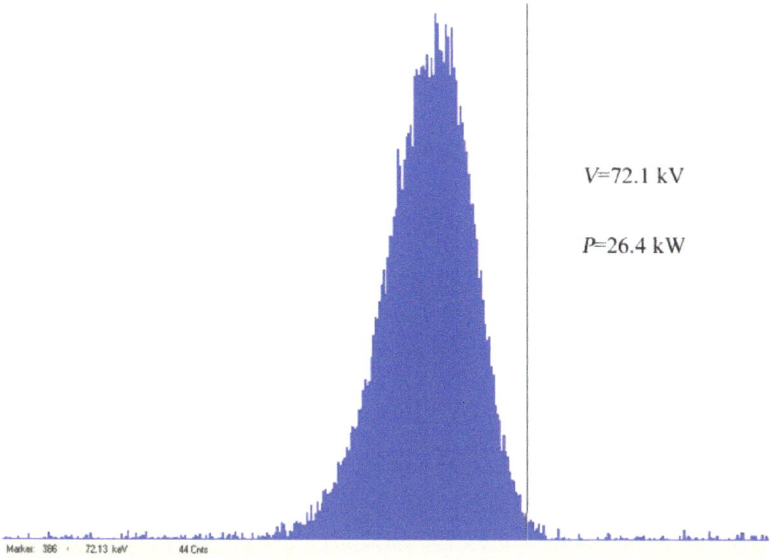

Fig. 16. Roentgen spectrum at 26.4kW rf power.

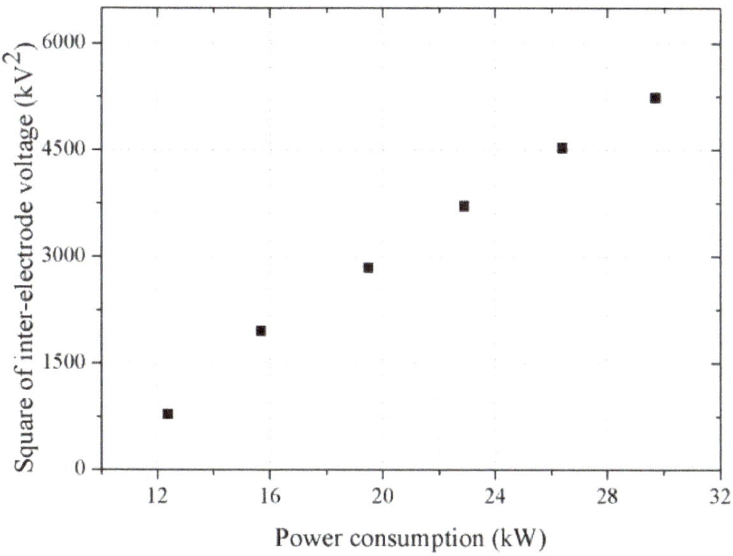

Fig. 17. Square of the inter-electrode voltage vs rf power.

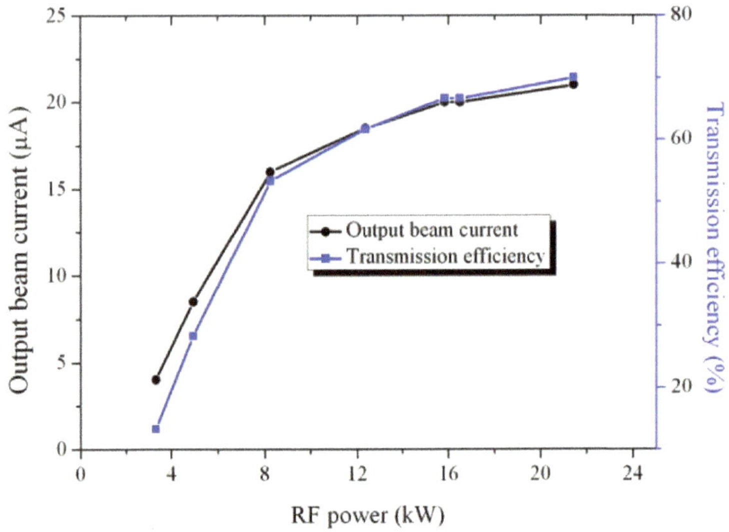

Fig. 18. The output current and transmission efficiency of the RFQ vs rf power at the extraction current 30μA($^4He^+$)

6. Summary

In conclusion, the RFQ accelerator is a competent ion implanter with many advantages such as high energy (MeV), high current (μA-mA), variable energy and ion species, compact size and so on. Further studies will be made on the practical usage of the RFQs in the ion implantation field. As for the ladder IH-RFQ accelerator of Peking University, after the further beam test is finished, some ion injection experiment will be made for the surface modification of materials.

7. Acknowledgement

The authors would like to thank Professor Dr. U. Ratzinger of IAP, J. W. Goethe University Frankfurt am Main, for the helpful discussions. Y. C. Nie would like to thank the China Scholarship Council (CSC) for the financial supports when he was in Frankfurt. This work was supported by the National Natural Science Foundation of China (Grant No. 10775009 and 11079001).

8. References

[1] M. Kapchinsky, V. A. Teplyakov, Linear Ion Accelerator with Spatially Homogeneous Strong Focusing, Prib. Tekh. Eksp., 1970, No.2, p.19.

[2] T. P. Wangler, Principles of RF Linear Accelerators, Wiley Series in Beam Physics and Accelerator Technology, 1998, p.225-257.

[3] H. Klein, Development of the Different RFQ Accelerating Structures and Operation Experience, IEEE Trans. Nucl. Sci., 1983, Vol. NS-30, p.3313.

[4] L. Young, 25 Years of Technical Advances in RFQ Accelerators, Proc. of PAC2003, p.60.
[5] A. Schempp, Research application of RFQ linac, Nuclear Instruments and Methods in Physics Research Section B, 1995, Vol.99, p.688-693.
[6] Ren Xiao-Tang, Lu Yuan-Rong, Yu Jin-Xiang, et al., Experimental research on simultaneous acceleration of positive and negative ions with equal q/m in an ISR RFQ, High Energy Physics and Nuclear Physics (in Chinese), 2000, Vol.24, p.347-351.
[7] U. Ratzinger, et al., The GSI 36 MHz high-current IH-type RFQ and HIIF-relevant extensions, Nuclear Instruments and Methods in Physics Research Section A, 1998, Vol.415, p.281-286.
[8] Y. R. Lu, C. E. Chen, J. X. Fang, et al., Investigation of high duty factor ISR RFQ-1000, Nuclear Instruments and Methods in Physics Research Section A, 2003, Vol.515, p.394-401.
[9] Kensuke Amemiya, Junya Ito and Katsumi Tokiguchi, Aluminum ion implantation using a variable energy RFQ implanter, Proc. of EPAC1998, p.2419.
[10] Byung-Wuek Lee, Jong-Won Kim and Holger Podlech, Design study of an RFQ for high-energy ion implantation, Journal of the Korean Physical Society, 2006, Vol.48, p.810-814.
[11] A. Schempp, et al., High duty factor nitrogen RFQ as a prototype high current implanter, Nuclear Instruments and Methods in Physics Research Section B, 1998, Vol.139, p.411-417.
[12] Zhiyu Guo, et al., Feasibility studies of RFQ based ^{14}C accelerator mass spectrometry, Nuclear Instruments and Methods in Physics Research Section B, 2007, Vol.259, p.204-207.
[13] NIE Yuan-Cun, LU Yuan-Rong, CHEN Jia-Er, et al., Theoretical design of a 104 MHz ladder type IH-RFQ accelerator, Chinese Physics Letters, 2010, Vol.27, p.112901.
[14] Fang Jiaxun, et al., RFQ accelerators for ion implantation, Nuclear Techniques (in Chinese), 2006, Vol.29, p.140-144.
[15] J. W. Staples, Reducing RFQ Longitudinal Emittance, Proc. of LINAC1994, p. 755.
[16] K. R. Crandall, R. H. Stokes, T. P. Wangler, RF quadrupole beam dynamics design studies, Proc. of LINAC1979, BNL-51134, p.205.
[17] X. Q. Yan, et al., Matched and equipartitioned design method for modern high-intensity radio frequency quadrupole accelerators, Nuclear Instruments and Methods in Physics Research Section A, 2007, Vol.577, p.402-408.
[18] www.cst.com
[19] H. Podlech, et al., Electromagnetic design of an 80.5 MHz RFQ for the RIA driver Linac, Proc. of EPAC2002, p.942-944.
[20] R. M. Hutcheon, An equivalent circuit model of the general 3-dimensional RFQ, IEEE Trans. Nucl. Sci., 1983, Vol. NS-30, p.3524-3526.
[21] J. X. Fang, and A. Schempp, Equivalent circuit of a 4-Rod RFQ, Proc. of EPAC92, p.1331-1333.
[22] J. Schmidt, et al., Tuning of the 4-Rod RFQ for MSU, Proc. of IPAC'10, p.762-764.
[23] P. Fischer and A. Schempp, Tuning of a 4-rod cw-mode RFQ accelerator, Proc. of LINAC06, p.728-730.
[24] P. G. Bricault, Methods for measurement of the specific shunt impedance of RFQ, TRIUMF DESIGN NOTE TRI-DN-94-30, 1994.

[25] Lu Yuan-Rong, Chen Wei, Nie Yuan-Cun, et al., Power test of the ladder IH-RFQ accelerator at Peking University, Chinese Physics Letters, 2011, Vol.28(7), p.072901.
[26] M. Zhang, S. X. Peng, Z. Z. Song, et al., Experimental results of an ECR oxygen source and a LEBT system for 1 MeV ISR RFQ accelerator upgrade project, Chinese Physics C, 2008, Vol.32(Suppl.I), p.220-222.

Section 2

Characterization

4

Spectroscopic Ellipsometry of Ion-Implantation-Induced Damage

Denis Shamiryan and Dmitriy V. Likhachev
Globalfoundries, Dresden
Germany

1. Introduction

In modern semiconductor manufacturing ion implantation requires precise control and such a control is impossible without adequate measurements of the implanted media. As the global trend of miniaturization dictates continuous decrease of the energy and dose of the implants, the measurement precision should match stricter requirements.

Well-established analysis techniques, including secondary ions mass-spectrometry (SIMS) (Benninghoven et al., 1987; Brundle et al., 1992; Alford et al., 2007), high-resolution transmission electron microscopy (HRTEM) (Hirsch et al., 1977; Horiuchi, 1994; Williams & Carter, 2009) allow direct observation of the effect of ion implantation, although they suffer from being destructive, time consuming and unusable for real-time in-line monitoring of ion implantation process or to measure uniformity. Some of the more common methods, like sheet resistance measurements (four-point probe, FPP) (Keenan et al., 1985; Johnson, 2001; Schroder, 2006) and thermal-wave technology (TW) (Smith et al., 1985; Guidotti & van Driel, 1985; Smith et al., 1986), are fast and relatively inexpensive, but require special efforts in order to get reliable information and exclude measurement errors. Ideally, the technique should be non-destructive to the sample, fast, accurate and precise to be used for routine process monitoring and control.

Among various characterization methods, ellipsometry is a fast, non-intrusive and informative technique that is sensitive to changes in optical properties and thicknesses of thin films. As ion implantation significantly changes optical properties of many materials (especially, crystalline), the ellipsometry is one of the most suitable technique for monitoring the results of ion implantation with no special sample preparation requirements. Ellipsometric analysis is able to detect and characterize the degree of crystallinity in buried layers as well as depth profiling in the sample. Besides the crystalline substrates, the ellipsometry is able to measure crust, formed on top of photoresist during masked ion implantation. The crust often poses a problem in post ion implantation photoresist strip and its characterization is important for optimization of strip processes.

Ellipsometry of implanted crystalline substrates is based on the fact that those substrates are amorphized and amorphous medium has very different optical properties from the crystalline one. In traditional models, ion-beam-induced amorphization occurs as a phase transition induced by adequate number of point defects created by individual ions (point-defect concentration exceeds some critical value; homogeneous amorphization) or the formation of continuous amorphous layer is due to overlapping of isolated damaged regions (heterogeneous amorphization) (Pelaz et al., 2004). Despite numerous studies, the mechanisms of amorphization by ion implantation are still under intensive investigations. Currently, various numerical simulations for ion implantation processes, like Monte-Carlo (MC) and molecular dynamics (MD) techniques, have been widely used to reconstruct the amorphization profiles (Sigmund, 2004; Ziegler et al., 2010; Nordlund, 1995; Beardmore & Grønbech-Jensen, 1998).

The amorphization depends on the crystalline medium, implanted elements, energy and dose of the implantation. There are number of papers describing dependence of the optical properties and depth of the amorphized layer on the processing conditions. However, the recent results indicate that even for rather low-energy low dose implants (with energy of several keV, dose around 10^{15} ions/cm^2), which do not cause amorphization, the changes of the optical properties of the implanted crystalline semiconductor are significant and can be detected by ellipsometry. The implantation depth measured by ellipsometry corresponds to direct observations of distorted crystalline lattice by TEM. These results allow extension of the ellipsometry capabilities toward the lower limits of energy and dose, where no amorphization occurs.

Ellipsometric measurements of the photoresist crust formation are based on the fact that energetic ions remove lighter photoresist fraction leaving behind graphite-like carbon-rich layer. Having optical properties different from those of the bulk photoresist, such a layer could be measured by ellipsometry. It should be noted, however, that with reduction of dose and energy of the ion implantation, the formation of this layer is less pronounced and it causes less issues in post ion implantation photoresist strip.

In the chapter, a short overview of ellipsometry principles is given in section 2. The application of ellipsometry to investigate ion implantation of crystalline and polycrystalline silicon is reviewed in section 3. The recent results obtained on very shallow low-dose implantation are also presented. Finally, summary and future perspectives are presented in section 4.

2. Fundamentals of ellipsometry

In this section, a very brief overview of the ellipsometry principles is presented. For much more extended and detailed up-to-date discussion on fundamental principles of ellipsometry, instrumentations, data analysis as well as multiple applications, see (Tompkins, 1993; Tompkins & McGahan, 1999; Tompkins & Irene (Eds.), 2005; Fujiwara, 2007; Losurdo et al., 2009; Azzam, 2010).

Ellipsometry (reflection polarimetry; single-wavelength as well as spectroscopic) is an optical measurement technique for evaluation geometrical and material properties of substrates, thin films and multilayer structures. Determination of fundamental optical properties (complex dielectric functions (ε) or complex refractive indexes (N)) gives an opportunity to characterize other important material properties such as composition, phase structure, doping, stress, uniformity, electrical properties, etc. The principles of this technique as well as first applications have been established in the late 19th century but it became widely utilized only in 1960s and 1970s due to significant developments in the instrumentation, computers and data analysis algorithms (Azzam & Bashara, 1977; Rzhanov et al., 1979; Theeten & Aspnes, 1981; Riedling, 1988; Azzam (Ed.), 1991).

Ellipsometry measures the changes in the polarization state of light upon reflection from a sample surface at non-normal (oblique) incidence (although transmission ellipsometry at normal incidence can be used for optically anisotropic samples) and those changes typically expressed in terms of two values (ellipsometric angles) called Psi (Ψ) and Delta (Δ). These represent an amplitude ratio and relative phase shift between p- and s-components of the polarized light[1] which are induced by reflection from the sample (see Fig. 1).

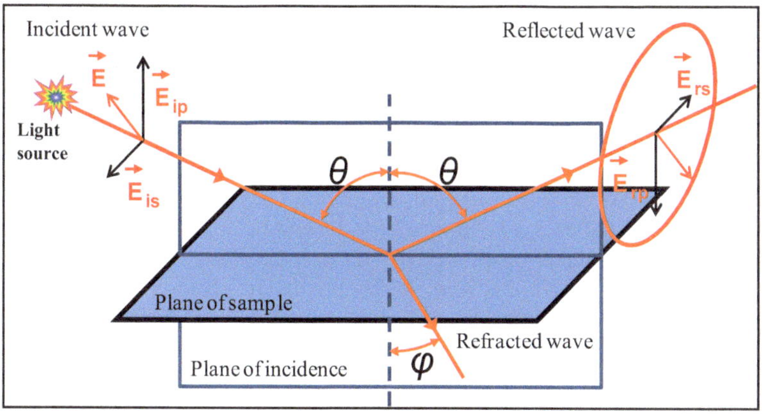

Fig. 1. Scheme showing the basic principle of ellipsometry: linearly polarized light with p- and s-components at oblique incidence is reflected and it becomes elliptically polarized. The parameters of the resulting ellipse depend on the initial direction of polarization, the angle of incidence and the optical properties of the surface. $\mathbf{E_{ip}}$, $\mathbf{E_{is}}$, $\mathbf{E_{rp}}$, and $\mathbf{E_{rs}}$ represent the p- and s-components of the incident (i) and reflected (r) light waves.

The measured quantities Ψ and Δ are described by the fundamental equation of ellipsometry

$$\rho \equiv \tan(\Psi)e^{i\Delta} \equiv \frac{R_p}{R_s} \equiv \left(\frac{E_{rp}}{E_{ip}}\right) \Big/ \left(\frac{E_{rs}}{E_{is}}\right), \tag{1}$$

where R_p and R_s are the Fresnel reflection coefficients for the p- and s-polarized light, respectively. The reflection coefficients are directly related to the optical properties of the sample. In the simple case of single thin film on substrate (see Fig. 2), ρ can be described as a function

$$\rho \equiv \tan(\Psi)e^{i\Delta} = f(N_0, N_1, N_2, \theta_0, d), \tag{2}$$

[1] The term "p-polarization" was taken from the German word *"parallel"* since this component of the electric field **Ep** is parallel to the plane of incidence. The component perpendicular to this plane, **Es**, was named "s-polarization" and derived from the German word *"senkrecht"* (perpendicular).

where $N = n - ik$ is the complex index of refraction.[2]

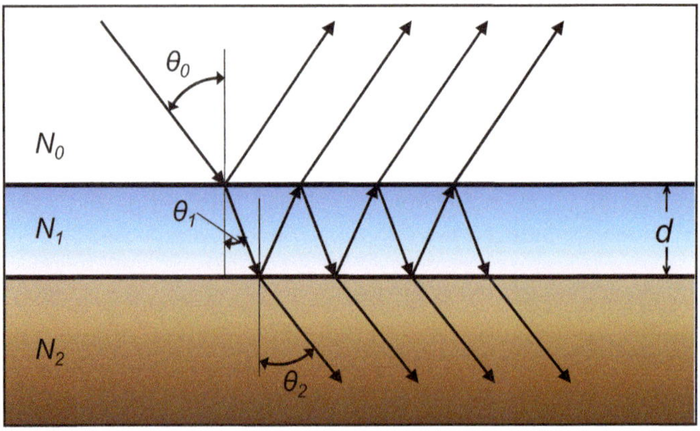

Fig. 2. Three-phase (ambient/film/substrate) optical model.

There are two major flavours of ellipsometric technique. Single-wavelength ellipsometry (SWE) uses a monochromatic light source (typically, high-intensity probe beam from HeNe laser with a wavelength of 632.8 nm) which results in only single set of ellipsometric parameters (Ψ,Δ). Multi-wavelength approach or spectroscopic ellipsometry (SE) involves measurements of Ψ and Δ as functions of wavelength (Tompkins & McGahan, 1999; Fujiwara, 2007). Therefore, the SE determines the complex index of refraction as a function of wavelength λ (material dispersion; $N = f(\lambda)$). This is definitely an advantage of the SE as compared with SWE since knowledge of the n and k spectra is very helpful in establishing the correlations to various physical and chemical properties of the materials (for instance, the ultraviolet (UV) range of the n and k spectra allows us to determine the degree of crystallinity in polycrystalline silicon or, together with visible range, to ascertain SiGe stoichiometry). In variable-angle spectroscopic ellipsometry (VASE), a variant of the SE, the ellipsometric angles Ψ and Δ are measured as functions of both wavelength and angle of incidence (AOI) which results in higher sensitivity and accuracy due to additional optical paths for various AOI's (Woollam, 1999 ; Woollam et al., 1999 ; Johs et al., 1999).

The real part of N, namely n, is the conventional refractive index. The imaginary part of N, namely k, is called the extinction coefficient and it is directly related to the absorption coefficient of the medium:

$$\alpha = \frac{4\pi k}{\lambda}$$

where λ is the free-space wavelength of light.

These two wavelength-dependent quantities are often called the "optical constants" of the material, although they are not constants in fact but depend, in addition to the λ (or light

[2] Here we follow the traditional ellipsometric convention which defines the imaginary part of N with a „minus" sign („The Nebraska Convention", 1968; see (Muller, 1969; Holm, 1991; Bennett, 2010)). In other areas of physics the complex index of refraction is defined with a „plus" sign.

frequency), also on material's conditions. The "optical constants" can be considered as *intrinsic* properties of the material which completely characterize material's optical response.

There is an equivalent approach to describe the optical properties in terms of the complex dielectric function ε (sometimes also called "complex dielectric constant"):

$$\varepsilon = \varepsilon_1 - i\varepsilon_2.$$

The complex index of refraction N and complex dielectric function ε are related to each other through the following relation derived from Maxwell's equations:

$$\varepsilon = N^2.$$

Therefore,

$$\varepsilon_1 = n^2 - k^2,$$
$$\varepsilon_2 = 2nk$$

and

$$n = \frac{1}{\sqrt{2}}\left(\varepsilon_1 + \sqrt{\varepsilon_1^2 + \varepsilon_2^2}\right)^{1/2},$$
$$k = \frac{1}{\sqrt{2}}\left(-\varepsilon_1 + \sqrt{\varepsilon_1^2 + \varepsilon_2^2}\right)^{1/2}.$$

The fact that ellipsometry measures the changes in polarization (rather than the absolute light intensity or absolute phase) makes it highly accurate, robust and very reproducible technique. Presence of "phase" information (Δ) also makes ellipsometric measurements extremely sensitive for the analysis of surfaces.

Experimentally, the ellipsometric parameters Ψ and Δ can be determined with very high precision. However, the next step, namely, extraction of required information (the optical constants n and k and film thickness(-es)) from those measured values, requires adequate optical model for calculations and it might be quit complicated to construct one in some cases. Thus, the ellipsometric technique is the *indirect* characterization method. [3] Then using linear regression analysis technique the film stack parameters are determined by minimizing fitting errors to the measured spectroellipsometric data using various error functions (for more details on data analysis procedure see, for instance, (Jellison, 1993, 1998; Fujiwara, 2007)).

3. Spectroscopic ellipsometry measurements on implanted silicon wafers

In the past few decades the spectroscopic ellipsometry was extensively used to investigate ion implantation (Si^+, Ge^+, B^+, P^+, As^+, Ar^+, Xe^+, and N_2^+) of crystalline and polycrystalline silicon (Ibrahim & Bashara, 1972; Adams & Bashara, 1975; Adams, 1976; Jellison et al., 1981; Ohira & Itakura, 1982; Lohner et al., 1983; Vasquez et al., 1985; Vedam et al., 1985; Nguyen &

[3] Direct n and k extraction from the ellipsometric parameters Ψ and Δ is only possible in case of flat, isotropic and homogeneous substrates (medium of semi-infinite thickness).

Vedam, 1990; Miyazaki & Adachi, 1993; Fried et al., 1992, 2004; Müller-Jahreis et al., 1995; Shibata et al., 1999, 2010; Giri et al., 2001; Tsunoda et al., 2002; Petrik et al., 2003; Yoshida et al., 2005; Stevens et al., 2006; Lioudakis et al., 2006a, 2006b; Petrik, 2008; Matsuda et al., 2010; Mohacsi et al., 2011). One of the motivations of that was to get non-contact, non-destructive and rapid measurement technique with high accuracy and sensitivity for industrial applications (in particular, for integrated circuits (IC) manufacturing). For instance, Vanhellemont et al. ([Van91], as cited in Petrik, 1999) claim that spectroscopic ellipsometry "can be considered as a non-destructive, cheap poor man's optical Rutherford backscattering spectrometer and even as a one-dimensional optical high-resolution microscope". With a proper optical model, calibrated to other analytical techniques, it is possible to non-destructively characterize damage depth profiles with various degrees of crystallinity as well as effects of subsequent annealing of ion-implanted Si layers.

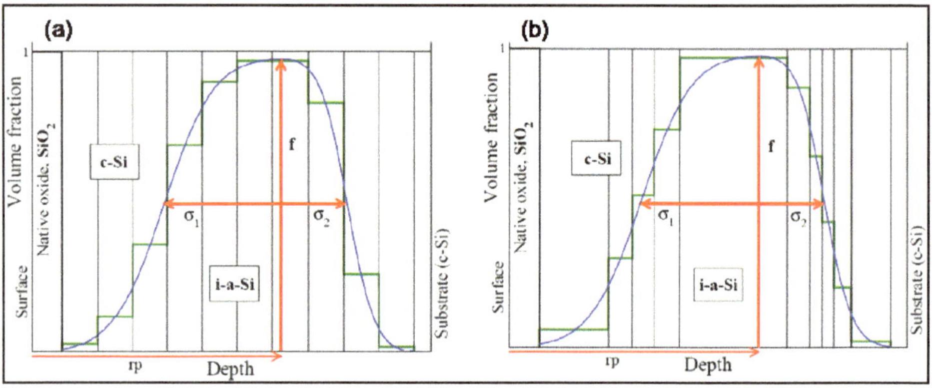

Fig. 3. Realistic damaged depth profile models where the damaged regions are divided into sublayers: (a) with fixed thicknesses (Fried et al., 1992); (b) with thicknesses inversely proportional to the slope of the profile (Petrik et al., 2003) (reprinted from (Petrik et al., 2008), with permission from Dr. P. Petrik).

The usual way of describing the ion-implanted media includes two constituents. At first, the parametrization of the amorphization profile needs to be established. For instance, typical layer structure for ion-damaged crystalline of polycrystalline Si consists of native oxide on the top, an extensively damaged region where amorphization exceeds critical amorphization density, and partially amorphous layer underneath. Those damaged regions can be divided in sublayers with various ratios for crystalline and amorphous components. Very sophisticated damaged depth profile models were introduced (Fig. 3) in which coupled half-Gaussian functions used to describe the damage levels in the sublayers with fixed thicknesses (Fried et al., 1992) or with thicknesses inversely proportional to the slope of the profile (Petrik et al., 2003). Secondly, it is necessary to describe the optical properties of the disordered layers or, in other words, parametrize the complex dielectric functions of such layers (sublayers). Typically, it can be modeled as a composition of crystalline (c-Si) and amorphous, ion-induced damaged (i-a-Si), phases using self-consistent Bruggeman effective medium approximation (B-EMA). Since dielectric functions of the amorphous silicon are not unique, various optical models can be selected to obtain its optical properties (for example, well-established Tauc-Lorentz (TL) optical model (Jellison & Modine, 1996a, 1996b) which has been the most widely used parametrization of the optical functions for amorphous materials).[4]

However, even a simple model of the ion-damaged Si layer could be used for characterization of ion-implanted Si for semiconductor manufacturing, where the robustness and the speed of measurements are the key factors. The measurements are based on the fact that the top part of the crystalline Si is amorphized by high energy ions and since amorphous Si (a-Si), crystalline Si (c-Si) and SiO_2 (which inevitably exists on top of Si after exposure to air) have very distinctive optical properties, as shown in Fig. 4.

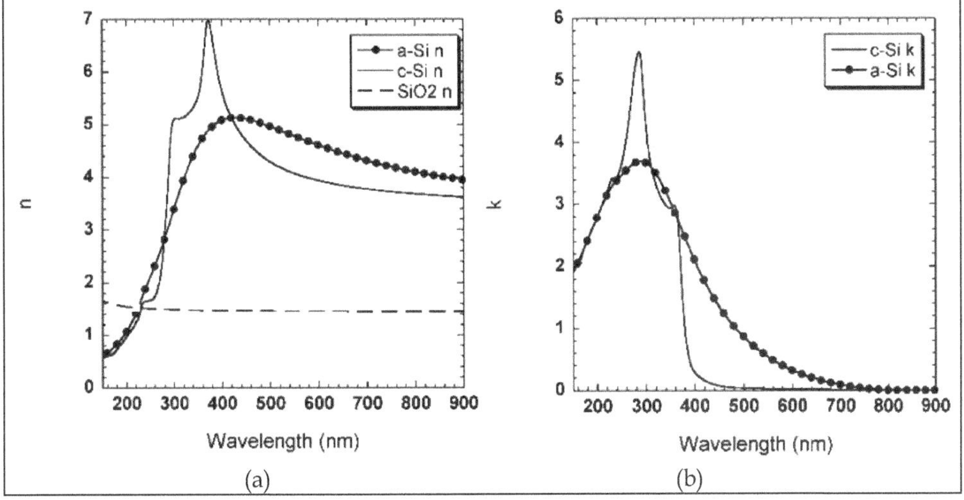

Fig. 4. Dispersions of the refractive index n (a) and the extinction coefficient k (b) of crystalline Si, amorphous Si and SiO_2. Extinction of SiO_2 is zero in the plotted range.

Significant difference in optical properties of those three layers allows thicknesses measurement of both the amorphized layer and SiO_2 on top of the crystalline Si substrate after ion implantation. It should be noted that the implantation energies in the aforementioned works are rather higher (10 keV or more) while modern ultra-shallow junctions require implantation energy close to 1 keV or even less. At such low energies Si might not be amorphized but only damaged. However, recently published papers (Shamiryan et al., 2010; Radisic et al., 2009, 2010) demonstrated that spectroscopic ellipsometry can be used for implanted Si measurements even in the case of low energy ion implantation, when no Si amorphization is observed.

Si implanted with B and As species at low energies (as specified in Table 1) was measured by spectroscopic ellipsometry with a spectral range of 150–895 nm. The optical model consisted of a Si substrate and two layers: SiO_2 on top and a damaged Si (d-Si) layer at the

[4] Recently, Ferlauto et al. suggested the Cody-Lorentz dispersion model which has a few advantages over TL model and better describes the optical properties of some amorphous materials (see details in (Ferlauto et al., 2002)).

bottom. The SiO_2 model was taken from library since the optical properties of SiO_2 were considered not modified by ion implantation as it has been reported that the change of refractive index of silica upon ion implantation does not exceed 1-2% (Bayly & Townsend, 1973; Webb & Townsend, 1976). This could be easily explained by the fact that, unlike Si substrate, the SiO_2 layer is already amorphous prior to ion implantation and energetic ions do not significantly change its state. The model for the damaged (implanted) Si layer was based on the a-Si model represented by a set of harmonic oscillators. In order to fit the model to the measured spectra, the thickness of both SiO_2 and d-Si were varied as well as optical properties of the d-Si layer. After the implantation, besides ellipsometry, the samples were also inspected by transmission electron microscopy (TEM). This technique allows direct observation of the crystallinity of the layers and determination of their thicknesses using interatomic distance as a reference.

Element	Energy, keV	Dose, cm^{-2}	Tilt, degrees
B	0.5; 1; 3	$1.5x10^{15}$	7
As	1; 1.5	$1.5x10^{15}$	7

Table 1. Ion implantation conditions.

After fitting of the two layer model to the ellipsometric data obtained after ion implantation it was found that the top damaged layer is indeed similar in optical properties to amorphous Si. Fig. 5 shows the *n* and *k* dispersions for As- and B-doped Si along with the dispersions for a-Si taken from the library. One can see that the dispersions of both implanted layers are closer to the a-Si than to c-Si (cf. Fig. 4).

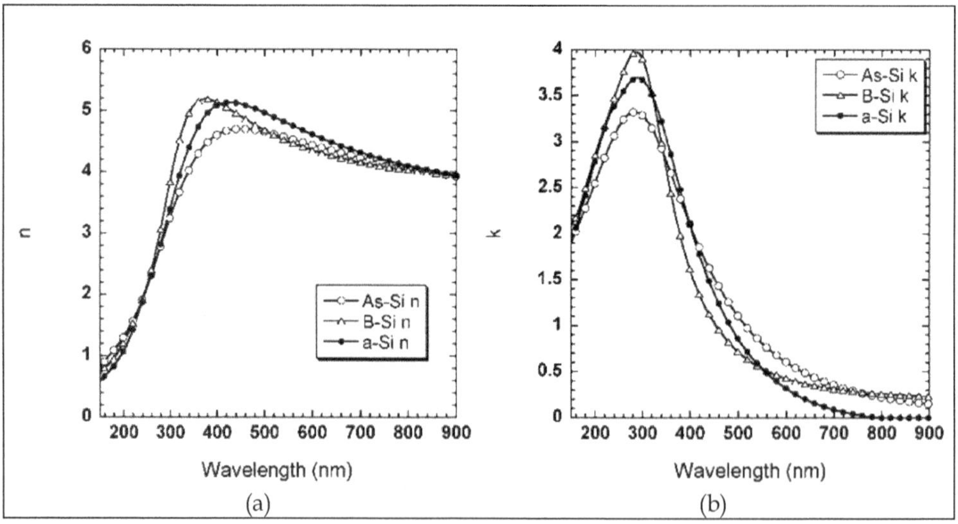

Fig. 5. Dispersions of *n* (a) and *k* (b) for a-Si (solid curve – taken from KLA-Tencor library) and for As (1 keV, $1.5x10^{15}$ cm^{-2})- and B (0.5 keV, $1.5x10^{15}$ cm^{-2}) -doped Si (circles and triangles, respectively) as measured by SE.

TEM micrographs (Fig. 6) reveal that for implantation of B with energies of 0.5 keV and 1 keV the Si substrate is not amorphized. However, the top implanted layer is still visible on TEM images due to strain induced by ion implantation. The thickness of this strained layer is in a good agreement with SE measurements. As implantation energy of B increases to 3 keV, TEM images show two layers: a bottom strained layer and a top amorphous layer. The total thickness of these two layers is in agreement with SE measurements. Similar results were obtained for As, but the top layer was amorphized even for the lowest studied implantation energy of 1 keV. Summary of TEM measurements and SE measurements is shown in Table 2. One can see that both d-Si and SiO_2 measurements are in a good agreement with TEM data.

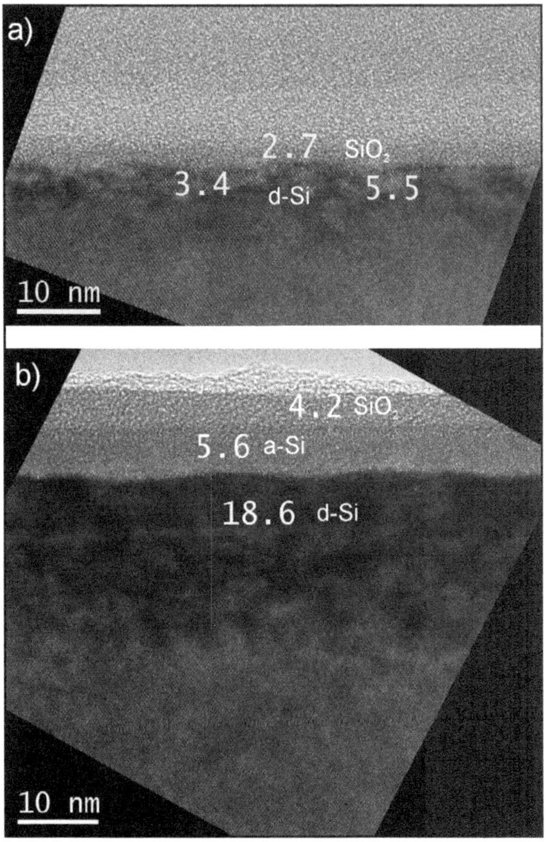

Fig. 6. TEM images of Si implanted with B at 0.5 keV (a) and 3 keV (b), the dose is $1.5x10^{15}$ in both cases. All measurements on the images are in nm.

From the comparison of the TEM and SE measurements we can make two important conclusions:

1. Even when the implanted Si is not amorphized, its optical properties change significantly, so it can be easily distinguished from c-Si.

2. Since the optical properties of the d-Si are close to those of a-Si (see Fig. 5) ellipsometry can hardly distinguish between those two layers when they both are present at higher implantation energies.

Therefore, it is possible to measure thickness of the implanted Si layer even though it might be impossible to tell whether the implanted layer is just strained or amorphized.

	B implantation energy, keV			As implantation energy, keV
Technique	0.5	1	3	1.5
	SiO_2 thickness (nm)			
TEM	2-2.7	2.4-2.6	2.1-2.9	2.3
SE	2.8±0.1	2.4±0.1	2.5±0.1	2.7
	Damaged Si thickness (nm)			
TEM	3.5-5.5	7-10	20-24	7-10
SE	4.7±0.1	7.4±0.1	23.5±0.2	6.0

Table 2. Summary of TEM and SE measurements of thickness of the implanted Si substrates.

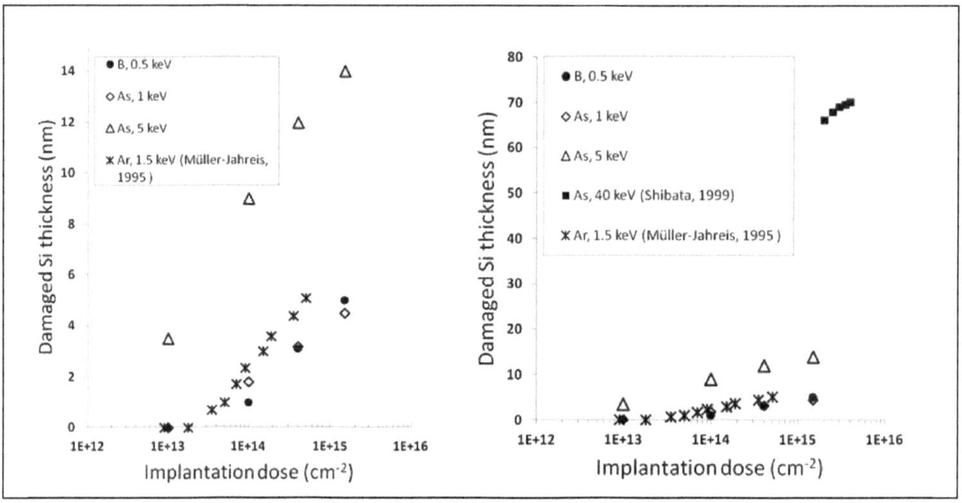

Fig. 7. Depth of the damaged Si as function of the implantation dose, as measured by ellipsometry. The left panel shows all data, the right one shows the data obtained for low energies (less than 10 keV).

The results of the ellipsometric measurements of ion-implanted monocrystalline Si are shown in Figures 7 and 8. One can see that there are detection limits above which the ellipsometry can detect damaged Si layer. The detection limits depend on the dose, energy and ion mass. The dependence of the damaged layer thickness on dose is logarithmic (the thickness is proportional to the logarithm of the dose; see Fig. 7), while the dependence of the damaged thickness d on the ion energy E exhibits power dependence in the form

$$d = A \cdot E^x ,$$

where A is a constant related to the implanted element, and x is the power factor in the range of 0.7-0.8 (see Fig. 8).

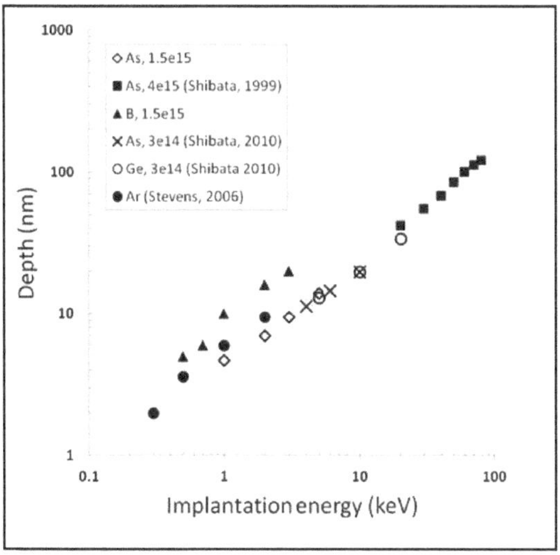

Fig. 8. Depth of the damaged Si as function of the ion implantation energy, as measured by ellipsometry.

4. Conclusion

In this chapter the basic principles of ellipsometry as well as application of spectroscopic ellipsometry to investigate ion implantation of crystalline and polycrystalline silicon have been reviewed. Spectroscopic ellipsometry can be used for characterization of ion-implanted crystalline substrates before anneal. Ion implantation creates a damaged Si layer, whether amorphized (for higher implantation energies) or just strained with a distorted lattice. SE can measure this damaged layer since its optical properties in both cases (amorphization or lattice distortion) are similar and significantly different from the crystalline Si. Due to similarity in optical properties, SE cannot distinguish between amorphized and distorted Si layer.

5. Acknowledgments

The authors would like to thank *imec* pilot line (Leuven, Belgium) for sample preparation, Dr. Andrey Zakharov for fruitful and spirited discussions and Dr. Manfred Mört for critical reading of the manuscript and comments.

6. References

Adams, J. R. & Bashara, N. M. (1975). Determination of the complex refractive index profiles in P^{+}_{31} ion implanted silicon by ellipsometry. *Surface Science,* Vol. 49, No. 2, (April 1975), pp. 441-458, ISSN 0039-6028

Adams, J. R. (1976). Complex refractive index and phosphorus concentration profiles in P^{+}_{31} ion implanted silicon by ellipsometry and auger electron spectroscopy. *Surface Science,* Vol. 56, (June 1976), pp. 307-315, ISSN 0039-6028

Alford, T. L.; Feldman, L. C., & Mayer, J. W. (2007). *Fundamentals of Nanoscale Film Analysis,* Springer Science+Business Media, Inc., ISBN 978-0-387-29260-1, New York, NY, U.S.A.

Azzam, R. M. A. & Bashara, N. M. (1977). *Ellipsometry and Polarized Light,* North-Holland Publishing Co., ISBN 978-0-720-40694-8, New York, NY, U.S.A.

Azzam, R. M. A. (Ed.) (1991). *Selected Papers on Ellipsometry,* SPIE Optical Engineering Press, ISBN 978-0-819-40571-5, Bellingham, WA, U.S.A.

Azzam, R. M. A. (2010). Ellipsometry, In: *Handbook of Optics, Vol.I. Geometrical and Physical Optics, Polarized Light, Components and Instruments,* 3rd Ed., M. Bass, (Ed.), pp. 16.1–16.25, McGrow-Hill Companies, Inc., ISBN 978-0-071-62925-6, New York, NY, U.S.A.

Bayly, A. R. & Townsend, P. D. (1973). Ellipsometric analysis of refractive index profiles produced by ion implantation in silica glass. *Journal of Physics D: Applied Physics,* Vol. 6, No. 9, (June 1973) pp. 1115-1128, ISSN 0022-3727

Beardmore, K. M. & Grønbech-Jensen, N. (1998). Efficient molecular dynamics scheme for the calculation of dopant profiles due to ion implantation. *Physical Review E,* Vol. 57, No. 6, (June 1998), pp. 7278–7287, ISSN 1539-3755

Bennett, J. M. (2010). Polarization, In: *Handbook of Optics, Vol.I. Geometrical and Physical Optics, Polarized Light, Components and Instruments,* 3rd Ed., M. Bass, (Ed.), pp. 12.3–12.31, McGrow-Hill Companies, Inc., ISBN 978-0-071-62925-6, New York, NY, U.S.A.

Benninghoven, A.; Rüdenauer, F. G., & Werner, H. W. (1987). *Secondary Ion Mass Spectrometry: Basic Concepts, Instrumental Aspects, Applications, and Trends,* John Wiley & Sons, Ltd, ISBN 978-0-471-01056-2, New York, NY, U.S.A.

Brundle, C. R.; Evans, C. A., Jr., & Wilson, S. (1992). *Encyclopedia of Materials Characterization: Surfaces, Interfaces, Thin Films,* Butterworth-Heinemann & Manning Publications Co., ISBN 978- 0-750-69168-0, Boston, MA, U.S.A.

Ferlauto, A.S.; Ferreira, G. M., Pearce, J. M., Wronski, C. R., Collins, R. W., Deng, X., Ganguly, G. (2002). Analytical model for the optical functions of amorphous semiconductors from the near-infrared to ultraviolet: Applications in thin film photovoltaics. *Journal of Applied Physics,* Vol. 92, No.5, (September 2002), pp. 2424-2436, ISSN 0021-8979

Fried, M.; Lohner, T., Aarnink, W. A. M., Hanekamp, L. J., & van Silfhout, A. (1992). Nondestructive determination of damage depth profiles in ion-implanted semiconductors by spectroscopic ellipsometry using different optical models. *Journal of Applied Physics,* Vol. 71, No.6, (March 1992), pp. 2835-2843, ISSN 0021-8979

Fried, M.; Petrik, P., Lohner, T., Khánh, N. Q., Polgár, O., & Gyulai, J. (2004). Dose-dependence of ion implantation-caused damage in silicon measured by ellipsometry and backscattering spectrometry. *Thin Solid Films,* Vol. 455-456, (May 2004), pp. 404-409, ISSN 0040-6090

Fujiwara, H. (2007). *Spectroscopic Ellipsometry: Principles and Applications,* John Wiley & Sons, Ltd, ISBN 978-0-470-01608-4, Chichester, UK

Giri, P. K.; Tripurasundari, S.; Raghavan, G.; Panigrahi, B. K.; Magudapathy, P.; Nair, K. G. M.; & Tyagi, A. K. (2001). Crystalline to amorphous transition and band structure evolution in ion-damaged silicon studied by spectroscopic ellipsometry. *Journal of Applied Physics,* Vol. 90, No.2, (July 2001), pp. 659-669, ISSN 0021-8979

Guidotti, D. & van Driel, H. M. (1985). Spatially resolved defect mapping in semiconductors using laser-modulated thermoreflectance. *Applied Physics Letters,* Vol. 47, No. 12, (December 1985), pp. 1336-1338, ISSN 0003-6951

Hirsch, P. B.; Howie, A., Nicholson, R. B., Pashley, D.W., & Whelan, M. J. (1977). *Electron Microscopy of Thin Crystals* (2nd Ed.), R. E. Krieger Pub. Co., ISBN 978-0-882-75376-8, Huntington, NY, U.S.A.

Holm, R. T. (1991). Convention confusions, In: *Handbook of Optical Constants of Solids II,* E. D. Palik, (Ed.), pp. 21–55, Academic Press, ISBN 978-0-125-44422-2, San Diego, CA, U.S.A.

Horiuchi, S. (1994). *Fundamentals of High-Resolution Transmission Electron Microscopy,* North-Holland Publishing Co., ISBN 978-0–444–88744–3, Amsterdam, London, New York, Tokyo

Ibrahim, M. M. & Bashara, N. M. (1972). Ellipsometric study of 400ev ion damage in silicon. *Surface Science,* Vol. 30, No. 3, (May 1972), pp. 632-640, ISSN 0039-6028

Jellison, Jr., G. E. (1993). Data analysis for spectroscopic ellipsometry. *Thin Solid Films,* Vol. 234, No.1-2, (October 1993), pp. 416-422, ISSN 0040-6090

Jellison, Jr., G. E. (1998). Spectroscopic ellipsometry data analysis: measured versus calculated quantities. *Thin Solid Films,* Vol. 313-314, (February 1998), pp. 33-39, ISSN 0040-6090

Jellison, Jr., G. E. & Modine, F. A. (1996a) Parametrization of the optical functions of amorphous materials in the interband region. *Applied Physics Letters,* Vol. 69, No. 3, (July 1996), pp. 371-373, ISSN 0003-6951

Jellison, Jr., G. E. & Modine, F. A. (1996b) Parametrization of the optical functions of amorphous materials in the interband region. *Applied Physics Letters,* Vol. 69, No. 14, (September 1996), p. 2137, ISSN 0003-6951

Jellison, Jr., G. E.; Modine, F. A., White, C. W., Wood, R. F., & Young, R. T. (1981). Optical properties of heavily doped silicon between 1.5 and 4.1 eV. *Physical Review Letters,* Vol. 46, No. 21, (May 1981), pp. 1414–1417, ISSN 0031-9007

Johnson, W. H. (2001). Sheet resistance measurements of interconnect films, In: *Handbook of Silicon Semiconductor Metrology,* A. C. Diebold, (Ed.), pp. 215–244, Marcel Dekker, Inc., ISBN 978-0-824-70506-8, New York, NY, U.S.A.

Johs, B. ; Woollam, J. A., Herzinger, C. M., Hilfiker, J., Synowicki, R., & Bungay, C. L. (1999). Overview of variable angle spectroscopic ellipsometry (VASE), Part II: Advanced applications, *Optical metrology, Proceedings of SPIE Conference,* Vol. CR72, pp. 29-58, ISBN 978-0-819-43235-3, Denver, CO, U.S.A., July 18-19, 1999

Keenan, W. A.; Johnson, W. H., & Smith, A. K. (1985). Advances in sheet resistance measurements for ion implant monitoring. *Solid State Technology,* Vol. 28, No. 6, (June 1985), pp. 143–148, ISSN 0038-111X

Lioudakis, E.; Christofides, C., & Othonos, A. (2006a). Study of the annealing kinetic effect and implantation energy on phosphorus-implanted silicon wafers using spectroscopic ellipsometry. *Journal of Applied Physics,* Vol. 99, No.12, (June 2006), pp. 123514-123514-6, ISSN 0021-8979

Lioudakis, E.; Nassiopoulou, A., & Othonos, A. (2006b). Ellipsometric analysis of ion-implanted polycrystalline silicon films before and after annealing. *Thin Solid Films,* Vol. 496, No.2, (February 2006), pp. 253-258, ISSN 0040-6090

Lohner, T.; Mezey, G., Kótai, E., Pászti, F., Manuaba, A., & Gyulai, J. (1983). Characterization of ion implanted silicon by ellipsometry and channeling. *Nuclear Instruments and Methods in Physics Research,* Vol. 209-210, Pt.2, (May 1983), pp. 615-620, ISSN 0168-9002

Losurdo, M.; Bergmair, M., Bruno, G., Cattelan, D., Cobet, C., Martino, A., Fleischer, K., Dohcevic-Mitrovic, Z., Esser, N., Galliet, M., Gajic, R., Hemzal, D., Hingerl, K., Humlicek, J., Ossikovski, R., Popovic, Z. V., & Saxl, O. (2009). Spectroscopic ellipsometry and polarimetry for materials and systems analysis at the nanometer scale: state-of-the-art, potential, and perspectives. *Journal of Nanoparticle Research,* Vol. 11, No.7, (October 2009), pp. 1521-1554, ISSN 1388-0764

Matsuda, A.; Nakakuboa, Y., Takaoa, Y., Eriguchi, K., & Ono, K. (2010). Modeling of ion-bombardment damage on Si surfaces for in-line analysis. *Thin Solid Films,* Vol. 518, No.13, (April 2010), pp. 3481-3486, ISSN 0040-6090

Miyazaki, T. & Adachi, S. (1993). Spectroscopic ellipsometry study of Si surfaces modified by low-energy Ar+-ion irradiation. *Japanese Journal of Applied Physics,* Vol. 32, No. 11A, (November 1993), pp. 4941-4945, ISSN 0021-4922

Mohacsi, I.; Petrik, P., Fried, M., Lohner, T., van den Berg, J. A., Reading, M. A., Giubertoni, D., Barozzi, M., & Parisini, A. (2011). Characterisation of ultra-shallow disorder profiles and dielectric functions in ion implanted Si. *Thin Solid Films,* Vol. 519, No. 9 (February 2011), pp. 2847-2851, ISSN 0040-6090

Muller, R. H. (1969). Definitions and conventions in ellipsometry. *Surface Science,* Vol. 16, (August 1969), pp. 14-33, ISSN 0039-6028

Müller-Jahreis, U.; Thiele, P., Bouafia, M., & Seghir, A. (1995). Determination of low-energy ion implantation damage parameters by an ellipsometric method. *Journal de Physique III,* Vol. 5, No. 5, (May 1995), pp. 575-584

Nguyen, N.V. & Vedam, K. (1990). Spectroscopic ellipsometry studies of crystalline silicon implanted with carbon ions. *Journal of Applied Physics,* Vol. 67, No.8, (April 1990), pp. 3555- 3559, ISSN 0021-8979

Nordlund, K. (1995). Molecular dynamics simulation of ion ranges in the 1-100 keV energy range. *Computational Materials Science,* Vol. 3, No. 4, (March 1995), pp. 448–456, ISSN 0927-0256

Ohira, F. & Itakura, M. (1982). Ellipsometric measurement of damage depth profiles for ion beam processed Si surface layer. *Japanese Journal of Applied Physics,* Vol. 21, No. 1, (January 1982), pp. 42-46, ISSN 0021-4922

Pelaz, L.; Marqués, L.A., & Barbolla, J. (2004). Ion-beam-induced amorphization and recrystallization in silicon. *Journal of Applied Physics,* Vol. 96, No.11, (December 2004), pp. 5947-5976, ISSN 0021-8979

Petrik, P. (1999). Characterization of polysilicon thin films using *in situ* and *ex situ* spectroscopic ellipsometry. *Ph.D. Thesis,* Technical University of Budapest and Research Institute for Technical Physics and Materials Science of the Hungarian Academy of Sciences, Budapest, 1999, p.65

Petrik, P.; Polgár, O., Fried, M., Lohner, T., Khánh, N. Q., & Gyulai, J. (2003). Ellipsometric characterization of damage profiles using an advanced optical model. *Journal of Applied Physics,* Vol. 93, No.4, (February 2003), pp. 1987-1990, ISSN 0021-8979

Petrik, P. (2008). Ellipsometric models for vertically inhomogeneous composite structures. *physica status solidi (a),* Vol. 205, No.4, (April 2008), pp. 732-738, ISSN 1862-6300

Petrik, P.; Lohner, T. , Polgár, O., & Fried, M. (2008). Ellipsometry on ion implantation induced damage, *16th IEEE International Conference on Advanced Thermal Processing of Semiconductors (RTP 2008),* pp. 93-101, ISBN 978-1-4244-1950-0, Las Vegas, NV, U.S.A., September 30 - October 3, 2008

Radisic, D.; Shamiryan, D., Mannaert, G., Boullart, W., Rosseel, E., Bogdanowicz, J., Goosens, J., Marrant, M., & Bender, H. (2009). Metrology for implanted Si substrate and dopant loss studies, *Cleaning and Surface Conditioning Technology in Semiconductor Device Manufacturing 11, 216th ECS Meeting,* pp. 367-374, ISSN 1938-5862, Vienna, Austria, October 4-9, 2009

Radisic, D.; Shamiryan, D., Mannaert, G., Boullart, W., Rosseel, E., Bogdanowicz, J., Goossens, J., Marrant, M., Bender, H., Sonnemans, R., & Berry, I. (2010). Metrology for implanted Si substrate loss studies. *Journal of the Electrochemical Society,* Vol. 157, No.5, (May 2010), pp. H580-H584, ISSN 0013-4651

Riedling, K. (1988). *Ellipsometry for Industrial Applications,* Springer-Verlag, ISBN 978-3-211-82040-7, New York, NY, U.S.A.

Rzhanov, A. V.; Svitashev, K. K., Semenenko, A. I., Semenenko, L. V., & Sokolov, V. K. (1979). *The Principles of Ellipsometry,* Nauka, Novosibirsk, U.S.S.R.

Schroder, D. K. (2006). *Semiconductor Material and Device Characterization* (3rd Ed.), John Wiley & Sons, Inc., ISBN 978-0-471-73906-7, Hoboken, NJ, U.S.A.

Shamiryan, D.; Radisic, D., & Boullart, W. (2010). In-line control of Si loss after post ion implantation strip. *Microelectronic Engineering,* Vol. 87, No.9, (November 2010), pp. 1669-1673, ISSN 0167-9317

Shibata, S.; Nambu, Y., Etoh, R., & Fuse, G. (1999). Evaluation of high dose ion implantation by spectroscopic ellipsometry, *Proceedings of 1998 International Conference on Ion Implantation Technology, Vol. 1,* pp. 465-467, ISBN 0-7803-4538-X, Kyoto , Japan, June 22-26, 1998

Shibata, S.; Kawase, F., Kitada, A., Kouzaki, T., & Kitamura, A. (2010). Evaluation of pre-amorphized layer thickness and interface quality of high-dose shallow implanted silicon by spectroscopic ellipsometry. *IEEE Transactions on Semiconductor Manufacturing,* Vol. 23, No.10, (November 2010), pp. 545-552, ISSN 0894-6507

Sigmund, P. (2004). *Stopping of Heavy Ions: A Theoretical Approach,* Springer Verlag, ISBN 978-3-540-22273-6, New York, NY, U.S.A.

Smith, W. L.; Rosencwaig, A., & Willenborg, D. L. (1985). Ion implant monitoring with thermal wave technology. *Applied Physics Letters,* Vol. 47, No. 6, (September 1985), pp. 584-586, ISSN 0003-6951

Smith, W. L.; Rosencwaig, A., Willenborg, D.L., Opsal, J., & Taylor M. W. (1986). Ion implant monitoring with thermal wave technology. *Solid State Technology,* Vol. 29, No. 1, (January 1986), pp. 85–92, ISSN 0038-111X

Stevens, A. A. E.; Kessels, W. M. M., van de Sanden, M. C. M., & Beijerinck, H. C. W. (2006). Amorphous silicon layer characteristics during 70–2000 eV Ar+-ion bombardment of Si(100). *Journal of Vacuum Science and Technology A*, Vol. 24, No.5, (September 2006), pp. 1933-1940, ISSN 0734-2101

Theeten, J. B. & Aspnes, D. E. (1981). Ellipsometry in thin film analysis. *Annual Review of Materials Science*, Vol. 11, No. 1, (August 1981), pp. 97-122, ISSN 0084-6600

Tompkins, H. G. (1993). *A User's Guide to Ellipsometry*, Academic Press, Inc., ISBN 978-0-126-93950-7, San Diego, CA, U.S.A.

Tompkins, H. G. & McGahan, W. A. (1999). *Spectroscopic Ellipsometry and Reflectometry: A User's Guide*, John Wiley & Sons, Ltd, ISBN 978-0-471-18172-9, New York, NY, U.S.A.

Tompkins, H. G. & Irene, E. A. (Eds.) (2005). *Handbook of Ellipsometry*, William Andrew Publishing/Noyes, ISBN 978-0-815-51499-2, Norwich, NY, U.S.A.

Tsunoda, K.; Adachi, S., & Takahashi, M. (2002). Spectroscopic ellipsometry study of ion-implanted Si(100) wafers. *Journal of Applied Physics*, Vol. 91, No.5, (March 2002), pp. 2936-2940, ISSN 0021-8979

Vasquez, R. P.; Madhukar, A., & Tanguay, A.R. (1985). Spectroscopic ellipsometry and x-ray photoelectron spectroscopy studies of the annealing behavior of amorphous Si produced by Si ion implantation. *Journal of Applied Physics*, Vol. 58, No.6, (September 1985), pp. 2337-2343, ISSN 0021-8979

Vedam, K.; McMarr, P. J., & Narayan, J. (1985). Nondestructive depth profiling by spectroscopic ellipsometry. *Applied Physics Letters*, Vol. 47, No. 4, (August 1985), pp. 339-341, ISSN 0003-6951

Webb, A. P. & Townsend, P. D. (1976). Refractive index profiles induced by ion implantation into silica. *Journal of Physics D: Applied Physics*, Vol. 9, No. 9, (June 1976) pp. 1343-1354, ISSN 0022-3727

Williams, D. B. & Carter, C. B. (2009). *Transmission Electron Microscopy. A Textbook for Materials Science* (2nd Ed.), Springer Science+Business Media, LLC, ISBN 978-0-387-76500-6, New York, NY, U.S.A.

Woollam, J. A. (1999). Ellipsometry, Variable Angle Spectroscopic, In: *Wiley Encyclopedia of Electrical and Electronics Engineering*, J. G. Webster, (Ed.), pp. 109–117, John Wiley & Sons, Inc., ISBN 978-0-471-35895-9, New York, NY, U.S.A.

Woollam, J. A.; Johs, B., Herzinger, C. M., Hilfiker, J., Synowicki, R., & Bungay, C. L. (1999). Overview of variable angle spectroscopic ellipsometry (VASE), Part I: Basic theory and typical applications, *Optical metrology, Proceedings of SPIE Conference*, Vol. CR72, pp. 3-28, ISBN 978-0-819-43235-3, Denver, CO, U.S.A., July 18-19, 1999

Yoshida, K. & Adachi, S. (2005). Rapid thermal annealing characteristics of P+-ion-implanted Si(100) wafers studied by spectroscopic ellipsometry. *Japanese Journal of Applied Physics*, Vol. 44, No. 2, (February 2005), pp. 802-807, ISSN 0021-4922

Ziegler, J. F.; Ziegler, M. D., & Biersack, J. P. (2010). SRIM – The stopping and range of ions in matter (2010). *Nuclear Instruments and Methods in Physics Research Section B: Beam Interactions with Materials and Atoms*, Vol. 268, No. 11-12, (June 2010), pp. 1818-1823, ISSN 0168-583X

5

Lattice Strain Measurements in Hydrogen Implanted Materials for Layer Transfer Processes

Caroline Moulet and Mark. S. Goorsky
University of California, Los Angeles
USA

1. Introduction

The ion implantation technique introduces atomic species in a material in order to modify its electronic, optical, and / or structural properties. The beam of species is accelerated through an electric field and thus bombards the target material with an electron-volt to mega-electron-volt energy range. Depending on the nature of the ions, their quantity and their energy, the properties of the target material are modified leading to different applications. This chapter is focused on one of these applications, which is the semiconductor slicing generated by ion implantation.

In 1977, Chu et al. realized the first ion cutting experiment by implanting hydrogen ions in silicon (Chu et al., 1977). The annealing of the samples led to the exfoliation of the silicon layer above the implantation zone. For the next two decades, this discovery remained unexploited until, Bruel et al. in 1995 bonded an implanted silicon wafer using similar implant conditions as Chu, to a second silicon wafer. After thermal treatment, the implanted wafer split along the implantation zone and left a thin layer of silicon on the second wafer. The layer transfer process also called "smart cut technology" (Bruel et al., 1995) or more commonly ion-cut process (Höchbauer et al., 1999) was thus created.

Because of the high crystalline quality of the transferred layer, and its controlled and uniform thickness, the layer transfer process was first dedicated to the fabrication of silicon on insulator substrates (SOI). Nowadays, this process offers multiple possibilities in the microelectronics field and constitutes an innovative approach for materials integration. For example, III-V semiconductors are particularly needed because of their optical and electronic properties, and their growth using epitaxy can generate a high density of defects in the layer impacting on the properties. The layer transfer process could be an effective and economical method to integrate these dissimilar materials (Hayashi et al., 2006).

However, the success of the layer transfer is not trivial. The implantation step and the choice of the implantation parameters are crucial to obtain a successful splitting of the implanted wafer. These parameters can be anticipated by analysing the implanted materials through X-Ray scattering measurements. Indeed, the implantation of ion generates strain in the lattice of the target materials, preceding the splitting. This strain can be quantified using High Resolution X-Ray Diffraction (HRXRD) and used as an indicator for atomic migration, as demonstrated in the second part of this chapter. Before that, a brief description of the layer transfer process is presented and the main mechanisms involved in the splitting process are reviewed. This first part contains explanations about the choice and the ranges of implantation parameters which lead to exfoliation. The second part of the chapter presents how to use HRXRD measurements to measure strain in implanted materials based on previous studies. Information from such measurements is extracted via simulation of the diffraction profiles. Finally, the chapter ends with various examples of HRXRD measurements of hydrogen-implanted materials. Analyses are realized on different implantation conditions and after annealing treatments. We thus demonstrated how HRXRD constitutes a fundamental tool in the study of implanted structures.

2. Ion implantation for semiconductor slicing

In this part, a general overview of the ion implantation technique used for semiconductor slicing is presented. As noted previously, ion implantation can be used for various purposes depending on the implant parameters. The idea here is to present the transfer layer process which involves specific conditions of ion implantation. The essential steps leading to the slicing are summarized based on the silicon case. This part of the chapter ends with an introduction of the parameter window to obtain exfoliation or semiconductor slicing.

2.1 Layer transfer process: From ion implantation to semiconductor slicing

The layer transfer process described in figure 1 consists in the direct bonding of an implanted wafer (wafer 1) to a second wafer called handle wafer (wafer 2). The annealing of the whole structure leads to the full splitting of the wafer along the implanted region. Thus, a thin layer of the implanted wafer is transferred on the handle wafer. Using chemical-mechanical polishing, the surface of the transferred layer can be very smooth and processed again. The thickness of the transferred layer can be adjusted depending on the implantation energy. The quality and properties of the transferred materials are equivalent to bulk materials. The layer transfer process was initially developed on silicon, and allowed the SOI production at lower cost. Nowadays it constitutes a generic process and can be used to transfer any materials on any substrates, such as Ge (Colinge et al., 2011) or III-V compounds (Hayashi et al., 2006).

To generate the splitting, species capable of forming gas such as hydrogen are generally implanted at a certain depth in the material. Strain and defects in the lattice are initially created in the material and evolve (under the appropriate thermal annealing conditions) into microcracks which generate the splitting of the material. Mechanisms explaining this evolution have been widely described in literature, particularly in the case of hydrogen implanted silicon. Although fewer investigations have been done on others materials, similar mechanisms have been demonstrated for Ge (Ferain et al., 2010) and III-V compounds (Hayashi et al., 2006). Here, we review the main steps of this evolution inspired from silicon studies, which can be generally transposed to others materials (not in details).

Based on different analysis techniques such as IR spectroscopy, Transmission Electronic Microscopy (TEM), Weldon et al. explain in five main steps the mechanism from defect formation to exfoliation in silicon (Weldon et al., 1997). First, there is the formation of hydrogenated point defects (Si_IH_y, I as Si interstitial) and vacancies (V_xH_y) near the implanted region. In the same step, specific defects due to H implantation called platelets are created. They are filled of hydrogen atoms either involved in Si-H or H-H bonds and measure between 50-100 Å. During annealing (350-450 °C), defects and vacancy collapse in formation of H_2, and platelets evolve in hydrogen-terminated surfaces. H_2 diffuses and gets trapped in these microvoids in which the pressure increases and induces the formation of microcracks. These fissures lead to blistering and exfoliation of the surface (figure 2).

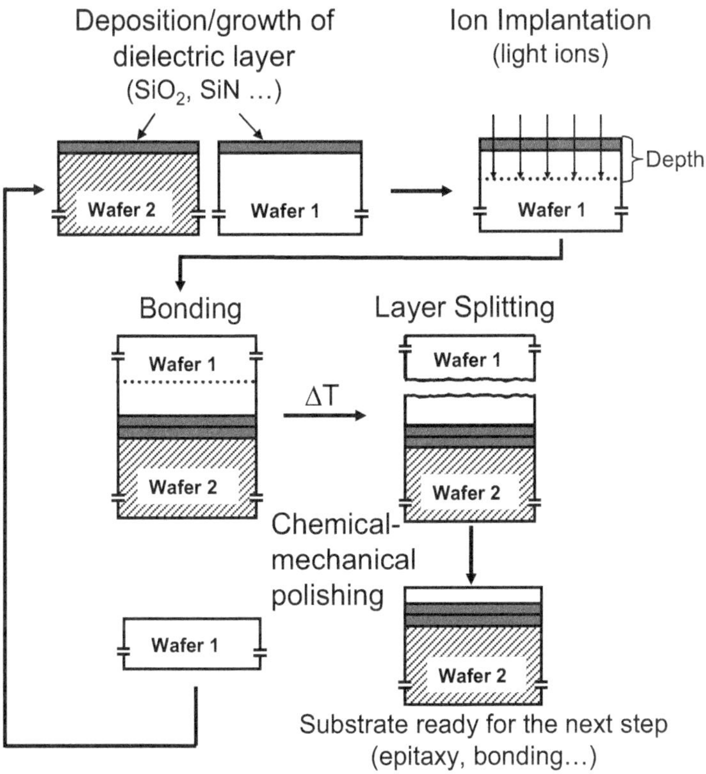

Fig. 1. Schematic of the layer transfer process which is involves ion implantation, direct wafer bonding and splitting of wafer induced by annealing.

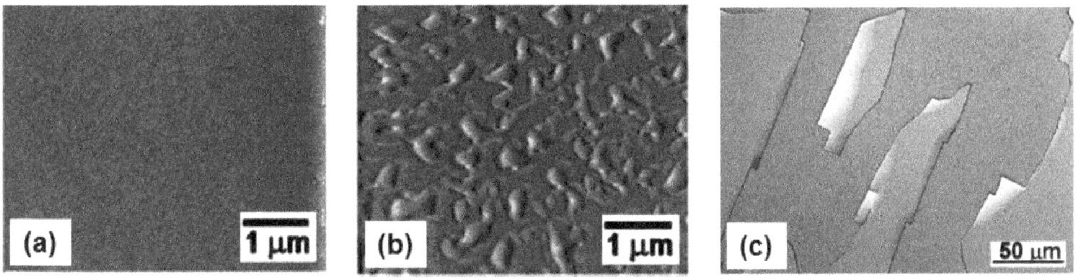

Fig. 2. Optical microscope image of hydrogen implanted InP showing (a) as-implanted surface, (b) blistering on the surface and (c) exfoliation of the blisters (Hayashi et al., 2004).

In the presence of handle wafer which acts as a mechanical stiffener, the top surface can not be distorted, so the cracks propagate parallel to the surface and induce the transfer of a thin film of the material.

2.2 Parameter window for exfoliation or slicing

2.2.1 Nature of implanted ions

Although a wide range of ions can be implanted in any material, the choice for semiconductor slicing is very limited for two main reasons. The first requires that the implanted species form a gas during annealing to create cracks in the bulk material and generate the full wafer splitting. The second reason is related to the atomic weight of the implanted species. During the implantation, ions are accelerated under an electric field and introduced into a target material. They interact with matter creating defects and loosing energy in collisions which finally stop them in matter. Two stopping mechanisms occur during their travel: electronic stopping and nuclear stopping. In electronic stopping, incident ions interact with the electrons of the target materials while in the nuclear stopping they collide with target nuclei (creating more damage in this case). The predominance of one of the stopping mechanisms depends on the atomic weigh of incident ions and their energy.

In the case of semiconductor slicing, the crystalline quality of the transferred layer must be preserved during the implantation and with the introduction of minimal damage. Consequently, the implant conditions are chosen to favour electronic stopping by using high implantation energy (detailed in the next paragraph) and light ions. Indeed, light ions transfer a small amount of energy during each collision, they are deflected through a large scattering angle and can travel deeply into the target material. On the other hand, heavy ions produce a large amount of energy transferred per interaction which generates many displacements of target atoms and so creates considerable lattice damage (see figure 3).

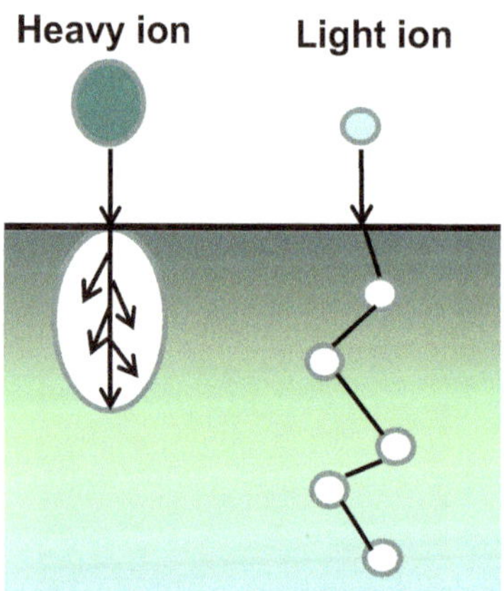

Fig. 3. Comparison between the travel of light ions and heavy ions in a target material during ion implantation.

Regarding the parameters discussed above, hydrogen and helium are the best candidates and remain the most common ions used for semiconductor slicing. Either hydrogen or helium can be implanted in a material, or they can be co-implanted depending on the target material to exfoliate. This is the case of GaAs for example, where the co-implantation of helium and hydrogen allow a better controlled surface blistering (Radu et al. 2002). However, because of its higher weight, helium produces more defects in the lattice and bigger exfoliated area. Hydrogen is more often used for a lot of semiconductors such as Si (Höchbauer et al., 1999), Ge (Zahler et al., 2007), SiC (Malouf et al., 2005), InP (Aspar et al., 2001), (Hayashi et al., 2006).

2.2.2 Implantation energy and dose

The implantation energy represents the energy acquired by ions under the acceleration of the electric field. As introduced in the previous paragraph, high implantation energy is required for semiconductor slicing. In this way, the electronic stopping is predominant and a local damage zone parallel to the surface is created in the target material where the slicing occurs. Standard values of implantation energy go from 10 keV up to 200 keV. Depending on this energy, ions penetrate into the target materials (~ 10 nm to 2 µm). The total distance that an ion travels is called the range R (see figure 4). However, the parameter usually studied is the projected range R_p which is the distance that an ion travels perpendicular to the surface and corresponds to the implantation depth. The projected straggle ΔR_p is a parameter which represents the standard deviation of the range also defined as the spread of the Gaussian-like strain distribution.

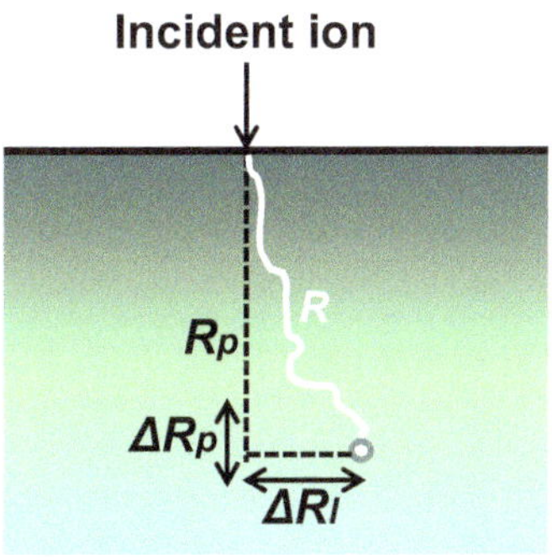

Fig. 4. Definition of the range R and the projected range R_p, and the projected straggle ΔR_p and lateral straggle ΔR_l.

The second important parameter during implantation is the dose or fluence. It is defined as the number of ions per area (usually cm^2) implanted into the sample. For semiconductor slicing, the dose is higher than doses required for others applications such as doping, and is generally between $2x10^{16}$-10^{17} cm^{-2}. This value is chosen according to the target material and the implantation energy.

The projected range, the straggle and the ion distribution in the target material can be estimated from TRIM simulations (Ziegler, 1985). This program based on quantum mechanical treatment of ion-atom collisions simulates the implantation of any element in any material. The implantation energy and the number of ions can be chosen for the simulation. In the following part, we demonstrate how these simulations are complementary of XRD measurements.

3. X-Ray diffraction analysis of implanted materials

3.1 Background and principle of the measurement

X-Ray scattering measurements appeared as a very sensitive technique to detect damage in single crystal materials in the 1980's. In 1982, Speriosu studied the implantation of silicon ions in gallium arsenide, silicon and germanium crystals for different implant doses (Speriosu et al., 1982). He investigated the modifications in the crystal structures with double-crystal X-Ray diffraction and showed that the diffraction profiles contain fringes after implantation. These fringes are due to a strain distribution in the crystal that is fairly well represented with a Gaussian profile. This was explained a few years later by Servidori who measured damage in ion implanted silicon using X-Ray Diffraction (XRD) (Servidori, 1987) and showed how these measurements can be a powerful diagnostic technique of structural investigation in the field of ion implantation.

Indeed, when ions are introduced in a crystal, interstitial type defects are generated leading to a distortion or strain of the lattice (figure 5). The distance between crystallographic planes changes and can be detected by longitudinal XRD measurements (i.e. $\omega/2\theta$ scans). In this geometry, the scattering planes are parallel to the surface and a positive strain perpendicular to the surface $\varepsilon^{\perp}$ is measured.

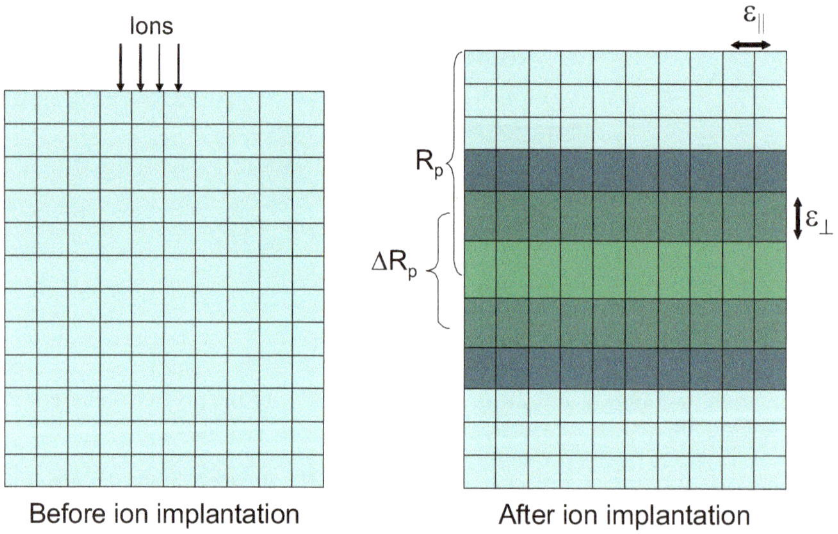

Fig. 5. Representation of the lattice strain in a single crystal after ion implantation.

For example, a <001> oriented crystal can be analyzed through the (004) Bragg reflection presented in figure 6. For angles lower than the Bragg angle, the curve for the as-implanted crystal presents additional scattered intensity with fringe pattern due to the change in the expansion of lattice spacing (Meda et al., 1989). The periodicity of the fringes is related to the width of the strained layer in the crystal and the most distant fringe from the substrate peak corresponds (approximately) to the maximum strain in the crystal. This diffraction profile is due to the strain profile from which information such as the maximum strain and the thickness of the strained layer can be calculated via simulation discussed in the next paragraph. For now, we just remind the equation that relates strain level to the strained lattice parameter **a** and to the Bragg angle.

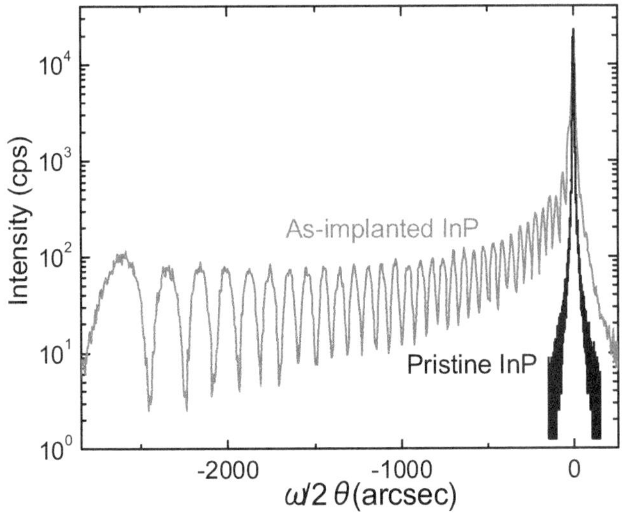

Fig. 6. Example of XRD measurements of non-implanted InP and after implantation in the conditions to obtain exfoliation.

The implanted material is a cubic substrate with a lattice parameter equals to a_{sub} and contains strained layer with a strained lattice parameter perpendicular to the substrate surface a_{layer}. Both the substrate and the layer follow the Bragg's law:

$\lambda = 2d_{sub}\sin(\theta_{B,sub})$ for the substrate;

$\lambda = 2d_{layer}\sin(\theta_{B,sub} + \Delta\theta)$ for the strained layer; of which the Bragg's angle is slightly different than the Bragg's angle of the substrate.

For a common reflection ((004) reflection for example):

$$2d_{sub}\sin(\theta_{B,sub}) = 2d_{layer}\sin(\theta_{B,sub} + \Delta\theta)$$

Replacing the plane spacing by a:

$$a_{layer} = a_{sub}\frac{\sin(\theta_{B,sub})}{\sin(\theta_{B,sub} + \Delta\theta)}$$

Subtracting by a_{sub} and then dividing by a_{sub}:

$$\frac{a_{layer} - a_{sub}}{a_{sub}} = \frac{a_{sub}\frac{\sin(\theta_{B,sub})}{\sin(\theta_{B,sub} + \Delta\theta)} - a_{sub}}{a_{sub}}$$

The strain ε is so expressed by:

$$\varepsilon = \frac{a_{layer} - a_{sub}}{a_{sub}} = \frac{\sin(\theta_{B,sub})}{\sin(\theta_{B,sub} + \Delta\theta)} - 1$$

For small angles:

$$\varepsilon = \frac{a_{layer} - a_{sub}}{a_{sub}} \approx -\Delta\Theta \cdot \cot\left(\theta_{B,sub}\right) \tag{1}$$

Thus, equation (1) allows the variation in strain to be calculated from the measured fringe spacing and the maximum strain corresponding to the fringe at the lowest angles. This calculation is in general carried out by simulating the strain profile, which is introduced in the next paragraph.

3.2 Simulation of the diffraction profile

The diffraction profiles contain information that is not so easily extracted without the use of simulations. Indeed, their determination is hindered by the "phase problem". Lattice displacements occur during ion implantation and affect the phase diffracted amplitude E, while the actual quantity measured is the intensity I equals to EE*. Consequently, the diffraction profile can not be obtained by a direct inversion of the diffracted intensity.

This issue is usually avoided by fitting the experimental data with a simulated profile based on the dynamical theory of XRD. The model generally includes depth-dependent strain and can also contain an atomic disorder parameter (Debye-Waller factor) discussed later. The formalism suggested by Taupin in 1964 about the dynamic theory in deformed crystals (Taupin, 1964) constitutes the fundamental concepts of all models utilized later. Belong them, Larson and Barhorst used the true dynamic model of XRD to simulate the strain profile of boron implanted in silicon crystals (Larson & Barhorst, 1980). The model was then employed by Wie et al. a few years later, who presents the dynamic theory in the layer approximation (Wie et al., 1986). In parallel, Kyutt developed a semikinematical model, which is an approximation of the dynamical theory and allows a strong reduction of the computation time (Kyutt et al., 1980). The semikinematical model was then used by Cembali and Servidori (Cembali et al., 1985), who went back to the dynamical theory based on Vreeland's work a few years later, for an extensive study of implanted materials.

Another model is also suggested to interpret diffraction profile in the early 1980's by Speriosu called the kinematical theory (Speriosu, 1981). This is actually an approximation of the dynamical theory in which all multiple scattering effects are neglected to simplify mathematical expressions and in order to save computation time. Speriosu demonstrated that the kinematical theory is valid for thin films (< 1μm) with a reflecting power lower than 6 %. The model incorporates the depth-dependent strain similarly to the dynamical theory, and also included a parameter u representing the atomic disorder. It corresponds to the displacement of the atoms from their lattice sites. A few years later, Servidori revealed its importance and showed that the best-fits of the diffraction profiles were obtained with the variation of u, an atomic disorder parameter (Cembali et al., 1985). He explained that this atomic disorder parameter begins to play an important role beyond certain strain values, which is demonstrated later in one of his reports (Balboni et al., 1995). In the calculation, u is taken into account in the static Debye-Waller factor (L_H), approximately described by a spherically Gaussian distribution:

$$L_H = 8[\pi \sin(\theta_B) / \lambda]^2 . < u^2 > \qquad (2)$$

where θ_B is the Bragg's angle, λ the wavelength and $<u^2>$ is the mean square displacement. The Debye-Waller factor is included in the expression of the structure factor and impacts the intensity of the reflected beam. In the same report, the authors demonstrated that in the case of heavy ion implantation (B, Si), L_H is an important parameter in the best fit of XRD curves for as-implanted structures and after annealing. However, in the case of hydrogen implantation, L_H is very small compared to the strain in as-implanted silicon.

As said previously when the studies of implanted materials with XRD emerge in the 1980's, the simulation of stain profile was carried out by resolving the Takagi-Taupin equation from the dynamical theory through iterations and numerical routines. Nowadays, with the drastic improvement in computational power, the calculation time has been considerably reduced and programs dedicated to the simulation of diffraction data have been developed. A very common one used for this application is Rocking-Curve Analysis by Dynamical Simulation (RADS) developed by Bede Scientific. Calculations are based on Takagi-Taupin model established from the dynamical theory of XRD. The stain profiles presented in this chapter are obtained from RADS simulation.

In the case of implanted layers, if we assume that ion implantation causes a strain profile similar in shape to the implanted ion concentration profile (obtained by SRIM simulation) the depth-dependent strain can be simulated by multiple layers with an increase of the strain as a function of the depth. This strain distribution can be linear, quadratic, exponential, etc, depending on the material. An example of simulated diffraction profiles is presented in figure 7 for two thicknesses of strained layer.

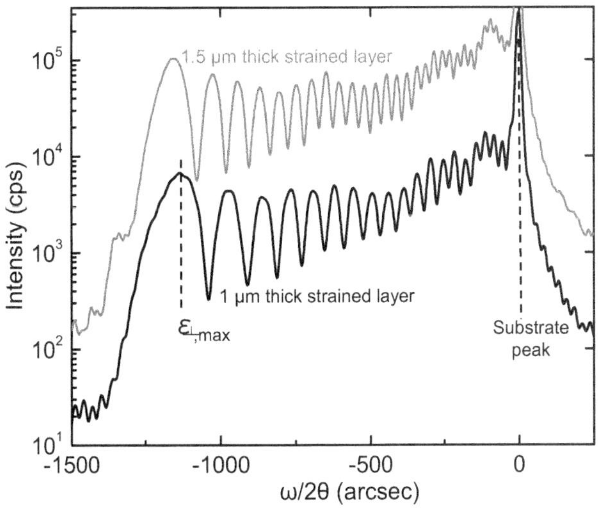

Fig. 7. Example of diffraction profiles simulated with RADS for two thicknesses, 1 µm and 1.5 µm of strained layer. The maximum strain is calculated from the last fringe.

The following figure 8 shows the diffraction profile of InP as-implanted with hydrogen.

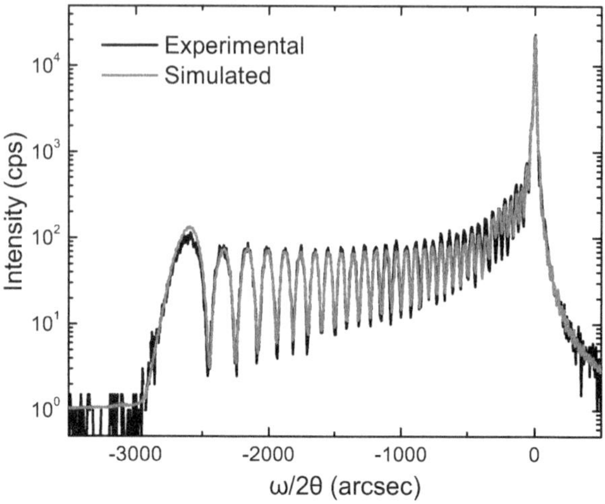

Fig. 8. Experimental and simulated diffraction profile obtained on InP after hydrogen implantation with cooling.

From the simulation of the profile, the strain as a function of depth can be plotted and then the maximum strain is determined (figure 9). This evolution can be compared to TRIM simulations (Ziegler, 1985) providing ions distribution as a function of depth for example, and facilitate the understanding of defects formation (Miclaus & Goorsky, 2003).

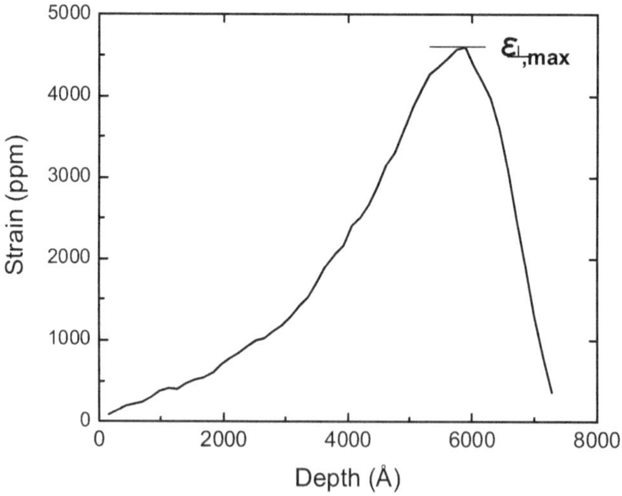

Fig. 9. Strain as a function of depth obtained from the simulation of the diffraction profile.

As introduced above, the Debye-Waller factor L_H which represents the atomic disorder, can be plotted versus depth similar to the strain distribution. This is realized by manual fitting of the diffraction profile as done by Balboni et al. (Balboni et al., 1995). In the case of light ions (H, He), it is expected that relatively small displacements are created compared to heavy ion implantation. Thus, perhaps this parameter can be neglected for these as-implanted cases. Additionally, in Balboni's report, authors discuss double axis XRD measurements and indicate the presence of significant Debye-Waller scattering for annealed structures; without triple axis measurements, it is not possible to determine whether the reduced intensity is simply due to reconstructed, tilted regions. Most of the measurements presented here are realized in triple axis XRD, which has less diffuse scattering (if any exists from the sample) compared with double axis XRD (Horng et al., 1994).

3.3 XRD analysis of different semiconductors

3.3.1 Strain in hydrogen implanted Silicon and Germanium

Because of the SOI production, hydrogen implanted in silicon has been widely studied and investigated by many characterisation techniques, such as IR spectroscopy (Weldon et al., 1997), SIMS and TEM analyses (Höchbauer et al., 2001). These techniques provide essentially chemical information about how hydrogen is bonded to the lattice, its quantification and distribution in the materials. XRD is a complementary technique providing physical information about strain in the crystal, and has the advantage compared to the others ones to be non-destructive. The information obtained is directly related to the implant conditions such as dose and implantation energy as demonstrated in this second part of the chapter.

Miclaus and Goorsky have performed XRD measurements on hydrogen implanted silicon for different doses, before and after annealing treatment (Miclaus & Goorsky, 2003). The strain profile is clearly different when the dose changes and evolves with the annealing temperature. In figure 10 is presented the diffraction profile of silicon implanted with hydrogen ($5x10^{16}$ cm^{-2}, H^+) at room temperature and at 140 keV. The diffraction profiles were obtained by doing a $\omega/2\theta$ scan on the (004) reflection.

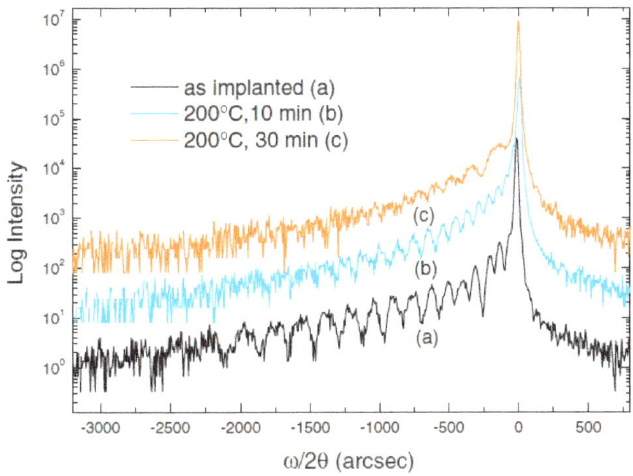

Fig. 10. Diffraction profiles ($\omega/2\theta$ scan) of hydrogen implanted <001> silicon (a) as implanted and after annealing at (b) 200 °C for 10 min and (c) 200 °C for 30 min.

With the annealing and particularly for longer time (30 min), we observe a reduction in the fringe contrast and fringe broadening in the vicinity of the substrate reflection. The reduction of the fringes is consistent with strain reduction due to the diffusion of the implanted hydrogen and related point-defects during the annealing at 200 °C.

The broadening can be best assessed through triple axis diffraction (TAD) ω scans (rocking curve) which reveal information on the change of mosaicity of the sample with high sensitivity (Horng et al., 1994). An example is presented in figure 11 for hydrogen implanted silicon measured for the same sample presented figure 10 (Miclaus & Goorsky, 2003).

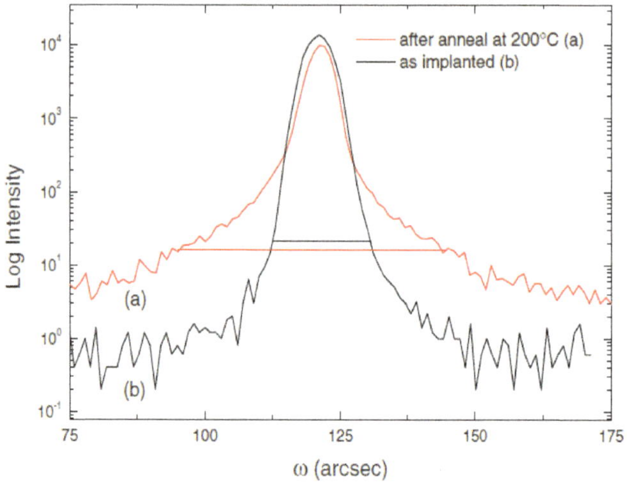

Fig. 11. Triple axis rocking curve of as implanted and after annealing at 200 °C silicon wafer with hydrogen.

The increase of the FW0.001M from 19 arcsec as implanted to 54 arcsec after annealing is very obvious. This increase is consistent with the presence of a thin layer of silicon that exhibits mosaic tilt and the increase of hydrogen platelet diameter. Indeed, previous studies have demonstrated that the diameter of such platelets increases with annealing (Grisolia et al., 2000), to finally form microcracks leading to the exfoliation of the wafer.

X-Ray scattering measurements have been also used to investigate hydrogen implantation in silicon with different crystallographic orientations (Sousbie et al., 2006). The authors measured the strain in <100>-, <110>- and <111>-oriented silicon wafers through ω/2θ scans, and support their results with TEM, IR spectroscopy and SIMS measurements. They showed that the strain profile is crystallographic orientation-dependent, meaning that implant conditions to obtain a successful exfoliation are also different for these three orientations. This can be related to the change in the Young Modulus E. Indeed, results showed that the maximum strain is proportional to E, as measured also by Ventosa-Moulet et al. on various semiconductor materials (Ventosa-Moulet et al., 2010).

Finally, XRD measurements can be also used to investigate co-implanted materials. Silicon has been co-implanted with helium and hydrogen by some authors (Duo et al., 2001),

(Reboh et al., 2010). They showed the sensitivity of the measurements to the annealing temperature, the dose and the nature of the implanted ions.

In the same IV-group of materials, germanium is gaining interest as a semiconductor material because bulk germanium has the highest hole mobility of all semiconductors. However, bulk germanium wafers is more expensive. With the layer transfer process, Ge can be used as a thin layer (GeOI substrates), and presents as good crystallographic and electronic properties as bulk germanium. The transfer or germanium on silicon wafer has been realized by Colinge et al. (Colinge et al., 2011). Those authors performed HRXRD measurements on hydrogen implanted Ge in order to estimate the optimum thermal process to induce exfoliation. Some of these measurements are presented in figure 12.

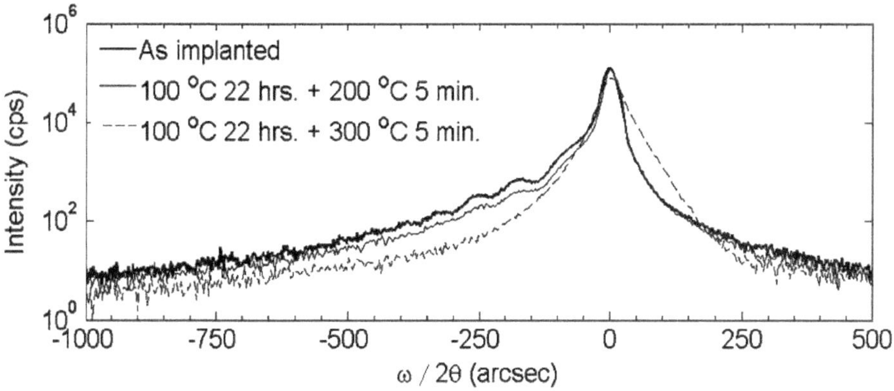

Fig. 12. Diffraction profiles (ω/2θ scan) of hydrogen implanted <100> germanium, as implanted and after annealing.

The as-implanted Ge substrate presents some fringes near the substrate peak indicating strain in the lattice. After annealing at 100 °C for 22 hours and 200 °C for 5 minutes, the fringe pattern is reduced. This indicates the hydrogen diffusion in the materials to form defect and nano-cracks parallel to the substrate surface similarly observed with TEM. At higher annealing temperature, the fringe pattern entirely disappeared because of the formation of micro-cracks suitable for exfoliation.

TAD ω diffraction scans have also been performed on the similar samples. Results are presented in figure 13. After annealing at 100 °C for 22 hours and at 300 °C for 5 minutes, an important broadening of the germanium is observed. This is attributed to local lattice deformations due to hydrogen Oswald ripening leading to exfoliation.

Through these two examples of hydrogen implantation in silicon and germanium, the sensitivity and the usefulness of X-rays scattering measurement have been pointed out. The strain in the implanted material can be quantified and allows us to know if the sample is suitable for exfoliation in a fast and non-destructive way. Additionally, these measurements are also sensitive to the implantation conditions such as dose, energy and implantation temperature. A good example of this is the case of hydrogen implantation in InP, that is why the next paragraph is fully dedicated to this material.

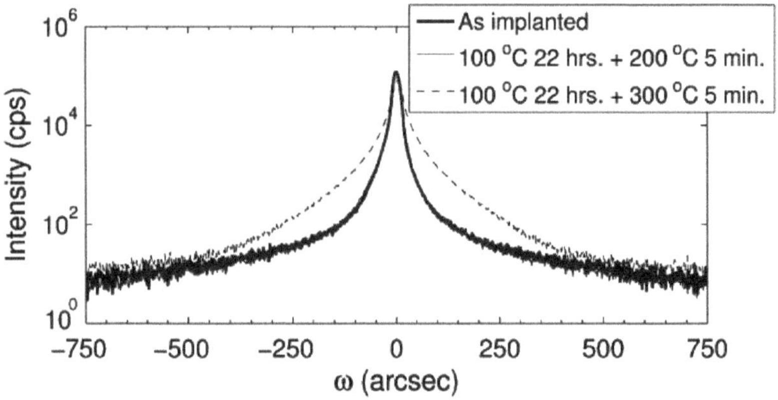

Fig. 13. Triple axis rocking curve (ω scan) of as implanted and after annealing germanium wafer with hydrogen.

3.3.2 Strain in InP: Temperature and dose dependence

Because of its remarkable electronic and optical properties, InP remains one of the most popular semiconductor materials among III-V compounds, and appears in many fields (electronics, optics, solar cells ...). For some applications, the use of InP in a thin layer configuration on the top of other materials can offer many benefits. The layer transfer process realized for the first time on InP by Tong et al. in 1999 (Tong et al., 1999), is an alternative way to produce thin layer of InP with qualities equivalent to bulk InP. However, those authors introduced the fact that the exfoliation of this material is not trivial. The implantation temperature plays a key role in the process. They indicated that implantation into InP must be conducted at ~ 150-250 °C to produce exfoliation although no explanation was proposed.

A few years later, Hayashi et al. presented an investigation of hydrogen implanted InP to obtain exfoliation (Hayashi et al., 2004). InP wafers were cooled to below room temperature (-20 °C) during implantation to assess blistering kinetics without any uncontrolled temperature excursions that may occur from the implantation process. XRD measurements were then performed on the wafers and are presented in figure 14.

With even modest cooling during implantation, fringes are clearly observable on the diffraction profile and reveal the presence of strain in the substrate. After specific annealing steps discussed later in the paragraph, this type of wafers implanted with cooling can be exfoliated. On the other hand, when the temperature is not controlled during implantation, the hydrogen diffuses rapidly during implantation, there is no strain in the InP wafer and therefore is not suitable for exfoliation.

XRD measurements also provide the strain evolution after annealing of implanted materials to study blistering kinetics as presented before in the cases of Si and Ge. In the particular case of InP, Hayashi et al. showed that an annealing at 150 °C was crucial to obtain a successful exfoliation before annealing at 300 °C or above (Hayashi et al., 2006), which is usually the temperature range for exfoliation. We can see on the diffraction profiles the evolution of the fringe pattern after annealing at 150 °C and 300 °C (figure 15). There is less strain relaxation for the 150 °C annealing compared to the 300 °C.

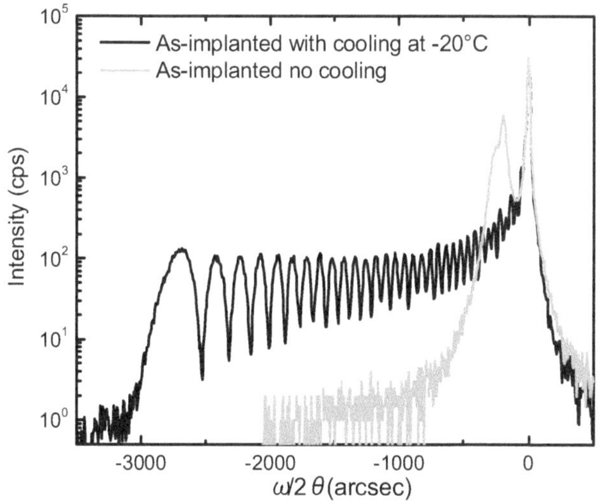

Fig. 14. Diffraction profiles (ω/2θ scan) of hydrogen implanted <100> InP as implanted. The black curve corresponds to implantation with cooling and the grey curve corresponds to implantation without temperature controlled during implantation.

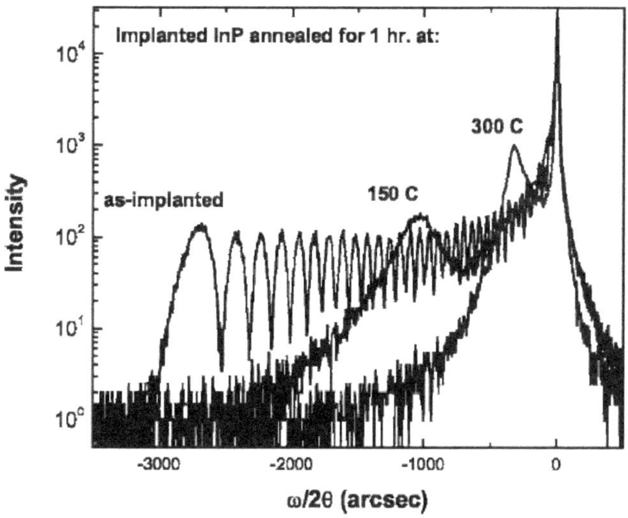

Fig. 15. Diffraction profiles (ω/2θ scan) of hydrogen implanted <100> InP as implanted and after annealing at 150 °C and 300 °C for 1h each.

Authors explain with the support of TEM observations that, the 150 °C annealing allows the nucleation of platelets necessary for exfoliation. This low temperature regime is favorable to the formation of hydrogen complexes trapped in defects while in the high temperature regime (300 °C) without a nucleation step, hydrogen may have enough thermal energy to diffuse away from the implanted region and escape defect traps.

Finally, the dose of hydrogen implanted is a good example to show how XRD measurements are sensitive to implant conditions (Hayashi et al., 2006). Figure 16 presents diffraction profiles of hydrogen implanted InP substrates in the same conditions, only the hydrogen dose is different. From the simulation of the curves, the strain profile and the maximum strain in the implanted region related to the position of the left-most interference peak are determined. It can be seen that the maximum strain is proportional to the implant dose. However, the straggles for each case have been determined from the simulation and they are the same, confirming that the straggle is dependent on the implantation energy rather than the dose.

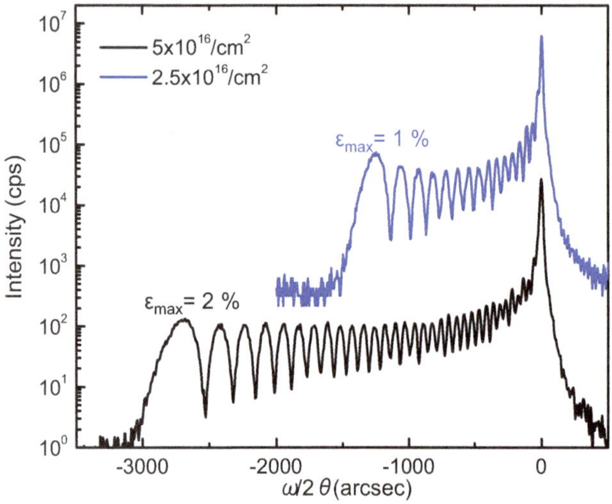

Fig. 16. Diffraction profiles (ω/2θ scan) of hydrogen implanted <100> InP as implanted with a dose of $5x10^{16}/cm^2$ and $2.5x10^{16}/cm^2$.

Thus this paragraph dedicated to InP, shows that XRD analyses are very sensitive to implant conditions and annealing steps after implantation. From these nondestructive measurements, the uniformity of the implant profile can be determined across a wafer and the reproducibility from wafer-to-wafer can be readily evaluated.

3.3.3 Strain in hydrogen implanted GaN

GaN and other III-N semiconductor materials are extremely interesting due to their optoelectronic and electronic properties. In theory, GaN can be alloyed with varying levels of indium, tuning its bandgap to cover the entire visible light spectrum. However, GaN substrates are expensive and currently limited in size. They are epitaxially grown on sapphire, SiC or Si substrates despite the significant lattice parameter and thermal mismatches. The layer transfer process is a promising solution to overcome these limitations. GaN can be transferred on substrates with similar coefficient of thermal expansion such as polycrystalline AlN, and the qualities of the transferred layer are conserved.

The splitting or exfoliation of GaN has been investigated by many authors in literature (Hayashi et al., 2003), (Tauzin et al., 2005), (Singh et al., 2006), (Moutanabbir et al., 2009). In these studies, the implant energies range from 10-200 keV and the implant temperature goes from -196 °C up to 800 °C. All show that the exfoliation requires higher doses of hydrogen (typically 0.5-5x10^{17}/cm^2 (H^+)) than doses usually implanted in others semiconductors. Diffraction profiles of each report present similarities with each other with a double peak shape, and a fringe pattern more or less marked according to the implantation temperature. Indeed, when the hydrogen implantation occurs at low temperature (-196 °C), the diffraction profile contains fringes due to strain in GaN, in agreement with measurements realized on InP (see previous paragraph). An example of such diffraction profile extracted from Padilla et al. is shown in figure 17 (Padilla et al., 2010).

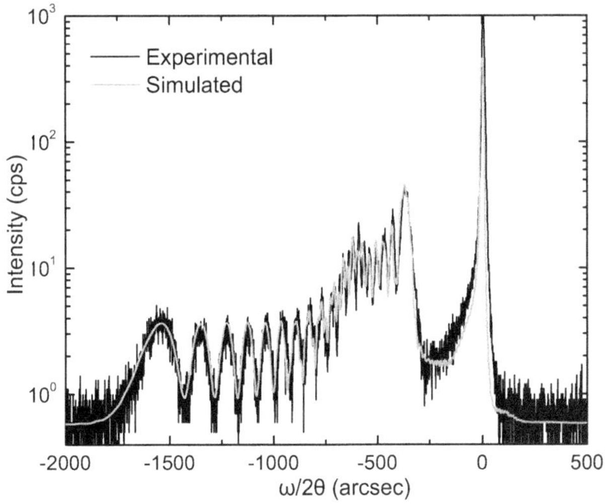

Fig. 17. Diffraction profile (ω/2θ scan) of hydrogen implanted <0001> GaN after implantation at -196 °C, 200 KeV and with a dose of 0.5-1.5x10^{17}/cm^2. The simulation of the profile has been realized with RADS.

Through a simulation of the diffraction profile with RADS, strain can be quantified and represented as a function of depth in GaN. In parallel, the ion distribution and the recoil energy are obtained from TRIM simulations. From these simulations and based on previous studies, the strain evolution can be accurately established (see figure 18).

Analysis and modeling of the strain distribution in as-implanted GaN show two distinct strain peaks, one near the surface and the second at the projected range. Such a double peak was not observed in the implant of standard III-V or Group IV materials. The strain near the surface is not explained by implant conditions. The strained surface layer is not present in either the GaN substrates or in the GaN/sapphire templates prior to implantation and is present in both after implantation.

Between the strained surface layer and the maximum strain, there is a region from around 2500 Å to 8000 Å where the strain is minimum. This could correspond to a damage band observed with TEM by Singh (Singh et al., 2006) and Moutanabbir (Moutanabbir et al., 2009). The damage band is located around 100-200 nm under the surface for implantation energies between 50-100 KeV. Here the implantation energy used is 200 KeV, which involves a deeper implantation and a possible shift of the damage band to higher depths.

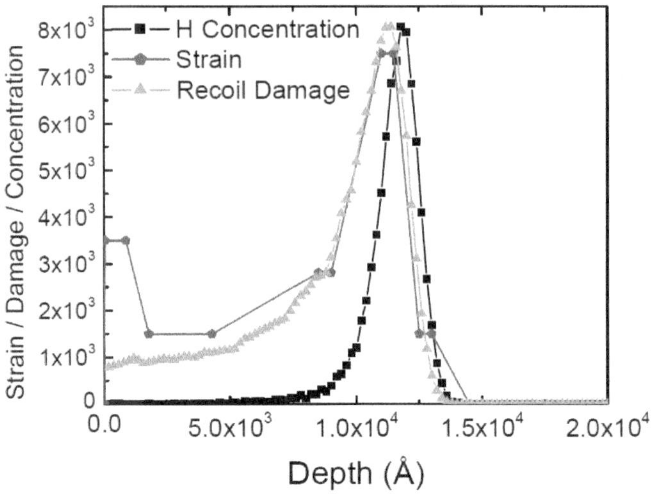

Fig. 18. The strain profile for the simulated diffraction scan in figure 12 is shown along with the energy transferred from the implanted hydrogen to the lattice as a function of depth (recoils) as well as the hydrogen ion distribution.

3.3.4 Strain and recoil energy

Earlier papers about implantation in semiconductor materials have related the measured strain to the energy lost in nuclear collisions F_D also called recoil energy (Paine et al., 1987). In this report, authors implanted heavy species into GaAs at 300 keV with relatively low doses. They showed the strain profile is the maximum where the energy loss by nuclear collisions is a maximum, such that

$$\varepsilon_{\perp,max} = K \cdot \Phi \cdot F_{D,max} \quad (3)$$

where Φ is the implant dose in ions/cm^2 (or ions/Å^2) and K is a constant representing the conversion of the recoil energy to strain. The K term also may be altered by defect migration and recombination that may reduce the strain - for example, by subsequent annealing.

The linear relationship between strain and F_D did not extend to higher doses and decreased for a dose greater than $2x10^{13}$ cm^{-2}. In the case of exfoliation, there are distinct differences with the earlier studies. First, hydrogen is light and each implanted hydrogen ion loses significant energy due to electronic stopping not included in F_D. Second, the doses typically encountered in an exfoliation process are in the 10^{16}-10^{17} cm^{-2} range, which is much higher than doses used for heavier ions. To verify if the linear relationship between strain and F_D exists also for exfoliation conditions, different semiconductor materials have been hydrogen implanted using the same implantation parameters. XRD measurements have been performed on the implanted wafers and they are presented in figure 19.

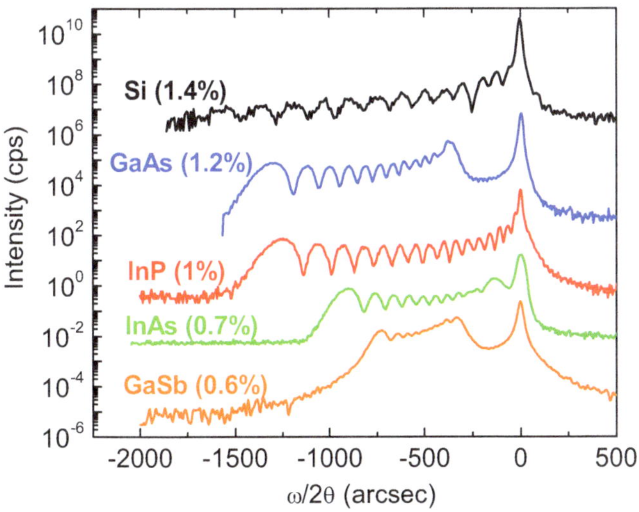

Fig. 19. (004) diffraction profiles (ω/2θ scan) of different cubic semiconductors implanted with hydrogen (H_2^+, dose $2.5x10^{16}$ cm^{-2}) at 77 K and 160 keV. The maximum strain values that correspond to the terminal fringe are noted in parentheses.

In parallel, F_D has been calculated using TRIM simulation for each semiconductor. For this, the simulation requires the implantation energy and the dose. The results obtained are presented in figure 20, showing the evaluation of F_D as a function of the depth in the implanted material. We observe that the evolution of F_D is very similar to the strain (see figure 6 for InP), and reach a maximum at 5400 Å versus 5800 Å for the strain.

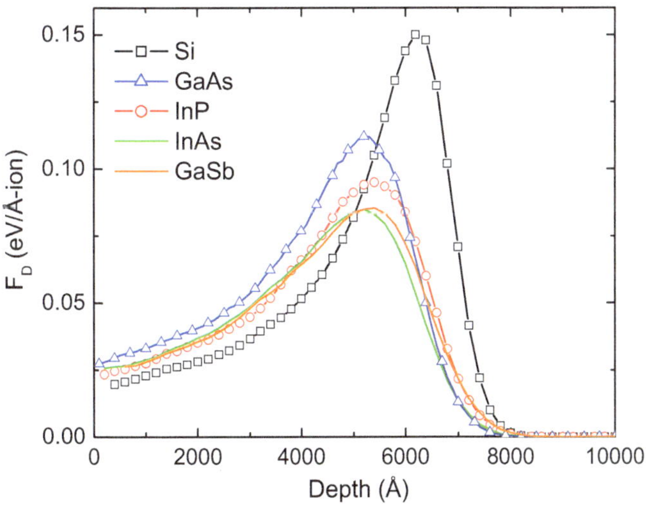

Fig. 20. Profiles of F_D as a function of depth for different semiconductors, obtained by TRIM simulations using the experimental implant parameters in figure 12.

On the other hand, a linear relationship between the maximum strain and F_{Dmax} is shown in figure 21. Also included in figure 14 is the data measured from implanted Si as well as from SiC and GaN.

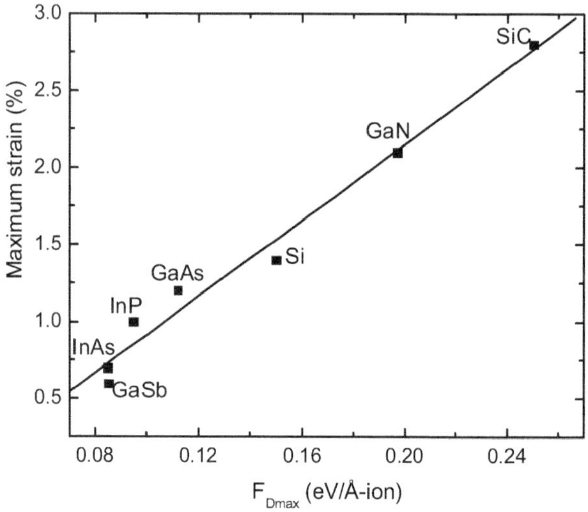

Fig. 21. The relationship between the maximum experimentally determined strain and the maximum recoil energy $F_{D,max}$ determined by TRIM simulation.

This graph demonstrates that, for the hydrogen implant doses and energies typically used for exfoliation and layer transfer, a linear relationship between F_D and strain exists. Whereas the earlier studies demonstrated this relationship only for much lower doses and heavier implant species. The observation that a linear relationship exists for high dose hydrogen implantation into these semiconductors indicates that the relatively low level of nuclear stopping associated with implanted hydrogen maintains a low enough level of damage to ignore higher order damage effects (such as amorphization).

A second interesting result is that the linear relationship between strain and F_D for a given dose indicates that the K value is the same (or very similar) for all of the implanted materials. This result contradicts earlier results which observed that there are significant (order of magnitude) differences in K for a given implant species into different targets (Paine et al., 1986). Noting that "defect motion, trapping, or recombination mechanisms are the most responsible for the variations of K between (*sic*) materials", the implantation at low temperatures effectively prevents significant defect motion, etc., and thus K becomes a more 'universal' value. Additionally, even at these high doses, doubling the dose doubles the strain (Hayashi et al., 2006), showing that the linear relationship between dose and strain exists at the high end of exfoliation implant conditions. Si implanted at an even higher dose produces the same strain - dose relationship (Sousbie et al., 2006) - i.e., K holds for those conditions. This result further supports the concept that the quantitative analysis of the implant-induced strain profile is suitable for a wide range of hydrogen-implanted materials used in exfoliation and layer transfer studies.

4. Conclusion

In this chapter, we have discussed the interest of HRXRD analyses in the ion implantation induced material splitting for layer transfer. This process is a combination of direct wafer bonding and ion implantation. First developed for the SOI production, it has recently been adapted to a wide a number of materials such as Ge, III-V semiconductors, SiC, LiBbO3, etc. Thus, thin layers of materials with qualities comparable to bulk materials can be used for different applications. However, the choice of the implantation parameter is crucial to obtain a successful splitting. Therefore, an analysis of the implanted structure prior splitting is a good way to optimize this choice.

HRXRD constitutes a fast and non-destructive technique to obtain a diagnosis of the implanted structure and determine if they are suitable for exfoliation. We have demonstrated that the ion implantation process induces strain and change in crystallographic planes in the target material. Because of this phenomenon, the diffraction profile measured by XRD presents interference fringes called diffraction pattern. The diffraction profile contains information about the strain in the structure accessible via simulation. The simulation of the diffraction profile can be realized manually by resolving the diffraction equations, or with a program named RADS used in this chapter. With both methods, a localisation and a quantification of the strain in the material is obtained.

We have shown that the strain profile evolves during annealing prior splitting, and HRXRD measurements constitute a good way to monitor this evolution. A diagnosis of the structure is obtained providing information about the stage of the process (hydrogen diffusion, platelet formation, etc). This analysis can help to refine the annealing conditions and obtain the exfoliation.

Finally, the strain profile is related to implantation parameters such as the dose and implantation temperature illustrated with the case of InP. The strain profile is also related to the target materials and their mechanical properties as demonstrated through several examples of semiconductor materials. This validates the fact that HRXRD is very sensitive to any changes in the material structure and constitutes a powerful tool in the study of implanted materials.

5. References

Aspar, B., Moriceau, H., Jalaguier, E., Lagahe, C., Soubie, A., Biasse, B., Papon, A.M., Claverie, A., Grisolia, J., Benassayag, G., Letertre, F., Rayssac, O., Barge, T., Maleville, C. & Ghyselen, B. (2001). The generic nature of the Smart-Cut((R)) process for thin film transfer. *Journal of Electronic Materials*, Vol.30, No.7, (July 2001), pp. 834-840, ISSN 0361-5235

Balboni, R., Milita, S. and Servidori, M. (1995). On the role of the static Debye-Waller factor in X-Ray rocking curve analysis. *Physica Status Solidi A-Applied Research*, Vol.148, No.1, (March 1995), pp. 95-105, ISSN 0031-8965

Bruel, M. (1995). Silicon on insulator material technology. *Electronics letters*, Vol.31, No.14, (July 1995), pp. 1201-1202, ISSN 0013-5194

Cembali, F., Servidori, M. & Zani, A. (1985). Double crystal X-Ray difraction analysis of low temperature ion-implanted silicon. *Solid-State Electronics*, Vol.28, No.9, (1985), pp. 933-943, ISSN 0038-1101

Chu, W.K., Kastl, R.H., Lever, R.F., Mader, S. & Masters, B.J. (1997). Distribution of irradiation damage in silicon bombarded with hydrogen. *Physical Review B*, Vol.16, No.9, (May 1977), pp. 3851-3859, ISSN 0163-1829

Colinge, C.A, Byun, K.Y., Ferain, I.P., Yu, R. and Goorsky, M.S. (2011). Low-temperature fabrication of germanium-on-insulator using remote plasma activation bonding and hydrogen exfoliation, In: *Semiconductor-on-insulator materials for nanoelectronics applications*, A. Zanarov, J.-P. Colinge, F. Balestra, J.-P. Raskin, F. Gamiz, V.S. Lysenko, pp. 31-46, SpringerLink

Duo, X., Liu, W.L., Zhang, M., Wang, L.W., Lin, C.L., Okuyama, M., Noda, M., Cheung, W.-Y., Wong, S.P., Chu, P.K., Hu, P.G., Wang, S.X. and Wang, L.M. (2001). Evolution of hydrogen and helium co-implanted single-crystal silicon during annealing. *Journal of Applied Physics*, Vol.90, No.8, (October 2001), pp. 3780-3786, ISSN 0021-8979

Ferain, I.P., Byun, K.Y., Colinge, C.A., Brightup, S. & Goorsky, M.S. (2010). Low temperature exfoliation process in hydrogen-implanted germanium layers. *Journal of Applied Physics*, Vol.107, No.5, (March 2010), pp. 054315-1-054315-5, ISSN 0021-8979

Grisolia, J., Ben Assayag, G., Claverie, A., Aspar, B., Lagahe, C. & Laanab, L. (2000). A transmission electron microscopy quantitative study of the growth kinetics of H platelets in Si. *Applied Physics Letters*, Vol.76, No.7, (February 2000), pp. 852-854, ISSN 0003-6951

Hayashi, S., Poust, B., Heying, B. & Goorsky, M.S. (2003). Exfoliation of GaN layers using hydrogen implantation, *Proceedings of Compound Semiconductors 2003, International Symposium*, ISSN 0-7803-7820-2, San Diego, August 2003

Hayashi, S., Bruno, D. & Goorsky, M.S. (2004). Temperature dependence of hydrogen-induced exfoliation of InP. *Applied Physics Letters*, Vol.85, No.2, (July 2004), pp. 236-238, ISSN 0003-6951

Hayashi, S., Goorsky, M.S., Noori, A. & Bruno, D. (2006). Materials issues for the heterogeneous integration of III-V compounds. *Journal of the Electrochemical Society*, Vol.153, No.12, (October 2006), pp. G1011-G1014, ISSN 0013-4651

Höchbauer, T., Nastasi, M. & Mayer, J.W. (1999). Hydrogen blister depth in boron and hydrogen coimplanted n-type silicon. *Applied Physics Letters*, Vol.75, No.25, (December 1999), pp. 3938-3940, ISSN 0003-6951

Höchbauer, T., Misra, A., Nastasi, M. & Mayer, J.W. (2001). Investigation of the cut location in hydrogen implantation induced silicon surface exfoliation. *Journal of Applied Physics*, Vol.89, pp. 5980-5990 ISSN 0021-8979

Horng, S.T., Goorsky, M.S., Madok, J.H. & Haegel, N.M. (1994). Strain and mosaic spread of carbon and gallium co-implanted GaAs. *Journal of Applied Physics*, Vol.76, No.4, pp. 2066-2069 ISSN 0021-8979

Kyutt, R.N., Petrashen, P.V. & and Sorokin, L.M. (1980). Strain profiles in ion-doped silicon obtained from X-Ray rocking curve. *Physica Status Solidi A-Applied Research*, Vol.60, No.2, (1980), pp. 381-389, ISSN 0031-8965

Larson, B.C. and Barhost, J.F. (1980). X-Ray study of lattice strain in boron implanted laser annealed silicon. *Journal of Applied Physics*, Vol.51, No.6, (1980), pp. 3181-3185, ISSN 0021-8979

Malouf, G., Poust, B., Hayashi, S., Yoshizawa, G. & Goorsky, M.S. (2005). Hydrogen-induced blistering of SiC: the role of post-implant multi-step annealing sequences. *Materials Science Forum*, Vol.527-529, (October 2006), pp. 855-858, ISSN 1662-9752

Meda, L., Cerofolini, G.F., Dierckx, R., Mercurio, G., Servidori, M., Cembali, F., Anderle, M., Canteri, R., Ottaviani, G., Claeys, C. & Vanhellemont, J. (1989). Hydrogen implantation into (100) silicon - A study of the released damage. *Nuclear Instruments & Methods in Physics Research section B-beam interaction with materials and atoms*, Vol.39, No.1-4, (March 1989), pp. 381-385, ISSN 0168-583X

Moutanabbir, O., Chabal, Y.J., Chicoine, M., Christiansen, S., Krause-Rehberg, R., Schiettekatte, F., Scholz, R., Seitz, O., Süßkraut, F. & Gösele, U. (2009). Mechanisms of ion-induced GaN thin layer splitting. *Nuclear Instruments & Methods in Physics Research B*, Vol.267, pp. 1264-1268

Nastasi, M., Höchbauer, T., Lee, J.K., Misra, A., Hirth, J.P., Ridgway, M. & Lafford, T. (2005). Nucleation and growth of platelets in hydrogen-ion-implanted silicon. *Applied Physics Letters*, Vol.86, No.15, (April 2005), pp. 154102, ISSN 0003-6951

Padilla, E., Jackson, M. and Goorsky, M.S. (2010). The role of the nucleation annealing temperature annealing on the exfoliation of hydrogen-implanted GaN, *Proceedings of Semiconductor wafer bonding 11: Science, Technology and Applications*, ISBN 978-1-56677-823, Las Vegas, October 2010.

Paine, B.M., Hurvitz, N.N. and Speriosu, V.S. (1987). Strain in GaAs by low-dose ion implantation. *Journal of Applied Physics*, Vol.61, No.4, (February 1987), pp. 1335-1339, ISSN 0021-8979

Radu, I., Szafraniak, I., Scholz, R., Alexe, M. & Gösele, U. (2002). Blistering and exfoliation of hydrogen and helium implanted (100) GaAs, *Proceedings of International Semiconductor Conference*, 0-7803-7440-1, Romania, October, 2002.

Reboh, S., Schaurich, F., Declemy, A., Barbot, J.F., Beaufort, M.F., Cherkashin, N. and Fichtner, P.F.P. (2010). On the microstructure of Si coimplanted with H^+ and He^+ ions at moderate energies. *Journal of Applied Physics*, Vol.108, No.2, (July 2010), pp. 023502-1-023502-6, ISSN 0021-8979

Servidori, M. (1987). Characterization of lattice damage in ion implanted silicon by multiple crystal X-Ray diffraction. *Nuclear Instruments & Methods in Physics Research section B-beam interaction with materials and atoms*, Vol.19-2, (February 1987), pp. 443-449, ISSN 0168-583X

Singh, R., Radu, I., Gösele, U. & Christiansen, S.H. (2006). Investigation of hydrogen implantation induced blistering in GaN. *Physica Status Solidi*, Vol.3, No.6, pp. 1754-1757

Speriosu, V.S. (1981). Kinematical X-Ray diffraction in nonuniform crystalline films: strain and ion distributions in-ion implanted garnets. (1981). *Journal of Applied Physics*, Vol.52, No.10, (1981), pp. 6094-6103, ISSN 0021-8979

Speriosu, V.S., Paine, B.M. & Nicolet, M-A. (1982). X-Ray rocking study of Si-implanted GaAs, Si and Ge. *Applied Physics Letters*, Vol.40, No.7, (1982), pp. 604-606, ISSN 0003-6951

Sousbie, N., Capello, L., Eymery, J., Rieutord, F. and Lagahe, C. (2006). X-Ray scattering study of hydrogen implantation in silicon. *Journal of Applied Physics*, Vol.99, No.11, (May 2006), pp. 103509-1-103509-7, ISSN 0021-8979

Taupin, D. (1964). The dynamic theory of the diffraction of X-Rays by deformed crystals. *Bulletin de la Societe Francaise de Mineralogie et de Cristallographie*, Vol.88, No.4, pp. 469-511

Tauzin, A., Akatsu, T., Rabarot, M., Dechamp, J., Zussy, M., Moriceau, H., Michaud, J.-F., Charvet, A.-M., Di Cioccio, L., Fournel, F., Garrione, J., Faure, B., Letertre, F. & Kernevez, N. (2005). Transfers of 2-inch GaN films on sapphire substrates using Smart Cut™ technology. *Electronics Letters*, Vol.41, No.11, pp. 668-670, ISSN 0013-5194

Tong, Q.-Y., Chao, Y.-L., Huang, L.-J. & Gösele, U. (1999). Low temperature InP layer transfer. *Electronics Letters*, Vol.35, No.4, pp. 341-342, ISSN 0013-5194

Ventosa-Moulet, C., Hayashi, S.L. and Goorsky, M.S. (2010). Lattice strain and strain evolution in hydrogen implanted materials: the role of mechanical properties and hydrogen diffusion, *Proceedings of Semiconductor wafer bonding 11: Science, Technology and Applications*, ISBN 978-1-56677-823, Las Vegas, October 2010.

Weldon, M.K., Marsico, V.E., Chabal, Y.J., Agarwal, A., Eaglesham, D.J., Sapjeta, J., Brown, W.L., Jacobson, D.C., Caudano, Y., Christman, S.B. & Chaban, E.E. (1997). On the mechanism of the hydrogen-induced exfoliation of silicon. *Journal of Vacuum Science & Technology B*, Vol.15, No.4, (July August 1997), pp. 1065-1073, IDS XT088

Wie, C.R., Tombrello, T.A. & Vreeland, T. (1986). Dynamical X-Ray diffraction from non uniform crystalline films: application to X-Ray rocking curve analysis. *Journal of Applied Physics*, Vol.59, No.11, (June 1986), pp. 3743-3746, ISSN 0021-8979

Zahler, J.M., Morral, A.F.I., Griggs, M.J. Atwater, H.A. & Chabal, Y.J. (2007). Role of hydrogen in hydrogen-induced layer exfoliation of germanium. *Physical Review B*, Vol.75, No.3, (January 2007), pp. 035309-1-035309-10, ISSN 1098-0121

Ziegler, J.F. (1985). *The stopping and range of ions in solids*, TarryTown, ISBN 008021603X, New York, USA

6

Determination of Near-Neighbour Bonding in the Mn-Implanted GaSb Crystals

Anna Wolska
Institute of Physics, Polish Academy of Sciences
Poland

1. Introduction

Ion implantation is a powerful technique for the production of new materials. However, in order to find an optimal way of obtaining compounds with desirable properties, the influence of many parameters must be tested. The characterization, among other things, should contain a very precise determination of the local neighbourhood of the implanted ions. This can be revealed by the X-ray absorption fine structure (XAFS) techniques. Their main advantage is elemental selectivity, which allows the experimenter to extract information on the atomic neighbourhood of the chosen species of atom even at very low concentration of this element.

2. X-ray absorption fine structure techniques

When the X-ray photon energy is equal to the binding energy of a core level of an atom in the material an abrupt increase in the absorption coefficient occurs. This phenomenon is known as the absorption edge. The edge energy is characteristic for the type of the excited inner shell and the type of the absorbing element.

The X-ray absorption spectrum is being usually divided into two regions: X-ray absorption near edge structure (XANES) up to around 50 eV above the edge and extended X-ray absorption fine structure (EXAFS) which starts around 50 eV and can reach even 1000 eV above the edge. However, the exact energy dividing both regions depends mainly on the investigated system.

Detailed description and interpretation of the XAFS processes can be found in many books and papers (Eisenberger & Kincaid, 1978; Koningsberger & Prins, 1988; Lee et al. 1981; Stohr, 1996; Teo, 1986; Teo. & Joy, 1981). Here only very simple explanation is provided. The photoelectron created during the X-ray absorption process can be represented as an outgoing wave which is scattered by the surrounding atoms. The outgoing and scattered waves interfere with each other which modifies the absorption coefficient in a way that depends on the type and spatial arrangements of the neighbouring atoms. Such pattern does not appear for an isolated atom.

In XANES multiple scatterings dominate. The shape of a spectrum in this region depends not only on the radial distances, the angles between the atoms, their type and number but also on their chemical bonding. Different chemical bondings (ionic, covalent, metallic) alter core-level binding energies producing absorption edge shifts and changing the distribution of the unoccupied states above the Fermi level. As an example the measurements of the manganese oxides at the Mn K edge can be considered. Fig. 1 presents the XANES spectra for Mn, MnO, Mn_2O_3 and MnO_2. It can be noticed that the edge position moves toward higher energy while the Mn atoms ionicity changes from 0 to IV. Studies presented by Croft et al. 1997 showed that in case of the Mn K edge the edge position changes linearly with the valence state. However, not only the edge position but also the shape of the spectra changes together with the rearrangements of the atoms around the absorber, reflecting the change in the local density of states.

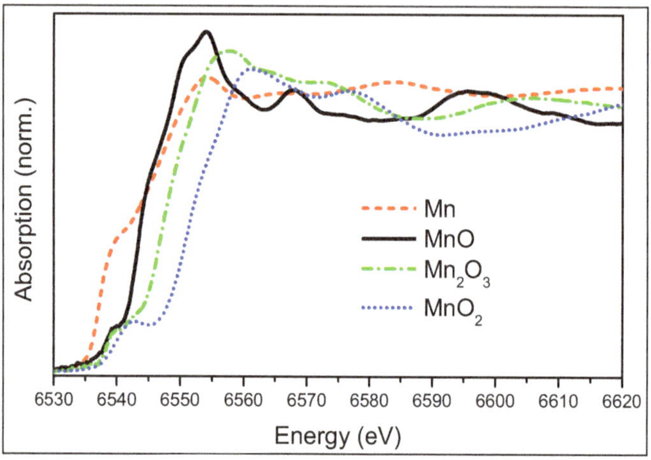

Fig. 1. XANES spectra at the Mn K-edge of the Mn foil and manganese oxides.

In the EXAFS region single scatterings dominate which simplifies the analysis. Fig. 2a shows the EXAFS spectrum of MnO. As can be noticed, at the end of the range, the oscillations are very weak. The absorption coefficient can be defined as $\mu(E) = \mu_0(E) * (1 + \chi(k))$, where $\mu_0(E)$ represents a smooth atomic background and $\chi(k)$ represents the EXAFS oscillations. By using this equation $\mu(E)$ can be converted into $\chi(k)$. The result is shown in Fig. 2b. These oscillations include information about the type, number and distances of the neighbouring atoms. The next step is to perform Fourier transformation (FT) from k (photo-electron wavenumber) to R (real) space. The $\chi(R)$ is a complex function; in the Fig. 2c the magnitude and the real part for MnO spectrum are shown. The positions of the peaks are related to the distances between the absorber and its neighbours. The amplitude and shape of the peaks depend on type and number of neighbouring atoms. E.g. the first peak in the plot consists of 6 oxygen atoms at a distance R = 2.2 Å, while in the second one 12 Mn (R = 3.14 Å) and 8 O (R = 3.85 Å) atoms are located. In order to analyse this kind of data a realistic model should be assumed first, and then the fitting of this model to the experimental data should be performed. The fitting parameters include the distance between given coordination shell and the absorbing atom; the number of the atoms in a given shell as well as static and thermal disorder. The thermal disorder can be reduced by performing the measurements at liquid nitrogen temperature.

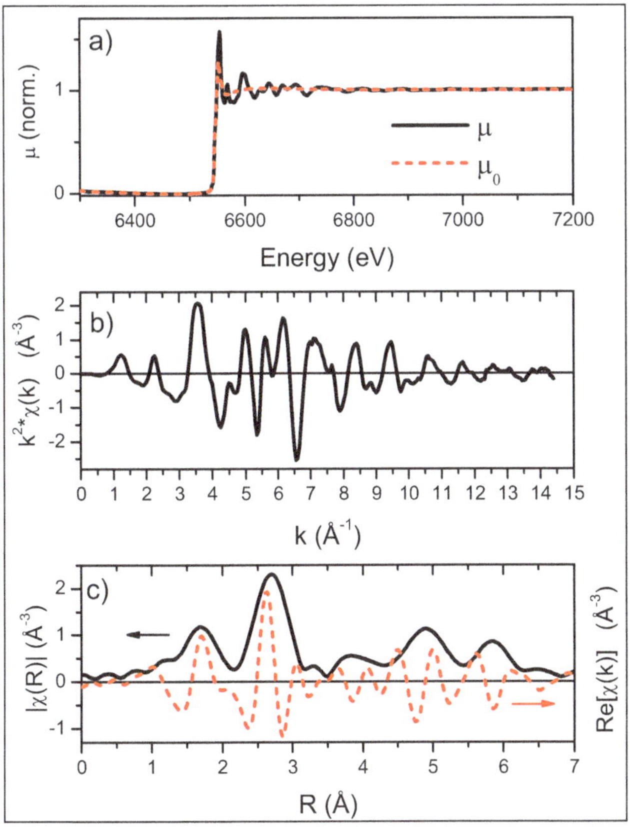

Fig. 2. a) EXAFS spectrum of MnO at the Mn K-edge; b) EXAFS oscillations extracted from a) after atomic background removal; c) the magnitude (black solid line) and the real part (dashed red line) of the FT EXAFS oscillations from b).

The fitting can be done using several accessible programs e.g. SixPACK, EXCURVE, WinXAS. The analysis presented in this chapter was conducted using IFEFFIT data analysis package with Athena and Artemis programs (Ravel & Newville, 2005).

The XAFS techniques are selective (only the selected element is investigated), local (only the closest surrounding of the absorbing atoms is investigated, up to 10 Å for perfectly ordered structures) and sensitive (even very low concentration of the element can be detected). Moreover, they can be used for samples in different states: solids (crystalline or amorphous), liquids, solutions or gaseous. However, it should be emphasized that if the investigated atoms form different compounds, the data obtained from the XAFS experiment is averaged over all those neighbourhoods.

3. Materials

The interest in granular materials composed of semiconductor matrix with nanoinclusions of magnetic materials is rapidly growing due to the demonstration of their usefulness in the construction of magnetic tunnel junctions with huge magnetoresistance. (Hai et al., 2009) In this kind of material, small ferromagnetic nanoparticles are immersed in the semiconductor host lattice providing a built-in local magnetic field. However, in case of MnAs precipitates, two crystalline structures with different magnetic properties were found: the cubic zinc blende type and hexagonal one. This obstacle does not exist for MnSb clusters since they have only hexagonal form. Another advantage of MnSb is a T_C well above room temperature. It was shown that bulk MnSb has a T_C of 587 K (Panchula et al., 2003) while for MnAs it is only 318 K. What's more, the magnetic force microscopy (MFM) as well as superconducting quantum interference device (SQUID) measurements revealed the existence of magnetic anisotropy for the MnSb inclusions grown by the MBE method. The X-ray magnetic circular dichroism (XMCD) experiment also confirmed that the direction of the easy magnetization axis in GaSb:MnSb granular layers grown on the GaSb(100) substrates was close to the in plane direction. The ratio of orbital to spin momentum depends on the shape and size of inclusions. (Lawniczak-Jablonska et al., 2011) In the case of small inclusions the enhancement of orbital momentum was observed which offers interesting opportunities for magnetic applications in a low dimensional limit. The observed change in the direction of the easy magnetization axis indicates on the possibility to control it by proper choice of the substrate and growth conditions. These results show that a compound consisting of MnSb inclusions and semiconductor matrix exhibits very interesting magnetic and structural properties which can be used in the construction of the spintronic devices. Finding the way of forming this kind of material by the implantation method could increase its variety of applications.

3.1 Implantation with low energy Mn^+ ions (10 keV)

The GaSb (100) crystals implanted with low energy Mn^+ ions were considered first. The implantation energy was equal to 10 keV and the doses of the Mn^+ ions were: $1x10^{16}$ Mn/cm^2 (#1 sample), $2x10^{16}$ Mn/cm^2 (#2 sample) and $3x10^{16}$ Mn/cm^2 (#3 sample). After the implantation all samples were annealed in a vacuum furnace at 650 °C for 10 min.

The depth profiles of the element distribution in sample #2 as measured by the secondary ion mass spectroscopy (SIMS) are shown in the Fig. 3. Only the region very close to the surface is disturbed by the implantation which, for this technique, is almost below the edge of detection. However, still it is an indication that the Mn atoms concentrate close to the surface as it was expected in this case.

The shapes of the XANES spectra after normalization look almost the same, which indicates that the Mn atoms are located in the similar positions independently on the applied dose. Since the aim of the implantation was to form the MnSb inclusions, the spectrum of powdered MnSb was measured as well. It can be seen in Fig. 4 that the spectrum of the standard MnSb sample is completely different from the spectra of the implanted samples. This excludes the possibility of MnSb formation in the investigated samples. On the other hand, the strong first peak is characteristic for MnO. The other maxima for the #1 sample in comparison to MnO are weaker and slightly shifted in energy which means that other type of compound dominates here. However, existence of some amount of manganese oxides cannot be excluded.

From EXAFS analysis more detailed information about the distribution of the atoms can be found. XANES spectra revealed that there is no MnSb, therefore, other cases should be considered. As it was predicted, the MnO structure does not dominate here. Oxygen can be only responsible for the tail visible between 1 and 2 Å (see Fig. 5). Therefore, other models should be checked.

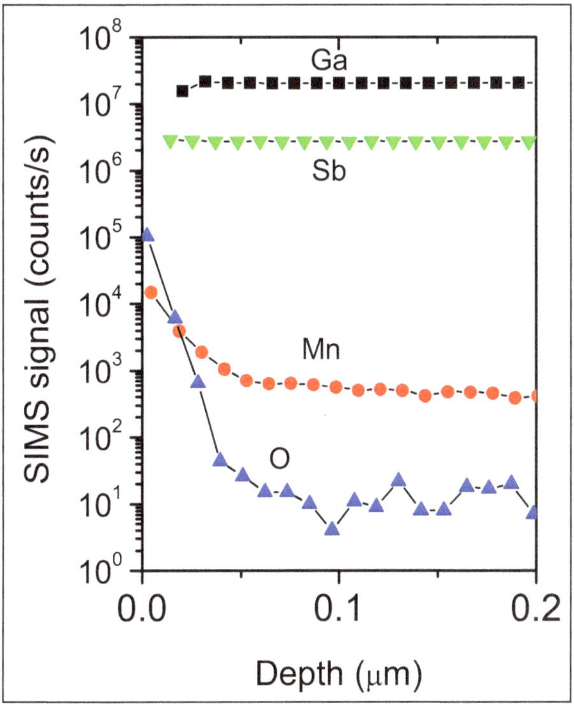

Fig. 3. SIMS depth profiles of the element distribution in the #2 sample.

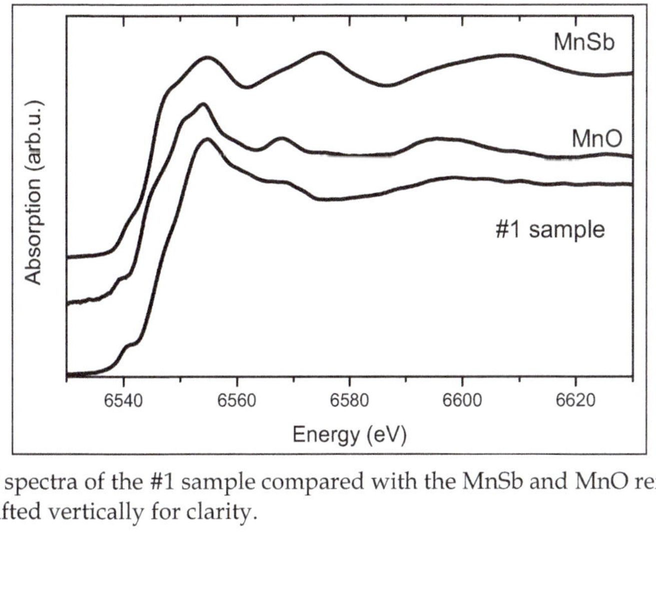

Fig. 4. XANES spectra of the #1 sample compared with the MnSb and MnO references. Spectra are shifted vertically for clarity.

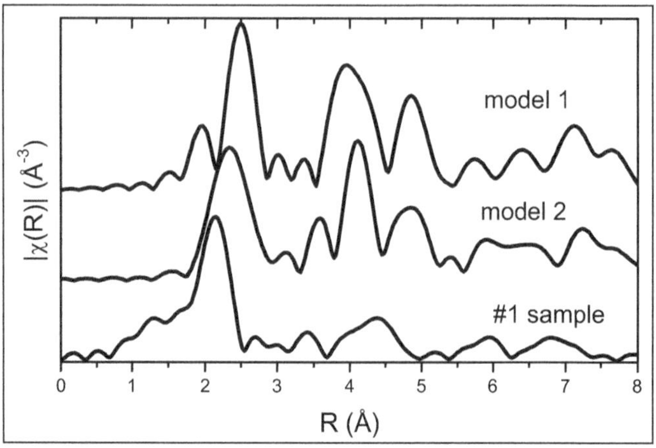

Fig. 5. FT EXAFS spectrum of the #1 sample and theoretical models based on the GaSb crystallographic data. Model 1: Mn substituting Ga (Mn_{Ga}); Model 2: Mn substituting Sb (Mn_{Sb}). Spectra are shifted vertically for clarity.

Since the Mn atoms are implanted into the GaSb crystal, the GaSb model seems to be the obvious choice. First, two cases have to be checked: 1) the Mn atom substituting the Ga atom in the GaSb structure (Mn_{Ga}) and 2) the Mn atom substituting the Sb atom (Mn_{Sb}). Fitting of both models returned unphysical parameters. Therefore, in Fig. 5, not the fitting results but the simulated spectra for the models are compared with the magnitude of FT EXAFS oscillations for the sample with the lowest dose (#1). Model 1 is a sum of 8 single scattering paths corresponding to the consecutive 4 Sb and 4 Ga coordination spheres in the GaSb crystal, where central Mn atom is located in the Ga position. The number of atoms in each coordination sphere was kept set according to the GaSb crystallographic data. Model 2 presents such sum for the Mn central atom located in the Sb position.

A closer look at the calculated spectra for models reveals the possibility that the Ga atoms can be located in the first shell but at shorter distance. Moreover, it seems that the Sb atoms are not located in the vicinity of the Mn atoms. It leads to model 3 also based on the GaSb structure, where only the Ga atoms are present. In Fig. 6 a sum of 4 single scattering Ga paths corresponding to four consecutive shells consisting of Ga atoms is shown. The result seems to be closer to the experimental spectrum but the distances to the subsequent shells are too large. However, with a lack of antimony atoms, it can be expected that remaining Ga atoms rearrange their positions and move closer to the central atom. This is shown in model 4 (Fig. 6), where the distances to each shell were modified to agree with the experiment.

The preliminary fitting of the experimental spectrum was performed using model 3. While defining the parameters, it has to be taken into account that the farther given shell is located in model 3 the more it has to be moved to agree with the experiment. Therefore, the ΔR parameter responsible for monitoring changes in the distances (R) was defined as $R^*\alpha$, where α is a parameter measuring lattice contraction. The result is shown in Fig. 7. The fit is not perfect because only single scattering Ga paths were considered, however in spite of that; the model works reasonably well even up to 8 Å. It can be noticed that the distances are getting significantly shorter. The comparison between the distances to the subsequent shells found from model 3 and fit is shown in Tab. 1.

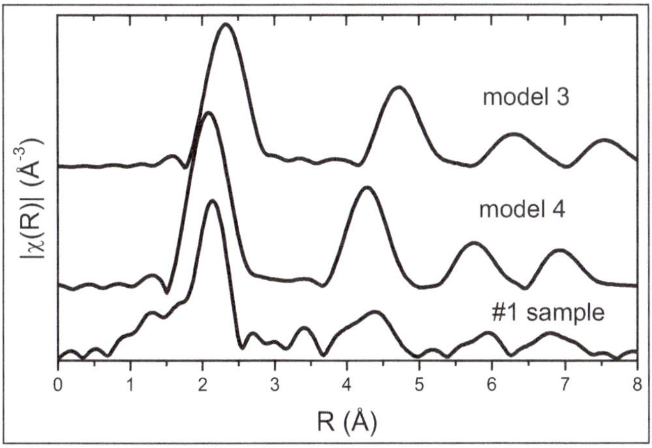

Fig. 6. FT EXAFS spectrum of the #1 sample and theoretical models based on the GaSb crystal with the Sb atoms removed. Model 3: distances to the Ga shells as in the GaSb crystal; Model 4: shortened distances to the Ga shells. Spectra are shifted vertically for clarity.

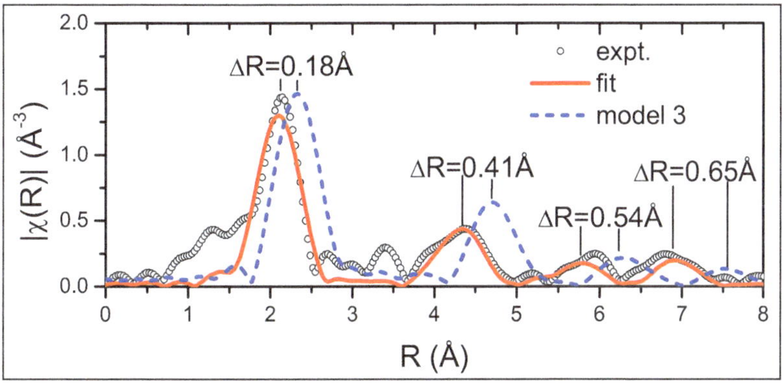

Fig. 7. Fitting of the FT EXAFS spectrum of the #1 sample using single scattering Ga paths for four consecutive Ga spheres in the GaSb crystal compared with the same paths as found from the model 3.

The structure above 5 Å in the spectra for higher doses was not well preserved which indicates that an increase of dose introduces higher disorder in the sample. Therefore, only two Ga shells were considered in the precise fitting. The first and second shells consist of gallium atoms. The tail below the first shell was fitted assuming presence of oxygen. MnO did not crystallize here; only in average 2-3 oxygen atoms at the distance of 2.02 Å were found. In order to further improve quality of the fits also some residual antimony atoms had to be introduced between the shells. In Fig. 8 the results of fitting for samples with all doses are shown.

In summary, the EXAFS analysis revealed that even low energy implantation and quick annealing lead to the significant deficit of Sb atoms around implanted Mn atoms. Some traces of the Mn-O bonds are detected, however, the majority of the Mn atoms is found to be located in the antimony positions in the GaSb matrix where the Sb sublattice is almost completely destroyed and the remaining Ga sublattice contracts significantly. It leads to the creation of the inclusions containing Mn-Ga material. In these inclusions Mn atoms are located within the contracted Ga sublattice preserving a crystalline order characteristic for the GaSb crystal with the F-43m space group (Wolska et al., 2010).

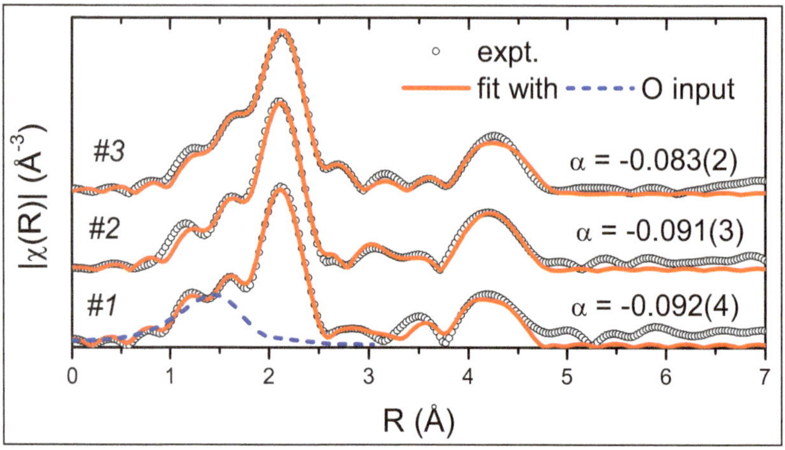

Fig. 8. Fitting results of the FT EXAFS spectra for the implanted samples up to the second coordination shell. Residual O and Sb atoms were also used in the fits. For the #1 sample the contribution from the oxygen atoms is shown (blue dashed line). The α parameter is a measure of lattice contraction. Spectra are shifted vertically for clarity.

	I shell (4 Ga)	II shell (12 Ga)	III shell (12 Ga)	IV shell (16 Ga)
R_{Ga} [Å] (model)	2.65	5.07	6.66	7.95
R_{Ga} [Å] (fit)	2.43	4.66	6.12	7.30

Table 1. The distances to the subsequent Ga shells found from the GaSb model and the fit.

3.2 Implantation with high energy Mn^+ ions (150 keV) at room temperature

The implantation of the low energy Mn^+ ions did not succeed in the creation of the MnSb inclusions. Therefore, in the next step higher energy of the Mn^+ ions was used. The implantation energy was equal to 150 keV and the dose was 1.7×10^{17} Mn/cm^2. One of the samples was left without any additional treatment. The other one was annealed in Ar atmosphere at 350 °C for 5 min.

The examination of XANES spectra of the samples revealed that this approach was not successful as well. MnSb inclusions were not formed. The XANES spectrum of the as implanted sample is quite shapeless which suggests that the Mn atoms' surrounding is amorphous (see Fig. 9). For the annealed sample weak maxima around 6570 eV and 6600 eV can be distinguished. They correspond to the maxima seen for the samples implanted with low energy ions (in Fig. 9 the #3 sample is shown as an example) described in the previous procedure. Both spectra are similar, except for the first maximum which is more pronounced for the #3 sample. Since it was attributed to the presence of oxygen, it would imply lack of manganese oxides in the new samples.

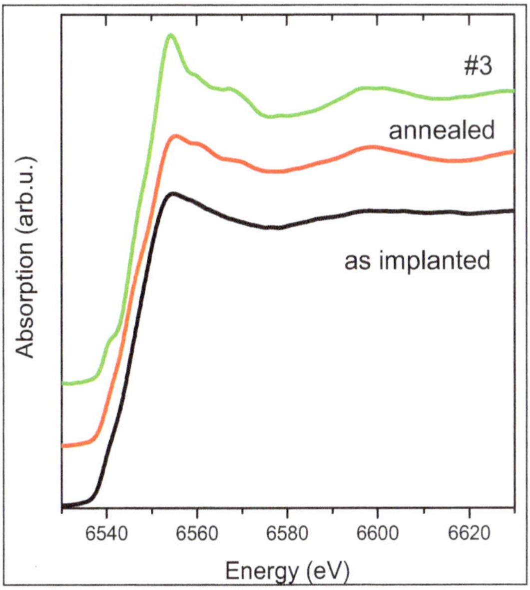

Fig. 9. XANES spectra of the as implanted and annealed samples compared with the #3 sample implanted with low energy ions described in previous subchapter. Spectra are shifted vertically for clarity.

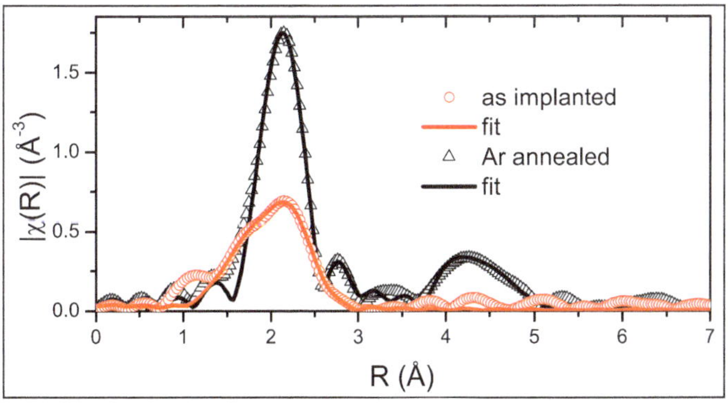

Fig. 10. Fitting results of the FT EXAFS spectra for the as implanted and annealed samples.

In order to obtain more precise information about the Mn location in the samples, the EXAFS analysis was performed. In the FT EXAFS of the as implanted sample only the first shell can be distinguished (Fig. 10). This confirms the conclusion drawn from XANES analysis that the Mn neighbourhood is amorphous here. This shell is mainly populated by the Ga atoms but some traces of the Mn atoms can also be fitted here. As was expected, the implantation with high energy ions leads to a lack of a long range order around the implanted ions.

After annealing in Ar atmosphere the amplitude of the first shell became higher and the second shell can be distinguished (Fig. 10). The annealing enabled to reconstruct the crystalline structure around the implanted ions; however, still the structure is not the MnSb or GaSb. The Mn atoms are no longer located in the first shell. The Ga atoms dominate in the near neighbourhood forming contracted Ga lattice as it was shown for the samples with low Mn doses. However, in this case there are no traces of the Mn-O bonds and still no Sb atoms were found around Mn.

In the attempt to reintroduce the Sb atoms into the crystal, next two implanted samples were annealed in vacuum with Sb vapour at 400 °C for 48 h and at 600 °C for 2h. The distributions of the elements as measured by SIMS are presented in Fig. 11. Clear attenuation in the depth profiles of the Ga and Sb atoms reflects the disturbed region in the crystal. Therefore, the ratio between the Sb and Ga atoms distributions is taken into consideration in the analysis while investigating the relative changes in the concentration of Sb atoms.

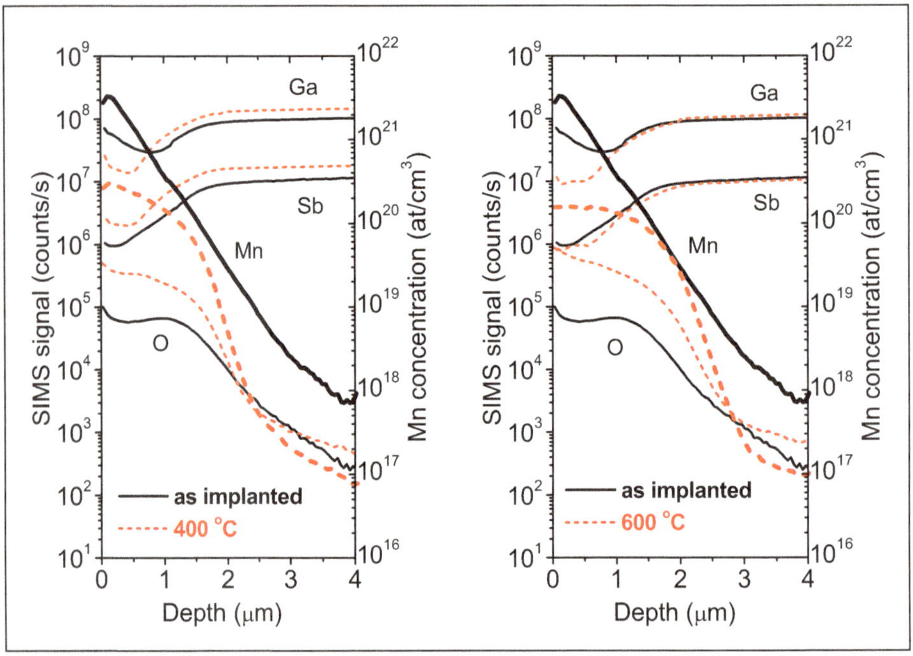

Fig. 11. SIMS depth profiles of the element distribution in the samples annealed in Sb vapour compared with the profiles for the as implanted sample.

SIMS data show that there is a depletion in the distribution of the Sb atoms in respect to the Ga atoms in the vicinity of the surface for the as implanted sample. For the samples annealed in Sb vapour the ratio of the Ga to Sb atoms stays almost constant. It confirms that such annealing reintroduces the Sb atoms into the implanted samples. On the other hand, together with antimony high amount of oxygen was also introduced during this procedure.

EXAFS analysis of the samples annealed in Sb vapour shows completely different atomic order around Mn atoms in comparison to previously investigated samples. It was found from SIMS data that high amount of oxygen was introduced during the heat treatment. And indeed, as seen from the EXAFS spectra in Fig. 12, the first shells show the distribution characteristic for oxides. However, in both samples the Mn atoms don't form stoichiometric MnO with long range order but the oxygen atoms are found in the first shell and different numbers of Mn atoms in the second one. What's more, the Mn-Mn distances differ between the samples annealed at 400 °C (R = 3.12 Å) and 600 °C (R = 3.55 Å) which means that different types of manganese oxides start to be forming.

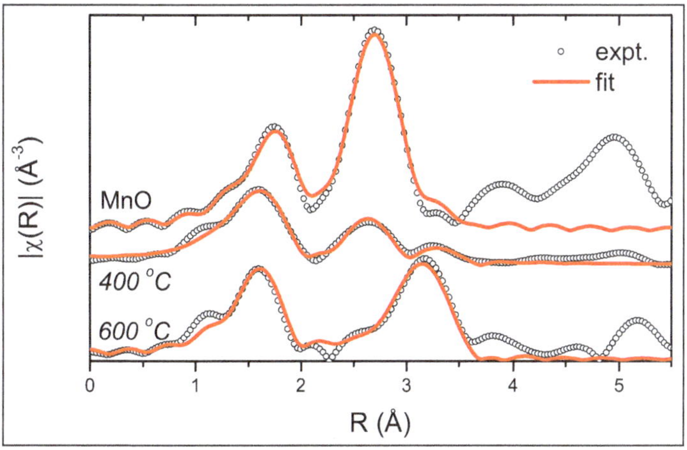

Fig. 12. Fitting results of the FT EXAFS spectra for the samples annealed in Sb vapour compared with the MnO reference. Spectra are shifted vertically for clarity.

In summary, the EXAFS results showed that in all considered cases the implantation removed the Sb atoms from the neighbourhood of the Mn atoms. After annealing in Ar atmosphere the Mn-Ga inclusions were formed. The annealing in Sb vapour reintroduced Sb atoms but together with oxygen which is more reactive and was bounded to the Mn atoms forming Mn oxides (Wolska et al., 2010).

3.3 Implantation with high energy Mn^+ ions (150 keV) at low temperature (80 K) with noble gases co-implantation

The attempts shown above revealed that, in order to establish the proper procedures, it is necessary to investigate processes with more steps. It was assumed that the cold and disordered matrix would prevent the escape of the Sb atoms. Therefore, the co-implantation of noble gases was considered. The He^+ and Ne^+ ions were used to make the matrix disordered. Then, the Mn^+ or the Mn^+ and Sb^+ ions were implanted. The substrates' temperature during the implantation processes was kept at 80 K. The energy and doses of Mn^+ and Sb^+ ions were chosen according to the depth profile ion distribution simulated by the SRIM2008 code (Ziegler et al., 1985) in order to locate the Mn^+ and Sb^+ ions at the distance from 50 to 150 nm below the surface. The implantation of noble gases was assumed not to influence the GaSb matrix density. The calculated ranges of ions are listed in Tab. 2.

Four series of implantation were prepared:

a. *NeMn* where the substrate was implanted first with the Ne^+ ions (250 keV, dose $5x10^{16}$ cm^{-2}) and then with the Mn^+ ions (150 keV, dose $9x10^{14}$ cm^{-2}).
b. *NeMnSb* where after the procedure described in point a) the Sb^+ ions (250 keV, dose $9x10^{14}$ cm^{-2}) were implanted
c. *HeMn* where the substrate was implanted first with the He^+ ions (80 keV, dose $5x10^{13}$ cm^{-2}) and then with the Mn^+ ions (150 keV, dose $9x10^{14}$ cm^{-2}).
d. *HeMnSb* where after the procedure described in point c) the Sb^+ ions (250 keV, dose $9x10^{14}$ cm^{-2}) were implanted.

	I step (noble gases)	II step (Mn^+ ions)	III step (Sb^+ ions)
NeMn	Ne+ GaSb Rp=369.1 nm ΔRp=149.2 nm	Mn+ GaSb Rp=94.82 nm ΔRp=48.11 nm	
NeMnSb	Ne+ GaSb Rp=369.1 nm ΔRp=149.2 nm	Mn+ GaSb Rp=94.82 nm ΔRp=48.11 nm	Sb+ GaSb Rp= 74.20 nm ΔRp= 33.70 nm
HeMn	He+ GaSb Rp= 433.6 nm ΔRp= 151.4 nm	Mn+ GaSb Rp= 94.70 nm ΔRp= 48.28 nm	
HeMnSb	He+ GaSb Rp= 433.6 nm ΔRp= 151.4 nm	Mn+ GaSb Rp= 94.70 nm ΔRp= 48.28 nm	Sb+ GaSb Rp= 78.10 nm ΔRp= 37.30 nm

Table 2. The calculated ranges of ions for each implanted series.

Each of the implanted samples was subsequently divided in two parts. One part was left as such (as-implanted), the second part was subject to rapid thermal annealing (RTA) for 5 min. in Ar atmosphere at the temperature 350 °C (RTA 350).

The SIMS measurements were conducted on the samples annealed at 350 °C. The results are shown in Fig. 13. There are clear differences in the distribution of the elements depending on the type of noble gas used in the co-implantation. What's more, in all cases a considerable amount of oxygen atoms was incorporated deeply into the matrix.

Close comparison of the Mn and Sb distributions found from SIMS measurements with those calculated by the SRIM code confirms that the results depend on the matrix preparation. In Fig. 14 the SRIM and SIMS data for the NeMn and NeMnSb samples are presented. In case of the Mn atoms there is a perfect agreement - the maximum of the concentration is located around 85 nm from the surface for the calculations and 90 nm for the measurements. The maximum for the Sb atoms is supposed to be around 70 nm. From SIMS data it is hard to determine the maximal point but comparison between the NeMn and NeMnSb clearly shows that additional Sb atoms are located within the same depth as the Mn atoms. This could suggest that at least in the NeMnSb sample the possibility of the MnSb formation exists.

In case of the samples co-implanted with the He+ ions the changes introduced to the matrix resulted in remarkable distortion from predicted Mn and Sb distributions. In Fig. 15a it can be seen that according to the calculations the maximum of the concentration should be around 85 nm for Mn and 70 nm for Sb. However, SIMS results (Fig. 15b) show that for the HeMn sample maximum of the Mn concentration is around 200 nm, while for the HeMnSb sample there is a long plateau slightly tilted down. The results are in correlation neither to the Mn atoms nor to the predictions.

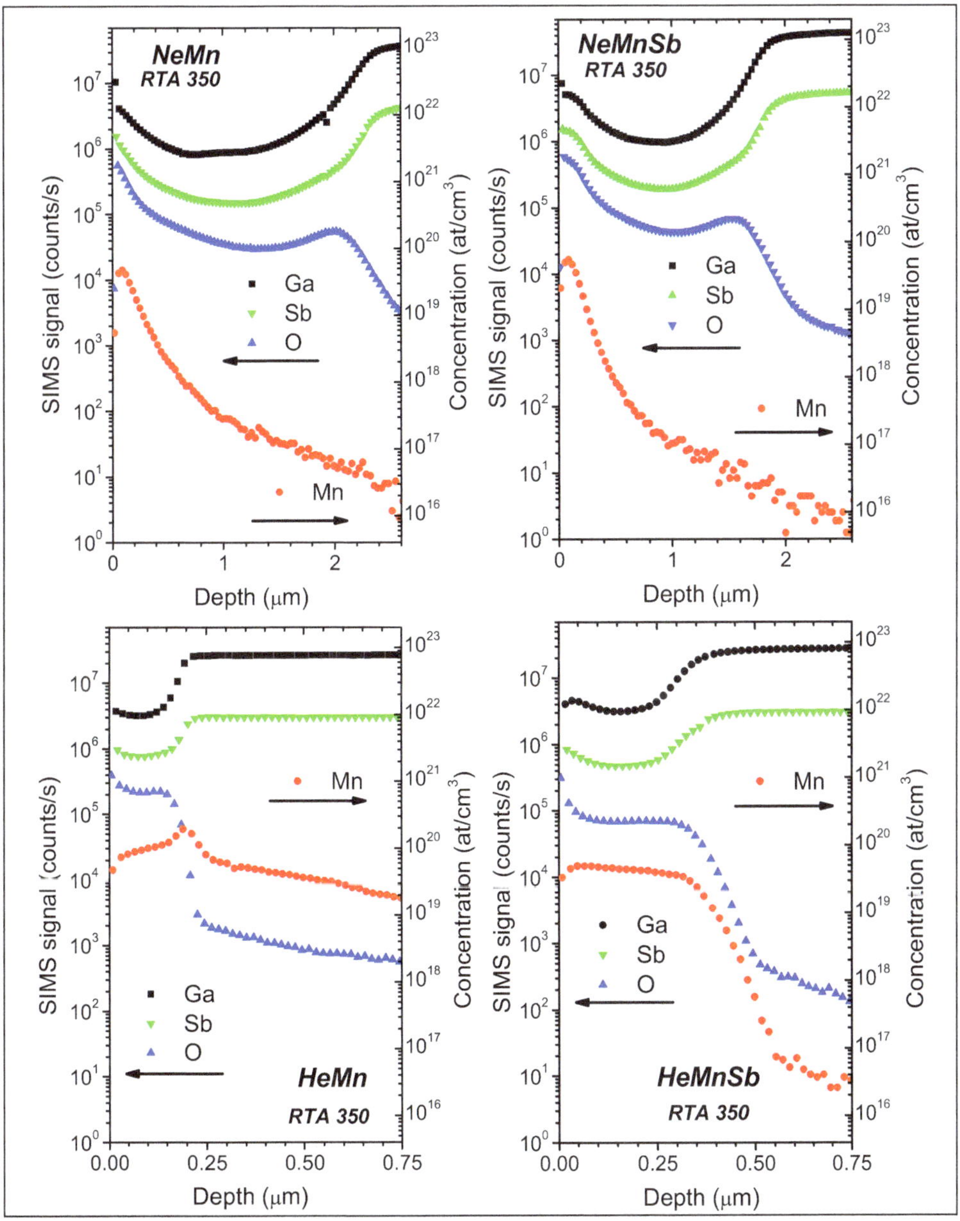

Fig. 13. SIMS depth profiles of the element distribution in the RTA 350 samples.

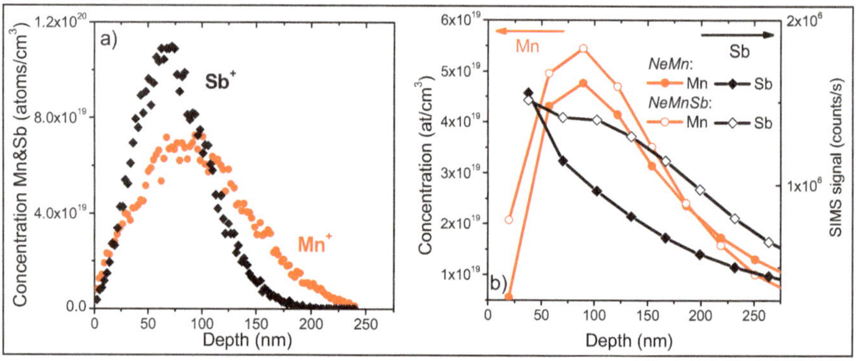

Fig. 14. a) Depth distribution of the implanted Mn and Sb atoms (after the Ne+ implantation) as calculated by SRIM 2008; b) Depth distribution of Mn and Sb atoms in the samples as measured by SIMS.

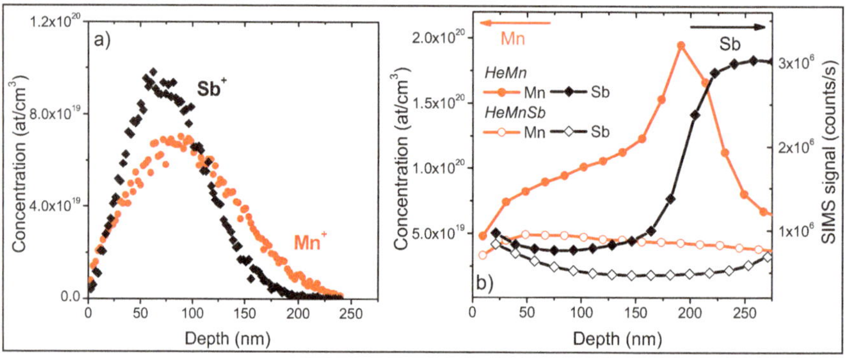

Fig. 15. a) Depth distribution of the implanted Mn and Sb atoms (after the He+ implantation) as calculated by SRIM 2008; b) Depth distribution of Mn and Sb atoms in the samples as measured by SIMS.

From the analysis conducted above it can be concluded that the distribution of the Mn ions after the implantation by the Ne^+ ions is close to the prediction of SRIM2008 simulation, while changes introduced to the matrix by the He^+ ions resulted in remarkable distortion from predicted Mn distribution. Taking into account that the SIMS results for the NeMnSb sample show that the co-implantation of the Mn^+ and Sb^+ ions tends to gather them in the same region one can suspect that in this sample the MnSb compound may be formed.

The XANES measurements of the as implanted and RTA 350 samples are presented in Fig. 16. The MnSb spectrum is shown for comparison. It is obvious that the MnSb inclusions were not formed, even for the prospective NeMnSb sample. What's more, spectra for all look very similar which indicates that the Mn neighbourhood doesn't change much for any of them.

In Fig. 17 the FT EXAFS spectra for the RTA 350 samples are presented (the spectra of the as implanted samples look very similar and are not shown here). For all of them only the first peak can be distinguished. It means that the neighbourhood of the Mn atoms is amorphous or highly disordered. The MnO reference spectrum is shown for comparison. Its first shell consisting of oxygen looks like the one in the samples. The second, Mn consisting shell, is not formed in the samples.

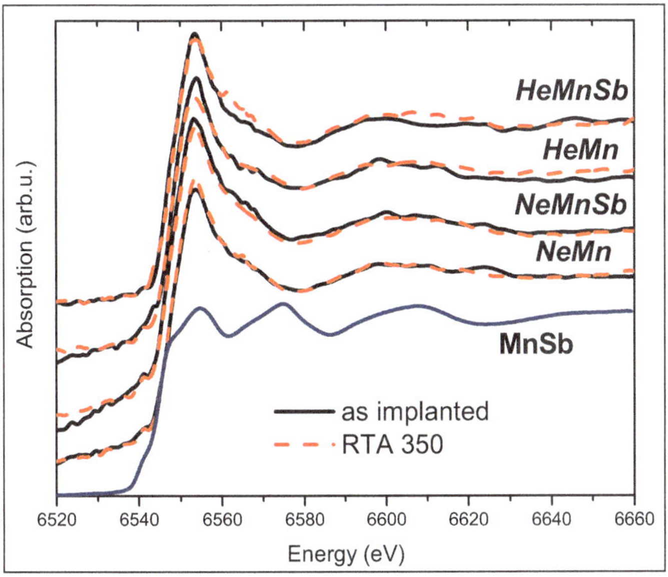

Fig. 16. XANES spectra of the as implanted and annealed samples compared with the MnSb reference. Spectra are shifted vertically for clarity.

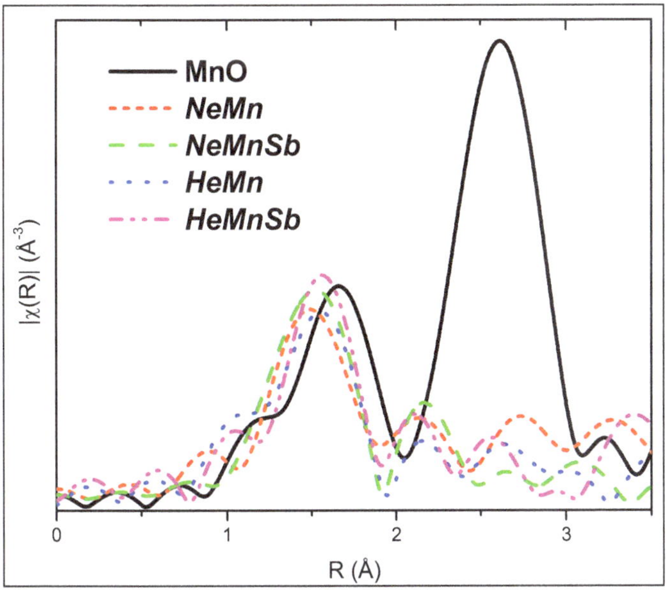

Fig. 17. FT EXAFS spectra for the RTA 350 samples compared with the MnO reference.

The model fitted to the spectra confirms that for all samples Mn-O bonds are formed. In all cases number of the O atoms is lower than for MnO (around 3 atoms instead of 6) and the Mn-O distances are shorter (around 2.03 Å instead of 2.22 Å).

In summary, the X-ray absorption analysis showed that independently on the co-implantation and post-implantation procedures only the Mn oxides were formed inside the GaSb matrix in the considered samples. The implantation into the cold and amorphous substrate indeed prevented the Sb atoms from escaping, however, at the same time through the formed frozen channels the oxygen atoms were incorporated deeply into the matrix.

3.4 Implantation with high energy Mn^+ ions (350 keV) into the Si_3N_4 capped samples

The procedures described above succeeded in preventing the Sb atoms from escaping but did not prevent the oxygen contamination. Therefore, in the next attempt the GaSb (100) crystals were capped with amorphous Sb and Si_3N_4 before the implantation of the Mn ions. The Si_3N_4 cap was supposed to protect the matrix from the introduction of oxygen while the Sb layer was providing additional Sb atoms. The implantation energy was equal to 350 keV and the dose was $1.3x10^{17}$ Mn/cm^2. The substrate temperature during implantation was kept below 50 °C. One of the implanted samples was left untreated; the other one was annealed at 500 °C for 20 minutes. Additionally, another not capped GaSb(100) crystal was submitted to the same implantation and post-implantation process in order to check the influence of the capping layer. Tab. 3 lists the names of the samples and the processes they were submitted to.

sample	not capped	capped
as implanted	210	210N
annealed	210V	210NV

Table 3. The names of the implanted samples and the processes they were submitted to.

The SIMS measurements show that indeed the presence of the Si_3N_4 capping layer introduces changes between the samples. In case of the capped samples, the depletion in the Ga and Sb distributions ends earlier that in the analogous uncapped ones. (Fig. 18)

For the as-implanted 210 and 210N samples the Mn and O distributions look rather similar. However, in case of the annealed ones the differences are quite significant. For the uncapped 210V sample the Mn and O distributions changed slightly the slope but still remained approximately parallel. For the capped annealed 210NV sample the O atoms were moved closer to the surface while the Mn atoms stayed deeper. It means that capping and annealing indeed had influence on the Mn and O atomic distributions.

Consideration of the SIMS data leads to the conclusion that in the 210NV sample new type of inclusions can be formed. However, in the XANES spectra only slight changes in the shapes are seen between the as implanted and annealed samples independently from the presence of the capping layer (see Fig. 19). The pronounced first peak characteristic for the manganese oxides dominated in the spectra.

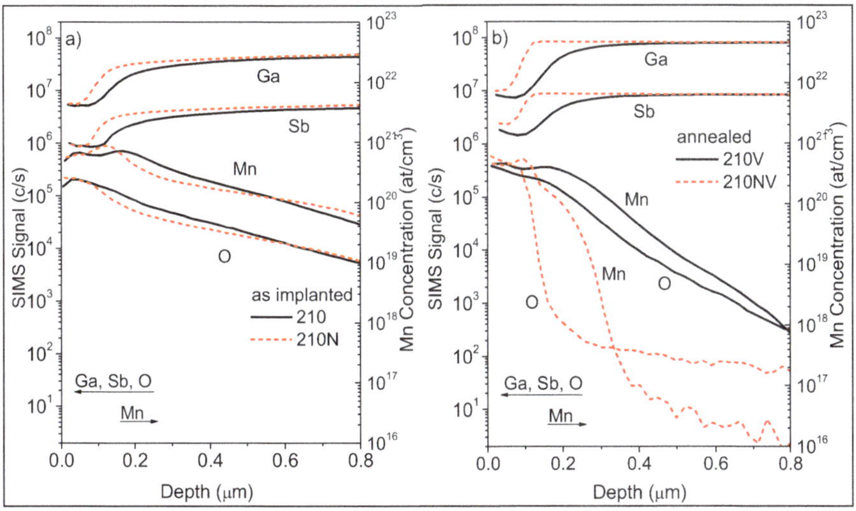

Fig. 18. SIMS depth profiles of the element distribution: a) in the as implanted samples; b) in the samples annealed in Ar atmosphere.

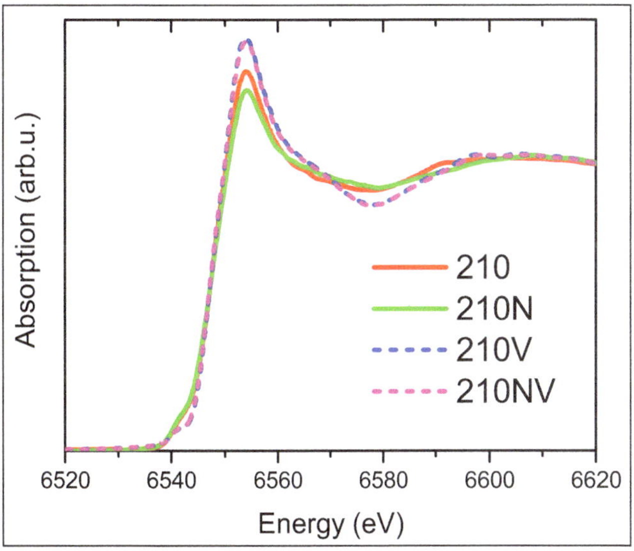

Fig. 19. XANES spectra of the as implanted and annealed samples.

FT EXAFS spectra reveal that in all samples an amorphous neighbourhood exists around the Mn atoms (see Fig. 20). Only first coordination shell is visible. Its amplitude differs between the as-implanted and annealed samples confirming differences noticed in the XANES spectra.

The model fitted to the spectra revealed that the Mn atoms' neighbourhood consists of O atoms. For the as-implanted samples there are in average 3 O atoms at the average distance of 2.09 Å, while for the annealed samples around 5 O atoms can be found at the same distance.

In summary, the SIMS measurements show that the capping layer influenced the distribution of the elements. However, the XAFS analysis reveals that the implantation temperature was too low to enable formation of the inclusions other than the manganese oxides.

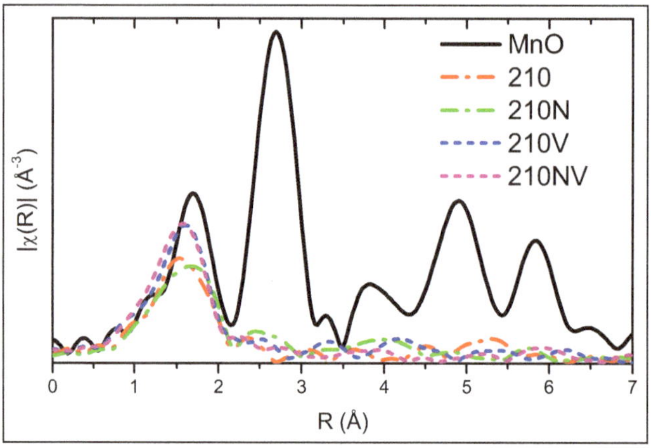

Fig. 20. Magnitude of FT EXAFS spectra for the samples compared with the MnO reference.

3.5 Implantation into warm substrate

Conclusions from the previous procedures suggested another approach - the implantation into the warm substrate. In this case the substrates were kept at 70 °C and 200 °C during the implantation. First, the Mn^+ ions (120 keV, dose $5x10^{16}$ cm^{-2}) were implanted and then the Sb^+ ions (240 keV, dose $5x10^{16}$ cm^{-2}). After implantation the samples were annealed in situ at 350 °C for 0.5 h.

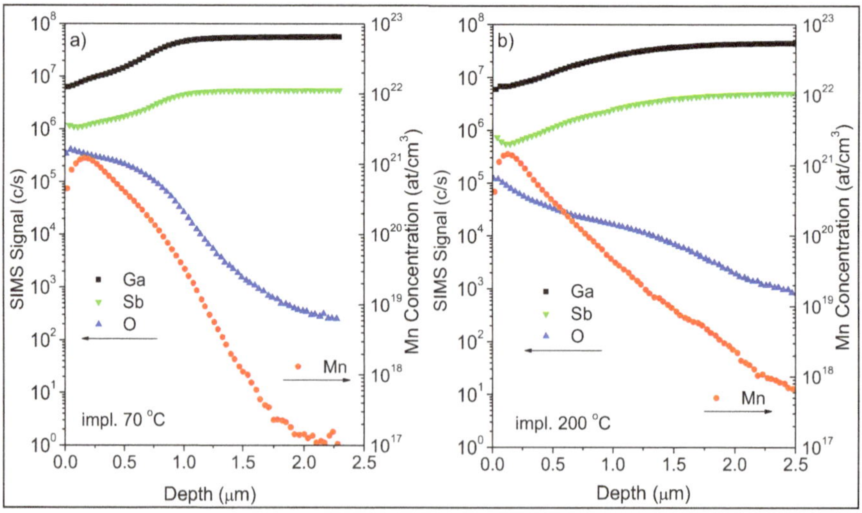

Fig. 21. SIMS depth profiles of the element distribution: a) in the sample implanted at 70 °C; b) in the sample implanted at 200 °C.

From the SIMS data the differences between the samples can be distinguished. For the 200 °C sample the maximum of the Mn atoms concentration is 50 nm closer to the surface than for the 70 °C sample and corresponds with the depletion in the Sb atoms distribution. For the 70 °C sample no depletion in the Sb profile in respect to the Ga one can be detected but the level of the O atoms is higher than in the 200 °C sample.

The XANES spectra confirm that there are differences between the samples (see Fig. 22). Comparison with the previously analysed various types of spectra leads to the conclusions that in the 70 °C sample some oxides can be expected while the 200 °C sample is probably the Mn-Ga type.

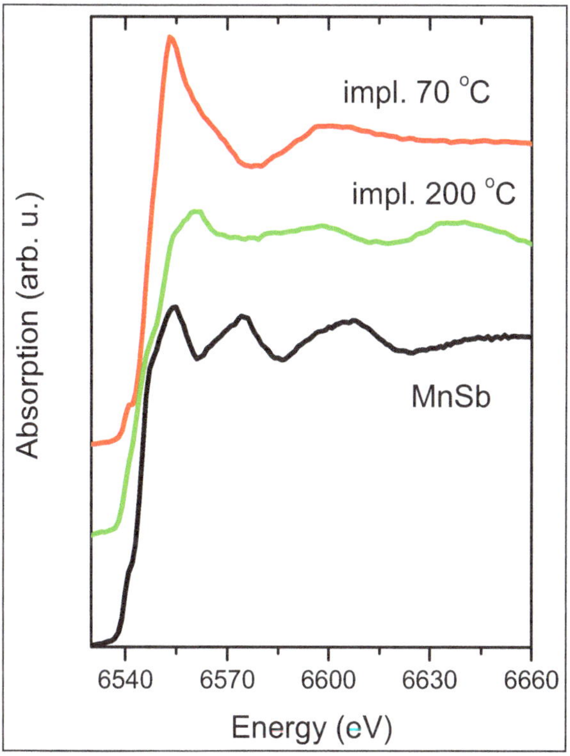

Fig. 22. XANES spectra of the samples implanted at 70 °C and 200 °C compared with the MnSb reference. Spectra are shifted vertically for clarity.

The EXAFS analysis confirms these conclusions. In the 70 °C sample two shells can be distinguished, see Fig. 23. The first one consists of the O atoms at the distance of 2.06 Å. In the second one the Ga atoms are present at the distance of 2.46 Å. In case of the 200 °C sample, according to predictions, two shells containing the Ga atoms are visible (Fig. 24).

It was assumed that the additional Sb atoms introduced into the matrix would facilitate the formation of the MnSb inclusions; however, the EXAFS analysis didn't confirm the presence of the Sb atoms in the Mn atoms' neighbourhood. In case of the sample implanted at 70 °C, the Mn oxides and Mn-Ga clusters coexist. For the sample implanted at 200 °C only the Mn-

Ga clusters are formed. It can be concluded that the thermal disorder during the implantation at higher temperature prevented the migration of the oxygen atoms due to the quick recrystallization of the matrix during cooling. However, the chemical affinity of Mn to Sb is lower than to Ga.

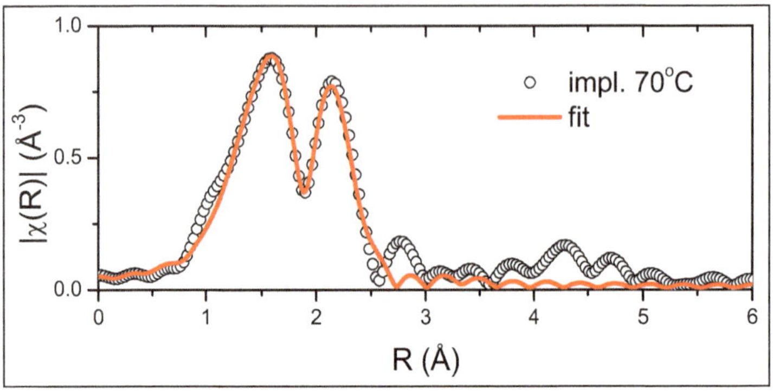

Fig. 23. Magnitude of the FT EXAFS spectrum of the sample implanted at 70 °C together with the fit. In the first shell the oxygen, in the second - the gallium atoms are located.

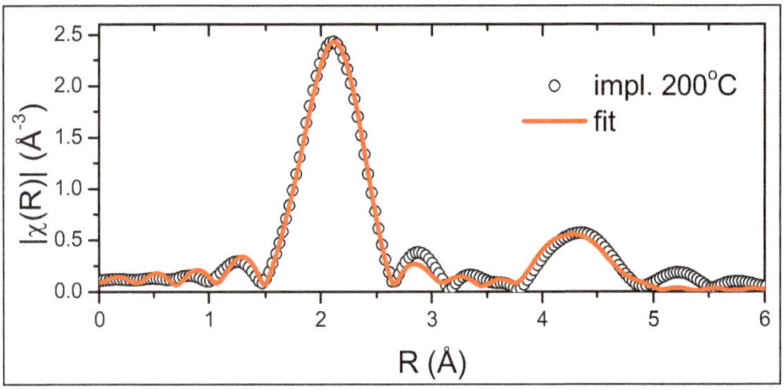

Fig. 24. Magnitude of the FT EXAFS spectrum of the sample implanted at 200 °C together with the fit. The first and second shells consist of the gallium atoms.

4. Conclusions

The examples of forming the MnSb inclusions in the GaSb crystals during the Mn^+ ion implantation shown in this chapter demonstrate that detailed information about the dopants' neighbourhood is a very important part of characterization of any new material. The attempts at obtaining the MnSb inclusions encountered several obstacles which were recognized by XAFS analysis. Implantation tends to remove Sb atoms from the neighbourhood of the Mn atoms even with the implantation temperature kept close to liquid nitrogen temperature. It was possible to reintroduce Sb atoms by the annealing in Sb vapour but together with oxygen which is more reactive and was bounded to the Mn atoms. Keeping the high temperature during the implantation prevented the migration of oxygen but still the chemical affinity of Mn to Sb is lower than to Ga and the Mn-Ga clusters are formed.

In summary, the theoretical calculations enable to determine the values of the implantation parameters in order to plan distribution of each element. Thanks to the SIMS measurements it can be checked if the element distributions are in agreement with the predictions. However, it is not enough to ensure that the implanted element is located in the right part of the material, it is also important to find out its chemical bonding which is possible with the XAFS techniques.

5. Acknowledgments

The author would like to acknowledge Dr. R. Jakiela from Polish Academy of Sciences in Warsaw for the SIMS measurements and Dr. M. Kulik from University Maria Curie-Skłodowska in Lublin for performing the SRIM calculations and some of the implantation procedures. The author would also like to thanks Dr. A. Hallen from Royal Inst. of Technology in Sweden and Dr. D. Arvanitis from Uppsala University in Sweden for advice and help in starting the GaSb implantation project.

This work was partially supported by national grant of Ministry of Science and High Education N202-052-32/1189. The synchrotron research leading to these results has received funding from the European Community's Seventh Framework Programme (FP7/2007-2013) under grant agreement n° 226716.

6. References

Croft, M.; Sills, D.; Greenblatt, M.; Lee, C.; Cheng, S.-W.; Ramanujachary, K.V. & Tran, D. (1997). Systematic Mn d-configuration change in the $La_{1-x}Ca_xMnO_3$ system: A Mn K-edge XAS study. *Phys. Rev. B,* Vol. 55, No. 14, pp. 8726–8732.

Eisenberger, P. & Kincaid, B.M. (1978). EXAFS: new horizons in structure determinations. *Science,* Vol. 200, No. 4349, pp. 1441-1447.

Hai, P.N.; Ohya, S.; Tanaka, M.; Barnes, S. E. & Maekawa, S. (2009). Electromotive force and huge magnetoresistance in magnetic tunnel junctions. *Nature,* Vol. 458, No. 7237, pp. 489-492.

Koningsberger, D.C. & Prins, R. (Ed.) (1988). X-ray Absorption. John Wiley and Sons Inc., New York.

Lawniczak-Jablonska, K.; Wolska, A.; Klepka, M.T.; Kret, S.; Gosk, J.; Twardowski, A.; Wasik, D.; Kwiatkowski, A.; Kurowska, B.; Kowalski, B. J. & Sadowski J. (2011). Magnetic Properties of MnSb Nano-inclusions Formed in GaSb Matrix Directly During MBE Process. *J. Appl. Phys.,* Vol. 109, No. 7, pp. 074308-1 - 074308-7.

Lee, P.A.; Citrin, P.H.; Eisenberger, P. & Kincaid, B.M. (1981). Extended x-ray absorption fine structure—its strengths and limitations as a structural tool. *Rev. Mod. Phys.,* Vol. 53, No. 4, pp. 769–806.

Panchula, A.F.; Kaiser, C.; Kellock, A. & Parkin, S.S. (2003). Spin polarization and magnetotransport of Mn–Sb alloys in magnetic tunnel junctions. *Appl. Phys. Lett.,* Vol. 83, No. 9, pp. 1812-1814.

Ravel, B. & Newville, M. (2005). ATHENA, ARTEMIS, HEPHAESTUS: data analysis for X-ray absorption spectroscopy using IFEFFIT. *J. Synchrotron Rad.,* Vol. 12, No. 4, pp. 537-541.

Stohr, J. (1996). NEXAFS spectroscopy. Springer-Verlag. ISBN 3-540-54422-4 Berlin, Heidelberg, New York.

Teo, B.K. (1986). EXAFS: Basic Principles and Data Analysis. Springer-Verlag, ISBN 0-387-15833-2 Berlin, Heidelberg, New York, Tokyo.

Teo, B.K. & Joy, D.C. (Ed.) (1981). EXAFS Spectroscopy. Plenum Press. ISBN 0-306-40564-3 New York and London.

Wolska, A.; Lawniczak-Jablonska, K.; Klepka, M.T.; Barcz, A.; Hallen, A. & Arvanitis, D. (2010). Study of the local environment of Mn ions implanted in GaSb. *Acta Phys. Pol. A,* Vol. 117, No. 2, pp. 286-292.

Wolska, A.; Klepka, M.T.; Lawniczak-Jablonska, K.; Sadowski, J.; Reszka, A. & Kowalski B.J. (2011). MnSb inclusions in the GaSb matrix studied by X-ray absorption spectroscopy. *Radiat. Phys. Chem.,* Vol. 80, No. 10, pp. 1026-1030.

Ziegler, J.F.; Biersack, J.P. & Litmar U. (1985). SRIM—The Stopping and Range of Ions in Solids. Pergamon Press, New York.

Section 3

Materials Applications

Ion Implantation in Phase Change $Ge_2Sb_2Te_5$ Thin Films for Non Volatile Memory Applications

Stefania Maria Serena Privitera
Consiglio Nazionale delle Ricerche (CNR)
Istituto di Microelettronica e Microsistemi (IMM)
Catania
Italy

1. Introduction

Phase change materials are characterized by the existence of at least two different phases, between which it is possible to reversibly and repeatedly switch in a very short timescale. Large variation of the electrical and optical properties occurs upon phase change. These unique properties have made phase change materials very attractive for the realization of nonvolatile electronic memories based on the concept of using the reversible amorphous-to-crystalline phase transition as writing and erasing mechanism. The two logic states are the high resistance, low reflectivity amorphous phase and the crystalline, high reflectivity, low resistance phase. The switching can be obtained either by using a laser pulse or an electric pulse.

The technological success of phase change materials was firstly linked to the realization of optical memories, and it was enabled by the discovery of a class of materials, belonging to the pseudobinary line GeTe-Sb_2Te_3, characterized by large variation of optical contrast by laser pulses. Thanks to the use of such materials, the rewritable optical storage technology has been developed, reaching its third generation of blu-ray disks with 100 Gb capacity. The most common used material for optical disks has composition $Ge_2Sb_2Te_5$ and it is also characterized by order of magnitude of resistance variation when changing from amorphous to polycrystal, with face centered cubic structure (*fcc*). Following the optical memory concept, the phase change materials have been proposed for the realization of nonvolatile memories based on the phase transitions driven by electric pulses. Phase change memories (PCM) employing $Ge_2Sb_2Te_5$ as recording media have been successfully manufactured, obtaining large electrical contrast, with orders of magnitude of difference between the resistance in amorphous and crystalline phase (Lai & Lowrey, 2001; Redaelli & Pirovano, 2011).

Many material aspects have to be taken into account for the realization of phase change electrical memories. One limits the application of phase change memories as multi-level memory, and it is the drift effect of the amorphous phase resistance, i.e. a resistance increase occurring even at room temperature. The second major issue is due to the low crystallization temperature of the amorphous phase, which may limit the data retention under operation conditions and therefore the reliability of phase change memories that should be stable in the amorphous phase for ten years at 80°C for embedded applications and at 150°C for automotive applications.

In this chapter we will show how ion beam can be employed as a useful tool to investigate and modify some physical properties of phase change materials. In particular we will focus on the stability of the amorphous phase, the data retention properties and the crystallization time, that represents the most limiting time constant.

2. Phase change memory

In phase change memories (PCM) the write and erasing mechanisms are based on the reversible amorphous-to-crystal transition, being the two logic states the amorphous and the *fcc* polycrystalline phase. Data reading is performed by measuring the resistance of the Ge-Sb-Te cell, without producing any phase transition.

The crystallization of the amorphous material (SET operation) is obtained through Joule heating by current pulses but it is enabled by a threshold switching mechanism which allows enough current to flow through the material to heat it above crystallization temperature. The threshold switching occurs at a certain threshold field, on the order of 10-100 V/μm (Krebs et al., 2009), depending on the material composition, and it is characterized by a sudden decrease of resistivity, with negative differential resistance, as schematically shown in figure 1. Several models have been suggested to explain this effect, such as thermal runaway caused by Joule heating (Owen et al. 1979) or energy gain of electrons in a high electric field, leading to a voltage-current instability (Ielmini, 2008). However, detailed experimental validations of the proposed models are further required to better understand the physical mechanisms.

It is experimentally observed that the threshold switching is a reversible phenomenon. If a short voltage pulse above the switching threshold is applied and quickly removed, after a short delay time, the memory cell returns to the high resistance value of the amorphous state, without memory switching. If the voltage pulse is long enough to heat the material above the crystallization temperature, the conversion into the crystalline phase occurs, and after pulse the cell has low resistance.

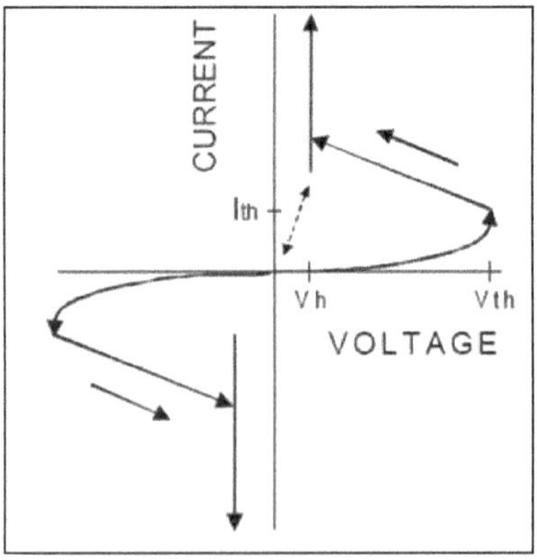

Fig. 1. Current-voltage characteristic of a phase change material, exhibiting electronic threshold switching.

The amorphous-to-crystal transformation is a first order phase transition governed by nucleation and growth and its investigation is crucial for memory applications since the most limiting time constant is the crystallization time and the reliability and retention properties strongly depend on the crystallization kinetics.

To increase the crystallization temperature (~150°C) adversely affecting the retention properties, nitrogen doped $Ge_2Sb_2Te_5$ (GST) has been proposed for PCM (Horii et al., 2003). In fact, it has been shown that nitrogen doping during deposition (Kojima et al. 1998, Jeong et al. 2000) increases the amorphous stability and reduces the grain size, improving the cell cyclability.

The erasing operation (RESET) is achieved through the crystal-to-amorphous phase transition. Amorphization of crystalline films can be obtained by melting and quenching of the melt fast enough that it solidifies in the amorphous state or by ion implantation at room temperature, producing films of various properties (Mio et al. 2010, Raoux et al. 2010). In nonvolatile electrical memory cells melt quenching is obtained by applying a high and short voltage pulse that melts the material, while in optical memories laser pulses are employed. In PCM devices, after melt quenching, the material usually undergoes structural relaxation leading to a resistance drift to higher values over time. This amorphous instability has not been yet understood and it has been explained by considering stress release (Pirovano et al. 2004, Boniardi et al. 2009), decrease of defect density (Ielmini, 2008), shift of the Fermi level or increase of the band gap (Pirovano et al., 2004). Indeed, it has been shown that amorphous phase change materials can have very different properties depending on the way they are produced (melt quenching, room temperature deposition or ion implantation) and on their thermal history (Lee et al., 2009, De Bastiani et al., 2008).

We will investigate the effect of ion beam on the electrical properties of amorphous $Ge_2Sb_2Te_5$ thin films by comparing the electrical properties of deposited amorphous materials to melt-quenched films obtained by laser pulses irradiation and to films amorphized by ion implantation.

Ion implantation has been also employed to introduce impurities belonging to different groups (nitrogen, oxygen and fluorine) into amorphous GST films, in order to produce films with different physical properties. The effects of doping on the electrical conduction of amorphous material and on the amorphous -to-crystal transition will be shown.

3. Amorphization by ion beam

In order to study the crystal-to-amorphous transition we have prepared crystalline films, 70 nm thick, by RF sputtering at 250°C from a single target, on a 500nm thick SiO_2 film, deposited by chemical vapour deposition on a Si substrate.

The film stoichiometry was measured by electron Auger spectroscopy and energy dispersion X-ray spectrometry (EDXS). X-ray diffraction (XRD) indicated the deposited samples were crystalline.

Amorphization was obtained either by ion implantation or laser melt quenching. For ion implantation, samples were obtained by implanting Ge at 130KeV with a dose of $9e13/cm^2$. This dose corresponds to 0.08 % of Ge in the film, and therefore it is low enough to not alter the film composition but it is effective to amorphize the material. Indeed, after implantation, the XRD spectrum was typical of an amorphous phase and the electrical resistivity was comparable to that measured in the as deposited amorphous film. The complete amorphization was detected by X-ray diffraction in glancing angle, to enhance the signal from the thin phase change layer. XRD measurements have been carried out using a SIEMENS D5005 diffractometer operating with Cu K_{α} (λ =0.15406 nm). The X-ray beam was incident at 0.7° on the sample surface and the detector angle was incremented with an angular step of 0.02°. Figure 2 shows in red the XRD spectra of 70 nm thick crystalline films, as deposited at 250°C, with typical rocksalt patterns, and, in blue, the same film after ion implantation of Ge. No XRD peak is visible anymore and the pattern is typical of amorphous materials.

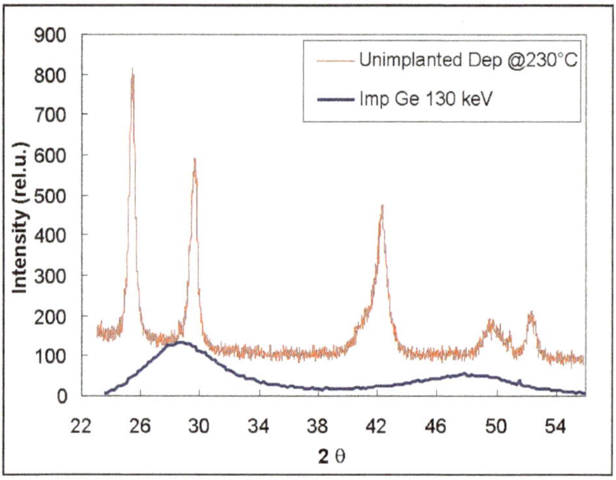

Fig. 2. XRD spectra of $Ge_2Sb_2Te_5$ deposited at 250°C, in the rocksalt structure (red), and after ion implantation with Ge at 130KeV, 9e13/cm2 (blue). After implantation the XRD peak is typical of an amorphous material.

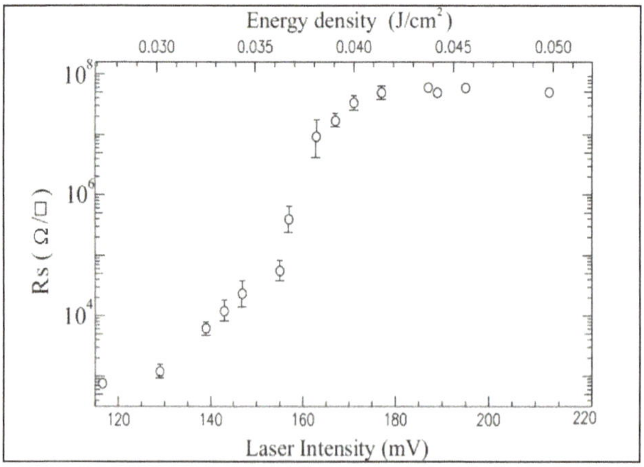

Fig. 3. Sheet resistance of melt quenched crystalline films versus laser intensity. The higher sheet resistance is measured in completely amorphized films.

Melt-quenched films were obtained by using a Ruby laser operating at 695nm. Samples were irradiated with a 30 ns laser pulse, employing a spot diameter of 3.5 mm. The energy density was varied and the effect of film irradiation was monitored by measuring the sheet resistance using a four point probe.

Figure 3 shows the sheet resistance as a function of laser intensity. Below 30 mJ/cm^2 the measured sheet resistance is around 1000Ohm/square, typical of crystalline GST. This value gradually increases with laser intensity and reaches a saturation value of 5E7 Ohm/square for energy densities above 40mJ/cm^2.

Data shown in fig. 3 can be explained by considering the depth of melt versus energy density. The heating and melting produced by the laser pulse can be described by considering the heat diffusion equation with the addition of a source term driven by the absorption of laser radiation (Baeri, 1982). If the film is spatially uniform, then away from the edges of the irradiated area any temperature variation will occur only along the axis perpendicular to the sample surface. This simplifies the problem to one dimension and it is described by the equation

$$\frac{\partial T}{\partial t} = \frac{\alpha}{\rho C_p} I(z,t) + \frac{1}{\rho C_p}\frac{\partial}{\partial z}\left(\kappa \frac{\partial T}{\partial z}\right) \tag{1}$$

where T is the temperature at time t and depth z, α is the absorption coefficient of the illuminated material, with specific heat capacity C_p and density ρ, κ is the thermal conductivity and I is the laser power density.

$$I(t,z) = I_0(t)\ (1-R)\ exp(-\alpha z) \tag{2}$$

For the same laser pulse duration, the melt region depth can be calculated by dividing the sample into slices of thickness Δz and discretising time into units Δt. The temperature of each slice is then calculated from the energy deposited into the material by absorption of laser light and heat transfer by diffusion. The parameters used for calculations are shown in table 1.

Melting Temperature (K)	891
Thermal Diffusivity (cm^2/s)	4.7e-3
Thermal conductivity (W/cm)	0.582e-2
Specific Heat (J/cm^3 K)	1.2
Density (g/cm^3)	6.2
Latent Heat of fusion (J/cm^3)	600

Table 1. Parameters used for numerical calculations of the depth of melting region under irradiation by a 30 ns laser pulse

The results of simulations are shown in figure 4, where the melt depth is plotted as a function of time for different energy densities. According to the measured sheet resistance no melting, which means no amorphization, is expected for energy density below 30mJ/cm^2. For energy densities above this threshold value, the melting depth increases as the energy density increases.

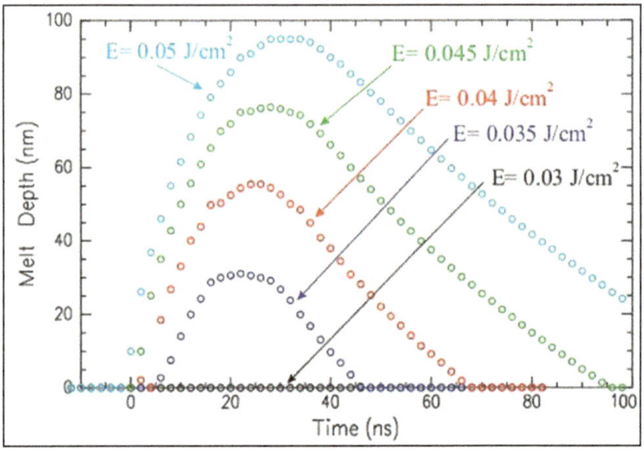

Fig. 4. Melted depth versus time as simulated for several energy densities.

Figure 5 shows the simulated melt depth as a function of laser energy density. The arrows indicate the energy density at which the thickness of melted material is expected to reach the film thickness of 70 nm. The simulated value of 40 mJ/cm^2 required to melt the entire film agrees very well with the energy density at which sheet resistance saturation is observed, indicating that all the material is converted into the amorphous phase by melt quenching.

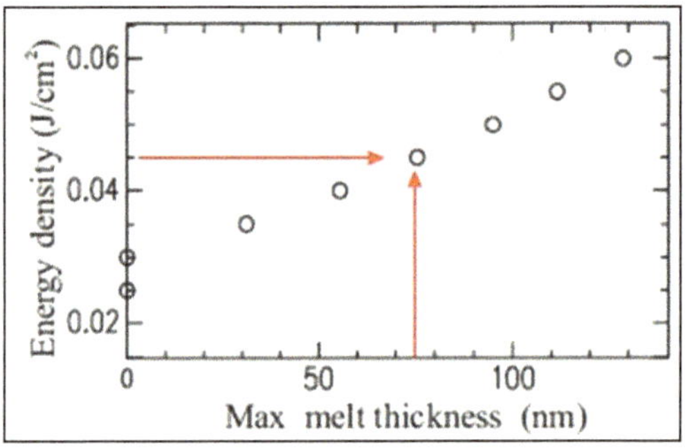

Fig. 5. Energy density versus melt thickness. For 70 nm thick film complete amorphization is expected above 40mJ/cm^2.

For comparison, amorphous films were also deposited at room temperature over SiO_2 covered Si substrates, from a single target by using RF sputter. Thickness of amorphous films was 50 nm. The stability of the amorphous resistance has been studied through sheet resistance measurements at different temperatures, in the range 40-120°C.

Figure 6 (a), (b) and (c) shows as black symbols the conductivity versus the reciprocal of temperature for as deposited, implanted and melt quenched amorphous films, respectively. The figure also shows as red symbols the conductivity obtained after annealing at temperatures below 140°C, i.e. not so high to produce the conversion into the crystalline phase. Four different samples have been measured for amorphous and as implanted material. For melt quenching, samples irradiated at two different laser intensities have been tested. Both laser intensities values are high enough to melt and quench the entire film thickness.

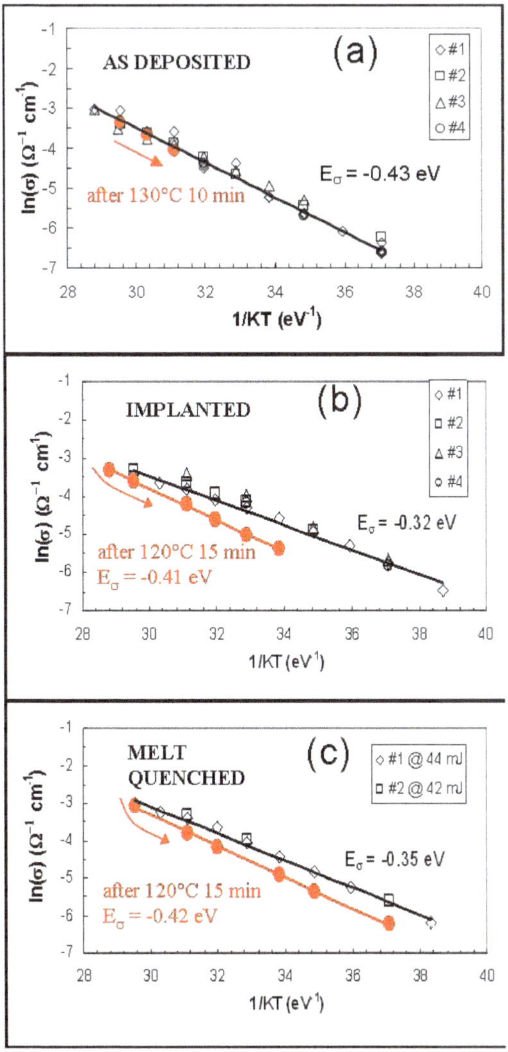

Fig. 6. Conductivity versus temperature for as deposited (a), implanted (b) and laser annealed (c) amorphous GST.

The carrier transport in amorphous GST in the temperature range we studied (40-120 °C), is dominated by the band conduction through extended states, in which the charge carriers hop between trap centers via excitations above the mobility edge. Therefore the conductivity of amorphous samples is exponentially dependent on the energy gap, whose value is about twice the activation energy for conduction (Mott & Davis, 1979).

In as deposited samples the mobility gap, measured by linearly fitting the data shown in Fig.6, is 0.86 eV and its value does not change after annealing at 130°C for 10 minutes. Therefore in as deposited GST we do not observe the resistance drift reported for PCM devices.

In films amorphized by ion implantation, instead, the conductivity decreases after annealing at 120°C. The mobility gap also changes from 0.7eV, obtained in fresh samples, to 0.82-0.84 eV, a value very close to the as deposited amorphous. A similar behavior is also observed for melt quenched amorphous films.

The drift of activation energy for conduction in amorphous GST has been already observed in amorphous cells and it has been explained considering two possible candidates: a change of the bandgap value due to mechanical strain; a change of the donor acceptor-like defect density (Pirovano et al., 2004). Under compressive strain the bandgap of several chalcogenide materials decreases by as much as 2eV/ relative volume variation (Kastner, 1972). It is known that the phase change cause a compression of 6%. Therefore melt-quenched films could exhibit lower bandgap due to accumulated mechanical stress. This stress can be then released, producing and increase of E_σ, thanks to the bandgap increase during stress release.

On the other hand, it has been observed that ion implantation can produce densification in phase change materials (Rimini et al., 2009), obtaining amorphous films of substantially higher mass density than as deposited amorphous films (Raoux et al., 2010). Therefore we can reasonably consider the observed modification of E_σ in both melt quenched and ion implanted film as due to compressive stress release.

Another possible explanation is given by recent calculations (Akola et al., 2010) and involves the role of Ge atoms bonding. Atomistic simulations of molecular dynamics performed on amorphous phase-change materials have shown that in melt quenched and as deposited amorphous GST, Germanium atoms have different environments: predominantly tetrahedral in as deposited GST and distorted octahedral in melt quenched. EXAFS data (Kolobov et al., 2004) indicate that in amorphous GST Ge atoms are prevalently tetrahedrally coordinated, with considerably shorter bond lengths (2.61-2.63 Å) than in the crystal, where Ge atoms occupy the octahedral position with Ge-Te bond length 2.83 Å. Therefore the decrease of activation energy for conduction observed in films amorphized by ion implantation and in melt quenched GST could be determined by a larger fraction of Ge atoms with distorted octahedral coordination.

4. Doping by Ion beam: Amorphous properties

Films doped with nitrogen, oxygen or fluorine have been obtained by ion implantation on amorphous $Ge_2Sb_2Te_5$. The energy of the ions was 15 keV, in order to have distributions peaked at the center of the film. Different fluencies were implanted, obtaining peak concentrations of 1.5 at.% and 10 at.% of nitrogen; 1.5 at.% and 6 at.% of oxygen; 1.5 at.% and 6 at.% of fluorine.

The amorphous properties were studied by electrical measurements. The sheet resistance was measured *in situ* using a four-point probe configuration, with the employment of a thermal chuck.

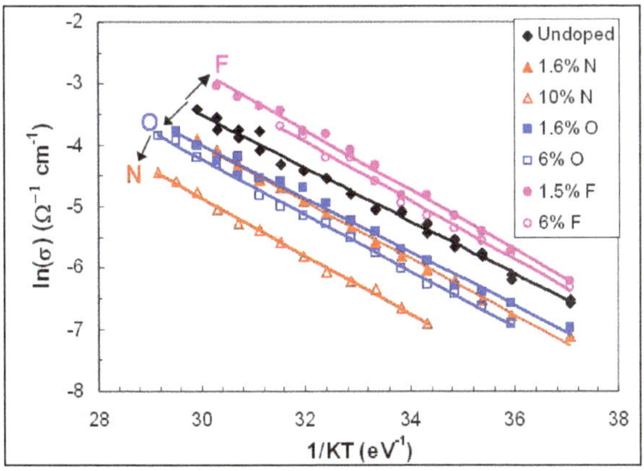

Fig. 7. Conductivity versus reciprocal temperature for undoped amorphous GST (diamonds) and samples doped with oxygen (squares), nitrogen (triangles) or fluorine (circles).

Figure 7 shows the conductivity of amorphous samples as a function of the reciprocal temperature, measured in undoped GST (black diamonds) and in samples with: 1.5 at.% of O (full squares), 6 at.% of O (open squares), 1.5 at.% of N (full triangles) 10 at.% of N (open triangles), 1.5 at.% of F (full circles) and 6 at.% of F (open circles). As shown by the arrows in Fig.7, in the case of nitrogen or oxygen doping, the conductivity decreases as the dopant concentration increases. In the case of GST doped with fluorine, instead, the conductivity increases.

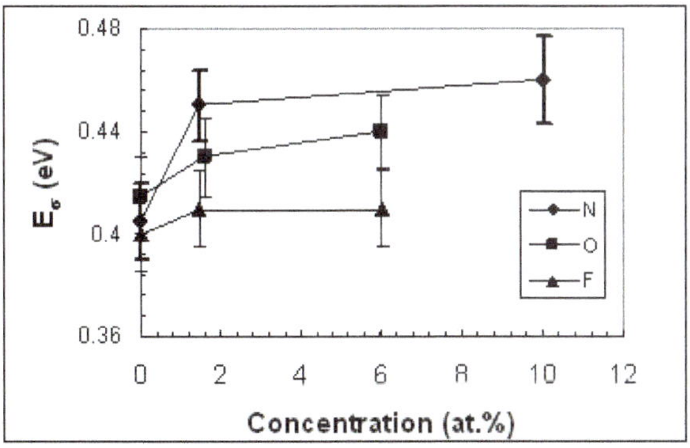

Fig. 8. Activation energy for conductivity as a function of doping concentration, obtained for nitrogen (diamonds), oxygen (squares) and fluorine (triangles).

Figure 8 shows the activation energy versus doping concentration, as obtained by linearly fitting data shown in Fig. 7.

In samples doped with nitrogen or oxygen the activation energy for conduction slightly increases, while for fluorine doping the activation energy remains almost the same. Largest variations have been obtained in the case of nitrogen: by doping GST with 10 at.% of N, the conductivity decreases of one order of magnitude and E_σ increases from 0.42 ± 0.01 eV, measured in undoped samples, up to 0.47 ± 0.01 eV.

These results are in agreement with changes in the optical band gaps of nitrogen doped GST films reported in literature (Kim et al., 2007), indicating that the increased band gap is closely related to the incorporation of nitrogen.

According to the results of numerical calculations shown by (Akola et al., 2010), an increased activation energy for conduction in nitrogen and oxygen doped samples could indicate the presence of more Ge atoms tetrahedrally coordinated.

5. Amorphous-to-crystal transition in undoped $Ge_2Sb_2Te_5$ films

Several experimental techniques have been employed to follow the amorphous-to-*fcc* transition in $Ge_2b_2Te_5$, such as optical or electrical measurements. However, only microscopical techniques allow to separate the contribution of nucleation and growth, obtaining valuable information for understanding of the involved physical mechanisms. For this reason we have studied by *in situ* transmission electron microscopy (TEM) the amorphous to crystal phase transition as a function of temperature and annealing time, obtaining a detailed measurement of all the physical parameters governing the transformation.

To render the sample transparent for TEM, specimens were ground, dimpled and etched in a HNO_3:HF:H_2O (4:1:1) solution. This procedure has been adopted because it assures no variation of the initial interfaces and film thickness, which could affect the nucleation and growth process (Ohshima, 1996, Friedrich et al., 2000). Indeed, the specimen under observation in the microscope does not contain a hole, but it is made by a continuous film over its SiO_2 substrate layer. Moreover, sample preparation by etching instead of conventional ion milling, can assures no undesired heating of the sample above the crystallization temperature, which could produce some pre-existing crystalline grains, therefore altering the kinetics of crystallization.

In situ TEM analyses have been performed using a JEOL 2010 TEM operating at 200 kV, with a sample holder equipped with a furnace around the specimen and a thermocouple. A magnification of 10000x was used for all the experiments. Since it has been shown the crystallization can be affected by the electron beam (Kooi et al., 2004), the beam has been switched off during annealing and fresh areas have been analysed, for each time step, in order to have a good statistical evaluation of the grain density and size. The transformation has been followed in the range of temperatures 130-143 °C.

Figure 9 (a) and (b) shows a TEM micrograph obtained for undoped GST after annealing at 135 °C for 50 min and 66 min, respectively. Grains appear as dark or bright circular regions and can be observed during the incubation regime (a), and after the onset of steady state nucleation (b). Grains with minimum detectable size of 7 nm are detected up to the complete crystallization, indicating that no saturation of nucleation sites occurs.

Figure 10 (a), (b), and (c) show the grain size distributions (GSDs) of samples annealed for various times, at 130 °C, 135°C and 140°C, respectively. GSDs have been obtained by assigning at each grain with surface area A an average radius $r = \sqrt{(A/\pi)}$, and by plotting, as a function of r, the density dN of crystalline grains with radius in the size interval $(r, r+dr)$, divided by dr. Figure 10 shows that all of the measured distributions smoothly decrease as a function of grain radius. At low grain size the distributions tend to reach the same value, independent on time. This value indicates the establishment of steady state nucleation condition at the lowest grain radii. Moreover, the largest detected grain size increases with time, following a linear law.

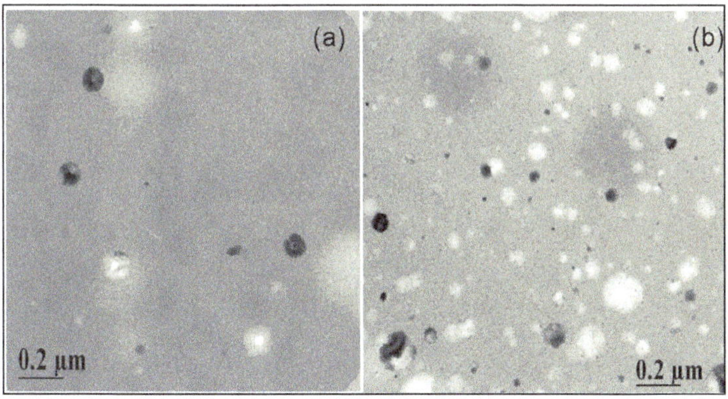

Fig. 9. TEM image of undoped amorphous GST after anneling at 135°C for 50 min (a) and 66 min (b).

From TEM images acquired at different annealing time, for several analyzed temperatures, it is possible to evaluate the evolution of the grain density ad the grain size distribution.

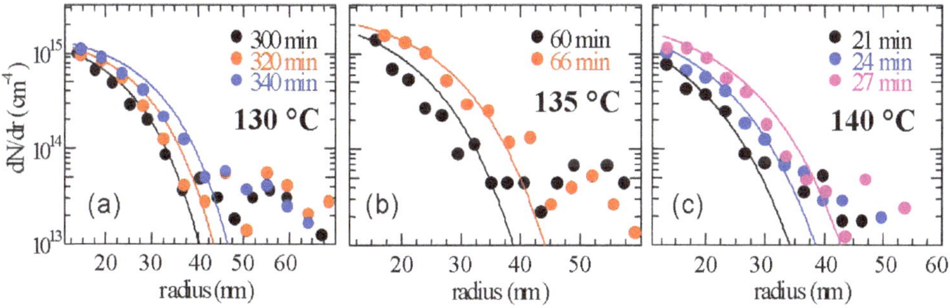

Fig. 10. Grain size distributions in undoped amorphous GST obtained after anneling at 130° C (a), 135°C (b) and 140°C (c). The distribution for each annealing temperature, have been obtained at different annealing time. Solid lines are fit to data.

According to a well known approach based on the capillarity approximation, the growth or shrinkage of a cluster is due to the attachment or detachment of atoms at the interface between the two phases. Assuming the transition is limited by the reaction at the interface, the net growth rate of individual nuclei can be approximated to (Turnbull &. Fisher, 1949):

$$v(r) = a f \Delta g_v/(KT) (1-r^*/r) \tag{3}$$

where $r^* = 2\ a\ \gamma/\Delta g_v$ is the critical radius, a is the lattice parameter, γ is the surface free energy at the interface, Δg_v is the free energy difference between the two phases per atom, $f = f_0\ exp[E_g/(KT)]$ is the thermally activated atomic jump frequency, and E_g is the activation energy for the atomic motion responsible for the single elementary crystallization step. Therefore, for grain size much larger than r^*, $v(r)$ tends to $v_g = a\, f\, \Delta g_v/(KT)$.

The nucleation process at any time t for a given cluster size i can be in general described considering the nucleation rate $J_{i,t}$ of clusters with size larger than i.

It can be shown that the steady-state value of the nucleation rate (Becker & Döring, 1935) is:

$$J_{ss}=N_a\, O_{i^*}\, f\, [\Delta\, g_v / (6\, \pi KT\, i^*)]^{\,1/2}\, exp[\, -\Delta G^* / (KT)] \qquad (4)$$

where N_a is the atomic density and $O_{i^*} = 4\pi\, [3/(4\pi)]^{2/3}\, i^{*\,2/3}$ is the number of atoms at the surface of a critical nucleus. Following this description, the steady-state nucleation rate J_{ss} exhibits an activated temperature dependence with an activation energy $E_n=E_g + \Delta G^*$.

The steady-state condition does not exist from the beginning of the transformation but it is established after a transient period during which the nucleation rate depends on size and time. All the data acquired from *in situ* TEM analyses as a function of annealing time and temperature, i.e. the grain size distributions, the transformed fraction and the grain density, have been simultaneously fitted following the model proposed by (Shneidman, 1987, 1988, 1991), as described in detail in (Privitera et al., 2007a). Four main parameters are used for the fitting procedure, the nucleation rate, the growth rate, the incubation time and the transient time τ, representing the time spent by the growing cluster in the critical region. Results of fitting have been shown in fig. 10 as solid lines.

The main parameters determined by fitting, i.e. the nucleation rate and the growth velocity, exhibit an activated temperature dependence, with activation energy of 2.6 eV and 2.3 eV, respectively (Privitera et al., 2007a).

In the classical theory of nucleation, the difference between the activation energy for the nucleation rate and that for the growth velocity is equal to the barrier energy for the nucleation ΔG^*, which corresponds to the energy necessary for the nucleation of a critical nucleus. Following this description, the value of ΔG^* here determined is less than 1eV and can be explained by considering the effect of heterogeneous nucleation. In this case, the model is also valid, provided that the atomic density N_a is replaced by the density of nucleation sites and the barrier energy ΔG^* is reduced according to the lower surface energy term. Indeed, it has been observed that the nucleation of the metastable *fcc* phase is dependent on the substrate type and different activation energies have been found for $Ge_2Sb_2Te_5$ over different substrates (Ohshima, 1996, Friedrich et al., 2000). To clarify this point in-situ isothermal annealing experiments of cross-sectional scanning-TEM (STEM) specimens of GST have been also performed (Lombardo et al. 2010). This analysis has shown the formation of crystalline grains preferentially occurs near the upper free interface, subsequently followed by the nucleation of crystal grain also at the GST / SiO_2 interface.

By fitting the experimental data the pre-exponential terms can be also determined. Estimates of these can be done according to Eqs. (3) and (4) considering that f_0 is of the order of 10^{13} s^{-1}, the lattice parameter a and the GST density N_a are known, and the critical radius r^* can be evaluated as the product of the measured v_g at large sizes, multiplied by the transient time τ. Exponential factors for J_{ss}, v_g obtained from best fit to data, however, result anomalously different from the above estimates. In order to reproduce the experimental value of v_g, the calculated value from Eq (3) should be multiplied by an adimensional factor equal to $\approx 7 \times 10^{13}$. For the steady-state nucleation rate J_{ss} we have to consider an additional factor equal to 0.3 which, however, is reasonably explained by the evidence of heterogeneous nucleation. So, data can be well explained by employing the classical model of grain growth and nucleation based on capillarity, provided that we consistently assume an anomalously large ($\approx 10^{14}$ times) factor multiplying the frequency of crystallization jumps.

This result has been extensively explained in (Privitera et al., 2007a) by considering the effect of maximum entropy production rate, based on the large disorder of the *fcc* crystal structure. The role of entropy in the grain growth during first order phase transformations has been put in clear evidence in literature (Hill, 1990), for example driving the transition from planar to dendritic growth. It is reasonable to expect that in our case the entropy production rate is extremely large. In fact, it has been shown that the *fcc* structure of GST is a combination of two *fcc* lattices, one made of Te atoms and a second made of Ge, Sb, and vacancies, in the ratio 2:2:1 (Yamada et al., 1991). The Ge:Sb:vacancy *fcc* sublattice is randomly occupied by the three species. Kolobov et al. (Kolobov et al., 2004) have proposed a crystalline structure made by short-range ordered building blocks that may rotate by 90 degrees in an arbitrary direction, giving rise to a random distribution of Ge, Sb and vacancies on a long-range scale. Based on the type of bonding experimentally observed by EXAFS in amorphous and *fcc* phases, they have also proposed that the amorphous to *fcc* transition is likely due to an atomic flip of the Ge, which jumps from sp^3 tetrahedral bonds with four Te atoms, to an octahedral configuration, again surrounded by Te first neighbors (eight) but with bonds of a prevalent p character. Kolobov proposes that the memory principle is based on this Ge flipping effect (from tetrahedral to octahedral). Since it involves at the ultimate level a single Ge atom and its immediate neighbors, this mechanism does possess a very large potential for memory scaling. However, this only effect, cannot explain the observed high transition rate, coupled with the high activation energy. What we have proposed is that the anomalous pre-exponential factors experimentally observed can be explained in terms of Ge flipping, but taking into account also the large configuration entropy of the *fcc* crystal.

Indeed, as a consequence of a Ge flip, it follows that the number of atoms crystallized in a single flip event is much larger than one. Referring to Fig. 5 in (Kolobov et al., 2004) this is at least equal to 6 Te atoms plus their first neighbors Ge, Sb, or vacancy. These amount to a total of about 18 atoms and vacancies crystallized for a single Ge flip. The Ge:Sb:vacancy *fcc* sublattice contains 11 positions, randomly occupied either by Sb atoms or by vacancies. Therefore it has a large number of different possible configurations. A rough estimate can be done considering that we have 11 allowed positions that can be occupied by either a vacancy or a Sb atom, each one with 4 different possible positions (Kolobov et al., 2004). This means that the total number of possible configurations is of the order of $8^{11} \approx 8.6 \times 10^9$.

So, according to (Hill, 1990) that has shown that the entropy generation rate is proportional to the growth velocity and that the growth mode in a phase transition follows the criterion of maximum entropy generation rate, the effective frequency of crystallization jumps should be multiplied by the factor $18 \times 8^{11} \approx 1.5 \times 10^{11}$, taking into account for the number of crystallized atoms per Ge flip (≈ 18) and for the configuration entropy term ($\approx 8^{11}$).

This estimate does not reach the experimentally found factor (1.5 × 10^{11} against 7 × 10^{13}) but it represents a lower limit, found assuming one single Ge flip, no effect of Sb, and that no other Ge neighbor is in the *fcc* crystal phase. If we consider the configuration entropy contribution due to the presence of the pre-existing amorphous / crystal interface, the estimate of the pre-exponential factor would be larger, so we believe that the estimate presented, strictly based on a single Ge flip, is satisfactorily close to the experimental result.

The discussion here presented can be considered a starting point, from which it is clearly necessary to further investigate and find other evidences in favor. In agreement with the proposed description, it has been observed that the crystallization speed is increased in GeTe-Sb_2Te_3 compounds by partially replacing Sb atoms with Bi atoms (Matsunaga & Yamada, 2004), or when a proportion of Ge is replaced by Sn (Kojima & Yamada, 2001). Since both Sn and Bi occupy the same (4b) position as Ge and Sb in the *fcc* lattice (Matsunaga & Yamada, 2004), their presence definitely increases the number of possible configurations and therefore the entropy factor in the growth velocity.

Recently, the disorder of phase change materials has been also taken into account to explain metal-semiconductor transitions occurring in crystalline phase change GeSbTe alloy (Siegrist et al., 2011). The strong disorder, usually associated only with amorphous solids, seems to be at the origin of the remarkable reproducibility of the resistance switching observed in phase change material and strongly suggests that the disorder could be one of the key ingredient that makes the physical properties of phase change materials so peculiar.

6. Amorphous to crystal transition in doped films

Once we have studied in detail the crystallization process in undoped materials, we have investigated the effect of doping with light atom on the amorphous-to-crystal transition. In order to evidence the modifications due to doping, we have plotted in Fig. 11 the sheet resistance measured after 12 minutes anneals as a function of annealing temperature.

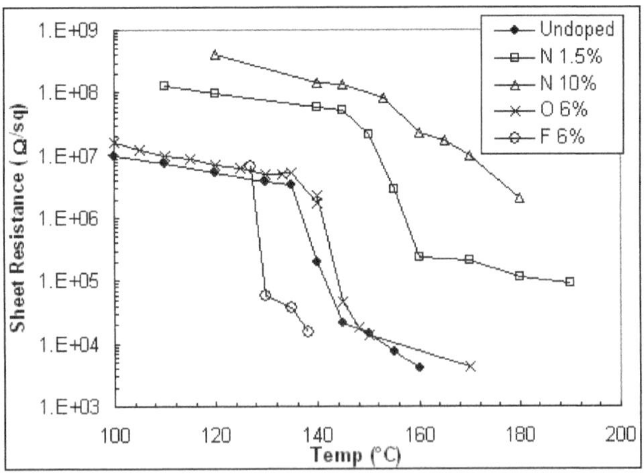

Fig. 11. Sheet resistance versus temperature as measured after 12 min isothermal anneals, for undoped and doped GST with oxygen (crosses), nitrogen (squares and triangles) or fluorine (circles).

Samples doped with nitrogen exhibit higher resistance in the amorphous phase and the crystallization occurs at much higher temperatures than undoped material. A small increase in the crystallization temperature is also observed by oxygen doping while, on the contrary, addition of fluorine produces a decrease in the crystallization temperature.

The dependence of the resistance as a function of time during isothermal anneals has been also studied by *in situ* resistance measurements using a four point probe. Figure 12 (a) and (b) shows the sheet resistance vs time as obtained during isothermal treatments at several annealing temperatures in the range 127-190°C, for undoped GST and for film doped by 6 at.% of fluorine, respectively. Fluorine doped samples are characterized by fast crystallization.

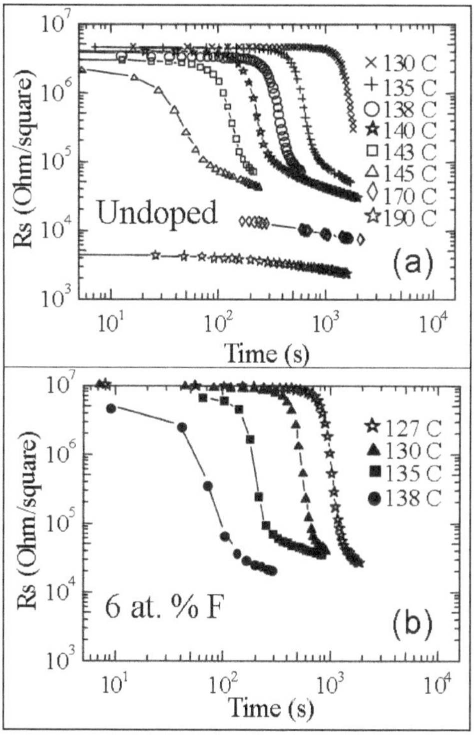

Fig. 12. Sheet resistance versus time as measured during anneal at various temperatures in the range 127-190°C, for undoped (a) and fluorine doped GST (b).

By taking the first derivative of the resistance versus time, the characteristic time for transition can be measured, as the time at which a minimum in the derivative is observed. The characteristic time has been then measured for all the studied samples and it has been used to determine the temperature dependence of the transformation. The activation energy obtained for undoped $Ge_2Sb_2Te_5$ is around 3 eV. The same value, within experimental errors, has been also determined for samples with low concentration of nitrogen and fluorine, as shown in Figure 13, reprinted from (Privitera et al. 2007b). Higher activation energy has been instead measured in GST doped with 6 at. % of oxygen.

Characteristic times observed in N and O doped samples are up to one order of magnitude higher than in pure GST, whilst the characteristic times observed in fluorine doped samples are lower.

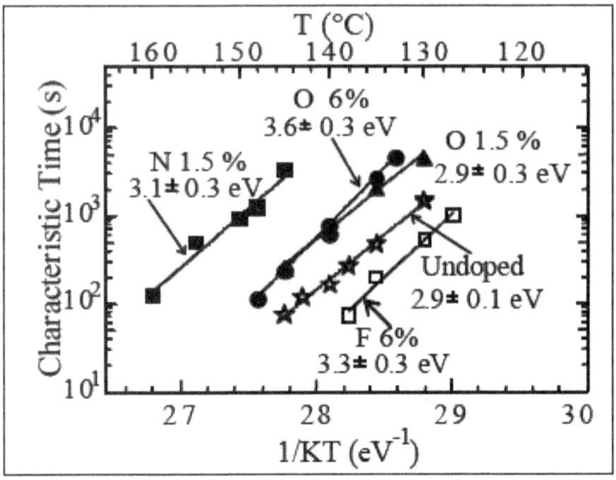

Fig. 13. Activation energy for crystallization in Undoped and Doped GST with O, N, or F. Reprinted from (Privitera et al., 2007b)

Results from electrical measurements have been related to the kinetics of the transformation through *in situ* TEM analyses. Figure 14 shows typical TEM plan view images obtained during the amorphous-to-*fcc* transition in different samples annealed at 130°C. Fig. 14 (a) has been acquired for undoped sample annealed at 130°C for 240 min. A few crystalline grains embedded in the amorphous matrix are visible as dark or bright regions. Fig. 14 (b) shows a sample doped by 6 at.% of oxygen, annealed at the same temperature and time of 130°C 240 min. The material is almost completely amorphous and only small crystalline grains can be detected. Such a result is a confirmation of the observed increase of incubation and characteristic times observed by electrical measurements, as shown in figures 12 and 13.

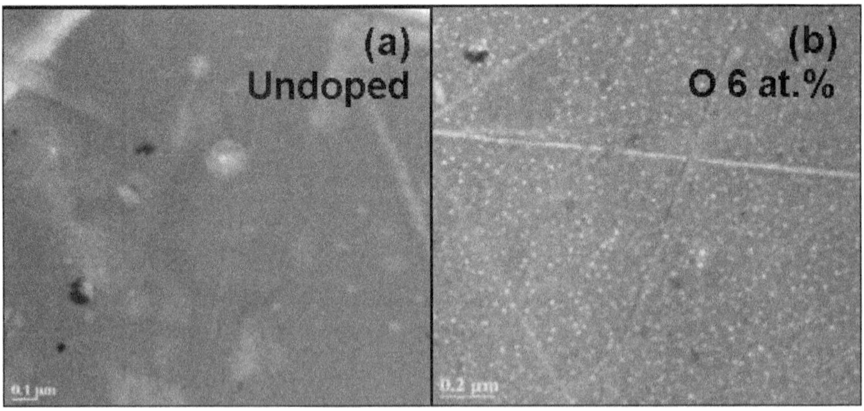

Fig. 14. Comparison between TEM image of amorphous GST after anneling at 130°C for 240 min for undoped (a) and oxygen doped GST (6at.%) (b).

Under annealing at higher temperature, samples doped with oxygen exhibit higher crystallization rate, compared to undoped GST. Figure 15 (a) and (b) shows TEM images of samples annealed in situ at 137°C for undoped and 6 at.% oxygen doped GST, respectively. Figure 15 (a) has been obtained in undoped material after 107 min annealing, while the image shown in fig. 15 (b) has been acquired on oxygen doped sample after only 70 minutes. Although annealed for shorter time, doped sample exhibits higher grain density and larger grain size.

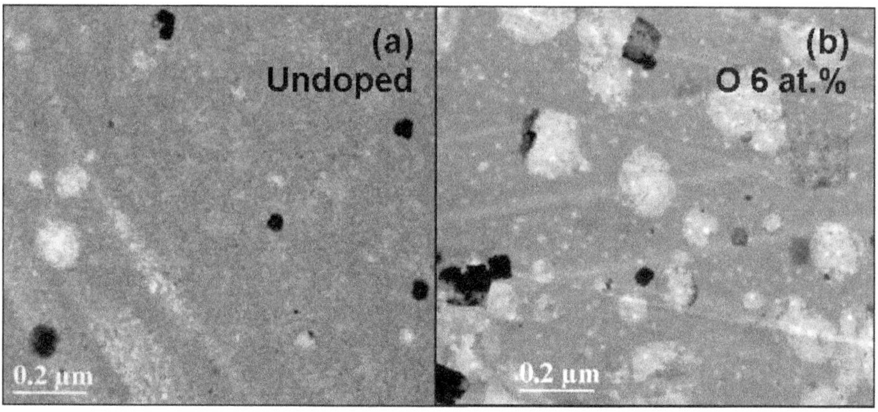

Fig. 15. TEM image of undoped amorphous GST after anneling at 137°C for 107 min (a) and of oxygen doped GST (6at.%) after annealing at 137°C for 70 min (b).

Such a results is in agreement with the higher activation energy obtained in O doped blanket material (Fig. 13) and in oxygen doped memory cells (Matsuzaki et al., 2005). Moreover, increase of the crystal grain size has been also reported in 6 at. % oxygen-doped phase change optical recording disk, improving the overwrite cyclability of the optical recording disk. Therefore, the increase of activation energy, coupled with the increase of incubation time, makes oxygen doped GST particularly interesting for cell endurance improvement, without decreasing the crystallization rate, i.e. without increasing the writing time.

The addition of 1.5 at.% of nitrogen by ion implantation produces strong variations in the kinetics of the transformation.

Figure 16 (a) and (b) shows the comparison between undoped sample (a) and sample with 1.5 at.% of N (b) after annealing at 140 °C for 75 min and 150 min, respectively. Nitrogen doped samples exhibit much longer incubation time compared to undoped film: after 150 min the material is still almost completely amorphous and only small crystallites are visible.

Even after annealing at higher temperature, the morphology of nitrogen doped films appears different from the one observed in undoped materials. As an example, Figure 17 shows a TEM image of a nitrogen doped sample (1.5 at.%) annealed at 150°C for 80 minutes. The film is crystalline but it is characterized by small crystalline size.

We have already observed in section 5 that the crystallization of undoped GST is heterogeneous and preferentially occurs near the upper free interface and, secondarily, at the GST / SiO_2 interface.

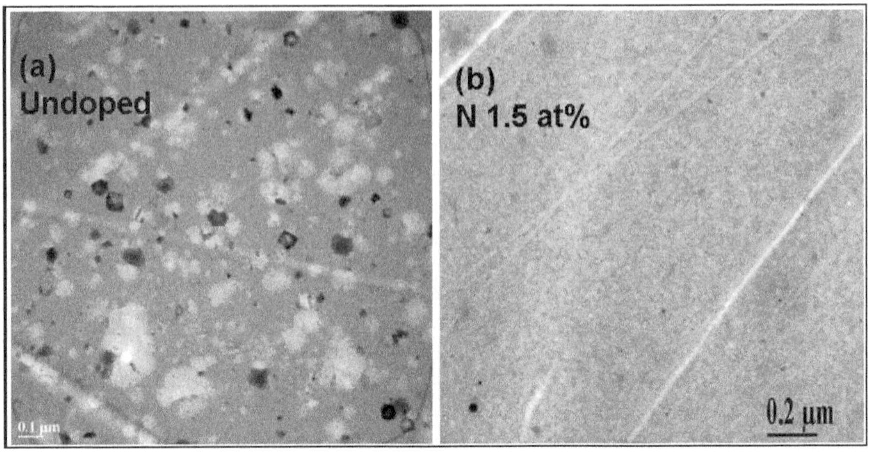

Fig. 16. TEM image of undoped amorphous GST after anneling at 140°C for 75 min (a) and of nitrogen doped GST (1.5 at.%) after annealing at 140°C for 150 min (b).

Since the ion implantation can modify the interfaces, causing some intermixing, we have performed in situ TEM analyses in cross section. Figure 18 (a) and (b) shows the amorphous GST film doped with 1.5 at.% of N as deposited and after annealing in situ at 150°C for 20 min, respectively. As observed for undoped GST (Lombardo et al., 2010) a continuous polycrystalline layer is formed. However, in contrast with observation on pure GST, in the case of nitrogen doping the nucleation has occurred preferentially at the GST/SiO_2 interface, with the addition of small dispersed crystalline regions in the amorphous material and at the upper interface. Therefore doping by ion implantation is also effective in modifying the nucleation sites at which heterogeneous nucleation preferentially occurs.

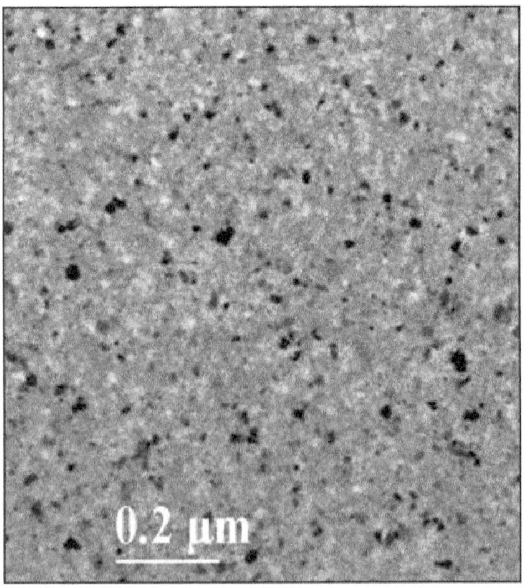

Fig. 17. TEM image of a sample doped by 1.5at.%of nitrogen annealed at 150°C for 80 min.

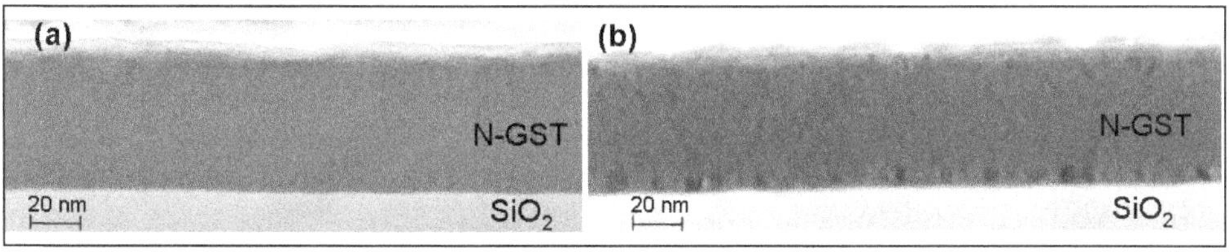

Fig. 18. Bright field TEM micrographs in cross section of a sample doped by N 1.5 at.% as-deposited (a), and annealed in situ at 150°C for 20 min (b). The first crystalline formation is clearly visible at the GST/SiO_2 interface, with additional crystalline grains nucleated also at the upper interface.

Therefore it appears clear from the experimental data that, at least in the range of compositions here investigated, the oxygen or nitrogen doping affects the crystallization process differently, the first one producing higher crystallization rate and larger grain size, while N doping gives rise to grain size refinement and longer incubation time.

The increase of the crystallization temperature in oxygen (Jeong et al., 2001) and nitrogen [Kojima et al., 1998, Jeong et al., 2000) doped GST has been already reported in literature, as well as the reduction of the grain size in samples containing nitrogen. Several models have been proposed (Jeong et al., 2001, Jeong et al., 2000) trying to simultaneously explaining the effect of both O and N in order to find the optimal dopant to engineer specific properties of GST.

First, it has been observed that the reduced grain size could be caused by dopant segregation near grain boundaries in the form of nitrides (Kojima et al., 1998). The resistance of the crystalline phase would increase due to additional electron scattering from grain boundaries (Horii et al., 2003). We cannot exclude, in the samples with 10 at.% of N, the formation of very thin nitrides layers at grain boundaries, which are probably responsible for the increased resistivity of the crystalline phase and for the smoothing of the shape of the resistance versus temperature shown in Fig. 11.

Regarding oxygen doping, the formation of Ge or Sb oxides has been reported for concentrations above 10 at.% (Jang et al., 2009, Morales-Sanchez et al., 2006). For lower concentration, as in the samples we have studied, no oxidation seems to occur. It has been therefore proposed that oxygen atoms, as well as nitrogen atoms at low concentration, are located at the tetrahedral interstitial sites (Jeong et al., 2000). Indeed, since nitrogen and oxygen have quite small atomic sizes compared to that of Ge, Sb and Te, they cannot substitute these atoms, filling the vacancies which are present in the structure. Therefore their effect is expected to be very different from the one observed in the case of Bi or Sn, (Matsunaga & Yamada, 2004), which substituting in the Ge:Sb:vacancy lattice, can increase the number of possible configurations and therefore the growth velocity. N or O atoms will be instead located at the tetrahedral interstitial sites distorting the unit cell and resulting in a strain field (Jeong et al., 2000). Since nitrogen has a larger atomic radius (0.071 nm) (Lide, 2004), compared to oxygen (0.066 nm) (Lide, 2004), a larger amount of strain is expected to accommodate nitrogen atoms in the structure.

Another factor to be taken into account together with the stress is the chemical bond between Ge and Te and its modifications due to the presence of nitrogen, or other dopants.

The main crystallization mechanism has been identified as an atomic flip of the Ge, which jumps from tetrahedral bonds with 4 Te atoms, to an octahedral configuration, surrounded by 8 Te first neighbors. Xray photoemission spectroscopy (XPS) analyses by employing synchrotron radiation have shown that when a GST film is crystallized in the *fcc* structure, the intensity of Germanium in tetrahedral position is decreased while the intensity of Ge octahedral is increased owing to the umbrella-flip change in Ge atoms from the tetrahedral to the octahedral position (Kolobov et al., 2004).

Both experimental data (Kim et al., 2007, Jung et al., 2007) and atomic *ab initio* calculations (Cho et al., 2011) performed on nitrogen doped GST have shown that the N atoms are prevalently bonded with the Ge atoms to form GeN_x, rather than bonding with the Te or Sb atoms. Moreover, data of Ref. (Kim et al., 2007) show that the formation of Ge-N bonds can disturb the umbrella-flip of Ge atoms between the tetrahedral and octahedral positions because, as nitrogen is incorporated into the GST film, the intensity of the peak relative to Germanium in the octahedral position is decreased, since nitrogen combines with Ge to form germanium nitrides. By reducing the portion of the peak of Ge octahedral nitrogen is therefore effective in suppressing the crystallization.

A similar behavior is expected also for oxygen atoms, with the formation of Ge-O bonds. However, nitrogen can form three bonds while oxygen only two. Moreover, N dopants increase the portion of tetrahedral Ge atoms, while, on the base of calculations reported in (Cho et al., 2011), O doped GST seems to be more similar to undoped material, without favoring tetrahedral sites respect to octahedral Ge atoms. This could explain the lower increase in the energy gap observed in the case of oxygen (Fig. 8).

Considering the heterogeneous character of the transformation, the proposed assumption that oxygen and a small amount of germanium oxide could act as nucleation centers during crystallization, increasing the crystallization rate (Morales-Sanchez et al., 2006) could be a good explanation.

The decrease of crystallization temperature observed for fluorine doping remains to be better investigated. It has been reported that amorphous GST can have fast crystallization under compressive stress. So the effect of F could be explained with the introduction of compressive stress in the amorphous material. More extensive studies however are needed to map out the relationship between the chemical nature of dopants and the induced modifications in the phase change structure, in order to better identify the optimal dopant in order to achieve the desired properties.

7. Conclusions

In this chapter we reviewed some applications of ion beam to the study of the physical properties of $Ge_2Sb_2Te_5$, the phase change material most widely employed for application as nonvolatile memories. Phase change memories are one of the most promising solutions as replacement of current FLASH technology, based on charge storage, but some material properties need to be optimized. We have shown that ion implantation can be successfully employed to modify the properties of the amorphous phase change material, obtaining conduction properties very similar to those of melt-quenched amorphous material, formed by electric pulses or laser in memory cells.

Ion implantation can be also adopted for doping of phase change materials, modifying the amorphous-to-crystal transition kinetics, upon which the reliability definitely depends. The crystallization is a fast process occurring at very low temperature, compared to other solid materials. The explanation of such a fast transition has been found in the peculiar structure of the *fcc* phase which contains a large amount of disorder and whose formation is driven by flips of Ge atoms. By nitrogen implantation it is possible to increase the crystallization temperature and the crystalline resistance improving the cell endurance, reducing the RESET current and enabling the use of PCM also for high temperature applications, such as automotive. The formation of Ge-N bonds has been identified as the main effect determining the increase of crystallization temperature, the increase of the band gap in the amorphous material and the reduction of crystalline grain size.

By oxygen implantation, a moderate increase of crystallization temperature has been obtained, maintaining short crystallization time, thanks to the increased activation energy. Decrease of crystallization temperature has been instead obtained by fluorine doping. The way these doping acts is still not well understood and more extensive studies are required to relate the chemical nature of dopants belonging to different groups in the periodic table to the induced modifications in the phase change structure, in order to optimize the dopant specie on the base of the desired properties.

8. Acknowledgment

The author wish to gratefully acknowledge Prof. Emanuele Rimini (University of Catania), Corrado Bongiorno (CNR), Romina Zonca (STMicroelectronics), Agostino Pirovano (Numonyx) and Roberto Bez (Numonyx) for the fruitful collaboration.

9. References

Akola, J.; Larrucea, J. & Jones, R. O. (2010), Polymorphism in amorphous $Ge_2Sb_2Te_5$: comparison of melt-quenched and as-deposited structures, *EPCOS 2010*, Milano, Italy, 2010. Available from http://www.epcos.org/library/papers/pdf_2010/Oral/F03-Akola.pdf

B.-S. Lee, G. W. Burr, R. M. Shelby, S. Raoux, C. T. Rettner, S. N. Bogle, K. Darmawikarta, S. G. Bishop, and J. R. Abelson, Direct observation of the role of subcritical nuclei in crystallization of a glassy solid, *Science*, Vol. 326, No. 5955, (November 2009), pp. 980-984, ISSN 0036-8075.

Baeri, P. & Campisano, S.U. (1982) Chapter 4: Heat Flow Calculations, In: *Laser Annealing of Semiconductors*, ed. by J. M. Poate & J. W. Mayer, pp. 75-110, Academic Press, ISBN 0125588208, New York.

Becker, R. & Döring, W. (1935) Kinetische behandlung der keimbildung in übersättigten dämpfen, *Annalen der Physik* (Leipzig), Vol. 24, N. 8 (1935), pp.719-752, ISSN 00033804.

Boniardi, M.; Redaelli, A.; Pirovano, A.; Tortorelli, I.; Ielmini, D.; & Pellizzer, F. (2009). A physics-based model of electrical conduction decrease with time in amorphous $Ge_2Sb_2Te_5$, *Journal of Applied Physics*, Vol. 105, No. 8, (April 2009), pp. 084506-084506-5, ISSN 0021-8979.

De Bastiani, R.; Piro, A. M.; Grimaldi, M. G.; Rimini, E.; Baratta, G. A. & Strazzulla, G. (2008). Ion irradiation induced local structural changes in amorphous $Ge_2Sb_2Te_5$ thin films, *Applied Physics Letters*, Vol. 92, No. 24 (June 2008), pp. 241925- 241925-3, ISSN 0003-6951.

Friedrich, I.; Weidenhof, V.; Njoroge, W.; Franz, P. & Wuttig, M. (2000), Structural transformations of $Ge_2Sb_2Te_5$ films studied by electrical resistance measurements, *Journal of Applied Physics*, Vol.87, No. 9, (May 2000), pp. 4130 -4134, ISSN 0021-8979.

Hill, A. (1990) Entropy production as the selection rule between different growth morphologies, *Nature*, Vol. 348, N. 6300 (November 1990) pp. 426-427. ISSN 0028-0836

Horii, H.; Yi, J.H.; Park, J.H.; Ha, Y.H.; Baek, I.G.; Park, S.O.; Hwang, Y.N.; Lee, S.H.; Kim, Y.T.; Lee, K.H.; U-In Chung; & Moon, J.T. (2003) A novel cell technology using N-doped GeSbTe films for phase change RAM, *2003 Symposium on VLSI Technology. Digest of Technical Papers*, pp.177-178, ISBN 4-89114-033-X, Kyoto, Japan, 10-12 June, 2003

Ielmini, D. (2008). Threshold switching mechanism by high-field energy gain in the hopping transport of chalcogenide glasses, *Physical Review B*, Vol. 78, No. 3 (July 2008), pp. 035308-035308-8, ISSN 1098-0121.

Im, J.; Cho, E.; Kim, D.; Horii, H,.; Ihm, J. & Han, S. (2010) Effects of pressure on atomic and electronic structure and crystallization dynamics of amorphous $Ge_2Sb_2Te_5$, Physical Review B, Vol. 81, No. 24, (June 2010), pp. 245211-245211-5, ISSN 1098-0121

Jang, M. H.; Park, S. J.; Lim, D. H.; Cho, M.-H.; Do, K. H.; Ko, D.-H. & Sohn, H. C. (2009) Phase change behavior in oxygen-incorporated $Ge_2Sb_2Te_5$ films, *Applied Physics Letters*, Vol. 95, No. 1 (January 2009), pp. 012102-012102-3, ISSN 0003-6951.

Jeong, T. H.; Kim, M. R. ; Seo, H. ; Park, J. W. & Yeon, C. (2000). Investigation of Crystallization Behavior of Sputter-Deposited Nitrogen-Doped Amorphous $Ge_2Sb_2Te_5$ Thin Films, *Japanese Journal of Applied Physics*, Vol. 39, No. 5A, (May 2000), pp. 2775-2779, ISSN 1347-4065.

Jeong, T. H.; Seo, H.; Lee, K. L. ; Choi, S. M.; Kim, S. J. ; & Kim, S. Y. (2001), Study of Oxygen-Doped GeSbTe Film and Its Effect as an Interface Layer on the Recording Properties in the Blue Wavelength, *Japanese Journal of Applied Physics*, Vol. 40, No. 3B, (March 2001), pp. 1609-1612, ISSN 1347-4065.

Jung, M.-C. Lee, Y. M. Kim, H.-D. Kim, M. G. Shin, H. J. Kim, K. H. Song, S. A. Jeong, H. S. Ko, C. H. Han, M. (2007) Ge nitride formation in N-doped amorphous $Ge_2Sb_2Te_5$, *Applied Physics Letters*, Vol. 91, No. 8 (August 2007), pp. 083514 - 083514-3, ISSN 0003-6951

Kastner, M. (1972). Compositional trends in the optical properties of amorphous lone-pair semiconductors, *Physical Review B*, Vol. 7, No. 12 (June 1973), pp. 5237-5252, ISSN 1098-0121.

Kim, Y.; Jeong, K.; Cho, M.-H.; Hwang, U.; Jeong, H. S. & Kim, K. (2007), Changes in the electronic structures and optical band gap of Ge_2Sb_2Te and N-doped Ge_2Sb_2Te during phase transition, *Applied Physics Letters*, Vol. 90, No. 17 (April 2007), pp. 171920-171920-3, ISSN 0003-6951.

Kojima, R. & Yamada, N. (2001), Acceleration of Crystallization Speed by Sn Addition to Ge–Sb–Te Phase-Change Recording Material, *Japanese Journal of Applied Physics*, Vol. 40 Part 1, No. 10, (October 2001) pp. 5930–5937, ISSN 1347-4065.

Kojima, R.; Okabayashi, S.; Kashihara, T.; Horai, K.; Matsunaga, T.; Ohno, E.; Yamada, N. & T. Ohta (1998) Nitrogen Doping Effect on Phase Change Optical Disks, *Japanese Journal of Applied Physics,* Vol.37, No. 4B, (April 1998), pp. 2098-2103, ISSN 1347-4065.

Kolobov, A. V.; Fons, P. ; Frenkel, A. I. ; Ankudinov, A. L. ; Tominaga, J. & Uruga, T. (2004) Understanding the phase-change mechanism of rewritable optical media, *Nature Materials*, Vol.3 (September 2004), pp.703-708, ISSN 1476-1122.

Kooi, B. J.; Groot, W. M. G. & De Hosson, J. Th. M. (2004) In situ transmission electron microscopy study of the crystallization of $Ge_2Sb_2Te_5$, *Journal of Applied Physics,* Vol.95, No. 3, (February 2004), pp. 924-932, ISSN 0021-8979.

Krebs, D.; Raoux, S. ; Rettner, C. T. ; Burr, G.W.; Salinga, M. & Wuttig, M. (2009). Threshold field of phase change memory materials measured using phase change bridge devices, *Applied Physics Letters,* Vol. 95, No. 8, (August 2009), pp. 082101-082101-3, ISSN 0003-6951.

Lai, S.; Lowrey, T., (2001). OUM-A 180nm Nonvolatile Memory Cell Element Technology for Stand Alone and Embedded Applications, *Electron Devices Meeting, 2001. IEDM Technical Digest. International,* pp. 36.5.1 - 36.5.4, ISBN 0-7803-7050-3, Washington, DC, USA, Dec 2-5, 2001

Lide, D. R. (2004) *CRC Handbook of Chemistry and Physics,* 85th Edition, CRC Press LLC, ISBN 0-8493-0485-7, Boca Raton, Florida, USA.

Lombardo S., Rimini, E.; Grimaldi, M.G. & Privitera, S. (2010) Amorphous-*fcc* transition in $Ge_2Sb_2Te_5$, *Microelectronic Engineering,* Vol.83, No.3, (March 2010), pp. 294-300, ISSN 0167-9317.

Matsunaga, T. & Yamada N. (2004) Crystallographic Studies on High-Speed Phase-Change Materials Used for Rewritable Optical Recording Disks, *Japanese Journal of Applied Physics,* Vol. 43, No. 7B, (July 2004), pp. 4704–4712, ISSN 1347-4065.

Matsuzaki, N.; Kurotsuchi, K.; Matsui, Y.; Tonomura, O.; Yamamoto, N.; Fujisaki, Y.; Kitai, N.; Takemura, R.; Osada, K.; Hanzawa, S.; Moriya, H.; Iwasaki, T.; Kawahara, T.; Takaura, N.; Terao, M.; Matsuoka, M. & Moniwa, M. (2005), Oxygen-doped GeSbTe phase-change memory cells featuring 1.5-V/100-μA standard 0.13μm CMOS operations, *Electron Devices Meeting, 2005. IEDM Technical Digest. International,* pp. 738 - 741, ISBN 0-7803-9268-X, Washington, DC, USA, Dec 5-7, 2005

Mio, A. M.; Carria, E. ; De Bastiani, R.; Miritello, M. ; Bongiorno, C.; D'Arrigo, G.; Spinella, C.; Grimaldi, M. G. & Rimini, E. (2010). Crystallization of ion amorphized $Ge_2Sb_2Te_5$ in nano-structured thin films, *Proceedings of Materials Research Society,* pp. 1251-H02-03, ISSN 0272-9172 , San Francisco, CA, USA, April 5-9, 2010.

Morales-Sanchez, E.; Gonzalez-Hernandez, J.; Herrera-Fierro, P.; Chao, B.; Kovalenko, Yu. & Prokhorov, E. (2006) Influence of oxygen on the crystallization process in Ge: Sb:Te:O films, 3rd International Conference on Electrical and Electronics Engineering, 2006, pp. 1-4, ISBN 1-4244-0402-9, Veracruz, Mexico, Sept 6-8, 2006.

Mott, N. F. & Davis, E. A. (1979) *Electronic processes in Non-Crystalline Materials,* Clarendon, ISBN 0198512880, Oxford.

Ohshima, N. (1996) , Crystallization of germanium–antimony–tellurium amorphous thin film sandwiched between various dielectric protective films, *Journal of Applied Physics,* Vol. 79, No. 11 (June 1996) , pp. 8357-8363, ISSN 0021-8979.

Owen, A. E.; Robertson, J. M. & Main, C. (1979). The threshold characteristics of chalcogenide-glass memory switches, *Journal of Non-Crystalline Solids*, Vol. 32, No. 1-3 (February-March 1979) pp. 29–52, ISSN 0022-3093.

Pirovano, A.; Lacaita, A.L.; Pellizzer, F.; Kostylev, S.A.; Benvenuti, A. & Bez, R. (2004). Low-field amorphous state resistance and threshold voltage drift in chalcogenide materials, *IEEE Transaction on Electron Devices*, vol. 51, No. 5, (May 2004), pp. 714–719, ISSN 0018-9383.

Privitera, S.; Lombardo, S.; Bongiorno, C.; Rimini, E. & Pirovano, A. (2007a). Phase change mechanisms in $Ge_2Sb_2Te_5$, *Journal of Applied Physics*, Vol. 102, No. 1 (July 2007) , pp. 013516 -1013516-5, ISSN 0021-8979.

Privitera, S.; Rimini, E. & Zonca, R. (2004), Amorphous-to-crystal transition of nitrogen- and oxygen-doped $Ge_2Sb_2Te_5$ films studied by in situ resistance measurements, *Applied Physics Letters*, Vol. 85, No. 15 (October 2004), pp. 3044-3046, ISSN 0003-6951.

Privitera, S.; Rimini, E.; Bongiorno, C.; Pirovano, A. & Bez, R. (2007b). Effects of dopants on the amorphous-to-fcc transition in $Ge_2Sb_2Te_5$ thin films, *Nuclear Instruments and Methods in Physics Research B*, Vol. 257, No. 1-2 (April 2007), pp. 352-354, ISSN 0168-583X.

Raoux, S.; Cohen, G. M.; Shelby, R. M.; Cheng, H.-Y. & Jordan-Sweet, J. L. (2010). Amorphization of crystalline phase change m aterial by ion implantation", *Proceedings of Materials Research Society*, pp. 1251-H02-06, ISSN 0272-9172, San Francisco, CA, USA, April 5-9, 2010.

Raoux, S.; Cohen, G. M.; Shelby, R. M.; Cheng, H.-Y.; Madan, A.; Ott, J. & Jordan Sweet, J. L. (2010). Effect of Ion Implantation on Crystallization Properties of Phase Change Materials, *EPCOS 2010*, Milano, Italy, 2010. Available from http://www.epcos.org/library/papers/pdf_2010/Oral/D01-Raoux.pdf

Redaelli, A.; Pirovano, A., (2011). Nano-scaled chalcogenide-based memories. *Nanotechnology*, Vol. 22, No. 25 (June 2011), pp. 254021-254025, ISSN 1361-6528.

Rimini, E.; De Bastiani, R. ; Carria, E. ; Grimaldi, M. G. ; Nicotra, G. ; Bongiorno, C. & Spinella, C. (2009). Crystallization of sputtered-deposited and ion implanted amorphous $Ge_2Sb_2Te_5$ thin films, *Journal of Applied Physics*, Vol. 105, No. 12 (June 2009) , pp. 123502-123502-6, ISSN 0021-8979.

Shneidman, V. A. (1987) Size-distribution of new-phase particles during transient condensation of a supercooled gas. *Soviet Physics Technical Physics*, Vol. 32, (1987), pp. 76-81, ISSN 0038-5662

Shneidman, V. A. (1988) Establishment of a steady-state nucleation regime. Theory and comparison with experimental data for glasses, *Soviet Physics Technical Physics*, Vol.33, (1988), pp.1338-1342, ISSN 0038-5662

Shneidman, V. A. (1991) Transient critical flux in nucleation theory, *Physical Review A* Vol.44, No.4 (August 1991), pp. 2609-2611, ISSN 1050-2947

Siegrist, T.; Jost, P.; Volker, H.; Woda, M. ; Merkelbach, P.; Schlockermann, C. & Wuttig M. (2011), Disorder-induced localization in crystalline phase-change materials, *Nature Materials*, Vol. 10, (January 2011), pp.208-210, ISSN 1476-1122.

Turnbull, S. & Fisher, J. C. (1949) Rate of nucleation in condensed systems, *Journal of Chemical Physics*, Vol. 17, No. 1 (January 1949), pp. 71-73, ISSN 0021-9606.

Yamada, N.; Ohno, E. ; Nishiuchi, K. ; Akahira, N. & Takao, M. (1991) Rapid-phase transitions of GeTe-Sb_2Te_3 pseudobinary amorphous thin films for an optical disk memory, *Journal of Applied Physics*, Vol. 69, No. 5, (March 1991), pp.2849 -2856 ISSN 0021-8979.

8

Si Nanocrystal Arrays Created in SiO_2 Matrix by High-Energy Ion Bombardment

Irina Antonova
A.V. Rzhanov Institute of Semiconductor Physics SB RAS
Russia

1. Introduction

In the next two decades, nanoscience and nanotechnologies will transfer into new products and processes. One emerging area where this challenge will be successfully met is the field of semiconductor nanocrystals (NCs) (see, for instance, Norris et al., 2008). The promises offered by dielectric layers with hosted semiconductor nanocrystals in applications include production of silicon-based light emitting systems (Iacona et al., 2006), wavelength-tunable lasers (Klimov et al., 2007), memory elements (Hu et al., 2009), solar cells (Gur et al., 2008), low-dimensional devices (for instance, single electron transistors operating at room temperature (Zhou et al., 2010)), and a wide spectrum of biotechnologies (for instance, bioimaging (Michalet et al., 2005)). The prospects for developing new electronic and optoelectronic devices on the basis of spatially ordered arrays of quantum dots (QDs) embedded in a semiconductor or insulator matrix have arouse currently growing interest in the production methods of such arrays and in the studies of their optical and electronic properties.

The idea to use non-overlapping tracks of high-energy ($E \geq 1$ MeV/amu) heavy ions for production of nanocrystalline (NC) systems is becoming increasingly attractive among the various methods for fabrication of nanodimensional structures (nanolithography, molecular nanoelectronics, organic nanostructures, etc.) (Fink et al., 2007; Rizza et al., 2009). Beams of heavy fast ions can be efficiently used not only for formation, but also for modification, of nanodimensional structures, in particular, metallic nanoclusters (Klaumunzer, 2006; D'Orleans et al., 2003). Avasthi et al. (Avasthi et al., 2010) have summarized experimental data on the formation of elongated metal nanoparticles, and suggested a model for the interaction of fast ions with such nanoparticles dependently on nanoparticle sizes. Studies of the effect of high-energy ion irradiation on the formation of semiconductor (silicon or germanium) nanoclusters have also appeared recently (Chaudhari et al., 2003; 2005; Arnoldbik et al., 2005). It was shown that, as a consequence of the high density of ionization energy losses in the material, silicon nanoinclusions could be formed in SiO_x layers without any additional thermal treatment given to the samples. Studies of the effect of high-energy ion irradiation on SiO_2 layers containing Si nanocrystals or an excess concentration of dissolved silicon (Si:SiO_2 layers) have made it possible to propose a new approach to formation of vertically ordered NC arrays (Antonova et al., 2008 a; 2009 a; 2009 b; 2010 a; 2010 b).

In the present review, we demonstrate the possibility of using irradiation of samples with high-energy gas ions (Kr, Xe, and Bi) for introduction and modification of Si nanocrystals in Si:SiO_2 and NC:SiO_2 layers, and also for formation of ordered arrays of Si nanocrystals embedded in a SiO_2 matrix. Due to the small size of nanocrystals synthesized in our samples (typically, ≤ 6 - 8 nm), in describing our results we will also use an alternative term 'quantum dots'. We will analyze the changes in the optical and electronic properties of NC systems and discuss possible applications of such systems. It should be emphasized here that a special feature of used irradiation, in contrast to conventional ion implantation, is huge ion-energy losses in the bulk of Si:SiO_2 layers and simultaneous creation of a relatively low defect concentration in such layers (Antonova et al., 2006a; Toulemonde et al., 2000; Meftah et al., 1997). The factor causing a low defect concentration in examined irradiated layers is the small thickness of those layers, which is normally much smaller than the projected ion range. Hence, while in practice defects are introduced in the substrate of film systems, the energy is mainly dissipated in the top Si:SiO_2 layer. The latter circumstance has allowed us to separate out the positive effect of ion irradiation (the huge energy losses in ion tracks) from the negative effect (defect formation).

2. Sample preparation and ion irradiation conditions

2.1 Nanocrystals in SiO_2 layers with varied concentration of excess Si

The Si:SiO_2 layers of thickness 400–1000 nm used in our study were sputtered from Si and SiO_2 sources onto silicon substrates. In the experimental arrangement used, the SiO_2 and Si sources were spaced ~100 mm apart. The large distance between the sources provided a possibility to form layers with varied concentration of excess silicon in the oxide. After annealing of samples and formation of nanocrystals, we could obtain layers with concentration and sizes of NCs varied in the lateral plane. That gave us an interesting opportunity to examine the properties of formed Si:SiO_2 layers as dependent on NC concentration and sizes. The excess Si content of the silicon-rich oxide layers ranged from 4 to 94 vol.%. The composition of deposited layers was estimated from the thickness of Si and SiO_2 layers obtained by individual deposition of films from the Si and SiO_2 sources onto different substrates (Shamin et al., 2010). Subsequent high-temperature annealing (at 1140°C for 40 min) used for part of samples resulted in the formation of silicon oxide layers containing Si nanocrystals (NC:SiO_2 layers), whose sizes only slightly depended on layer composition and whose concentration increased with coordinate along the sample by several orders of magnitude (Sa'ar et al., 2005; Antonova et al., 2006b). Experimental results are presented below as a function of excess Si concentration. This parameter is used to characterize the composition of our Si:SiO_2 layers since its value was inherently defined by the evaporation procedure, whereas other parameters, such as NC concentration and sizes, were complicated functions of layer composition, and also of thermal treatments given to the samples.

2.2 High-energy ion irradiation: Specific features

For irradiation of Si:SiO_2 and NC:SiO_2 layers obtained as described above, Kr, Xe, or Bi ions respectively with energies 90, 130-250, and 65 - 670 MeV were used. The ion fluences ranged from $3x10^{11}$ to $3x10^{13}$ cm^{-2}. The irradiations of our samples were performed with the use of accelerators available in the Laboratory of Nuclear Reactions, Joint Institute for Nuclear

Research, Dubna. For the ion energies used, the ion penetration range was 15–23 μm (see Table 1); i.e. it was substantially larger than the thickness of silicon-rich oxide layers. This means that almost all ions penetrated into the substrate, and the defects generated in elastic scattering channels were also predominantly introduced in the substrate, whereas prevailing energy losses in the layers were the electron stopping losses.

With reference to SRIM data, the electron stopping losses due to ionization dE_e/dx in the oxide layers were expected to vary from 7.1 (Kr), 10.2 (Xe), and 22.0 keV/nm (Bi) for the SiO_2 end of the sample to 9.5 (Kr), 12.3 (Xe), and 24.3 keV/nm (Bi) for its Si end. The ionization losses of ion energy lead to ionization of atoms in the material under irradiation. In a time interval of 10^{-13} - 10^{-12} s the excitation energy was transferred into anisotropic heating of the oxide layers due to thermalization process. Using the value of ionization losses, the dependence of ion-track sizes in α-SiO_2 on ion energy, and the thermal spike model of (Toulemonde et al., 2000; Meftah et al., 1997), we could estimate the width of ion tracks in our samples, which turned out to be about 4-6 nm. Results for the time evolution of temperature inside ion tracks calculated by TSRIKE-02 software (Toulemonde et al., 2000) are shown in Fig.1a. TSPIKE-02 simulates the electron-lattice interaction (transfer of the ionization losses of ions into the heating of the target lattice). The variable parameter of the ion-lattice interaction is the radius of the shells surrounding individual ion trajectories in which the main portion of ion energy is lost for ionization. As it follows from Fig.1a (see also Fig.15 a), the local transient temperature in ion track cores in our experiments could reach 1700 - 2500°C, the ion-track heating time being about 10^{-11} - 10^{-10} s. It means that almost all inter-atomic bonds along ion tracks could be broken to free the atoms and make them capable of displacing from their regular lattice sites. As a result, new bonds could readily form in the material, this process leading to subsequent amorphization or crystallization of part of previously existing Si nanoparticles as well as to the formation of new NCs in the oxide layer along ion tracks.

One more effect that was taken into consideration for materials subjected to the irradiation with high-energy ions (see also Section 6) is ion-beam-induced anisotropic plastic deformations (Van Dillen et al., 2004; 2005). The strain anisotropy is imposed by the direction of the ion beam: normally, the material expands in the direction normal to the ion beam and contracts parallel to it, while maintaining their volume. According to the developed model, the strain ε (determined as the relative change of lattice constant) versus time goes through a maximum and then reaches a constant value. The anisotropic strain rises with ion energy and fluence. It should be emphasized here that up to fluences 10^{12} cm^{-2} the ion tracks in the material were independent of each other, whereas at higher fluences they can become overlapping.

The ionization losses of the ions and the anisotropic heating of the material are higher in Si in comparison with SiO_2. The anisotropic deformations found in amorphous silicon are typically 1.5 orders of magnitude lower than the anisotropic deformations in SiO_2 (Van Dillen et al., 2005; Meftah et al., 1997). This means that the anisotropic strain and the anisotropic heating should demonstrate considerable changes with increasing concentration of Si nanocrystals embedded in the SiO_2 layer. In the case of variable NC concentration we have to expect a strong anisotropic-strain effect for SiO_2 layers with low Si-phase content and a more pronounced anisotropic heating effect for SiO_2 layers with high excess Si phase content.

Sample, thickness of SiO_x layer, nm	Annealing regime before irradiation	Ion energy, MeV	Fluence, cm^{-2}	R_p in Si, mm	$\Delta E_e/\Delta x$, keV/nm	Annealing after irradiation
NC-10, 700	1140°C, 40 min					
NC-11, 700	1140°C, 40 min	Kr, 90	$1x10^{12}$	16	7.1	
NC-11a, 700	1140°C, 40 min	Kr, 90	$1x10^{12}$	16	7.1	800°C, 20 min
NC-20, 1000	1140°C, 40 min					
NC-21, 1000	1140°C, 40 min	Kr, 90	$1x10^{12}$	16	7.1	-
NC-22, 1000	-	Xe, 130	$1x10^{12}$	17	10.2	1140°C, 40 min
NC-23, 1000	1140°C, 40 min	Xe, 130	$1x10^{12}$	17	10.2	
NC-23a, 1000	1140°C, 40 min	Xe, 130	$1x10^{12}$	17	10.2	800°C, 20 min
NC-24, 1000	1140°C, 40 min	Xe, 130	$1x10^{13}$	17	10.2	800°C, 20 min
NC-30, 400	1140°C, 40 min					
NC-31, 400	1140°C, 40 min	Kr, 250	$1x10^{12}$	33.1	10.2	
NC-31a, 400	1140°C, 40 min	Kr, 250	$1x10^{12}$	33.1	10.2	800°C, 20 min
NC-32, 400	1140°C, 40 min	Kr, 250	$2x10^{13}$	33.1	10.2	
NC-32a, 400	1140°C, 40 min	Kr, 250	$2x10^{13}$	33.1	10.2	800°C, 20 min
NC-33, 400	1140°C, 40 min	Bi, 65	$1x10^{12}$	10	8.9	800°C, 20 min
NC-34, 400	1140°C, 40 min	Bi, 130	$1x10^{12}$	16	13.2	800°C, 20 min
NC-35, 400	1140°C, 40 min	Bi, 670	$3x10^{11}$	43	18.0	800°C, 20 min
NC-36, 400	1140°C, 40 min	Bi, 670	$1x10^{12}$	43	18.0	800°C, 20 min
NC-37, 400	1140°C, 40 min	Bi, 670	$8x10^{12}$	43	18.0	800°C, 20 min
NC-38, 400	1140°C, 40 min	Bi, 670	$3x10^{13}$	43	18.0	800°C, 20 min

Table 1. Parameters of structures used, and annealing and irradiation regimes. R_p is the projected ion range in Si calculated by SRIM. $\Delta E_e/\Delta x$ values are the ionization losses of the ions in SiO_2.

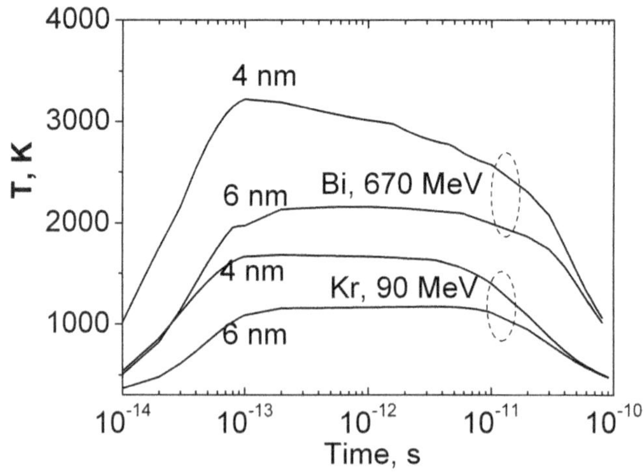

Fig. 1. Simulated evolution of temperature in time in α-SiO_2 (quartz) samples for increasing radial distances from the ion track axis. The simulations were performed using the TSPIKE-02 software (Toulemonde et al., 2000). For times shorter than 10^{-13} sec the temperature should be considered as quasi-temperature.

3. Structural modification of Si nanocrystals embedded in SiO_2 matrix

3.1 Ordered nanocrystal arrays in ion-modified layers

In the initial (non-irradiated) layers, the spherically shaped Si nanocrystals were distributed randomly over the insulator matrix; the NC spheres were 3–5 nm in diameter for excess Si contents of $Si:SiO_2$ layers 40-70 vol.%, with the atomic planes in the nanocrystals having random orientation. After ion irradiation with ionization losses over 6 keV/nm, in the oxide layers with a relatively high excess Si content (above 35%) vertically ordered NC arrays were formed along ion tracks (Fig.2) (Antonova et al., 2009 a; 2009 b). The morphology of NCs has also changed: the nanocrystals became extended along the ion tracks. As shown in Fig. 2 b, this effect is accompanied by another effect: the atomic planes in the formed NCs were aligned in the direction along ion tracks, with deviation $\pm$ 20°. It should be emphasized here that no other method of forming Si NCs allows production of vertically ordered arrays of nanocrystals with identically oriented atomic planes.

The TEM images presented in Fig.2 were obtained for samples NC-11 and NC-21 irradiated with 90 MeV Kr ions; the excess Si content of the oxide layer in those samples was 37–45%. The Kr ion had a minimal mass and a relatively low energy. We believe that in the case of using for irradiation heavier ions with a higher energy such effects will become essentially more pronounced.

In irradiated $Si:SiO_2$ structures which initially contained no NCs (sample NC-22), introduction of Si nanocrystals with a similar spatial and size distribution was revealed after additional high-temperature thermal treatments given to the samples.

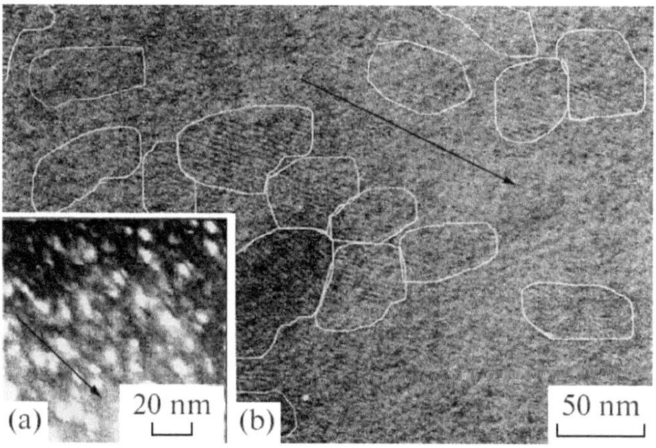

Fig. 2. Cross-sectional TEM and HRTEM images of $NC:SiO_2$ layers in samples NC-21 irradiated with 90 MeV Kr ions to a fluence of 10^{12} cm^{-2}. The Si phase contents of the samples were about 37 (a) and 73 vol.% (b). The arrows show the direction of ion tracks. Note that in the irradiated sample (b) the atomic planes inside nanocrystallites are aligned along ion tracks (with deviation $\pm$ 20°).

3.2 Amorphization and crystallization processes studied by Raman spectroscopy

Figure 3 a, b compares Raman spectra taken from initial and irradiated samples in areas with two different Si-content values, 51% and 62% (Antonova et al., 2009 a; 2010 a). The

Raman line due to the single-crystal Si substrate is detected at 521 cm^{-1}. After the samples were irradiated with Kr or Bi ions, the amorphous 480 cm^{-1} line has increased significantly. In the case of Bi irradiation, an NC-related line at 500-520 cm^{-1} was also observed in the spectra. To demonstrate this effect, we have plotted in Fig. 2a the difference between the spectrum of a sample irradiated with Bi ions and the spectrum of the same sample before irradiation. An NC-related peak at 500-520 cm^{-1} is also observed in the differential spectrum; this fact points to the formation of new NCs in samples subjected to irradiation. An increase in the irradiation dose (> 10^{12} cm^{-2}) and a decrease in ion energy and/or ion mass lead to domination of the amorphization process. Raman measurements performed on samples with higher Si contents (> 70 vol.%) have shown that, in the latter case, only amorphization of Si NCs was observed. Following an additional anneal at 800°C for 20 min, no substantial changes were observed in the Raman spectra.

It was found that irradiation of Si:SiO_2 structures that initially contained no previously formed NCs also brought about the formation of new Si nanocrystals in the course of irradiation. This effect was observed practically for all ions used in the present study. In this case, the formation of new nanocrystals also prevails at relatively low Si contents of Si:SiO_2 layers (below 50–60%), whereas the formation of amorphous inclusions prevails at higher Si contents. An increase in irradiation dose (over 10^{12} cm^{-2}) and a decrease in the specific ionization losses of the ions yield a net increase in the amount of optically active nanocrystals available in the samples (Antonova et al., 2010 b).

3.3 Evolution of NC sizes under ion irradiation

It was also shown that a study of Si NCs by means of charge deep-level transient spectroscopy (Q-DLTS) makes it possible to observe the capture of charge carriers by, and their emission from, dimensional quantization levels in NCs located close to the SiO_2/Si interface (Antonova et al., 2009 c; 2011). The Q-DLTS method was also employed to investigate into variations of energy barriers involved in the emission process of charge carriers from NC levels in NC:SiO_2 layers modified with high-energy ion irradiation.

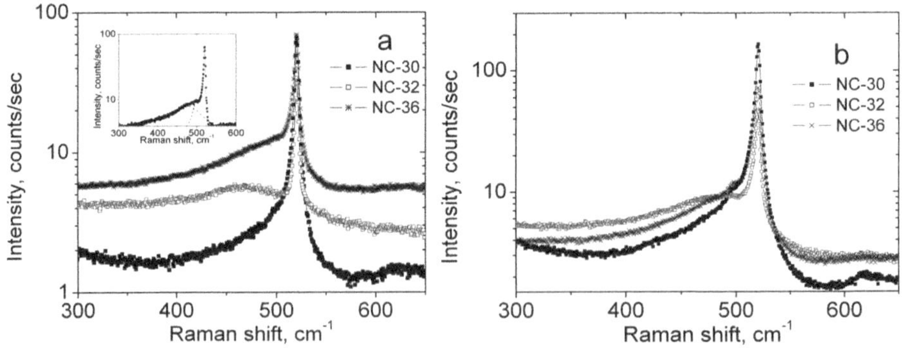

Fig. 3. Raman spectra of NC-SiO_2 layers on Si substrates for initial (non-irradiated) samples with (a) 51% and (b) 62% of the Si phase and for samples irradiated with Kr (NC-32) and Bi ions (NC-36). In order to demonstrate the added intensity around the "amorphous" 480 cm^{-1} line and the NC-related 500-520 cm^{-1} line due to Bi^+ irradiation, in the insert to the figure we have plotted the difference between the spectra measured after and before the irradiation of the sample with 670 MeV Bi^+ ions (NC-30 and NC-36).

Samples NC-36 and NC-37 to be discussed in the present section were irradiated with Bi ions of about 670 MeV energy (the projected ion range 43 nm) to fluences in the range 10^{12} – 10^{13} cm^{-2}. Those ions, having a maximal mass and energy, were expected to provide the most significant effect on Si NCs. The specific ionization losses of 670 MeV Bi ions in the SiO_2 layers were 24 keV/nm, which value far exceeded the threshold ionization energy (~2 keV/nm) required for the formation of latent tracks in SiO_2 (Toulemonde et al., 2000). Prior to performing Q-DLTS measurements, the radiation defects introduced in the substrate were removed from it giving the samples an additional anneal at 800°C during 20 min.

As a rule, Q-DLTS spectra of non-irradiated and irradiated samples involved a single peak. In the temperature range 80 to 300 K, this peak features two slopes in the Arrhenius dependence (Fig. 4a). The activation energies determined from the measurements made at several points on a sample with different compositions of the NC:SiO_2 layer are listed in Table 2. A hypothesis was put forward that the levels detected by Q-DLTS in the NC:SiO_2 layers were quantum confinement levels in Si nanocrystals. In the latter case, the energy interval between the levels could be expected to be dependent on NC sizes. Assuming that $\Delta E_i = E_{n=2}-E_{n=1}$ (here, $E_{n=2}$ and $E_{n=1}$ are the energies of the first excited and ground state of an electron in the quantum dot), we were able to evaluate the typical NC size value W from the theoretical relation between E_n and W. Starting with the effective mass approximation (Moskalenko et al., 2007), we have obtained, from the theoretically estimated values of ΔE_i, the W values given in the last column of Table 2. Trying then to approximate the dependence $E_n(W)$ with the simpler well-known expression (see, for example, Gaponenko, 1998),

$$E_n = \frac{\hbar^2}{2m} \cdot \left(\frac{\pi^2}{W} \right) \cdot n^2, \quad (1)$$

where E_n is the energy of the n-th confined level and m is the electron effective mass in crystalline Si (0.26 in units of the free electron mass), we have obtained W values given in the last column of the table as bracketed values. We see that the estimated values of W fall into the 3-6 nm range for the reference (non-irradiated) sample NC-30. These values of W show a good correlation with TEM data. Q-DLTS measurements yield practically identical sets of levels for various points measured along samples, this finding being in turn indicative of an identical set of NC sizes. As it follows from Table 2, the irradiation brings about a decrease of NC sizes. A further decrease in NC sizes was observed in samples irradiated to higher fluences. In the latter case, formation of smaller Si nanocrystals might proceed due to three different reasons. The first reason is the formation of new small Si nanocrystals in the samples under irradiation. The second reason is possible decrease in the sizes of already existing NCs, which could be a result, for instance, of oxidation of the nanocrystals. It should be noted here that dissolution of metallic nanoinclusions (in particular, Au inclusions) in samples irradiated with high-energy ions was observed by Rizza et al. (Rizza et al., 2007 a; 2007 b). The third reason, operating in a situation in which the shape of NCs could be altered by ion irradiation from spherical to ellipsoidal one, extended along ion tracks, is that the energies of quantum confinement levels were primarily defined by the smallest NC size. In the latter case, a decrease in NC sizes could be indicative of an elongation of nanocrystals and of a decrease of their effective diameter. Photoluminescence (PL) measurements performed on the same structures yields for the blue shift of the NC-related PL peaks observed in the spectrum (see Section 4.1) a value consistent with the decrease of NC sizes as estimated from Q-DLTS data.

Figure 4 b shows the density of deep traps detected by Q-DLTS as a function of the composition of the NC:SiO_2 layer. It is seen that, as a result of ion irradiation, the trap density has decreased by approximately one order of magnitude. The decrease in the density of nanocrystals detected by the Q-DLTS method in irradiated samples in comparison with pristine samples NC-30 correlates with the formation of conducting NC chains in ion tracks, which phenomenon will be discussed in more detail below. Only isolated Si nanocrystals located in the outside region of conducting chains and being therefore able to retain charge made a contribution to the Q-DLTS signal.

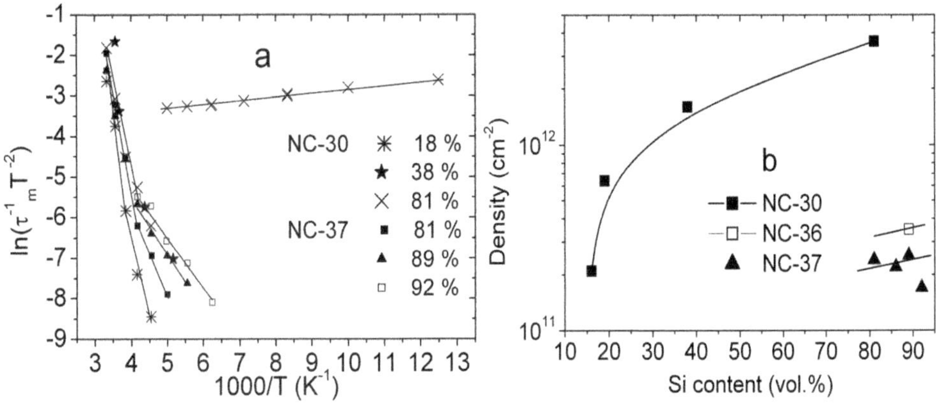

Fig. 4. (a) Arrhenius plots for Q-DLTS peaks measured in regions with different Si-phase contents along the initial (non-irradiated) sample NC-30 and along the sample NC-37, whose NC:SiO_2 layer was modified by irradiation of the sample with Bi ions. (b) The sheet concentration (density) of traps in the NC:SiO_2 layers versus the Si-phase content of these layers for three sets of samples, NC-30, NC-36 and NC-37.

A schematic energy diagram of a nanocrystal embedded within a SiO_2 matrix in a vicinity of the substrate is shown in Fig.5 for the regime of capture of charge carriers by NCs during Q-DLTS filling pulse (a) and for the regime in which the emission process is observed (b). A higher energy position of dimensional-quantization levels in the nanocrystal in comparison with the edge of the conduction band of silicon and an additional increase in the energy of the levels by the energy of NC charging with electrons suggest that, here, direct tunneling of NC-emitted charge carriers into the Si substrate is possible. The observation of NC-related levels by the Q-DLTS method and the evaluation of the activation energies related to the emission of charge carriers from NCs into substrate show that the emission process could proceed due to a combination of thermal activation and tunneling processes. The latter is possible if the emission of charge carriers proceeds via some intermediate level E^* available at the NC/SiO_2 interface. In such a situation, the activation energies experimentally determined from the treated Q-DLTS spectra present the energy differences $E^* - E_i$. The presence of an intermediate level E^* at the NC/SiO_2 interface was often postulated in interpretation of NC-related photoluminescence (*e.g.* Godefroo et al., 2006). The fact that the activation energy for the emission of charge carriers from NCs proved to be practically invariable with a strong increase of excess Si content in our samples suggests that the role of the matrix is not too much important. This suggests that the level E^* involved in the tunneling process was located at the NC/matrix interface rather than at the interface between the SiO_2 layer and the Si substrate. Most likely, in real samples Si nanocrystals embedded within the SiO_x matrix are coated with a SiO_2 shell.

From the position of the Q-DLTS peak in the Q-DLTS spectra, the characteristic time of emission of charge carriers from NC-related levels could be estimated. Figure 6 shows the temperature dependences of this time. Comparing the data shown in Fig.5 for the initial and irradiated samples, we may conclude that the substantial rearrangement in the NC system having occurred as a result of ion irradiation of the samples had no appreciable effect on the characteristic times of recharging of individual isolated nanocrystals in the temperature range from 220 to 300 K; those times amounted to 10^{-2}–10^{-4} s. On the average, at lower temperatures the times of emission of charge carriers from NC levels in irradiated samples proved to be shorter than the same times in pristine structures.

The average values of the activation energies for emission of charge carriers from nanocrystals in non-irradiated samples were found to be E^* - 0.24 and E^* - 0.43 eV. After irradiation with ions to a fluence 10^{12} cm^{-2}, those energies have decreased to E^* - 0.13 and E^* - 0.35 eV and, as the fluence was further increased, those energies have further decreased to E^* - 0.11 and E^* - 0.34 eV. It can be seen from a comparison of the values obtained that ion irradiation generally leads to a decrease in the activation energies. This decrease also

Samples	Si content, vol.%	E_a, eV	ΔE_i , eV	W, nm
NC-30	16	0.30 0.16	0.14	5.6 (3.6)
	19	0.43 0.24	0.19	4.8 (3.1)
	38	0.43 0.23	0.20	4.7 (3.0)
	81	0.43 0.27 0	0.16	5.1 (3.3)
NC-36 Bi, 670 MeV, $1x10^{12}$ cm^{-2}	89	0.35 0.13	0.22	4.4 (2.7)
NC-37 Bi, 670 MeV, $8x10^{12}$ cm^{-2}	82	0.44 0.12	0.32	3.6 (2.3)
	87	0.37 0.03	0.34	3.3 (2.0)
	89	0.33 0.10	0.23	4.3 (2.6)
	92	0.43 0.10	0.33	3.6 (2.2)

Table 2. The energies of deep traps E_a as extracted from the Q-DLTS spectra of samples with different Si contents of the NC:SiO_2 layer. The ΔE_i values are the energy separations between the levels E_a for a given Si content. Also indicated in the table are the quantum-dot sizes (diameters), W, as estimated through comparison of experimentally determined ΔE_i values with the theoretical predictions of ref. (Moskalenko et al., 2007) and from formula (1) (bracketed values).

manifests itself as a decrease of the emission times for charge carriers at relatively low temperatures. A decrease in the activation energy can be attributed to a decrease in NC sizes, or to an increase in the energies of dimensional quantization levels.

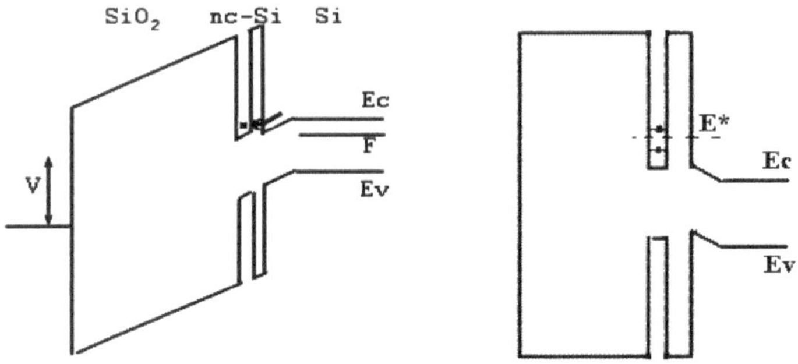

Fig. 5. Energy diagrams illustrating the NC charging process (left) and the process of electron emission from NC-related levels (right) observed during and after the filling pulse in the Q-DLTS procedure. For simplicity, only one NC, located in the vicinity of the SiO_x/Si substrate interface, is depicted in the diagrams.

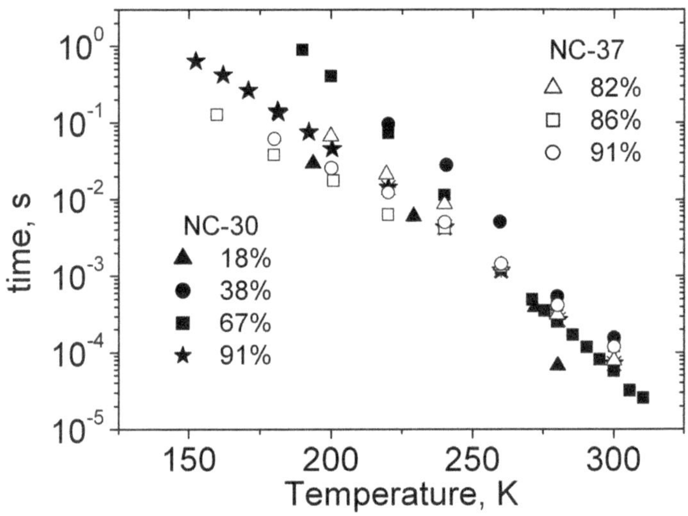

Fig. 6. Temperature dependence of the characteristic time of emission of charge carriers from quantum confinement levels in nanocrystals for samples NC-30 and NC-37. The emission time for sample NC-36 has intermediate values.

4. Optical and electrical properties of ion-modified arrays of nanocrystals

4.1 Photoluminescence of irradiated NC arrays

The capability of Si nanocrystals to emit light in the visible spectral region attracts a lot of interest, thus stimulating the studies of the optical properties of NC systems aimed at enhancement of the PL intensity coming from nanocrystals. For the initial (non-irradiated) NC-SiO_2 layers it was found that the PL signal due to NCs could be observed only in a certain range of layer compositions (Fig. 7 a). The PL intensity decreased with increasing concentration of Si NCs (or excess Si content of NC-SiO_2 layers) when the nanocrystals started to cluster together and, thereafter, percolation conduction appeared in the NC system (Antonova et al., 2006 b; 2008 b). The possibility of migration of charge carriers over extended NC clusters substantially diminishes the probability of radiative recombination of electrons and holes optically generated in NCs, thus resulting in a decreased PL intensity.

The electronic losses of the ions being relatively low (Kr, 90 MeV, dE_e/dx = 7.1 keV/nm), the ion irradiation induces no substantial changes in the PL intensity coming from NC-SiO_2 layers (Fig. 7 a, NC-10, NC-11) (Antonova et al., 2008 a). Following irradiation, an insignificant change in the spectral position of the NC-related PL peak was only observed (see Fig. 7 b). We have suggested that such a change could be a result of a change of NC sizes. The higher was the Si content of the NC-SiO_2 layers, the larger NCs were observed in such layers after ion irradiation. As estimated from the position of the PL peak in the spectrum and from the relation of the latter position with typical NC sizes in the NC systems (see Bitten et al., 2004), the average size of Si nanocrystals in sample NC-10 before irradiation was 3.6 nm. The latter value well correlates with the NC sizes that were estimated from Raman measurements and TEM observations. After ion irradiation (sample NC-11), based on the blue shift of the NC-related PL peak the typical NC size could be estimated as 3.5 nm for 27% and 3.9 nm for 57% Si content of NC-SiO_2 layer. It should be noted here that an additional peak at about 570 nm appeared in the PL spectra of irradiated NC-SiO_2 layers. This peak was attributed to the formation of small Si chains in irradiated samples (Kachurin et al., 1998).

Increase in the ionization losses of ions used for irradiation of samples (Xe, 130 MeV, dEe/dx = 10.2 keV/nm) widened the range of compositions of the NC-SiO_2 layers for which an appreciable NC-related PL signal could be detected (Fig. 7 a, NC-22). Samples NC-23 and NC-23a exhibited PL spectra similar to the PL spectrum of sample NC-22. Here, the intensity of the NC-related PL decreases in value with increasing the Si content of NC-SiO_2 layer over 40 vol.% while a more intense photoluminescence was observed in samples with lower Si contents. The spectral position of the PL peak maximum observed in the PL spectra of examined samples is indicated in Fig. 7 a. It can be seen that the PL peak position exhibits a distinct blue shift with increasing Si content of the NC-SiO_2 layers. Observation of NC-related photoluminescence in part of samples with low Si contents points to the fact that, here, new nanocrystals were formed right during irradiation. An additional peak with a maximum located at about 550 nm was observed in irradiated samples. The observed blue shift in the spectral position of NC-related peaks points to the formation of an increased number of smaller Si NCs.

Further increase in ionization losses (Bi, 670 MeV, dEe/dx = 18 keV/nm) demonstrates a possibility of reaching even greater PL intensity in NC systems irradiated with high-energy heavy ions (Antonova et al., 2009 a). The photoluminescence spectra that were taken from the reference sample NC-30 and from several irradiated samples with indicated Si content values are shown in Fig. 8 a. The NC-related PL peak in the reference sample NC-30 was observed at ~ 785 nm; the spectral position of this peak in the PL spectra of non-irradiated samples was roughly independent of the excess Si content of the Si:SiO_2 layers. In samples irradiated with relatively low ion fluences (10^{12} cm^{-2}, sample NC-36 in Fig.8 b), a very strong increase in the PL intensity coming from NCs in the NC:SiO_2 layers was found, although the main spectral characteristics of the NC-related PL peak (spectral position and half-width) were the same as in the PL spectra of non-irradiated samples. Increase in the Bi ion fluence to $8x10^{12}$ cm^{-2} (NC-37) leads to a very strong decrease of the registered PL intensity; simultaneously, there emerges a dependence of the spectral position of the NC-related PL peak on the excess Si content of the oxide layers. The data on the NC-related PL intensity considered as a function of the Si content of NC-SiO_2 layers are summarized in Fig. 8 d. These data suggest that an additional concentration of Si crystallites, resulting in PL intensity enhancement, could be formed in samples irradiated with relatively low ion fluence. Considering the data obtained in (Antonova et al., 2006 b; 2008 b), which show that the NC-related PL emission comes only from individual, geometrically isolated NCs, it can be inferred that in ion-irradiated NC:SiO_2 layer a high intensity of NC-related PL should be observed for relatively low excess Si contents of NC:SiO_2 layers. The PL quenching observed in the samples irradiated with higher ion fluences should be attributed either to formation of NC clusters or to amorphization of Si crystallites.

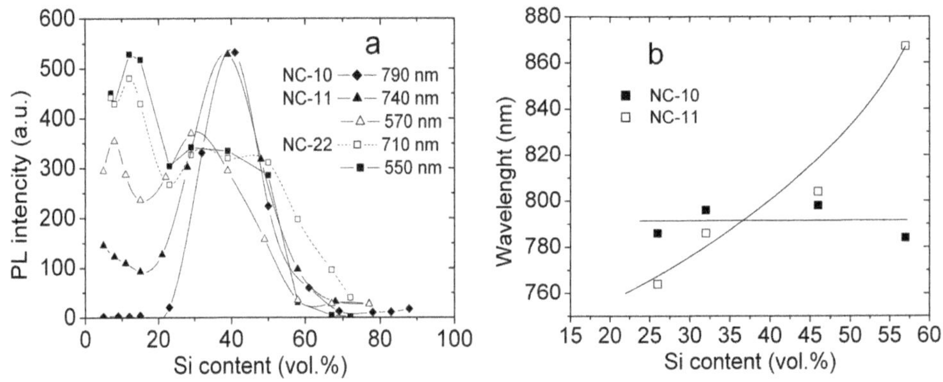

Fig. 7. (a) The dependence of the maximum PL intensity on the Si phase content in samples NC-10 and NC-11 irradiated with Kr ions; those samples contained a certain amount of Si NCs formed before irradiation. The sample NC-22 irradiated with Xe ions (10^{12} cm^{-2}) was subjected to a post-irradiation high-temperature anneal (1140°C, for 40 min). (Antonova et al., 2008 a) (b) Spectral position of the PL peak maximum in the PL spectra of samples NC-10 and NC-11 versus the Si content of the NC-SiO_2 layers.

The radiation effects observed in our samples can be summarized as follows:

i. At fixed ion dose, in samples irradiated with ions producing relatively weak ionization in the material (Kr, ionization energy losses dE_e/dx = 7.1 keV/nm) and in samples irradiated with relatively high ion fluences (≥ $8x10^{12}$ cm^{-2}), the spectral position of the NC-related PL peak proved to be dependent on the excess Si content of the NC-SiO_2 layers. The effect was attributed to a change of NC sizes.
ii. In the case of higher ionization losses (i.e. in the case of Xe and Bi ions with dE_e/dx ≥ 10.2 keV/nm) the range of excess Si contents for which an intense NC-related PL was observed was found to be extended towards lower Si contents. A maximal PL intensity

was found in samples containing 17 - 20 vol.% of excess Si. The PL intensity in samples irradiated with low ion fluence (~ $1x10^{12}$ cm^{-2}) could be a few times greater than the PL intensity coming from non-irradiated samples. Hence, at low fluence an efficient formation of NCs due to high-energy ion irradiation is observed in samples with relatively low Si content. Such a formation of Si NCs was found to occur both in silicon oxide layers with preliminarily formed NCs and in silicon oxide layers that contained just dissolved excess Si.

iii. Increase in the ion fluence quenches the NC-emitted PL due to formation of extended NC clusters in the NC-SiO_2 layers and/or due to amorphization of already formed Si crystallites.

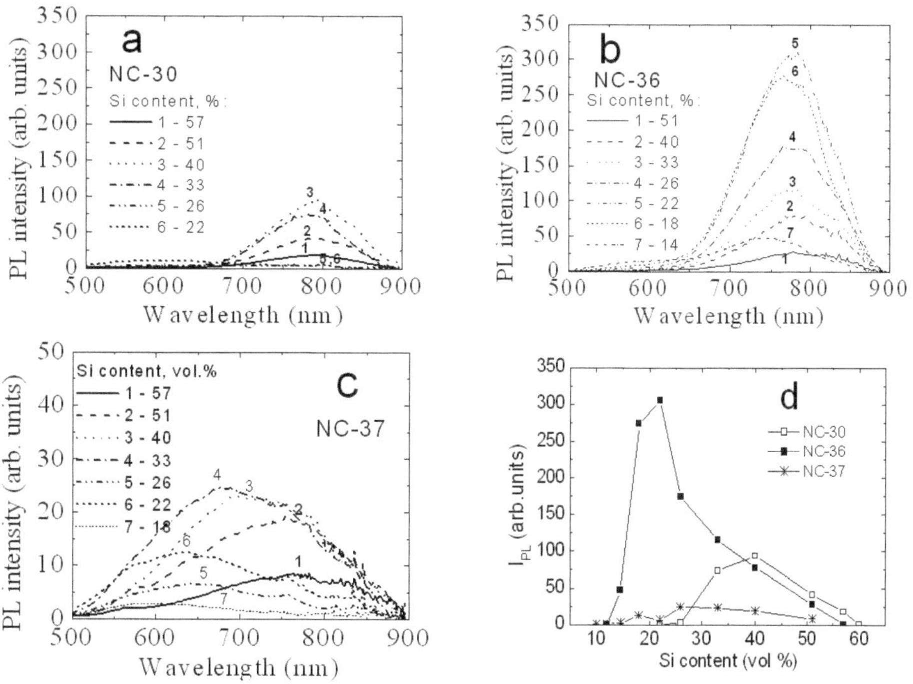

Fig. 8. PL spectra taken from samples with different Si contents: NC-30 (a), NC-36 (b) and NC-37(c). The results are summarized in graph (d), which shows the intensity of the most intense NC-related PL peak versus the Si content of oxide layers in samples NC-30 (785 nm), NC-36 (785 nm), and NC-37 (780 – 580 nm). (Antonova et al., 2009 a)

4.2 Charge storage capacity of irradiated NC:SiO_2 layers

The density of fixed oxide charges in the initial (non-irradiated) structure NC-30 and in the NC-SiO_2 layers of samples irradiated with Bi ions at different ion energies (NC-33, NC-34, NC-36) and at different ion fluences (NC-36, NC-37) versus the excess Si content of NC-SiO_2 layers is shown in Fig. 9 a, b. The data were extracted from measured C-V characteristics of MOS capacitors considering the flat-band voltage values. Note that prior to performing C-V measurements the irradiated samples were annealed at 800°C for 20 min with the aim to remove irradiation-induced electrically active defects from the substrate. The sign of the oxide charge (capture of electrons or holes) was found to be dependent on the substrate conductivity type, which was altered during irradiation. The fixed charge in the NC-SiO_2 layer of sample NC-35 irradiated with the lowest ion fluence was low ($< 10^{12}$ cm^{-2}) due to the high compensation of the substrate.

An important observation discussed in Refs. (Antonova et al., 2006 b; 2008 b) consists in that a relatively high density of electrons trapped and stored at nanocrystals in NC-SiO_2 layers presents a peaking function of the excess Si content of the layers. A decrease in the fixed oxide charge that was observed in NC-SiO_2 layers with high Si phase contents was attributed to the formation of NC clusters in the oxide layer, resulting in impossibility for individual NCs to retain trapped charges. Further increase in NC concentration leads to the formation of conductive NC chains in turn giving rise to percolation conductivity. The peaking dependence of charge storage capacity of NC-SiO_2 layers on the excess Si content of these layers simply reflects the fact that the density of isolated NCs capable of retaining trapped charges goes through a maximum if considered as a function of excess Si content of the layers. The most striking finding here is that, within experimental accuracy, the maximum charge storage capacity of NC-SiO_2 layers was observed at exactly the same Si contents for which a maximum PL intensity coming from the NC-SiO_2 layers was registered. Variations in the deposition and/or annealing conditions of Si-rich oxide layers alter the proportion between the amounts of the excess silicon contained in Si nanocrystals and Si amorphous inclusions; as a result, the maximum in the curve of stored charge and versus the Si content of NC-SiO_2 layers and the maximum in the curve of NC-related PL intensity both exhibit shifts. It is significant, however, that in all cases a maximum charge storage capacity of NC-SiO_2 layers was observed in the same samples in which a maximal NC-emitted PL emission was registered. In (Antonova et al., 2008 b), a concept of delocalization threshold for an ensemble of NCs was introduced, this threshold being associated with the onset of substantial migration of charge carriers within NC clusters. This delocalization threshold is determined by both carrier migration and inter-cluster tunneling processes, and it corresponds to the excess Si content at which the PL and charge storage maxima are observed. Thus, a remarkable matching between the onset of the formation of quantum dot clusters, the above-discussed charge-storage and PL properties of NC systems, percolation-clusters theory and intra-cluster carrier migration processes was revealed.

We have found that in irradiated structures, in agreement with the suggested delocalization threshold concept, the maximum charge trapping at nanocrystals was observed exactly in the same region of excess Si contents in which a most intense PL emission from NCs was detected (compare Figs. 9 and 7, 8). Irradiation of the NC-SiO_2 layers with high-energy heavy ions results in widening of the range of layer compositions for which a substantial charge trapping by NCs is observed and in shifting the position of the charge storage maximum towards lower excess Si contents (Antonova et al., 2010 a; 2010 b). With the increase in ion energy, the range of Si contents of NC-SiO_2 layers in which additional charge could be accumulated at NCs also exhibited a shift towards lower Si contents. This finding, proving that additional NCs were formed in irradiated samples, well correlates with our PL and Raman data.

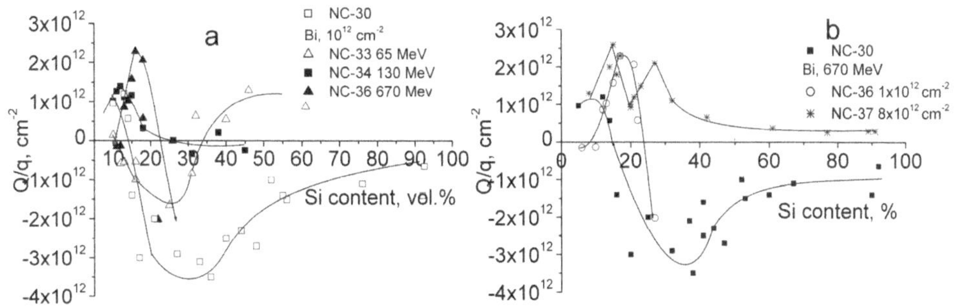

Fig. 9. The density of the NC-trapped charge as extracted from the flat band voltages of MOS capacitors versus the Si phase content of 400-nm thick NC-SiO_2 layers irradiated with Bi^+ ions: with variation of (a) the ion energies for the ion fluence of 10^{12} cm^{-2}, and (b) the ion fluencies for energy of 670 MeV. Irradiated regimes are indicated in the figures.

Observation of two peaks in the graph of Fig. 9 b for the sample irradiated with the highest ion dose correlates with the emergence of two peaks in Fig. 8 d in the curve for the same sample. Note that the high magnitude of the charge stored at NCs implies a high concentration of Si NCs formed due to irradiation in the range of relatively low Si contents of NC-SiO_2 layers. Thus, the higher the fluence, the higher is the NC concentration in the mentioned range of Si contents. With reference to Fig. 9, it can be hypothesized that the concentration of irradiation-induced Si NCs formed in NC-SiO_2 layers with a relatively low content of excess Si increases with increasing the ion fluence. The appearance of a dip at ~30% Si content in the dependence for the sample NC-37 (with simultaneous emergence of two peaking features on either side of the dip) can be attributed to the formation of NC clusters in the NC-SiO_2 layer with an increased concentration of irradiation-induced NCs. It should be noted here that the maximum value of the NC-stored charge for the data of Fig.9 is ~$3x10^{12}$ cm^{-2}; this value complies with the maximal density of isolated nanocrystals with sizes ~ 3-5 nm in NC:SiO_2 systems. One more interesting conclusion that can be drawn from a comparison of data in Figs. 9 b and 8 d is that the NC-related PL exhibits a more pronounced decrease with increasing NC concentration in comparison with NC charge storage capacity.

4.3 Transport properties and inter-cluster migration effects

In part of our samples with high Si content, percolation conduction was observed. For evaluating the conductivity of NC-SiO_2 layers, we measured I-V characteristics of reference and irradiated samples versus the Si-phase content. From such measurements, the curves of resistivity R of our NC:SiO_2 layers versus the Si content of those layers were deduced; these curves are shown in Fig.10. For reference (non-irradiated) samples, a percolation transition with a percolation threshold at Si contents 33 - 40 vol.% was revealed; at this threshold, a sharp decrease in the resistivity of NC:SiO_2 layers amounting to about six orders of magnitude was observed. The position of the percolation threshold was found to be dependent on the proportion between the contents of crystalline and amorphous Si inclusions in the SiO_2 layer. The 33% position of the threshold, consistent with theoretical predictions (Shklovskii & Efros, 1984), was observed in NC:SiO_2 layers deposited at low rates (Antonova et al., 2006 b). It was found that the percolation conduction threshold shifts to higher contents of excess Si in SiO_x layers after irradiation of the samples with high ion fluences (~10^{13} cm^{-2}) or with high-energy ions producing relatively weak ionization in the material. In those cases, as a rule, the conductivity of oxide layers in the interval of x-values above the percolation threshold decreased in magnitude in irradiated samples because of partial amorphization of Si nanocrystals, as supported by Raman data. In samples irradiated with relatively low fluences ($1x10^{12}$ cm^{-2}) of 670 MeV Bi ions the resistivity of the NC:SiO_2 layers in the region over the percolation threshold decreased by two orders in comparison with non-irradiated samples.

The above results can be self-consistently interpreted as follows. The conductivity in the range of composition of NC-SiO_2 layers above the percolation threshold depended weakly on the Si-phase content of our layers both in irradiated and non-irradiated samples. This finding is consistent with the predictions of percolation theory (Shklovskii & Efros, 1984). However, the conductivity of irradiated sample NC-36 was almost two orders of magnitude higher than that of non-irradiated sample NC-30. This is believed to be due to a much higher conductivity of the percolation network due to NC chains formed along ion tracks and having identical orientations of atomic planes within the ordered NC chains.

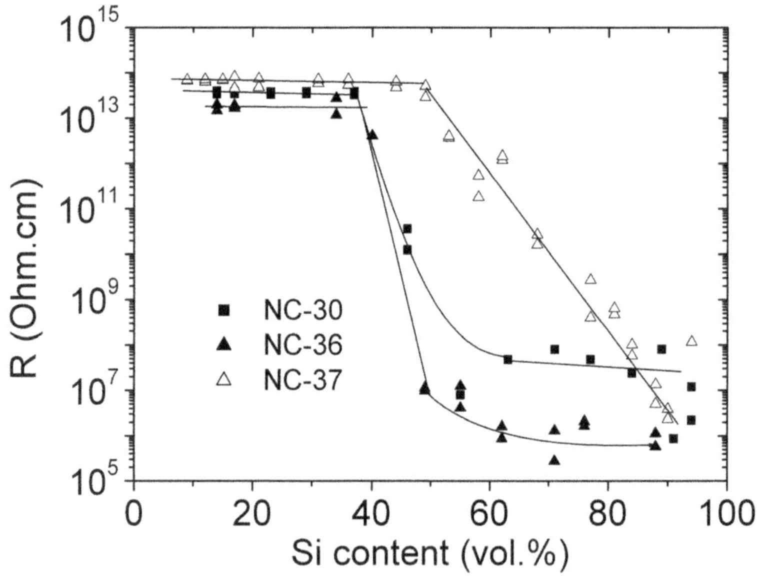

Fig. 10. The resistivity of NC:SiO_2 layers (measured at applied bias voltage 5 V) versus the Si phase content of the layers (vol.%) for reference sample NC-30 and for samples irradiated with 670 MeV Bi ions to fluences 10^{12} cm^{-2} (NC-36) and $8x10^{12}$ cm^{-2} (NC-37). (Antonova et al., 2009 a)

In view of a higher conductivity of crystalline Si phase in comparison with amorphous Si phase, a narrower percolation transition can be expected in samples containing a larger amount of Si crystallites formed in the oxide layer with a concomitant reduction of a-Si phase content as evidenced by TEM images and Raman data. Thus, crystallization is found to predominantly proceed in samples with relatively low contents of excess Si. For the other part of our samples, with a higher Si content (about and above the percolation threshold), and for higher ion fluences, amorphization of Si nanocrystallites becomes a dominating process. However, the formation of a larger amount of Si crystallites fails to offer a reasonable explanation to the lower resistivity of sample NC-36 in comparison with sample NC-30 since the average conductivity of NC chains at high Si contents (above the percolation threshold) decreases weakly with increasing NC concentration.

5. Low-dimensional effects in ion-modified 3D NC arrays

Modification of NC-containing films (NC-11a, NC-23a, and NC-31a) with high-energy ion irradiation was found to lead to the appearance of capacitance peaks in the low-frequency capacitance-voltage characteristics of MOS capacitors formed on our samples. The *C–V* characteristics of reference and irradiated samples measured at different frequencies are shown in Fig. 11. As it is seen, two distinct capacitance peaks appear in the low-frequency C-V curve after irradiation. Such peaks were observed only in irradiated structures that contained 60–80% of excess silicon in 400-1000 nm thick NC:SiO_2 layer (Antonova et al., 2009 b). The amplitudes of those peaks increased with decreasing the measurement frequency, while the voltage position of the peaks remained essentially the same for all frequencies at which our measurements were performed. Moreover, the bias voltages at which the peaks were observed were quite reproducible in measurements performed at different points on the surface of one and the same sample. In particular, in all investigated samples one of the observed peaks was located at $V \approx 0$.

The amplitude of the capacitance peaks (peaks 1 and 2 in Fig. 11 b) plotted as a function of frequency is presented in Fig. 12. This frequency dependence of capacitance C is given by

$$C(\omega) = \frac{C_{max}}{\sqrt{(1+(\omega\tau)^2}}, \quad (2)$$

where $\omega = 2\pi f$ (Shi et al., 2002; Punchaipetch et al., 2006). Hence, the frequency dependence of the peak amplitude allows one to estimate the time constant τ of the NC charging process. The τ values obtained by fitting the experimental $C(\omega)$ data (Fig. 12) with formula (2) were found to lie around $1.5x10^{-5}$ s and $5x10^{-6}$ s respectively for peaks 1 and 2. The value $1.5x10^{-5}$ s is in a fairly good correlation with the characteristic time of carrier emission from the ground state of NCs estimated for 300 K by means of Q-DLTS (see Fig. 6).

The bias-voltage dependences of the real (G) and imaginary (C) parts of the admittance of NC-23a films, which were irradiated with Xe ions and measured at 10-kHz frequency are shown in Fig. 13. As it follows from Fig. 11, this frequency is low enough for the capacitance peaks which we attribute to the charging of Si nanocrystallites, to emerge. The data in Fig. 13 suggest the possibility of dividing the observed peaks into two groups according to their bias positions (I and II). The voltage difference between the two groups is ~2.6 V, and the voltage difference between neighboring capacitance peaks within each group is roughly ~0.5 V. The G-V curves exhibit only two less pronounced peaks observed at the same voltages.

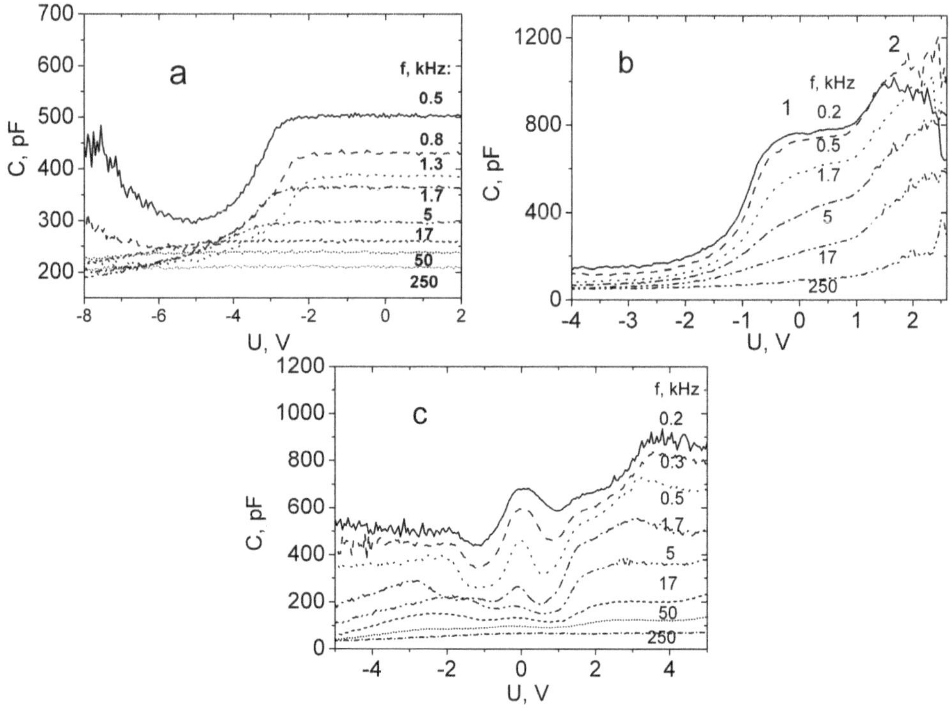

Fig. 11. C-V characteristics of NC-SiO_2 structures measured at room temperature and various frequencies for Si-phase contents 72% (a, b) and 79% (c) of oxide layer (initial sample NC-10 (a) and two samples from the sample series NC-11a irradiated with Kr ions (b, c). (Antonova et al., 2009 b)

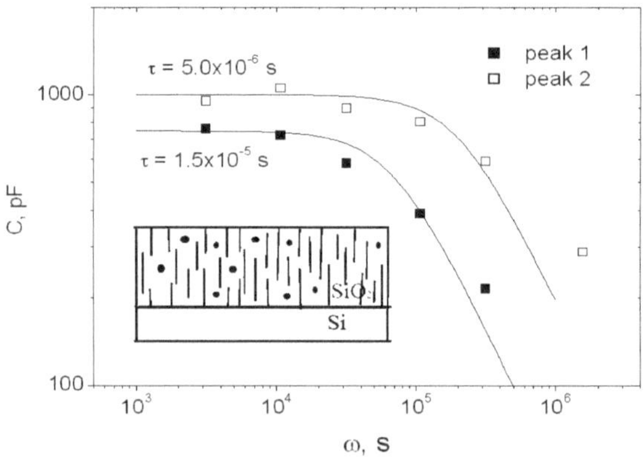

Fig. 12. The amplitudes of capacitance peaks 1 and 2 (Fig. 11 b) versus frequency. The peak numbers refer to the two main peaks in Fig. 11 b. The curves are the dependences that fit the τ values estimated from Q-DLTS data (symbols) with formula (2). The inset illustrates our model of NC chain formation in NC-SiO_2 layers subjected to high-energy ion irradiation. (Antonova et al., 2009 b)

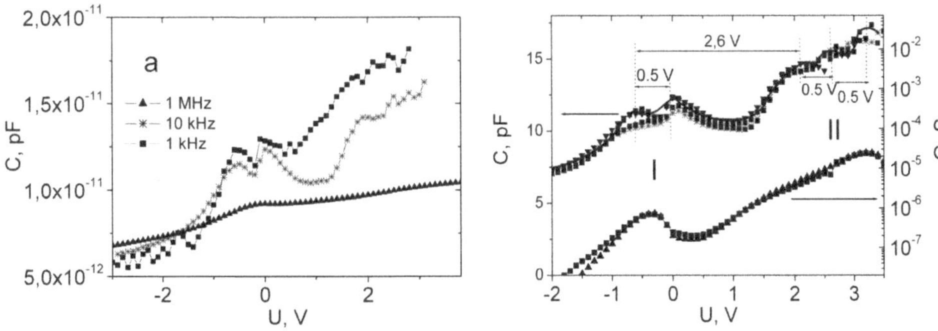

Fig. 13. Typical C-V and G-V characteristics of a MOS capacitor with Si NCs in the oxide layer (Si-phase content 63 vol.%) irradiated with Xe ions. Different points refer to different repeatedly performed measurements.

Similar peaks with similar characteristics were previously observed in C-V characteristics of thin (3 -10 nm) oxide layers with an embedded two-dimensional array of Si NCs; the latter peaks were attributed to the charging process of electronic levels in NCs, and the voltage spacing between them (in V) was associated with the energy spacing (in eV) between the quantum confinement levels in NCs (Wu et al., 2004; Yu et al., 2005). It is significant that in our samples similar effects were observed for thick (400–1000 nm) NC-containing layers. We have suggested that the peaks could also be due to resonance tunneling of charge carriers to electronic levels in nanocrystals. This interpretation leans upon the fact that at high frequencies the electrons participating in the tunneling processes could not follow the voltage modulation, while at low frequencies they could respond to it; as a result, tunneling of charge carriers through the oxide into NCs could give rise to the capacitance peaks. A qualitative model to interpret the nature of the experimentally observed effect in ordered three-dimensional NC arrays is described below.

Taking into consideration our data (TEM, Raman spectroscopy, transport and C-V measurements, Q-DLTS), we suggest that radiation-induced NC chains were formed within ion tracks in our samples and that the peaks found in the C-V characteristics possibly reflected the formation of those NC chains. Considering the fact that the Si-phase content of the structures on which C-V measurements were performed (> 60%) was in a notable excess of the Si-phase content of the structures that were used in TEM study (35 - 40%) while the number of ion tracks in both cases was roughly identical, we suppose that, very likely, more pronounced NC chains were formed in our layers with the higher Si content. The lengths of those chains were approximately identical, defined by the thickness of the NC-SiO_2 layer. In samples with a Si content exceeding the percolation threshold those chains formed a conducting network. The NCs forming the chains were unable to retain the captured charge due to the high electric conduction existing between the NCs. So, only isolated NCs in non-conductive cells in the conducting network formed by NC chains could retain charge, the recharging of those NCs proceeding through the "bottlenecks" located between the isolated NC and neighboring chains. It is the charge trapping at such isolated Si nanocrystals that was probably responsible for the emergence of additional peaks in the *C–V* characteristics of irradiated samples. The effect due to resonance tunneling could be observed in such NC systems because, here, the conducting network played the role of a "substrate" that created, together with a particular isolated NC, a two-barrier tunneling system in which the individual Si nanocrystal was separated from the "substrate" with a tunneling gap. Since the probability of tunneling through a barrier strongly (exponentially) depends on the barrier width, it can be assumed that the spread of tunnel-barrier widths involved in the phenomenon was rather small. Irrespective of the exact position of "bottlenecks" in the conducting NC chains, for statistical reasons the resistances of such chains and the resistances of the rest part of the NC cluster for different NC chains had close values and, as a result, the total bias voltage applied to the oxide was divided between each NC chain and the "bottleneck" in roughly identical proportions. Statistically, all the parallel NC chains involved tunneling gaps of roughly identical widths; as a result, NCs in different non-conductive cells can be expected to be recharged at roughly identical voltages. A sketch of QD chains formed along ion tracks in an irradiated sample is shown in the inset of Fig. 12. Charge carriers are captured by NCs at bias voltages at which resonant tunneling conditions are realized. In a real situation, the concentration N_{NC} of isolated NCs contributing to measured capacitance presents only a small fraction of the total NC concentration. In our case, all NC chains are connected in parallel and the total induced capacitance is therefore equal to the product $C_{NC}N_{NC}$. The magnitude of the capacitance peak ΔC is ~ 6 pF for the first group of NC-related capacitance peaks (Fig. 12). The capacitance of one NC is estimated as 8 aF (see below). Thus, we have $N_{NC} \approx 8\text{x}10^8$ cm^{-2}, this value constituting approximately 10^{-3} of the total number of all ion tracks (10^{12} cm^{-2}).

A more difficult problem here is interpretation of the bias voltage values of the C-V peaks. Let us try to correlate these values with characteristics of the resonant tunneling process and with the charging energies of a single NC. For that purpose, consider the τ values that were extracted from the C(ω) curves shown in Fig. 12. The probability T of carrier tunneling through the rectangular barrier between NCs as a function of the energy position of the levels in NC can be estimated using the expression

$$T = \exp(-\frac{2}{\hbar}\int_{x_2}^{x_1} p dx) \sim \exp(-\frac{d}{\lambda}), \tag{3}$$

where $p = \frac{h}{\lambda}$ is the carrier momentum, $\lambda = \frac{h}{\sqrt{2mE}}$ is the electron wavelength, and $d = x_1 - x_2$ is the width of the tunneling barrier between an isolated NC and the substrate or between an isolated NC and the adjacent NC chain, h is the Planck constant, and m is the electron effective mass in silicon. The tunneling being a dominating process, the tunneling time, or NC recharging time, is $\tau \sim T^{-1}$. In view of formula (3), the ratio τ_1/τ_2 (here, $\tau_1 = 1.5\text{x}10^{-5}$ s and $\tau_2 = 5\text{x}10^{-6}$ s, see above) can expected to be a function of the quantum confinement energies in the NCs, so that

$$\ln(\frac{\tau_2}{\tau_1}) = \frac{h}{d\sqrt{2m}}(E_1^{-1/2} - E_2^{-1/2}) . \tag{4}$$

Considering equation (1), we obtain a certain relation between τ_2/τ_1 and W. For the above-indicated values of τ_1 and τ_2, the estimated NC diameter W is 3.3 nm assuming that the tunneling distance is d = 3 nm. The value of W is in a very good agreement with the TEM values, 3-5 nm. This, then, suggests that, indeed, the capacitance peaks that we observed in the C-V characteristics can be associated with capture of electrons into quantum confined levels in NCs. Following the presented analysis, the first and the second groups of capacitance peaks in Figs. 11 and 13 can be associated with the charging of respectively the ground-state level and the first exited level in the NCs, whereas the relatively small peaks inside each group can be attributed to charging effects (Coulomb blockade energies) of NCs.

We have tested our model for consistency by performing an analysis typically used for NCs (Yu et al., 2005, Huang et al., 2003). The position of the photoluminescence peak observed in examined samples yields for the QD size a value 3.3 nm. The energy distance between the ground level and the first excited electronic level in such a QD is estimated to be ΔE = 103 meV according to equation (1). The voltage difference between the two groups of peaks can be represented as $\Delta V = eK\Delta E$, where K is a coefficient depending on the geometry of the QD/matrix system. In the present case, we have ΔV = 2.6 eV which value translates into K=0.048. The NC charging energy can be estimated from the relation $\Delta E_e = K\Delta V_e$; with the above value of K, this yields ΔE_e = 24 meV. On the other hand, we have $\Delta E_e = e/C_{NC}$, where C_{NC} is the NC capacitance. The capacitance of an individual NC in our NC-SiO_2 layers was estimated as 8 aF, this value being quite a reasonable one for an NC ~ 3 nm in size.

6. Interaction of high-energy ions with silicon nanocrystals in dielectric matrix

Consider the processes that can be responsible for the radical transformations of NC arrays in ion-irradiated samples. First of all, it should be noted that the diffusivity of silicon in SiO_2 at temperatures 1110–1400°C is rather low, 10^{-21}–10^{-17} $cm^2 \cdot s^{-1}$ (Mathiot et al., 2003; Tsoukalas et al., 2001; Takahashi et al., 2003). According to the data of (De Almeida et al., 2000; Aleksandrov et al., 2008), oxygen possesses a much higher diffusivity in SiO_2 amounting to 10^{-9}–10^{-7} $cm^2 \cdot s^{-1}$. That is why the formation processes of Si nanocrystals in Si-rich oxide layers are usually described with consideration given to the fact that those processes should involve migration of oxygen atoms from the region of the growing silicon nanoparticle (see, for instance, Khomenkova et al., 2007). In addition, in their Monte-Carlo simulations, Muller et al. showed that the formation process of Si nanocrystals in Si:SiO_2 layers with a low excess Si content involves nucleation and growth of Si nanocrystals, whereas in Si:SiO_2 layers with a high excess Si content Si NCs are formed due to spinodal decomposition with separation of the Si and SiO_2 phases (Muller et al., 2002). Thus, it follows from the above consideration that, due to the local release of energy in ion tracks and the corresponding heating of the material there (see Fig.15 a, and Section 2.2, Fig.1 a), oxygen is bound to leave track regions, where Si nanocrystals may form. It should be mentioned here that the high ionization of material in track regions can be expected to most likely enhance the diffusivity of all atoms.

All of the above-mentioned processes can be enhanced by the anisotropic strain fields produced by anisotropic heating and by the atoms leaving the track regions (Van Dillen et al., 2005). It is believed that, on irradiation, the material expands in the direction normal to the ion track and contracts in the direction parallel to the track axis. The steady-state (residual) strain ε_{st} (the relative change in lattice distance) can be estimated as (Van Dillen et al., 2005)

$$\varepsilon_{st} = exp(AF) - 1, \tag{5}$$

where A is the rate of strain accumulation in a material irradiated with a particular type of ions, and F is the ion fluence. The rate of strain accumulation can be estimated as

$$A = 0.463\frac{1+\nu}{5-4\nu}\cdot\frac{\beta\frac{dE_e}{dx}}{\rho C}, \tag{6}$$

where ν is the Poisson's ratio, β is the coefficient of thermal expansion, ρ is the material density, and C is the specific heat. As follow from (6), the rate of strain accumulation A is proportional to the electronic losses of ions dE_e/dx. Invoking the well-known constants for SiO_2 (ν = 0.2, $\beta = 5.5x10^{-7}$ K^{-1}, $\rho = 2.2x10^3$ kg/m^3, $C = 10^3$ J/kgK), we have estimated the strain accumulation rate A for the interval of ionization losses of used ions (see Fig. 14 a). For silicon, the quantity A has one and a half order of magnitude smaller value. For the case under consideration, using formula (5), we have also estimated the steady-state strain ε_{st} for ionization losses of ions ranging from 5 to 20 keV/nm. Those results are given in Fig.14 b. So, the strains in the layer are estimated as $3x10^{-5}$–$8x10^{-3}$ depending on the ion used and on the ion fluence. This value is high enough for provoking significant strain-induced effects. The strain decreases with increasing the Si content in the NC:SiO_2 layer. This means that, at low Si contents, the diffusion flux of oxygen from track regions and, consequently, the formation process of nanocrystals should be more pronounced. Our experimental data fairly well agree with such considerations. In particular, the high strain has to limit the phase separation in ion tracks, thus decreasing the size of formed NCs. This effect was clearly observed by means of PL and Q-DLTS measurements (see Sections 3.3 and 4.1): the higher are the ionization losses, the higher is the strain, and the smaller is the typical NC size.

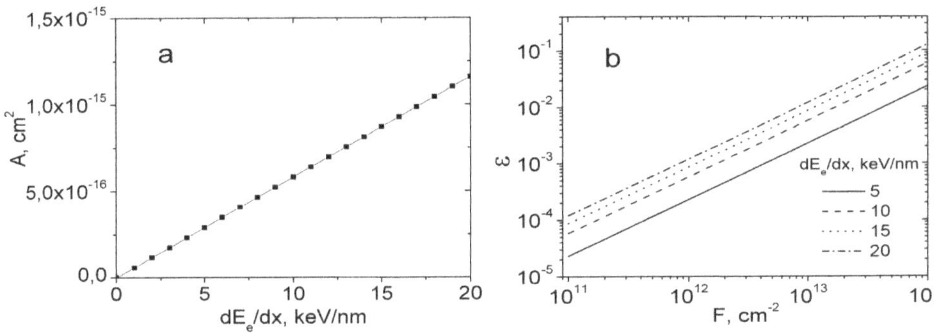

Fig. 14. (a) The strain accumulation rate A estimated for SiO_2 by formula (6) using constants $\nu = 0.2$, $\beta = 5.5x10^{-7}$ K^{-1}, $\rho = 2.2x10^3$ kg/m^3, and $C = 10^3$ J/kg·K. (b) The steady-state strain ε_{st} estimated from (5) for different ionization losses of ions.

If the Si content of the SiO_2 layer is high, other processes control the interaction of high-energy ions with Si nanocrystals previously formed in SiO_2. In this case, the interaction is defined by the relation between the nanocrystal diameter and the track diameter. With the example of Ge QDs, Schmidt et al. (Schmidt et al., 2007; 2009) showed that, if the nanocrystal size is such that the particles do not melt at a particular fluence, then the nanoparticles change their spherical shape for an ellipsoidal one with increasing the ion fluence, and the sizes of nanoparticles in an ion track decrease in value along the track length. The mechanism of such a modification of nanoparticle shape is associated with the above-discussed anisotropic deformations of the particles. At the same time, the Si nanocrystals get amorphized and partially dissolved, as it was observed in (Mishra et al., 2007) and in the present study for samples with a high Si content. Nanoparticles smaller than, or comparable with, track dimensions are stretched along ion tracks. The threshold electronic losses corresponding to the onset of material melting in ion tracks in SiO_2 are thought to be 6 keV/nm (Meftah et al., 1997). Figure 15 presents a sketch illustrating the interaction of a high-energy ion with a nanocrystal. The crossing length of the NC of radius R by the ion track is $L_{nc} \leq 2R$. The nanocrystal will melt if the energy lost by the ion over the length L_{nc} exceeds the energy needed for melting the nanocrystal,

$$\frac{dE_e}{dx} L_{nc} \geq \frac{4\pi}{3} R^3 H_m \rho_{nc}, \tag{7}$$

where H_m is the melting heat and ρ_{nc} is the NC density. The maximal radii R of the nanocrystals in our samples which could be melted by the ions used in the present study are given in Table 3. For Kr, Xe, and Bi ions with energies 1–3 MeV/nucleon, melting occurs in all cases. As a result, nanocrystals elongated along ion tracks retain their crystal structure, whereas nanocrystals located in between ion tracks become amorphous. Integrally, these effects result in general ordering of nanocrystals along ion tracks. The same processes proceed in the case of metal inclusions embedded in an insulator matrix and, as it was shown recently (Mishra et al., 2007; Giulian et al., 2010; Dawi et al., 2011; Kumar et al., 2011), irradiation with sufficiently high ion fluences and energies provides a means for forming long wire-shaped nanoparticles in ion tracks.

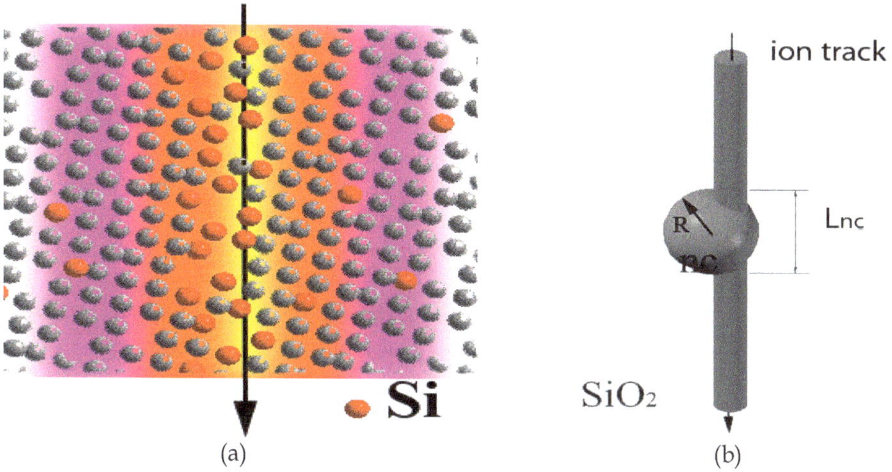

Fig. 15. (a) Sketch of an ion track in a composite crystal. Color gradient of the background is an illustration of the heating gradient. Non-uniform distribution of excess Si atoms (red balls) reflects the out-diffusion of oxygen from the track zone. (b) Sketch of the interaction of a high-energy ion with a nanocrystal.

Ion	Energy, MeV	dE_e/dx in Si, keV/nm	dE_e/dx in SiO_2, keV/nm	R, nm
Kr	90	8.0	7.1	≤40
Xe	130	12.3	10.5	≤43
Bi	65 - 670	9.1-23	7.9-18	≤50-68

Table 3. Maximal radii R of Si nanocrystals that could be melted by the used ions in ion tracks. Estimations were made with the use of inequality (7) and parameter values ρ_{nc} = 2.33 g/cm^3 and H_m = 164 J/g.

7. Conclusions and outlook

As shown in this chapter, irradiation of Si:SiO_2 and NC:SiO_2 layers with high-energy heavy ions under the above-described conditions results in the formation of ordered low-dimensional NC systems that possess properties untypical of three-dimensional arrays of Si nanocrystals randomly distributed in the bulk of oxide layers. We have found that irradiation of our layers with heavy ions could either enhance or retard the formation of silicon nanocrystallites in the host matrix according to the irradiation dose used. We have also showed that both the electrical and optical properties of ion-irradiated layers could be improved by irradiating samples with low ion fluences. The increase of NC-emitted PL intensity and the extension of the interval of excess Si contents in which trapping of charges at NCs and emission of NC-related PL become detectable show much promise in applications of Si NCs. This suggests that, using a combination of properly chosen annealing and irradiation conditions, one can optimize the properties of NC:SiO_2 layers according to posed requirements. The ordering of Si nanocrystallites along ion tracks in oxide layers of irradiated samples leads to the formation of NC chain arrays. Moreover, the formed NCs have identically oriented atomic planes. Such NC systems cannot be obtained by other known NC formation methods. Other important features demonstrated by samples with NC chains thus formed in comparison with non-irradiated samples include: (1) high conductivity of the NC chains and (2) identical lengths of those chains, approximately equal to the oxide-layer thickness. Due to these features, the initial 3D system of QDs acquires properties of a low-dimensional array of QDs. Thus, irradiation of NC-SiO_2 layers with high-energy heavy ions enables the formation of new nanostructured materials exhibiting nontrivial structural and electronic properties not observed in materials produced by other technologies. The irradiation of NC-SiO_2 layers with high-energy heavy ions provides a useful tool that makes it possible to optimize the properties of NC arrays and fabricate vertically aligned nanostructures with low-dimensional properties.

Systems involving Si nanocrystals are of interest for a wide spectrum of applications, from solar cells to biomedical applications. The understanding of fundamental mechanisms that control the conductivity, the charge storage capacity, and the photoluminescence of nanocrystalline films modified with high-energy ions gives us the key to altering the properties of nanoscale materials in a desirable controlled manner and, in particular, the key to solving the various problems encountered in applications of such materials. High-energy ion irradiation can be used as a tool for 3D structuring of composite materials allowing control of NC shape, size, and/or atomic plane orientation.

8. Acknowledgements

This study was supported by the Russian Foundation for Basic Research (Grant No. 11-02-92477).

9. References

Aleksandrov, O.V., & Dusj, A.I., (1997). Model of Thermal Oxidation of Silicon at the Volume-Reaction Front. *Semiconductors*, Vol. 42, No. 11, (November 2008), pp. 1370-1376, ISSN 1063-7826.

Antonova, I.V., Shaimeev, S.S. & Smagulova, S.A. (1997). Transformation of Electrically Active Defects as a Result of Annealing of Silicon Implanted With High-Energy Ions. *Semiconductors*, Vol. 40, No.5 , (May 2006 a), pp. 543-548, ISSN 1063-7826.

Antonova, I. V., Gulyaev, M.B., Yanovitskaya, Z. S., Volodin, V. A., Marin, D. V., Efremov, M. D., Goldstein, Y., & Jedrzejewski, J., (1997). Electrical Properties and Photoluminescence of SiO_x Layers with Si Nanocrystals in Relation to the SiO_x Composition. *Semiconductors*, Vol. 40, No. 10, (October 2006 b), pp. 1198-1203, ISSN 1063-7826.

Antonova, I.V., Gulyaev, M.B., Skuratov, V.A., Marin, D.V., Zaikina, E.V., Yanovitskaya, Z.S., Jedrzejewski, J., & Balberg, I., (1988). Modification of Silicon Nanocrystals Embedded in an Oxide by High Energy Ion Implantation. *Solid State Phenomen.*, Vol.131-133, (2008 a), pp.541-546, ISSN 1012-0394.

Antonova , I.V., Gulyaev, M.B., Savir, E., Jedrzejewski, J., & Balberg I. (1983). Charge Storage, Photoluminescence and Cluster Statistics in Ensembles of Si Quantum Dots. *Phys. Rev.B*, Vol. 77, No.12, (2008 b), 125318(5), ISSN 1098-0121.

Antonova, I.V., Gulyaev, M.B., Volodin, V. A., Cherkov, A.G., Marin, D.V., Skuratov V.A., Jedrzejewski, J., & Balberg, I., (1990). Modification of Si Nanocrystallites Embedded in a Dielectric Matrix by High Energy Ion Irradiation. *Nanotechnology*, Vol.20, No.2, (Fabruary 2009 a), 095205(5), ISSN 0957-4484.

Antonova, I.V., Kagan, M.S., Cherkov, A.G., Skuratov, V.A., Jedrzejewski, J., & Balberg, I., (1990). Low-Dimensional Effects in a Three-Dimensional System of Si Quantum Dots Modified by High-Energy Ion Irradiation. *Nanotechnology*, Vol.20, No.2, (April 2009 b), 185401(5), ISSN 0957-4484.

Antonova, I.V., Neustroev, E.P., Smagulova, S.A., Jedrzejewski, V.A., & Balberg, I. (1931). Charge Spectroscopy of Si Nanocrystallites Embedded in a SiO_2 Matrix, *J. Appl. Phys.*, Vol. 106, No 6, (September 2009 c), 064306(6), ISSN 0021-7989.

Antonova, I.V., Marin, D.V., Volodin, V. A., Skuratov, V.A., Jedrzejewski, J., & Balberg, I., (1988). Anisotropic Heating, Anisotropic Strain – Anisotropic Heating Engineering for Silicon Nanocrystals in SiO_2 . *Solid State Phenomen.*, Vol.156-158, (2010 a), pp.523-528, ISSN 1012-0394.

Antonova, I.V., Skuratov, V.A., Jedrzejewski, J., & Balberg, I., (1997). Ordered Arrays of Si Nanocrystals in SiO_2:Structural, Optical, and Electronic Properties, *Semiconductors,* Vol. 44, No. 4, (April 2010 b), pp. 482–487, ISSN 1063-7826.

Antonova, I.V., Smagulova, S.A., Neustroev, E.P., Skuratov, V.A., Jedrzejewski, V.A., Savir, E., & Balberg, I. (1997). Charge Spectroscopy of SiO_2 Layers with Embedded Silicon Nanocrystals Modified by Irradiation with High_Energy Ions, *Semiconductors,* Vol. 45, No. 5, (May 2011), pp. 582–586, ISSN 1063-7826.

Arnoldbik, W. M., Tomozeiu, N., van Hattum, E. D., Lof, R. W., Vredenberg, A. M., & Habraken, F. H. P. M. (1893). High-Energy Ion-Beam-Induced Phase Separation in SiOx Films. *Phys. Rev.B,* Vol. 71, No. 12, (March 2005), 125329 (7), ISSN 1098-0121.

Avasthi, D.K., Mishra, Y.K., Singh, F., & Stoquert, J.P. (1983). Ion Tracks in Silica for Engineering the Embedded Nanoparticles. *Nucl. Instrum. Methods Phys. Res. B,* Vol. 268, No.19 , (October 2010) , pp. 3027-3034, ISSN 0168-583X.

Bitten, J.S., Lawis, N.S., Atwater, H.A., Polman, A. (1962). Size-Dependent Oxygen-Related Electronic States in Silicon Nanocrystals. *Appl. Phys. Lett.,* Vol. 84, No. 12, (June 2004), pp. 5389-5391, ISSN 0003-6951.

Chaudhari, P. S., Bhave, T. M., Kanjilal, D., & Bhoraskar, S. V. (1931) Swift Heavy Ion Induced Growth of Nanocrystalline in Silicon Oxide. *J. Appl. Phys.,* Vol. 93, No 6, (March 2003), pp. 3486-3489, ISSN 0021-7989.

Chaudhari, P. S., Bhave, T. M., Pasricha, R., Singh, F., Kanjilal, D., & Bhoraskar, S. V., (1983). Controlled Growth of Nanocrystallites in Silicon Oxide Matrix Using 150 Mev Ag Ion Irradiation. *Nucl. Instrum. Methods Phys. Res. B,* Vol.239, No.3, (August 2005) , pp.185- 190, ISSN 0168-583X.

D'Orleans, C.; Stoquert, J. P.; Estournes, C.; Cerruti, C.; Grob, J. J.; Guille, J. L., Haas, F.; Muller, D.; & Richard-Plouet, M. (1893). Anisotropy of Co Nanoparticles Induced by Swift Heavy Ions *Phys. Rev.B,* Vol. 67, No. 22, (June 2003), pp. 220101(4), ISSN 1098-0121.

Dawi, E.A.; Vredenberg, A.M.; & Rizza, G.; Toulemonde, M. (1990). Ion-induced elongation of gold nanoparticles in silica by irradiation with Ag and Cu swift heavy ions: track radius and energy loss threshold. Nanotechnology, Vol. 22, No. 21, (May 2011), pp. 215607-215619, ISSN 0957-4484.

De Almeida, R.M.C., Goncalves, S., Baumvol, J.R., & Stediler, F.C., (1893). Dynamics of Thermal Growth of Silicon Oxide films on Si. *Phys. Rev. B,* Vol. 61, No. 19, (May 2000), pp.12992-12999, ISSN 1098-0121.

Fink, D.; Chadderton, L. T. ; Hoppe, K.; Fahrner, W. R. , Chandra, A.;& Kiv A, (1983). Swift-Heavy Ion Track Electronics (SITE). *Nucl. Instrum. Methods Phys. Res. B,* Vol.261, No.1-2 , (August 2007) , pp.727- 730, ISSN 0168-583X.

Gaponenko, S.V., (1998). *Optical Properties of Semiconductor Nanocrystals.* Cambridge University Press, ISBN 0521582415, Cambridge, UK.

Giulian, R., Kremer, F., Araujo, L. L., Sprouster, D. J., Kluth, P., Fichtner, P. F. P., Byrne, A. P., & Ridgway, M. C. (1893). hape transformation of Sn nanocrystals induced by swift heavy-ion irradiation and the necessity of a molten ion track. *Phys. Rev. B,* Vol. 82, No. 11, (September 2010), 113410 (4) , ISSN 1098-0121.

Godefroo, S., Hayne, M., Jivanescu, M., Stesmans, A., Zacharias, M., Lebedev, O. I., Van Tendeloo, G., & Moshchalkov, V. V. (2006). Classification and Control of the Origin of Photoluminescence from Si Nanocrystals. *Nature Nanotechnology*, Vol.3, No. 3, (March 2008), pp.174-178, ISSN 1748-3387.

Gur, I., Fromer, N. A. , Geier, M. L, Alivisatos, A. P. (1880). Air-Stable All-Inorganic Nanocrystal Solar Cells Processed from Solution. *Science*, Vol.310, (October 2008), pp.462-465, ISSN 0036-8075.

Hu, C.-W., Chang, T.-C., Tu, C.-H., Chiang, C.-N., Lin C.-C., Lee, S.-W., Chang, C.-Y., Sze, S.M., & Tseng, T.-Y. (1874). Enhancement of NiSi Nanocrystal Formation by Incorporation of Ge Elements for Nonvolatile Memory devices. *J. Elecrochem Soc.*, Vol.156, No 9, (July 2009), pp. H751 - H755, ISSN 0013-4651.

Huang, S., Banerjee, S., Tung, T.T., Oda, S. (1931) Quantum Confinement Energy In Nanocrystalline Silicon Dots from High-Frequency Conductance Measurement. *J. Appl. Phys.*, Vol. 94, No.11, (September 2003), pp.7261-7665, ISSN 0021-7989.

Iacona, F., Irrera, A., Franz, G., Pacifici, D., Crupi, I., Miritello, M. P., Presti, C. D., & Priolo, F. (1995). Silicon-Based Light-Emitting Devices: Properties and Applications of Crystalline, Amorphous and Er-Doped Nanoclusters. IEEE Journal of Selected Topics in Quantum Electronics, Vol.12, No 6, (Nov.-Dec. 2006) pp. 1596 - 1606, ISSN 1077-260X.

Kachurin, G.A., Rebohle, L., Skorupa, W., Yankov, R.A., Tyschenko, I.E., Floeb, H., Boehme, T., Leo, K. (1997). Short-wavelength photoluminescence of SiO2 layers implanted with high doses of Si+, Ge+, and Ar+ ions. *Semiconductors*, Vol. 32, No.4, (1998), pp. 392–396, ISSN 1063-7826.

Kim, S.-K., Cho, C.-H., Kim, B.-H., Park, S.-J., & Lee, J. W. (1962). Electrical and Optical Characteristics of Silicon Nanocrystals Solear Cells. *Appl. Phys. Lett.*, Vol.95, No 14, (October 2009), 143120 (3), ISSN 0003-6951.

Khomenkova, L. , Korsunska, N., Stara, T., Venger, Ye., Sada, C., Trave, E., Goldstein, Y., Jedrzejewski, J., & Savir, E. (1967). Depth Redistribution of Components of SiOx Layers Prepared by Magnetron Sputtering in the Process of Their Decomposition. Thin Solid Films, Vol. 515, No.17, (June 2007), pp.6749 - 6753, ISSN 0040-6090.

Klaumunzer, S. (1983). Modification of Nanostructures by High-Energy Ion Beams. *Nucl. Instrum. Methods Phys. Res. B*, Vol.244, No.1 , (March 2006) , pp.1- 7, ISSN 0168-583X.

Klimov, V. I., Ivanov, S. A., Nanda, J., Achermann, M., Bezel, I., McGuire, J. A. , & Piryatins, A. (1869). Single-Exciton Optical Gain in Semiconductor Nanocrystals, Nature Vol. 447, (May 2007) pp.441 - 446, ISSN: 0028-0836.

Kumar, H., Ghosh, S., Avasthi, D.K., Kabiraj, D., Mücklich, A., Zhou, S., Schmidt, H., & Stoquert, J.-P. (2006). Ion beam-induced shaping of Ni nanoparticles embedded in a silica matrix: from spherical to prolate shape. *Nanoscale Research Letters*, Vol. 6, No.1, (February 2011), 155 (9), SSN: 1556-276X.

Mathiot, D., Schunck, J.P., Perego, M., Fanciulli, M., Normand, P., Tsamis, C., & Tsoukalas, D., (1931). Silicon self-diffusivity measurement in thermal SiO_2 by

$^{30}Si/^{28}Si$ isotopic exchange. *J. Appl. Phys.*, Vol. 94, No.3, (May 2003), pp.2136-2138, ISSN 0021-7989.

Meftah, A., Brisard, F., Constantini, J.M., Dooryhee, E., Hage-Ali, M., Hervieu, M., Stoquert, J.P., Studer, F. & Toulemonde, M. (1893). Track Formation in SiO_2 Quartz and the Thermal-Spike Mechanism. *Phys. Rev.B*, Vol. 49, No. 18, (May 1997), pp. 12457-12463, ISSN 1098-0121.

Michalet, X., Pinaud, F. F., Bentolila, L. A., Tsay, J. M., Doose, S., Li, J. J., Sundaresan, G., Wu, A. M., Gambhir, S. S. & Weiss, S., (1880). Quantum Dots for Live Cells, in Vivo Imaging, and Diagnostics. *Science*, Vol. 307, (January 2005), pp.538-544, ISSN 0036-8075.

Mishra, Y.K., Singh, F., Avashthi, K., Pivin, J.C., Malinovska, D., & Pippel, E. (1962). Syntensis of elonged Au nanoparticles in silica matrix by ion irradiation, *Appl. Phys. Lett.*, Vol. 91, No. 6, (August 2007), 063103(3), ISSN 0003-6951.

Moskalenko, A.S., Beracdar, J., Prokofiev, A.A., & Yassievich, I.N. (1893). Single-Particle States in Spherical Si/SiO_2 Quantum Dots. *Phys. Rev.B*, Vol. 76, No.8, (2007), 085427(9), ISSN 1098-0121.

Muller, T., Heining, K.-H., & Moller, W. (1962). Size and Location Control of Si Nanocrystals at Ion Beam Synthesis in Thin SiO_2 Films. *Appl. Phys. Lett.*, Vol. 81, No 16, (August 2002), pp. 3049-3051, ISSN 0003-6951.

Norris, D.J.; Efros, A.L.; & Erwin S.C. (1880). Doped Nanocrystals. *Science*, Vol.319, (March 2008), pp.1779-1779, ISSN 0036-8075.

Punchaipetch, P., Ichikawa, K., Uraoka, Y., Fuyuki, T., Tomyo, A., Takahashi, E., & Hayashi, T. (1983). Experimental Investigation of Tunnel Oxide Thickness on Charge Transport Through Si Nanocrystal Dot Floating Gate Memories. *Vac. Sci. &Technol. B*, Vol. 24, No.3, (April 2006), pp. 1271-1277, ISSN 1071-1023.

Rizza, G., Ramjauni, Y., Gacoin, T., Vieille, L., & Henry, S. (1893). Chemically Synthesized Gold Nanoparticles Embedded in a SiO_2 Matrix: a Model System to Give Insights into Nucleation And Growth Under Irradiation. *Phys. Rev. B*, Vol. 76, No.24, (December 2007 a), 245414 (9) , ISSN 1098-0121.

Rizza, G., Chevery, H., Gacoin, T., Lamasson, A., & Henry, S. (1931) Ion Beam Irradiation of Embedded Nanoparticles: Toward an *in Situ* Control of Size and Spatial Distribution. *J. Appl. Phys.*, Vol. 101, No.1, (January 2007 b), 014321(7), ISSN 0021-7989.

Rizza, G.; Dawi, A. E. A.; Vredenberg, M., & Monnet, I. (1962). Ion Engineering of Embedded Nanostructures: from Spherical to Facetted Nanoparticles. *Appl. Phys. Lett.*, Vol.95, No.4, (July 2009), 043105(3), ISSN 0003-6951.

Sa'ar, A., Reichman, Y., Dovrat, M., Krapf, D., Jedrzejewski, J., & Balberg, I. (2001). Resonant Coupling between Surface Vibrations and Electronic States in Silicon Nanocrystals at the Strong Confinement Regime. *Nano Lett.* Vol.5, No.12, (November 2005), pp.2443-2447, ISSN 1530-6984.

Schmidt, B., Mucklich, A., Rontzsch, L., & Heining, K.-H. (1983). How Do High Energy Ions Shape Ge Nanoparticle Embeded into SiO_2?, *Nucl. Instrum. Methods Phys. Res. B*, Vol. 257, No.1-2 , (April 2007) , pp. 30-32, ISSN 0168-583X.

Schmidt, B., Heining, K.-H., Mucklich, A., & Akhmadaliev, C., (1983). Swift-heavy-ion-induced shaping of spherical nanoparticales into disks and rods. *Nucl. Instrum. Methods Phys. Res. B,* Vol. 267, No.8-9 , (May 2009) , pp. 1345-1348, ISSN 0168-583X.

Shamin, S. N., Galakhov, V. R., Aksenova, V. I., Karpov, A. N., Shvartz, N. L., Yanovitskaya, Z. Sh., Volodin, V. A., Antonova, I. V., Ezhevskaya, T. B., Jedrzejewski, J., Savir, E., & Balberg, I. (1997). X-Ray and Infrared Spectroscopy of Layers Produced by Co-Sputtering of Spatially Separated SiO_2 and Si Sources. *Semiconductors,* Vol. 44, No. 4, (April 2010), pp. 531-536, ISSN 1063-7826.

Shi, J., Wu, L., Huang, X., Liu, J., Ma, Z., Li, W., Li, X., Xu, J., Wu, D., Li, A. & Chen, K. (1963). Electron and Hole Charging Effect of Nanocrystalline Silicon in Double-Oxide Barrier Structure. *Solid State Commun.* Vol.123, (August 2002), pp. 437-442, ISSN: 0038-1098.

Shklovskii, B.I., & Efros, A.L. (1984). *Electronic Properties of Doped Semiconductors.* Springer, ISBN-10 0387129952, Berlin, Germany.

Takahashi, T., Fukatsu, S., Itoh, K., M., Uematsu, M., Fujiwara, A., Kageshima, H., Takahashi, Y., & Shiraishi, K., (1931). Self-Diffusion of Si in Thermally Grown SiO_2 under Equilibrium Conditions, *J. Appl. Phys.,* Vol. 93, No.6, (March 2003), pp. 3674-3676, ISSN 0021-7989.

Toulemonde, M., Dufour, Ch., Meftah, A., Paumier, E. (1983). Transient Thermal Processes in Heavy Ion Irradiation of Crystalline, Inorganic Insulators. *Nucl. Instrum. Methods Phys. Res. B,* Vol.166-167, (May 2000) , pp.903-912, ISSN 0168-583X.

Tsoukalas, D., Tsamis, C., & Normand, P., (1931). Diffusivity Measurements of Silicon in Silicon Dioxide Layers Using Isotopically Pure Material. *J. Appl. Phys.,* Vol. 89, No.12, (March 2001), pp.7809 - 7813, ISSN 0021-7989.

Van Dillen, T., Polman, A., Onck, P.R., & Van der Giessen, E. (1893). Anisotropic Plastic Deformation by Viscous Flow in Ion Tracks. *Phys. Rev.B,* Vol. 71, No. 2, (January 2005), 024103(12) , ISSN 1098-0121.

Van Dillen, T., de Dood, M.J.A., Penninkhof, J.J., Polman, A., Roorda, S., & Vredenberg, A.M, (1962). Ion Beam-Induced Anisotropic Plastic Deformation of Silicon Microstructures. *Appl. Phys. Lett.,* Vol. 84, No 18, (May 2004), pp.3591-3593, ISSN 0003-6951.

Wu, L., Dai, M., Huang, X., Li, W., & Chen, K. (1983). Size-Dependent Resonant Tunneling and Storing of Electrons in a Nanocrystalline Silicon Floating-Gate Double-Barrier Structure. *J. Vac. Sci. &Technol. B,* Vol. 22, No.2, (March 2004), pp. 678-681, ISSN 1071-1023.

Yu, L.W., Chen, K.J., Wu, L.C., Dai, M., Li, W., & Huang, X.F. (1893). Collective Behavior of Single Electron Effects in a Single Layer Si Quantum Dot Array at Room Temperature. *Phys. Rev. B,* Vol. 71, No. 24, (June 2005), 245305(5) , ISSN 1098-0121.

Zhou, X., Uchida, K., & Oda, S. (1962). Current Fluctuations in Three-Dimensionally Stacked Si Nanorystals Thin Films. *Appl. Phys. Lett.*, Vol. 96, No 9, (March 2010), 092112(3), ISSN 0003-6951.

9

Neon and Manganese Ion Implantation into AlInN

Abdul Majid
Department of Physics, University of Gujrat
Pakistan

1. Introduction

This chapter deals with the experimental study of ion implantation into wurtzite AlInN, grown by metal organic chemical vapor deposition (MOCVD) technique. III-nitride alloys (AlGaN, InGaN and AlInN) are the materials which provide tunable energy band gap, effective mass of the carriers, lattice and dielectric constants depending upon the alloy composition. Unlike AlGaN and InGaN, little attention has been paid to the study of AlInN because of the growth related problems. The first ever AlInN layer in the form of polycrystalline film was grown and reported by Starosta in 1981 when he fabricated such layers by means of radio-frequency (RF) reactive multi-target sputtering [1]. Later, Kubota et al reported high quality growth of $Al_{1-x}In_xN$ layers by RF magnetron sputtering in 1989 and demonstrated the compositional dependence of band gap and lattice constant in the whole composition range x = 0 to 1 [2]. Following these earliest efforts, many groups engaged themselves for the growth and characterization of AlInN layers [3, 4].

AlInN is a potential direct wide band gap compound semiconductor material for high power, high frequency and high temperature applications in electronic and optoelectronic devices due to its exceptional physical and chemical properties [5, 6]. The property which makes III-V nitrides prominent over the other compound semiconductors is the availability of wide band gap range from 0.7 eV (InN) to 6.2 eV (AlN), covering emission wavelengths from infrared (IR) to UV [7]. AlInN is the only III-nitride alloy which has ability to be grown perfectly lattice matched to GaN for indium content of 17-18 % [2]. In lattice matched composition, AlInN has larger band gap and higher refractive index with respect to GaN which makes this alloy attractive for several photonic applications [8, 9].

The fabrication of many of the devices mentioned above requires efficient, controlled and selective area doping. Although doping during the growth of materials can be performed but doping by ion-implantation offers several advantages, which can not be achieved by *in-situ* doping of III-nitrides. The low solubility of transition metal ions into III-V semiconductors (>10^{18}cm^{-3}) has put limits on their doping for achieving variety of magnetic semiconductor materials. Fortunately, ion implantation equipped with several added advantages has potential to act as an alternate technique [10]. Incorporation of any dopant atom at any desired depth of material with known concentration above solid solubility limits and selective area doping are advantages of doping by ion-implantation [6]. Furthermore it offers additional processing benefits like, electrical isolation, dry etching, quantum well intermixing and ion cut etc. Along with several unique benefits of ion implantation there are disadvantages like lattice damages and production of new defects [11]. A well established remedy to remove the implantation induced damages is the thermal annealing of the samples. The optical, electrical and magnetic activation of dopants is also achieved by the thermal annealing. There are several parameters that have to be taken into account for successful optimization of annealing to recover the lattice, e.g. annealing time, annealing temperature, annealing atmospheres, sample thickness, cap layers and implantation parameters.

In order to enhance the functionality of devices, scientists have shown a great enthusiasm to exploit the spin of electrons [12]. The communication between microelectronic devices occurs by the flow of charges; however the movement of spin can also be used to carry information among devices [13]. This has opened the possibility to utilize the charge and spin degrees of freedom simultaneously, to realize a new generation of electronic devices known as spin-electronic or spintronics [14]. The diluted magnetic semiconductors (DMS) are a class of materials in which small quantity of magnetic ions is introduced into normal semiconductors, are found suitable for spintronic device applications [15]. The most investigated DMS are (GaMn)As and (InMn)As but these suffer from the limitation of Curie temperature and disqualify for use in the practical spintronic devices [16, 17]. Dietl in year 2000 have predicted the possibility of Curie temperature above room temperature in GaN based DMS and provided new basis for DMS research [18]. Extremely high Curie temperature of around 940 K has been reported in 2002 for wurtzite GaN doped with Mn [19]. Unlike other nitride alloys, only few reports are available in the literature, on ion implantation into AlInN [20, 21]. The production and variation of residual strain caused by ion implantation in semiconductor crystals has been observed [22]. Lorenz et al have reported that the strain state of $Al_xIn_{1-x}N/GaN$ switches from tensile to the compressive when ε^T shifts from negative to positive on increasing the indium content beyond 17% [23, 24] have reported the electrical isolation of implanted AlInN but no report is available on the other areas like doping, microstructural changes and defect studies in this implanted material.

2. Background

2.1 Group III-Nitride semiconductors

Group III-Nitride semiconductors are the compounds of metals from group-III with nitrogen (group-V) which offer the binaries GaN, InN, AlN ternaries AlInN, AlGaN, InGaN and quaternary AlInGaN. The strong polarity of III-N bond makes these materials chemically stable and physically tolerant for the devices to work at high temperature, high frequency and in hostile environment [25]. These materials have earned the extensive research and industrial interest due to their promising technological capabilities for the electronic and opto-electronic devices [26]. Current applications of III-nitrides include ultraviolet/visible laser diodes, ultra-bright LEDs, UV detectors, high temperature electronics, high-density optical data storage, aerospace and automobiles technologies [6]. A variety of growth techniques have been used to achieve III-nitrides since 1862 when AlN powder was first made [27].

The choice of a lattice matched and thermally compatible substrate is a critical requirement for the homoepitaxial growth of semiconductors. Unfortunately, no such substrates are available for III-nitrides which have made heteroepitaxial growth, a practical necessity that puts limits on the quality of these materials. Along with the lattice and thermal compatibility, substrate should have some other qualities like, reasonably flat surface with minimum size of two inch diameter and ability to withstand high concentration of ammonia and hydrogen at elevated temperatures. Many substrate materials like Si, NaCl, GaP, InP, W, ZnO, $MgAl_2O_4$, TiO_2, MgO, SiC, Al_2O_3 etc have been tested for the heteroepitaxial growth of III-nitride [5]. However, sapphire and silicon carbide are the most popular materials for III-nitride heteroepitaxy on the basis of the pre-suppositions mentioned above.

III-nitrides can be crystallized into wurtzite, zincblende and rocksalt structures. However, under ambient growth conditions, wurtzite is the thermodynamically stable crystalline structure for these materials. The wurtzite structure has hexagonal unit cell containing 6 atoms of each type, with in-plane and out-of-plane lattice parameters 'a' and 'c' respectively. The space group of wurtzite structure is $P6_3mc$ (C_{6v}^4). The wurtzite crystal structure consists of two interpenetrating hexagonal closed pack sub-lattices, off-set along c-axis by 5/8 of the cell height. Each group-III atom is coordinated by four nitrogen atoms in tetrahedral arrangement.

2.2 Aluminum indium nitride (AlInN)

Aoming III-nitride alloys AlGaN, InGaN and AlInN, little attention has been paid to the study of AlInN because of the growth related problems. AlInN layers suffer crack formation and phase segregation due to large mismatch between InN and AlN. Following these earliest efforts [1, 2] many groups engaged themselves for the growth and characterization of AlInN layers [3, 28] A schematic structural sketch of AlInN is shown in figure 1.

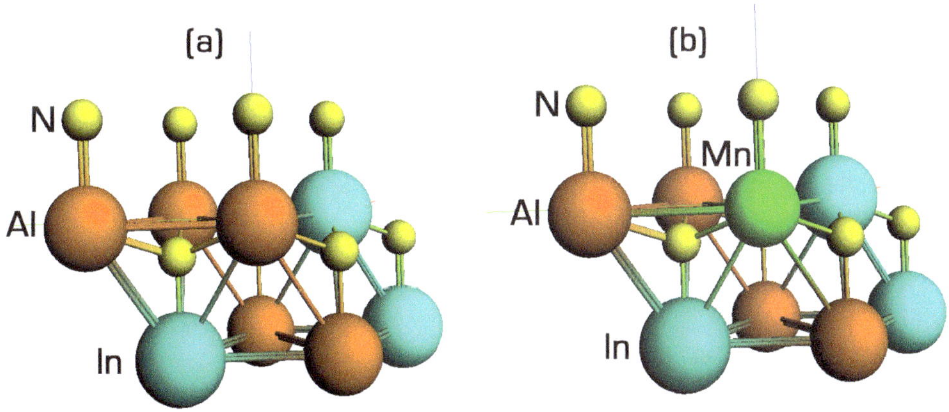

Fig. 1. Crystalline structure of AlInN (a) pure (b) Mn doped

The energy band gap of AlInN can be tuned at any value between the band gaps of InN (0.63 eV) [ref. 29] and AlN (6.2 eV) [ref. 30] which makes the material a promising candidate for several electronic and optoelectronic applications. Owing to the unique qualities, AlInN based structures are expected suitable for AlInN/GaN superlattice that can be used for applications in near IR devices based on inter-sub-band transitions [31], distributed Bragg reflectors (DBRs) [8], high quality factor microcavities [32] and high electron mobility transistors (HEMTs) [31]. The composition 'x' of $Al_{1-x}In_xN$ alloy is usually found from a linear interpolation between the relaxed bulk lattice constants of pure AlN (a=3.112 Å, c=4.982 Å) and InN (a=3.548 Å, c=5.760 Å) by assuming a linear relationship between the ternary composition and its lattice parameters. This method is known as Vegard's law [33]. However, it is suggested to use the Vegard's law carefully due to the phenomenon of band gap bowing, which gives the alloy composition lower than that calculated from Vegard's law [34]. The wurtzite $Al_{1-x}In_xN$ is a material having direct energy gap E_g of $0.7 < E_g < 6.2$ eV at 300 K [29]. The lattice parameters 'a' and 'c' lie in the range $3.112 < a < 3.548$ Å and $4.982 < c < 5.760$ Å respectively. The band gap and lattice parameters can be tuned by changing the indium mole fraction 'x' in the alloy. The band gap bowing phenomenon observed due to the deviation from Vegard's law is usually quantified by a bowing parameter 'b' which can be estimated by using the relation [35].

$$E_g(eV) = xE_{g,InN} + (1-x)E_{g,AlN} - bx(1-x)$$

High resolution X-ray diffraction has been proved to be successful in strain measurements of epitaxial films [36, 37]. The 'a' and 'c' lattice constants are usually found from skew-symmetric and symmetric XRD reflections respectively. Both the biaxial in-plane and out-of-plane strain can be determined by using the relations $\varepsilon^{\parallel} = \frac{a - a_o}{a_o}$ and $\varepsilon^{\perp} = \frac{c - c_o}{c_o}$ respectively, where a_o and c_o are the respective in-plane and out-of-plane lattice constants of unstrained relaxed bulk material [38].

2.3 Ion implantation into III-nitrides

Ion implantation into semiconductors offers a multipurpose processing tool used for the selective area doping, electrical isolation, dry etching, quantum well intermixing and ion cut etc [39]. William Shockley was the first who introduced ion implantation for doping of desired atoms into semiconductor materials. He got his technique registered as a patent in 1954 and later first commercial Ion-implanter was launched to the market during 1970s. Figure 2 highlights the suitable elements in the periodic table for implantation into semiconductors [10].

2.4 Ion implantation into AlInN

Unlike other nitride alloys, AlInN has not gain a considerable research interest from the implantation point of view, possibly due to its poor crystalline quality. The change in strain caused by the implantation and thus variation of the tetragonal distortion ($\varepsilon^T = \varepsilon^{\parallel} - \varepsilon^{\perp}$) provides valuable information about the structural properties of implanted material. The production and variation of residual strain caused by ion implantation in semiconductor crystals has already been observed [22]. Lorenz et al have reported that the strain state of $Al_xIn_{1-x}N$/GaN switches from tensile to the compressive when ε^T shifts from negative to positive on increasing the indium content beyond 17% [23]. Pearton et al [24] have reported the electrical isolation of implanted AlInN but no report is available on the other areas like doping, microstructural changes and defect studies in implanted material.

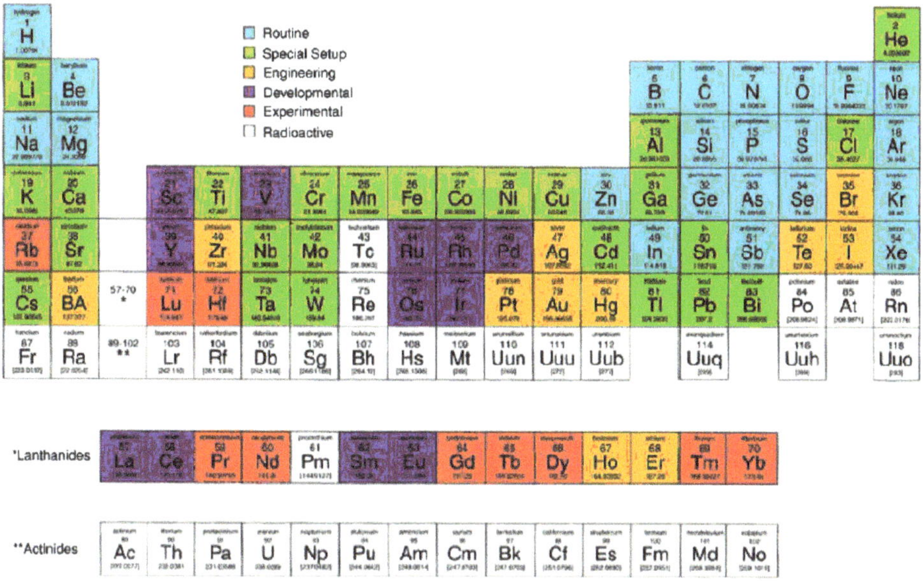

Fig. 2. Elements of periodic table used for ion implantation [10]

Despite promising benefits of ion implantation, this technique is still far away from its applications in GaN based devices. Prior to the use of implantation for a certain purpose, one has to optimize the implantation and thermal annealing conditions. Besides this, the implantation of different ions has shown different effects on different materials, therefore detailed investigations of implantation effects on AlInN are needed to explore the possibilities of the expected advantages.

A technologically well known phenomenon is the impurity gettering which has become a mature technique for silicon industry to getter the unwanted transition metal impurities. It is well known that inert gases produce cavities or bubbles in solids which contain high density of dangling bonds that exhibit high affinity for metallic contaminants to act as gettering centers. The formation of inert gas cavities in GaAs, SiC, InP etc urged us to implant the neon ions into AlInN which is still lacking in literature.

Until now, appropriate annealing conditions have not been established for the treatment of the implanted AlInN; therefore, different researchers have reported different results for same material and implantation conditions. The magnetic properties of the implanted semiconductors strongly depend upon annealing conditions. Manganese is the mostly used magnetic impurity in semiconductors to realize the DMS. Although few report are available for ion implantation into AlInN but no one has focused the study of effects of annealing on its properties. Therefore, a lot of work is still needed to optimize the annealing parameters for ion implanted AlInN.

Many groups have reported ferromagnetism in GaN, InN, AlN and related alloys AlGaN and InGaN by incorporating magnetic ions but there is no report on AlInN based DMS. To realize all possible benefits of nitride based DMS, ferromagnetic properties of all III-Nitride members should be known. AlInN is expected to be a promising DMS material after doping of magnetic ions and possibly the future magnetic DBR for polarization stability in the spin VCSELs. This has motivated us to implant magnetic ions into AlInN to fabricate the room temperature DMS based on AlInN.

3. Experimental details

This section describes the fabrication/processing of the samples and the characterization techniques used during this study. First section of this chapter deals with the fabrication of samples which consists of epitaxial growth by using metal organic chemical vapor deposition (MOCVD) along with the post fabrication processing which comprises high energy ion implantation and thermal annealing. After fabrication, the samples were studied in detail by means of the structural, optical, electronic and magnetic characterizations with variety of techniques described in the second section of this chapter.

The preparation of the samples comprises three major steps; fabrication using MOCVD technique, implantation of energetic ions into the grown materials and thermal annealing of implanted samples.

3.1 MOCVD growth of samples

All the samples used in this study are grown in a low pressure MOCVD system. Some times this technique is also referred as Metal Organic Vapor Phase Epitaxy (MOVPE) or Organo Metallic Vapor Phase Epitaxy (OMVPE). During the fabrication of the samples, a mixture of gases (called precursor) containing the materials required for deposition of the sample is passed over the heated substrate. The chemical decomposition (Pyrolysis) of precursor gases takes place in high temperature chamber and the atoms (needed for deposition) then deposit onto the substrate thereby making the bonding to build a crystalline layer. The growth process is sketched below in figure 3.

MOCVD growth has several advantages over other growth techniques (e.g. MBE). The growth process in MBE is controlled only by substrate temperature and molecular flow rates, whereas, MOCVD offers a variety of parameters (chamber pressure, gas flow and wide range of precursors) which make it more versatile. The precursors consist of organic materials which are less toxic, liquid at room temperature and inexpensive for industrial mass production. Moreover, no ultrahigh vacuum is needed like in MBE.

Samples were grown in a horizontal MOCVD reactor manufactured by Thomas Swan Scientific Equipment Limited. Hydrogen was used as a carrier gas for the growth of the samples. The c-plane sapphire wafers (Al_2O_3) of 2 inch diameter ware used as substrates for growth of the materials. The samples used during this study were hexagonal thin films of GaN and AlInN fabricated in the form of GaN/sapphire and AlInN/GaN/sapphire respectively. During the fabrication of GaN, the first step was the growth of a low temperature thin GaN buffer layer on a sapphire substrate. Next step was the growth of a GaN epilayer without any doping (so called i-GaN) i.e. GaN template. In case of doped samples, silicon and magnesium were incorporated to produce n- and p-GaN respectively. Sketches of AlInN samples used in this study are given in figure 4. For the fabrication of AlInN samples the first step was the growth of GaN template as described earlier. After that AlInN epilayer was grown without any doping. The samples were then classified into two groups. One group was left as it is, for further processing whereas other group of the AlInN samples was capped with GaN cap-layer as shown in fig. 4 (b).

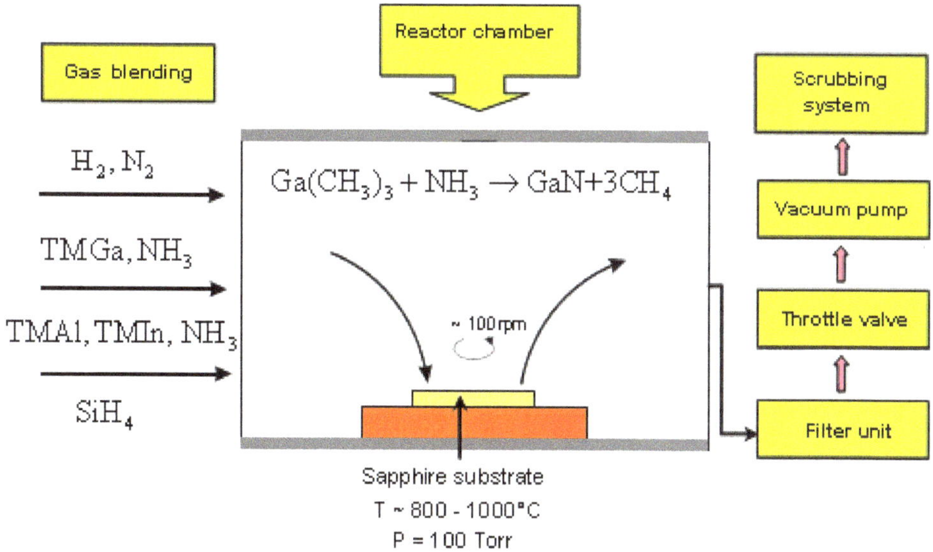

Fig. 3. Schematic diagram of MOCVD growth process.

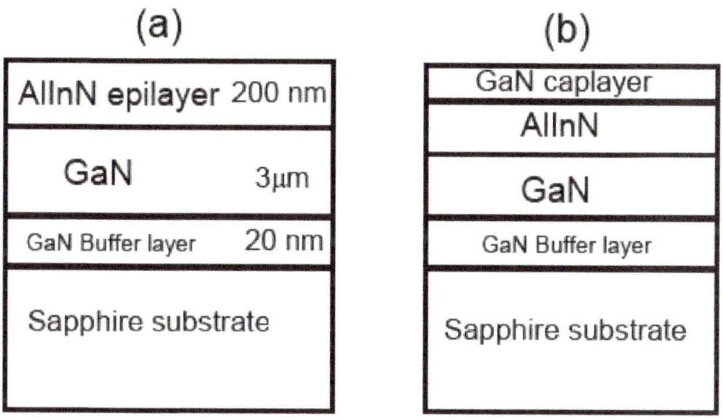

Fig. 4. Layered sketches of the AlInN samples used in this study.

Precursors for the MOCVD growth of III-nitrides consist of organic compounds involving elements from groups III and V along with a dopant precursor (if needed). A list of precursors used for growth of different samples is given below in table 1.

The samples with best structural and electrical properties were grown after the optimization of growth parameters. The growth conditions involve the chamber temperature, chamber pressure, precursors, precursor's flow rates, substrate rotation rate and growth time etc. Growth conditions of the samples used in this study are given in table 2.

Hall and X-Ray diffraction measurements were carried out to select the high quality wafers for further study. The samples of 1x1 cm^2 size were then carefully scribed from wafers to obtain different sets of samples and each set of samples scribed from the same wafer was used separately for implantation of different ions. The scheme of ion implantation is given in the next section.

Precursor	Chemical symbol	Advantages	Disadvantages
TMGa Tri-methyl gallium (Group III-Precursor)	$(CH_3)_3Ga$	liquid, high vapor pressure	Volatile and dangerous for contact to human body
TMIn Tri-methyl indium (Group III-Precursor)	$(CH_3)_3In$	solid, good vapor pressure for MOCVD	low vapor pressure
TMAl Tri-methyl aluminum (Group III-Precursor)	$(CH_3)_3Al$	liquid, good vapor pressure, good long term stability	oxygen contamination
Ammonia (Group V-Precursor)	NH_3	good stability, only practically available nitrogen precursor	high Pyrolysis temperature
Silane (n-dopant)	SiH_4	used as n-dopant for electronic devices	gaseous, flammable, high Pyrolysis temperature
Cp_2Mg Bis- (cyclopentadienyl) magnesium (p-dopant)	$(C_5H_5)_2Mg$	used as p-dopant and has low volatility	Volatile and dangerous for contact to human body

Table 1. List of precursors used for growth of samples.

No.	Sample and Purpose	Growth steps	Time (sec)	Temperature (°C)	Pressure (Torr)	NH_3 (ccm)	TMGa (ccm)	SiH4 (ccm)	cp2Mg (ccm)	TMIn (ccm)	TMAl (ccm)
1	AlInN (Used for Mn implantation)	1.GaN	850	1040	200	3500	51	0.05	----	----	----
		2.AlInN	5900	820	100	5200	N.A	N.A	N.A	200	20
2	AlInN (Used for Ce implantation)	1.GaN	850	1040	200	3500	51	0.05			
		2.AlInN	3550	820	60	5500	----	----	----	410	20
		3.GaN cap-layer	180	800	200	3000	30	----	----	----	----
3	AlInN (Used for Ne Implantation)	1.GaN	850	1040	200	3500	51	0.05	----	----	----
		2.AlInN	5200	820	90	4000	N.A	N.A	N.A	470	20
		3.GaN cap-layer	180	800	200	3000	30	----	----	----	----

Table 2. Growth conditions of AlInN samples used in this study.

3.2 Ion implantation

Ion implantation is a process in which impurity ions are projected into the target material to modify its structural, optical and electronic properties. It is relatively a simpler technique for introducing the ions into a semiconductor for doping, electrical isolation of active regions and device applications. Some of the several advantages of doping by ion implantation over the doping during growth are listed below.

1. It is an inexpensive and quick method to achieve the required material.
2. Any species of ion can be incorporated into any host material.
3. Ion can be introduced at the desired depth by the control of implantation energy.
4. Ions can be implanted into the desired areas of the sample for the selective area doping by the use of a proper mask system.
5. The required concentration of the ions can be implanted into the material after monitoring the implantation dose which is measured by the ion current.
6. Dopants can be incorporated above the solid solubility limits.
7. It is insensitive to the host properties, i.e. sample geometry, lattice structure, lattice defects etc.
8. It is a low temperature process.

The main parts of an ion implanter are the ion source, mass spectrometer, high voltage accelerators, scanning system and a target chamber. The material to be doped into the samples via implantation is filled in a cylindrical chamber in the form of gas. An electric potential of about 100 V is applied across the body of the chamber and the filament to generate electrons. The electrons emitted out of the filament of chamber ionize the gas atoms by impact ionization. The positive ions are accelerated towards the exit slit which is set at high negative potential of about 25 kV. The positive ions then enter the mass spectrometer which separates the ions of different masses and charge states to allow the desired ion species to pass through the exit slit of the spectrometer.

The ions selected by the spectrometer are accelerated by passing through the accelerating columns at high voltages. These ions are then passed through the pair of X and Y plates of a scanner system to produces a beam for uniform dose of implantation into the whole area of wafer. The scanned beam is passed through the defining aperture and then projected on to the wafer. Electrical contacts between the wafer and the metallic target holder facilitates the electron flow to neutralize the implanted ions received by the wafer. The ions received by the wafer are counted by the charge integrator which measures the time averaged swept beam current according to the relation,

$$Q = \int_0^t \frac{I}{nqA} dt$$

Where 'A' is the wafer area, 'n' is 1 for singly and 2 for doubly ionized atoms, and 'I' is the electron current integrated over total time of implantation. The wafer is biased at small positive voltage to minimize the secondary electron's dose errors.

The energy with which the ions are accelerated towards the samples depends upon the applied electrostatic potential and is known as implantation energy. It depends upon the implanter set-up and is normally in the range, 1 to 2000 keV. The number of ions received by the target per unit area (ions/cm^2) is known as dose and is found electronically by using a charge integrator.

The incident ion on striking the crystal surface loses its energy through scattering events during coulombs interactions with target atoms. By ignoring the complications of many-body interactions, the process can be best appreciated by the supposition of binary collision. The projected ion collides with a target atom, which in turn recoils and collides with other lattice atoms to start a collision cascade. The route of penetrating ion mainly consists of point defects (vacancy, interstitial, substitutional, antisite etc). The energy loss mechanism can be classified in two ways: the elastic or nuclear energy loss (coulomb interaction) and inelastic or electronic energy loss (electron-electron energy transfer, excitation and ionization phenomenon) [6].

The total distance covered by the incident ions normal to the target surface before coming to rest is called range of ions. It is usually denoted by R and is given by

$$R = \int_0^E \frac{dE}{NS(E)}$$

Where N is the atomic density, dE is the energy loss of ion and S(E) is the stopping cross-section.

The mean penetration depth of the ions below the target surface is an important parameter and is known as projected range. This is denoted by R_p and it is smaller than the actual distance R traveled by ions. The relation between R and R_p is approximated as

$$R_P \approx \frac{R}{1 + \frac{m_2}{3m_1}}$$

Where m_1 and m_2 are the masses of incident and target atoms respectively.

The distance traveled by the ions along the normal to the incidence direction is referred as the lateral range and is denoted by $R_\perp$, whereas the spread in ion range caused by the fluctuations in energy loss mechanism is known as straggling range. Some ions stop at depth less than R_p and some at depth greater than R_p. Therefore, standard deviation or straggling (usually denoted by ΔR_p) is used to find the distribution around R_p. The range and distribution of implanted ions into the target have been described by Lindhard, Scharff and Schiott (LSS) theory [40], according to them the distribution of implanted ions follows Gaussian function.

The concentration of implanted ions at a certain depth x can be written as

$$N(x) = N_{\max} e^{-\frac{(x-R_P)^2}{2\Delta R_P^2}}$$

Where N_{max} is the peak concentration of implanted ions and is given by

$$N_{\max} = \frac{\Phi}{\sqrt{2\pi}\Delta R_P} \approx \frac{0.4\Phi}{\Delta R_P}$$

Where Φ is the ion dose in ion/cm^2, using this value of N_{max} we can find N(x) as,

$$N(x) = \frac{\Phi}{\sqrt{2\pi}\Delta R_p} e^{-\frac{(x-R_p)^2}{2\Delta R_p^2}}$$

A more exact and quick way to find the range, distribution and damage analysis of implanted ions into target is the computer simulations. Transport and range of ions in matter (TRIM) is a comprehensive Monte Carlo binary collision computer simulation based on full quantum mechanical treatment of ion-atom collisions [41]. It is capable of providing the distribution of the ions as well as all kinetic phenomena associated with the ion's energy loss (target damage, sputtering, ionization, and phonon production) in the multilayered complex target materials.

Though Gaussian distribution is a reasonable approximation of the depth profile of implanted ions but this is true mostly for the amorphous targets. Due to higher crystallinity of semiconductor crystals, incident ion if implanted parallel to the crystal axis, moves through the crystal planes without facing any collision and most probably channel into the substrate. To avoid such channeling effect implantation is usually carried out in such a way that target crystal is tilted 7° off the direction of incident ions.

No.	Sample name	Material	Implanted ion	Implantation energy (KeV)	Implantation angle (Degree)	Beam current (μA)	Implantation dose Ions/cm²
1	TH070207E	AlInN	Neon	250	7	0.4	$1x10^{14}$
						1.8	$2x10^{15}$
						1.8	$5x10^{15}$
						1.8	$9x10^{15}$
2	TH070206C	AlInN	Manganese	250	7	0.6	$1x10^{14}$
						5.5	$2x10^{15}$
						7	$5x10^{15}$
						10	$9x10^{15}$
						14	$2x10^{16}$
						14	$5x10^{16}$

Note: Implantation is carried out at room temperature.

Table 3. Implantation conditions used for this study.

LC-4 high energy ion implanter installed at Institute of semiconductor, Chinese academy of science, Beijing, China was used for the implantation of the samples used for the work reported in this thesis. This implanter is manufactured by Institute No. 48, China Electron Science and Technology group. Implantation conditions for different samples are given in the table 3.

3.3 Thermal annealing

The energetic ion-implantation into semiconductor lattice leads to the production of defects, damages, local amorphous regions and disordered crystalline structures. In spite of the several benefits, ion implantation damages the lattice that causes quenching of the luminescence, band filling, band gap narrowing and band tailing effects. Such effects severely affect the electrical and optical properties of the devices. A necessary step after implantation of the samples is the thermal annealing, which not only recovers the lattice but also moves the implanted ions to a suitable place for the electrical, optical and magnetic activation of the dopants.

Rapid thermal annealing (RTA) is a process in which semiconductor samples are heated up to very high temperatures (up to 1000°C or even higher) for a short time (several seconds) in a controlled way. Rapid heating rates (1000°C per minute) and thus short time processing offered by RTA is its major advantage over the conventional furnace annealing (takes several hours). Moreover, RTA ensures comparatively high security, when sample decomposition is taken into account since sample is exposed to the high temperature for a very short interval of time. High temperature annealing of III-nitrides is usually performed in the nitrogen ambient to counter the loss of nitrogen from the surface of the samples. Rapid thermal annealing has been found an effective way to improve the crystal quality and repair the implantation induced damages in III-nitrides.

An RTA processor mainly consists of three major components, RTA chamber, a heating system and a temperature sensor.

RTA has been performed in an RTP-300 rapid thermal processor that uses 13 tungsten halogen lamps (1250 W each) as a heat source and a nitrogen gas flowing ambient. The temperature sensor used during RTA is K-type thermocouple embedded in the sample stage. The samples to be annealed have been placed inverted on another GaN wafer which is then put on the 4 inch silicon wafer positioned on the sliding sample holder in side the RTA chamber. After inserting the sample holder inside the chamber, window is tightly closed by the screws. RTA is programmed for the specified temperature and time.

Usually semiconductors are annealed up to the maximum temperature equal to 2/3 of the melting point of bulk martial. This suggests the use of annealing temperature of about 1800°C for GaN whose melting point is 2791°C. However, this temperature is normally higher than the growth temperature of the epitaxial semiconductors which may degrade the material due to its decomposition. Therefore, annealing temperature and time needs to be carefully optimized to keep a balance between the lattice recovery and the sample degradation. Many experiments were carried out by using the different annealing conditions for the search of optimized values of annealing time and temperature. In this study we have used the maximum annealing temperature of 1000°C for GaN and 850°C for AlInN with annealing time of 30 seconds.

3.4 Characterizations techniques

Several techniques were used for the characterization of as-grown and implanted samples to study the effects of ion implantation on the structural and magnetic properties of the materials. Following section describes the characterization techniques used for this study. Ion implantation produces modifications in the lattice of the materials which can be studied by performing the structural characterizations of the samples. In this study, X-Ray diffraction and Rutherford backscattering spectroscopy have been used for the study of the structural changes induced by the implantation and annealing of the samples.

3.4.1 X-ray diffraction

X-Ray diffraction (XRD) is very useful technique for the structural analysis of the crystalline samples. It offers an efficient way to identify the crystal structure and explore the lattice modifications in the implanted/annealed samples.

X-rays emitted out of the material as a result of inter shall transitions are used as probe for analysis in X-ray diffractometer. These rays are produced by the bombardment of high energy electrons on a copper target and then filtered to give a monochromatic beam of Cu-K_α radiation. When X-rays are incident onto a crystalline material they are diffracted from the planes of the crystal. The diffraction peak is obtained only when the distance traveled by the rays after reflection from the successive crystal planes differs by an integral multiple of wavelengths. This process takes place in accordance with the well known Bragg's equation, given below.

$$2d\sin\theta = n\lambda$$

Where d is interplaner spacing, θ is the incident angle, λ is the wavelength of incident X-rays and n is the order of diffraction. By changing the angle θ, the Bragg's conditions are satisfied and a strong reflection or XRD peak is obtained. The plot of the angular positions verses the intensities of the diffracted peaks produces a pattern, which is a characteristic of the material. The position of the peaks recorded in XRD spectrogram is compared with the peaks of known materials for the phase analysis of the samples.

XRD analysis of the samples was performed using Cu-K_α source of X-rays at room temperature by Rigaku SLX-1A X-ray diffractometer. Crystallinity of GaN and AlInN samples was studied in detail by performing $\omega/2\theta$ scans using double axis and triple axis diffraction. Furthermore, broadening of the peaks, tilt and twist features were also studied by the measurements of FWHM from ω-scans of high resolution XRD. Powder diffraction XRD has been carried out in 2θ range of 20 to 100° to study the phase analysis and detection of secondary phases in implanted samples.

3.4.2 Rutherford Backscattering Spectroscopy

Rutherford Backscattering Spectroscopy (RBS) is a versatile technique for finding the concentration of the atoms, thickness of the thin film, depth distribution of the atoms and the characterization of implantation induced damages in the lattice [23, 39]. RBS in angular mode also provide information about the substitution of dopants on regular sites of the lattice [42].

This technique works on the principle of Rutherford scattering. An energetic beam of helium ions ($^4He+$) is incident on the sample and then backscattered. The collision probability of incident alpha particle with the target atoms is determined by Rutherford cross-section. The elastic collision between incident alpha particles and target atoms suggests that the ratio of energy of the particles before and after collision is equal to the ratio of the masses of both particles. The energy of the backscattered alpha particles is analyzed, which contains the information about the atoms from which alpha particles are backscattered. According to the single scattering theory, an alpha particle faces only one large angle scattering before reaching the detector. This approximation helps to convert the energy scale into the depth scale of the sample within energy resolution of the detector. Lower is the energy of the backscattered alpha particles, deeper will be the detected atoms.

RBS measurements were performed in School of Physics, Peking University, Beijing, China. An ion beam of energy 2 MeV having diameter of 1 mm was used for this study. The samples were mounted on a two axis goniometer which has ability to align the sample at any desired angle with the incident beam. Chamber has two silicon detectors, one for the detection of particles backscattered at angle near to the incident beam and other for detection of particles backscattered at glancing angle to the sample surface. Glancing detector of resolution 18 keV was set at 165° relative to incident beam direction. This detector have aperture diameter of 5 mm and placed 80 mm away from the samples. The data obtained from RBS measurements was analyzed by using SIMNRA simulation.

3.4.3 Magnetic characterizations

The implantation of magnetic ions into AlInN is expected to make them diluted magnetic semiconductors. Therefore, magnetic characterizations of the magnetic-ion-implanted samples were carried out to search the room temperature ferromagnetism. Vibrating sample magnetometer and Superconducting quantum interference device were used for this purpose.

3.4.4 Vibrating sample magnetometry

This is a conventional technique for the magnetic characterization of the samples in which magnetic moments as a function of applied magnetic field and temperature can be recorded.

To perform the measurements, sample mounted on sample holder is placed in a uniform magnetic field and set to the mechanical vibrations at a known frequency through a vibrator. The vibrations produce a change in magnetic flux of the sample which induces a voltage in the pick up coils. This induced voltage is proportional to the magnetic moment of the sample.

DMS Model 4HF Vibrating Sample Magnetometer (VSM) of ADE Technologies having detection sensitivity 10^{-5} emu, was used to investigate the magnetic properties of the implanted samples. Magnetic hysteresis loops were recorded at room temperature for all the implanted samples using magnetic field 12000 Oe (Oe=oersted unit) in-plane and out-of-plane applied magnetic field of to saturate the samples. It was found that an easy axis for samples is the direction when the magnetic field is applied parallel to the sample plane (in-plane).

3.4.5 Superconducting quantum interference device magnetometry

Superconducting quantum interference device (SQUID) is a resourceful tool for the magnetic characterization of the samples. It offers a sophisticated way to detect the magnetic ordering by studying temperature dependence of magnetization (MT), field dependent magnetization (MH) and other properties of the samples. SQUID has very high sensitivity ($\sim 10^{-9}$ emu) for the detection of very weak magnetic moments. SQUID works on the principle of the electron-pair wave coherence and Josephson Effect. The flow of current across two superconductors separated by an insulated layer is known as Josephson's effect and current flowing is known as Josephson's current. A Josephon's junction consists of two superconducting coils, which are separated by a very thin insulating barrier, so that electrons can pass through it. SQUID magnetometer consists of a superconducting ring containing two Josephson's junctions in parallel, placed in magnetic field. In the presence of a magnetic field, a current starts to flow through the superconducting loop. When a ferromagnetic sample is placed between the superconductors in the presence of applied field the magnetic flux appears to change accordingly. This change of magnetic flux induces a current which changes the current circulating through the coil. The variation in the current helps to detect the magnetic moment of the material.

SQUID was used to study the MT and MH behaviors of the magnetic ions implanted samples. During MH analysis, hysteresis loops were recorded at 100 K and 300 K for implanted samples. MH loops were also recorded for un-implanted samples and this data was subtracted from that of implanted samples. For MT measurements, the sample was cooled up to 5 K without applied magnetic field. Then magnetic field of 500 Oe was applied and scan was run up to 350 K to record zero field cooled (ZFC) trace. The sample was then again cooled down to 5 K with the field of 500 Oe and magnetization was measured to get the field cooled (FC) trace. During magnetization measurements plane of sample was parallel to the applied magnetic field.

4. Neon implantation

4.1 Neon implantation into AlInN

The samples used for this study were ~200 nm thick AlInN thin films grown on a 4μm thick GaN template. AlInN films were finally capped with a very thin GaN layer to avoid the sample decomposition during high temperature annealing. The cap layer also helps to limit the penetration depth of incident ion within AlInN layer and to avoid their interaction with the buffer layer. Sketch of the samples is shown in figure 4 (b).The chamber temperatures during growth of GaN template, AlInN layer and GaN cap layer were 1040, 820 and 800°C respectively. The detailed growth parameters of these samples are given in table 2. The crystal quality and composition of the as-grown samples was checked by X-ray diffraction. FWHM value of $Al_xIn_{1-x}N$ peak measured by X-ray diffraction was 406 arcsec indicating its high crystalline quality. Vegard presented a way to estimate the composition of ternary alloys by assuming linear relation between composition and lattice parameters [33]. The composition of $Al_xIn_{1-x}N$ is usually found by Vegard's law by linear interpolation between relaxed bulk lattice constants of AlN and InN. The value of indium content in AlInN layer of as-grown sample estimated using the Vegard's law is 8.3 %.

Neon ions were implanted at 250 keV on four samples scribed from same wafer, with doses 10^{14}, $2x10^{15}$, $5x10^{15}$ and $9x10^{15}$ ions/cm^2. The samples were implanted at room temperature and incident angle of beam was set at 7° to the c-axis of samples. The implanted samples were annealed by RTA processor at 750°C for 30 sec in nitrogen ambient. Although, annealing temperature of 400°C above the growth temperature has been reported to be safe in implanted AlInN [20], but in general, annealing temperature below the growth temperature is preferred. We have annealed the neon implanted AlInN at temperature below the growth temperature.

4.2 Implantation induced strain in epilayers

Neon implanted AlInN thin films were characterized by XRD before and after annealing to study the effects of implantation and annealing on the structural properties of material [69].

Figure 5 shows the (0002) XRD curves of as-grown and neon implanted AlInN/GaN samples before annealing. The peak at $2\theta = 34.56^{o}$ is due to the signal from the GaN buffer layer and peak appearing at $2\theta = 35.51^{o}$ is due to the signal coming from AlInN epilayer of as-grown samples. After implantation, it has been observed that intensity of AlInN peak decreases with increasing dose. The position of the peak shifts systematically toward the lower angle with increase in dose up to $5x10^{15}$ cm^{-2}. In the scans measured after maximum dose $9x10^{15}$ cm^{-2}, AlInN peak is broad and its intensity is too small to measure its exact peak location. However it seems shifted slightly towards higher angle side as compared to its position for the samples implanted with dose $5x10^{15}$ cm^{-2}. An additional peak structure is also observed on lower angle side of GaN peak in the scans of all implanted samples. This peak structure broadens and becomes less intense with increasing dose of implantation, indicating the expansion of GaN lattice.

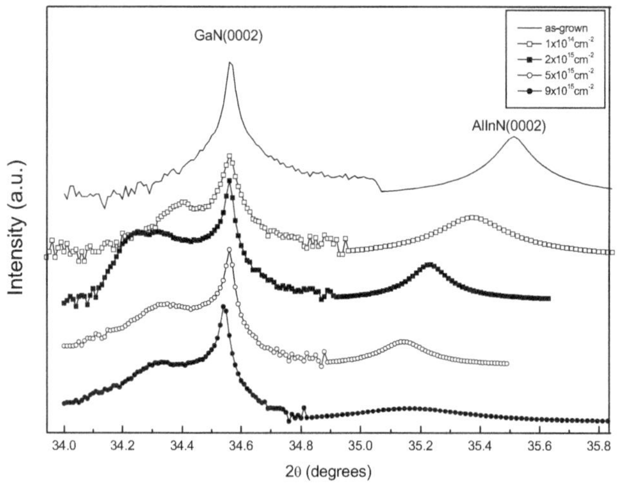

Fig. 5. X-Ray diffraction spectra showing (0002) peaks of GaN and AlInN for as-grown and neon implanted AlInN/GaN samples before annealing. AlInN peaks are shown Lorentz fitted.

Figure 6 shows the XRD spectra of same samples after annealing at 750°C. The post annealing spectra show a similar trend of GaN and AlInN peaks as observed before annealing. However the peak separation in each scan increases after annealing as compared to that of the separation in respective scans before annealing. This indicates that the position of AlInN related peak approaches to that of as-grown samples which points to lattice recovery after annealing. The broadness of the peak structure appearing on the left side of the main GaN peak is observed suppressed and shifted towards the higher angle side after annealing. The value of out-of-plane strain ($\varepsilon^{\perp}$) in AlInN layer of as-grown samples found from $\omega / 2\theta$ scan of (0002) symmetric XRD is $-2.66x10^{-2}$. In-plane strain calculated using relation $\varepsilon^{\|} = -(\frac{1-\nu}{2\nu})\varepsilon^{\perp}$ [ref. 43] is 2.7x10^{-2} (where ν is the Poisson ratio taken as 0.33). The percentage value of tetragonal distortion ($\varepsilon^{T} = \varepsilon^{\|} - \varepsilon^{\perp}$) calculated from in-plane $\varepsilon^{\|}$ and out-of-plane $\varepsilon^{\perp}$ strains is 5.36 ± 0.06.

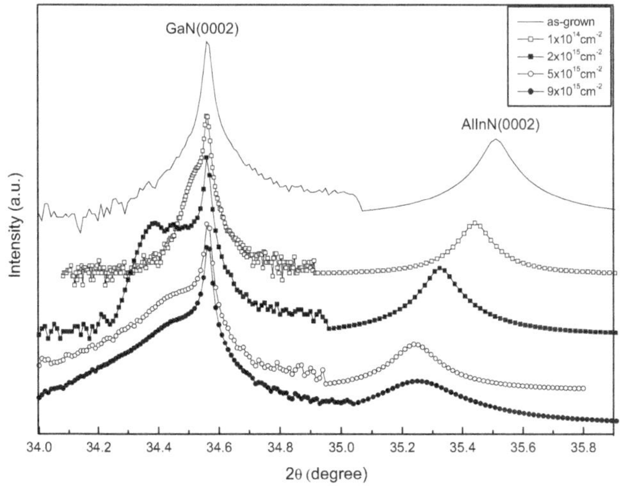

Fig. 6. X-Ray diffraction spectra showing (0002) peaks of GaN and AlInN for as-grown and neon implanted AlInN/GaN samples after annealing at 750°C for 30 sec. AlInN peaks are shown Lorentz fitted.

The positive value of ε^{T} indicates the presence of tensile strain [23]. Lorenz et al reported that the strain state of $Al_xIn_{1-x}N/GaN$ switches from tensile to the compressive when ε^{T} shifts from negative to positive with the increase in indium content beyond 17%. The out-of-plane lattice constant of AlInN layer calculated by $c^{\perp}_{AlInN} = (1+\varepsilon^{\perp})c_{GaN}$ for as-grown samples was found to be 5.047Å. The strain status (strained or relaxed) of hetro-epitaxial layers can be used to determine the indium content of $Al_xIn_{1-x}N$ samples [9].

Fig. 7 shows the comparison of dose dependence of absolute perpendicular strain of AlInN layer before and after annealing. The strain-dose curves display similar trend before and after annealing, but annealed samples exhibited comparatively less strain that suggests the partial relaxation caused by the redistribution of defects during annealing.

A consistent decrease in the intensity of AlInN peak was observed with increase in dose. This reduction of the intensity of Bragg reflections can be explained by Kinematic theory of scattering according to which the increase in lattice defects due to implantation is a consequence of decrease in scattering intensity [44]. Along with the intensity reduction, a clear shift in peak position AlInN was seen that gives information about the strain in the epilayer. To explain the strain accumulation we can divide the strain-dose curve into two regimes; (1) when strain is accumulating (2) when strain is dispersing. In the first regime, during lowest implantation dose of 10^{14}cm^{-2}, point defects (vacancies, interstitials) dominate due to atomic displacements. These defects are distributed within the lattice mostly away from one another for this dose. The tracks of penetrating ions will have mostly (interstitial, vacancy) pairs due to displacement of Al and In atoms from their regular sites. Vacancies are not only produced during implantation but also exist in as-grown samples. Most of the Ne ions are expected to be captured by vacancies at the end of their journey. The clustering of such occupied trapping sites leads to the formation of gas filled cavities called bubbles [45]. Defect density increases with the increase in dose which enhances the probability of interaction of defects with each other. This interaction facilitates the clustering of individual defects to produce the vacancy and interstitial clusters. Bubble size increases in this region due to addition of more gas atoms. All such events will increase the perpendicular strain and cause the lattice to expand accordingly. This situation persists until the dose approaches to $5x10^{15}cm^{-2,}$ where strain saturates. Interaction probability of newly produced defects with already existing defects is maximized at some particular defect density. Size of the biggest bubble has reached to a maximum possible value before breakup on further addition of gas atoms. In thermodynamic equilibrium internal gas pressure is compensated by the surface tension of the surrounding matrix [46]. At this stage, system has acquired a critical state and further increase in dose disturbs the equilibrium. The second regime starts in this situation, therefore $5x10^{15}cm^{-2}$ can be assumed as a critical dose for the system under study. This implantation dose seems to be the critical for providing a driving force for annihilation of vacancy and interstitial defects. Increase in the number of neon ions in the bubble increases the gas pressure and thus bubble gains critical size that triggers the bubbles to collapse; as a result strain reduces. The reduction in strain leads to the partial relaxation.

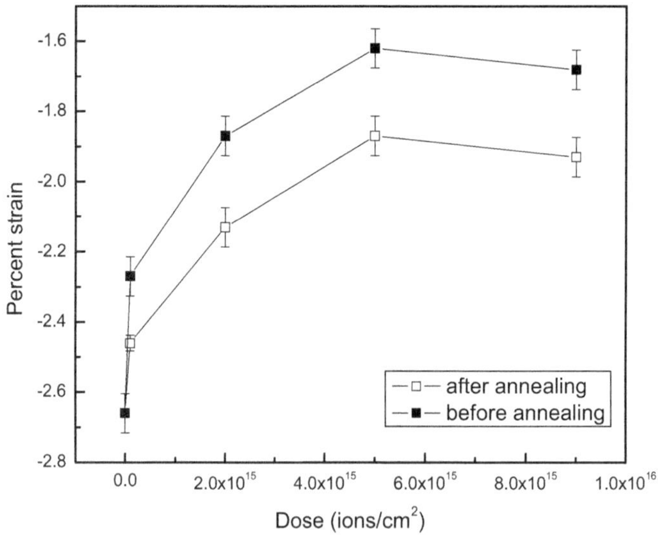

Fig. 7. Dose dependence of out-of-plane strain in neon implanted AlInN layer before and after annealing steps.

Defects become mobile during annealing which facilitates the annihilation of defects of opposite nature (e.g. vacancy and interstitial) and thus minimizes the thermodynamic free energy of the system. High temperature annealing also causes the bubbles to grow and they approach a critical size earlier than in the un-annealed samples [47]. This preferential growth of bubbles towards a critical size and their collapse is attributed to the low level of strain in the post implantation annealed samples as compared to as-grown samples. It is likely for the inert gas to out-diffuse from the bubbles during annealing and convert the gas filled bubbles into nano-sized cavities or voids [48]. These cavities have special significance to capture the mobile impurities and structural defects which results in the reduction of strain. It has been reported that cavities can be used as an effective source of the strain relaxation of semiconductors hetrostructures [49]. Annealed samples presented enhanced strain relaxation in comparison to that of un-annealed ones, which is attributed to the capturing of defects by cavities during gettering process [50].

4.3 Migration of indium atoms through interfaces

One would expect alloying at GaN/AlInN/GaN interfaces due to the atomic diffusion across both the interfaces caused by high energy implantation process that may lead to the formation of InGaN, AlGaN or InAlGaN in the top and bottom interface regions. If such new materials have formed and related lattices are thick enough to produce XRD signal then their signal may have superimposed on the AlInN related peak. The increase in width of AlInN related peak gives a clue of range of lattice parameters extended lattices. In case of alloying one would expect new peak/peaks which are not observed in the XRD measurements of our sample. It also appears unlikely that the signal due to such peak is hidden in the AlInN related peak in such a way that it behaves similar to AlInN peak with increase in dose. Similarly, one would also suspect that the peak appearing at lower angle side of GaN peak is due to possible alloying resulting from atomic diffusion across the interfaces. However, the similarity of such additional XRD peaks appearing in XRD spectra of neon implanted GaN/AlInN/GaN structures to that of neon implanted GaN samples having same implantation doses guides to rule out the possibility of such alloying. Hence it can be concluded that this peak is not due to alloying but purely due to expansion of GaN lattice. This observation also points that expected alloying has no contribution in broadening and shift of AlInN related peak. The observed features are due to implantation induced strain in the AlInN lattice.

5. Manganese implantation

A detailed study of manganese implantation into AlInN samples along with realization of diluted magnetic semiconductors (DMS) based on these materials is described in this chapter.

5.1 Manganese implantation into AlInN

The samples used in this study were AlInN/GaN/sapphire structures with ~200 nm thick AlInN epilayer and ~ 4 μm thick GaN templates. A sketch of layered structure of the samples is shown in figure 4 (a). The growth parameters of the samples are given in table 2. Mn ions were implanted at 250 keV into six samples scribed from the same wafer, with doses $1x10^{14}$, $2x10^{15}$, $5x10^{15}$, $9x10^{15}$, $2x10^{16}$ and $5x10^{16}$ ions/cm^2 at room temperature. Mn concentrations in implanted samples corresponding to these doses are ~ 0.01%, 0.2%, 0.5%, 0.9%, 2% and 5% respectively. Implantation was carried out at 7 degree with c-axis of the samples to avoid channeling into the substrate. Simulation of RBS measurements of highest dose samples indicated that implanted ions were limited in the top AlInN (200 nm thick) layer. After implantation, the samples were rapid thermal annealed (RTA) isochronically at 750 °C and 850 °C for 30 sec in the nitrogen ambient. During annealing samples were placed inverted on a GaN wafer to avoid the possible loss of nitrogen due to decomposition.

Mn implanted AlInN thin films were characterized by X-ray diffraction, Rutherford backscattering spectroscopy and superconducting quantum interference device magnetometer to study the structural and magnetic properties of material [70, 71]. The experimental results obtained from such measurements are described in the following sections.

5.2 Structual modifications

XRD (2θ = 20 –100°) measureemtns were carried out for phase analysis of the materials. Figure 8 shows the XRD spectra of as-grown and Mn implanted AlInN/GaN samples exhibiting the characteristic Bragg peaks from AlInN, GaN and sapphire substrate. This indicates the absence of any possible second phase in our implanted samples.

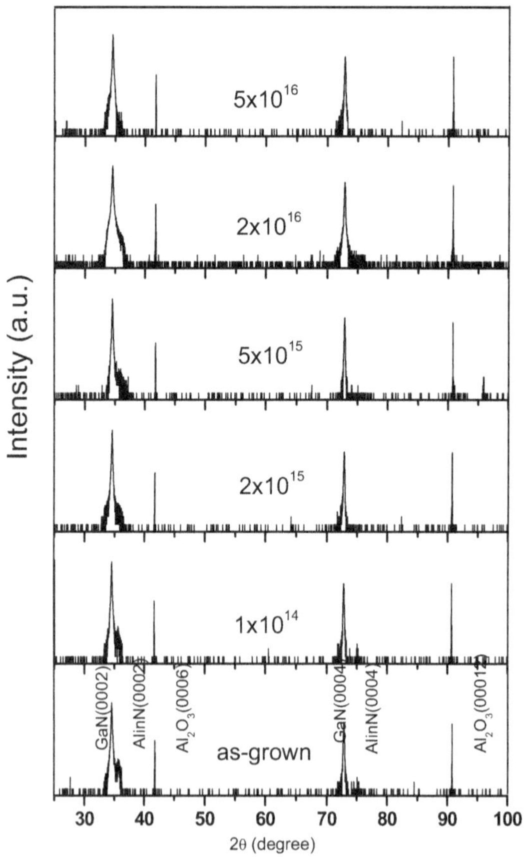

Fig. 8. (ω-2θ) XRD scans showing the peaks of GaN, AlInN and sapphire for as-grown and Mn implanted AlInN/GaN/sapphire samples after annealing at 750°C.

Figure 9 shows the (0002) symmetrical XRD spectra of as-grown and Mn implanted AlInN/GaN samples after annealing at 750 °C for 30 sec. GaN peak for all the samples appeared at 2θ=34.56°. AlInN peak was found at 2θ = 35.59° for as-grown samples but it appeared at different positions for the implanted samples. This AlInN peak shifted towards the lower angle side with increase in dose up to $5x10^{15}$ cm^{-2} and then it shifted back towards higher angle for higher doses. All the AlInN peaks, given in XRD spectra, were Lorentz fitted to extract the peak position.

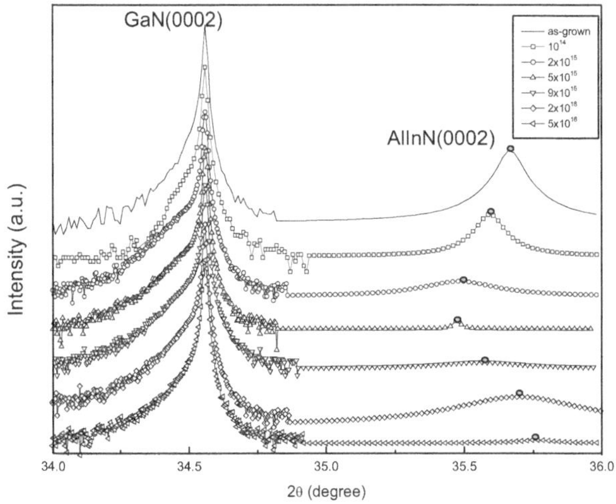

Fig. 9. X-Ray diffraction spectra from (0002) symmetric (ω-2θ) scans showing GaN and AlInN peaks of as-grown and Mn implanted AlInN/GaN samples after annealing at 750°C. AlInN peaks are shown Lorentz fitted.

Figure 10 shows the XRD spectra of as-grown and Mn implanted AlInN/GaN samples after subsequent annealing at 850 °C for 30 sec. A trend of shift in the position of AlInN peak, similar to the previous annealing, was observed for all the implanted samples. However, the increase in X-ray intensity of the peaks related to the GaN and AlInN was noted, which indicates the improved lattice recovery. No sizeable additional peak was seen on the lower angle side of GaN peak that indicates that either no or very small number of implanted ions have reached the buried GaN layer.

The perpendicular strain of implanted AlInN lattice relative to that of as-grown samples was calculated from the data extracted from the symmetric (0002) XRD scans by using the angular separation of substrate (GaN) and epilayer (AlInN) peaks. Figure 11 shows the perpendicular strain of AlInN layer for both the annealing steps. After first annealing at 750°C the strain increases with dose until it reaches to the maximum value for the sample implanted with dose of $2x10^{15}$ cm^{-2}. It remains constant for further increase in dose up to $5x10^{15}$ cm^{-2} and then starts decreasing with the increase in dose. The strain follows the same trend after second annealing at 850 °C with a minor difference; instead of being constant it starts decreasing immediately after reaching to a saturation value. Another difference observed is the higher strain for samples implanted with doses $2x10^{16}$ cm^{-2} and $5x10^{16}$ cm^{-2} after annealing at 850 °C as compared to that noted after annealing at 750 °C.

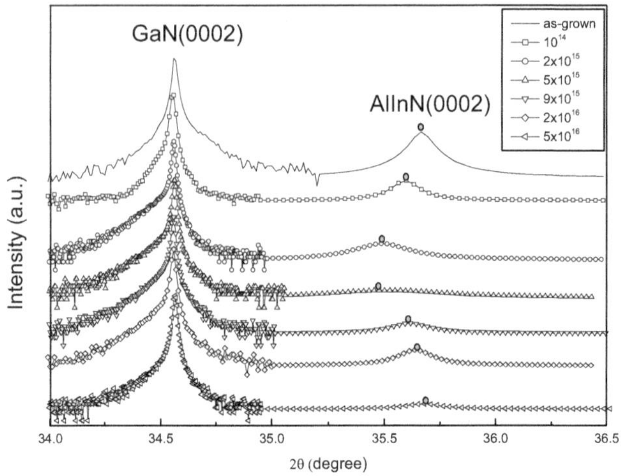

Fig. 10. X-Ray diffraction spectra from (0002) symmetric (ω-2θ) scans showing GaN and AlInN peaks for as-grown and Mn implanted AlInN/GaN samples after annealing at 850ºC. AlInN peaks are shown Lorentz fitted.

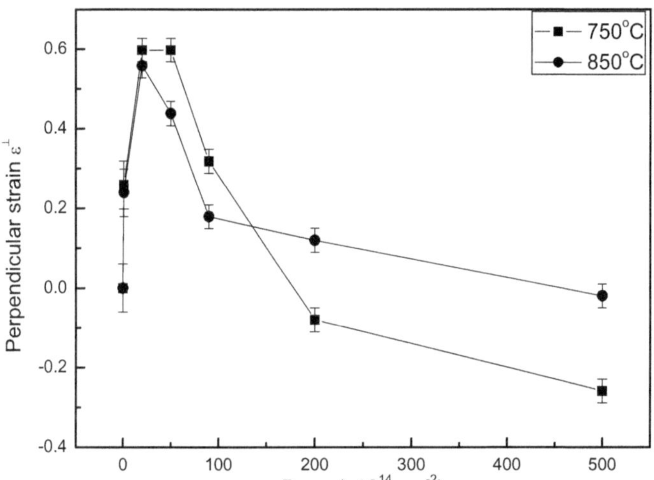

Fig. 11. Perpendicular strain extracted from XRD for as-grown and Mn implanted AlInN/GaN samples after both annealing steps of 750ºC and 850ºC. Solid lines are guides to eye.

5.3 Thermal annealing degradation of AlInN

RBS measurements were carried out to find the depth profile of elements of the samples. Figure 12 shows the random RBS spectra of as-grown and Mn implanted AlInN/GaN samples measured after annealing at 750 ºC. The energy of alpha particles backscattered from the interface is indicated by an arrow labeled by Ga. The arrows labeled Al and In indicate the energy of alpha particles backscattered from the respective atoms of aluminum and indium. The Mn signal was detected only in the spectrum of the samples implanted with the highest dose.

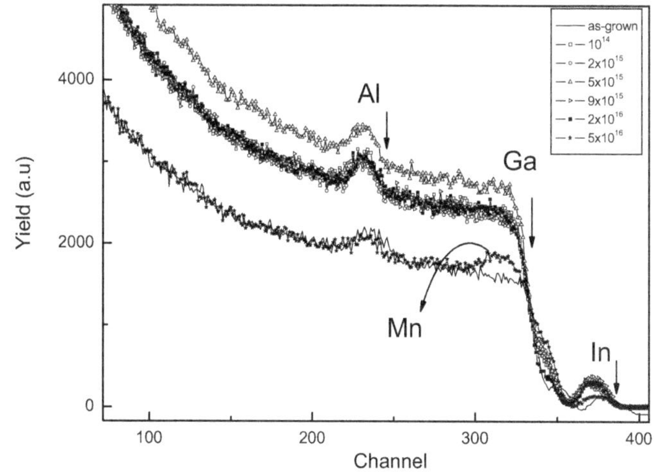

Fig. 12. RBS spectra of as-grown and Mn implanted AlInN after annealing at 750°C. Dose for all the samples is shown in cm^{-2}.

RBS measurements were again performed on the same samples after next annealing at 850 °C for 30 sec to study the thermally activated atomic migration of the constituents of the samples. The RBS random spectra of as-grown and such post implanted annealed samples are given in figure 13.

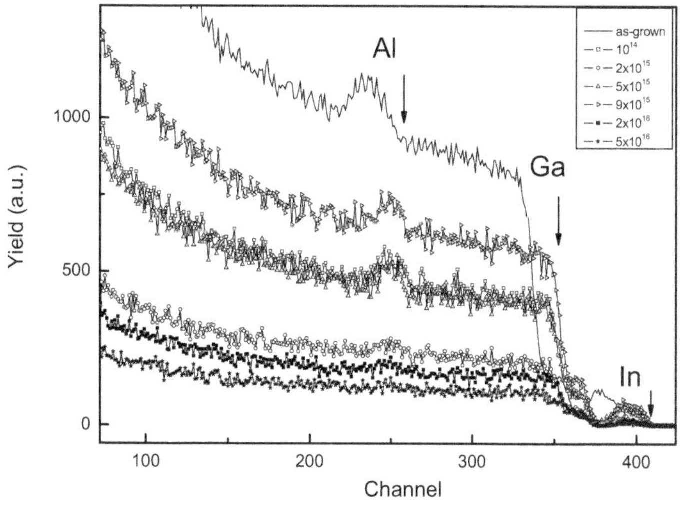

Fig. 13. RBS random spectra for as-grown and Mn implanted AlInN samples after annealing at 850°C. Dose for all the samples is given in cm^{-2}.

The comparison of indium and aluminum peaks after both anneals presents an interesting behavior that highlights the effect of annealing on atomic migration in implanted samples. Figure 14 shows the blown up RBS spectra of indium peak for as-grown and implanted samples after both anneals. The vertical arrows show the energy position of alpha particles backscattered from indium atoms located in the region near to the surface.

Indium peak after annealing at 750 °C is found shifted to the lower energy side when compared to that of as-grown samples as shown in figure 14 (a). The RBS spectra of indium peak after 850 °C annealing is shown in figure 14 (b). The spectra are found quite different from those measured after first annealing. In this case the indium peak is found shifted to the high energy side with reduction of yield as compared to that of the as-grown samples. This observation indicates the out-diffusion of indium atoms towards the surface after this annealing. The behavior of the peak with increasing dose is similar to that of observed after first annealing but reduction in yield is high in this case.

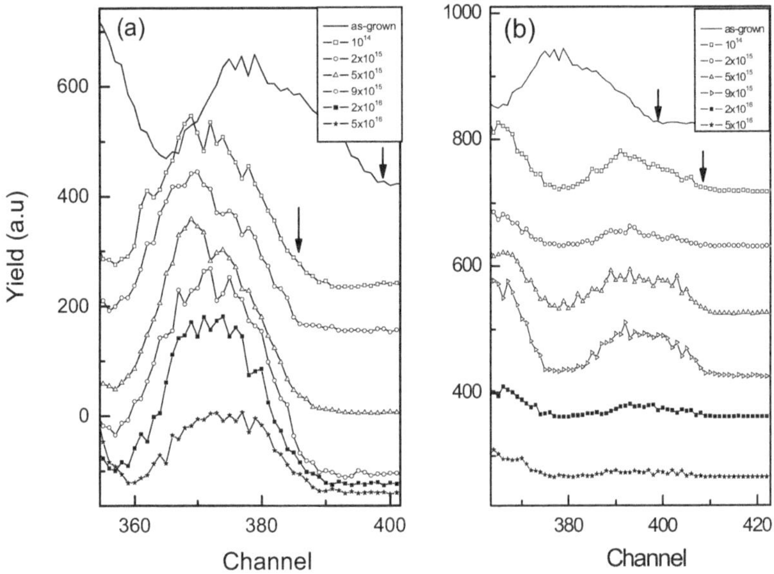

Fig. 14. RBS spectra of as-grown and Mn implanted AlInN samples showing In peak after annealing at (a) 750°C (b) 850°C.

The comparative behavior of Al signal after both anneals is shown in the figure 15. After the first annealing, Al peak is observed truncated from the higher energy side for the samples irradiated with lower doses as compared to that of as-grown samples. The decrease in yield with increase in dose is observed, which is prominent for samples implanted with highest dose. The comparison of scans of as-grown and implanted samples given in figure 15 (b) shows a clear shift in the aluminum signal towards higher energy side for implanted samples. A drastic decrease in the yield of Al peak can also be noted after second annealing as compared to that of first annealing.

Figure 16 shows the blown up RBS spectra of interface region measured after both anneals. The presence of Ga peak in between the interface and indium peak is clearly noticed in the

RBS spectra of as-grown samples, which is found suppressed for the implanted samples with the increase in dose. The position of the interface edge for the implanted samples remained the same after first annealing but it was found shifted toward higher energy side after second annealing.

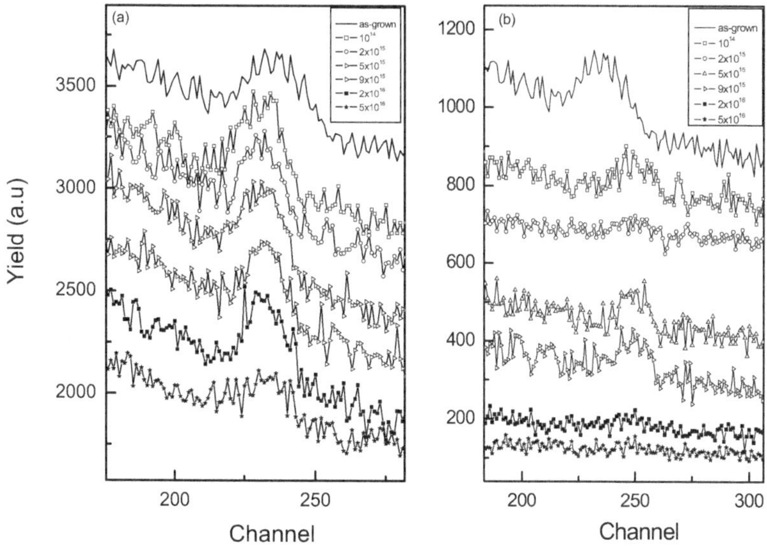

Fig. 15. RBS spectra of as-grown and Mn implanted AlInN samples showing Al peak after annealing at (a) 750°C (b) 850°C.

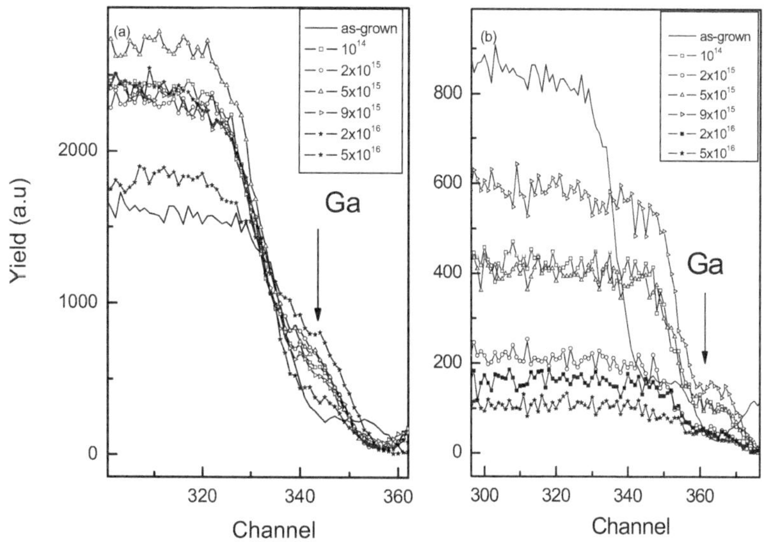

Fig. 16. RBS spectra of as-grown and Mn implanted AlInN samples showing Ga migration at the interface for annealing at (a) 750°C (b) 850°C

Figure 17 shows the comparison of RBS spectra in the region of Mn signal that appears clearly in between channel numbers 300 and 330 in RBS spectra of AlInN sample implanted with highest dose of 5×10^{16} cm^{-2} after annealing at 750°C but it could not be detected after annealing 850°C.

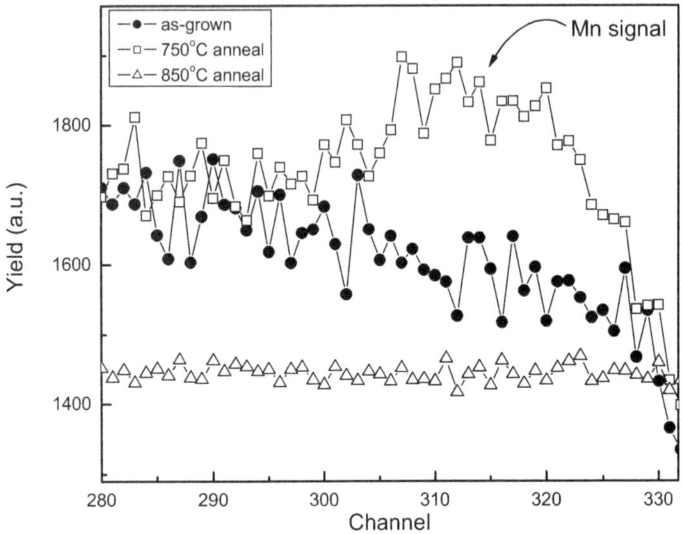

Fig. 17. RBS spectra showing the Mn signal in Mn implanted AlInN samples having dose $5 \times 10^{16} cm^{-2}$ after both anneals. Spectrum of as-grown sample is also given for comparison.

5.4 AlInN based diluted magnetic semiconductor

Magnetic properties of Mn implanted AlInN samples were studied using SQUID magnetometer by recording magnetization as a function of temperature and applied field. Hysteresis loops were recorded at 100 K and 300 K for implanted samples. Magnetic measurements were carried out on the samples implanted with doses 1×10^{14}, 2×10^{15} and $2 \times 10^{16} cm^{-2}$ and subsequently annealed at 750° for 30 sec. The temperature dependence of magnetization (MT curves) of as-grown and implanted samples having different Mn concentrations (0.01, 0.2 and 2 at. %) is shown in figure 18. As-grown samples exhibited very weak signal with a paramagnetic behavior, presumably due to the impurities in the substrate. There is a clear difference of FC and ZFC magnetization observed up to 235 K for the samples having Mn concentration of 0.01 at. %. The samples having Mn concentration of 0.2 at.% show a strong magnetization with a difference in FC and ZFC magnetizations that extend up to 285 K. Similar trend was observed in the MT curves of the sample having the highest Mn concentration of 2 at.%. An apparent transition observed in MT curves of this sample at around 260 K which points to the presence of at least two magnetic phases with one having Curie point T_c at 260 K and the other above room temperature. The highest magnetization has been observed for this sample with difference in FC and ZFC magnetization curves that persists up to nearly room temperature.

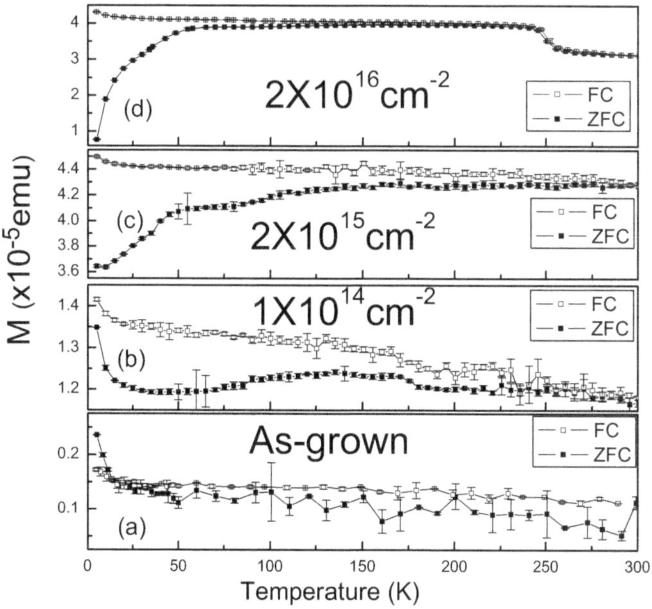

Fig. 18. FC and ZFC magnetization curves at applied field of 500 Oe for (a) as-grown and Mn implanted AlInN samples having Mn concentration of (b) 0.01 at. %, (c) 0.2 at. % and (d) 2 at. %.

Figure 19 shows the room temperature magnetization measurements as a function of applied field (MH loops). Clear hysteresis behavior is observed for all the samples. The values of saturation magnetization (M_s) have were to be 5.83, 25.4 and 27.92 emu/cm^3 for AlInN samples having Mn atomic concentrations of 0.01%, 0.2% and 2% respectively. The values of effective magnetic moment per Mn atom (p_{eff}) were calculated to be 113, 24.6 and 2.67 μ_B/ion respectively. This shows a trend of increase in M_s and decrease in p_{eff} in AlInN samples with increase in Mn concentration. The difference in FC and ZFC magnetization values is an established tool to exclude the possibility of paramagnetic and diamagnetic contribution in the observed magnetic properties. For all the samples, the difference between FC and ZFC along with well defined hysteresis loop at 300 K is consistent with presence of ferromagnetic ordering at room temperature.

In order to explore the effects of Mn implantation on the structural properties of the material, we conducted detailed XRD and RBS studies. Using the shift in position of AlInN related XRD peak, we studied the strain in AlInN epilayer as a function of implantation dose and annealing temperature. After annealing at 750 °C, the strain increases with dose until it reaches a short plateau and then decreases with further increase in dose. Similar dose dependent behavior of the strain variation has been reported by Partyka et al for the argon implanted AlGaAs/GaAs [51].

The variation of strain with dose in our samples can be divided into three regimes (1) strain accumulation (2) strain saturation and (3) strain dispersion. In the first regime, during lowest implantation dose of 10^{14}cm^{-2}, point defects (vacancy, interstitial, substitutional, and antisite) dominate due to the atomic displacements. These defects are spread within the lattice, mostly away from one another. The interstitial-vacancy pairs accumulate around the route of the penetrating ion due to displacement of Al and In atoms from their regular sites. Most of the Mn ions are expected to rest on some of the interstitial sites during this implantation dose. Projectile Mn ions encounter In atoms rarely in their way due to low indium concentration in the samples but In_i (interstitial indium atoms) are still expected to be produced due to collective motion of the cascade atoms. With increase in the dose the defect density increases, which enhances the probability of interaction of new defects with old ones. This favors the clustering of individual defects causing the increase in the level of lattice deformation. All such events will increase the perpendicular strain and cause the lattice to expand accordingly. This situation will persist until the dose of 2x10^{15} cm^{-2} at which the strain saturates. Interaction probability of newly produced defects with existing defects is maximized at a particular defect density in this situation. This saturation state is maintained in dose range 2x10^{15} to 5 x10^{15} cm^{-2}. The implantation dose of 2x10^{15} cm^{-2} seems to be critical for providing a driving force, which causes the annihilation of (vacancy, interstitial) defects and prevent the increase in strain. Mn ions have more probability to occupy substitutional sites in this atmosphere. The competition between defect creation and annihilation limits the net increase in the strain in this region. With further increase in dose, third regime starts where a decrease in strain is observed. This decrease of strain is attributed to the migration of defects into sinks (cavities, dislocation loops, grain boundaries etc). Defect clusters after attaining a critical size may have collapsed to minimize the thermodynamic free energy of the system. This collapse may produce either the increase of substitutional efficiency or capturing of defects by cavities or conversion of defects into local amorphous zones in the lattice. The strain relaxation will be the consequence of all such events. Most of Mn ions are expected to occupy the substitutional sites into the cationic sublattice of Al and In atoms in this regime of dose.

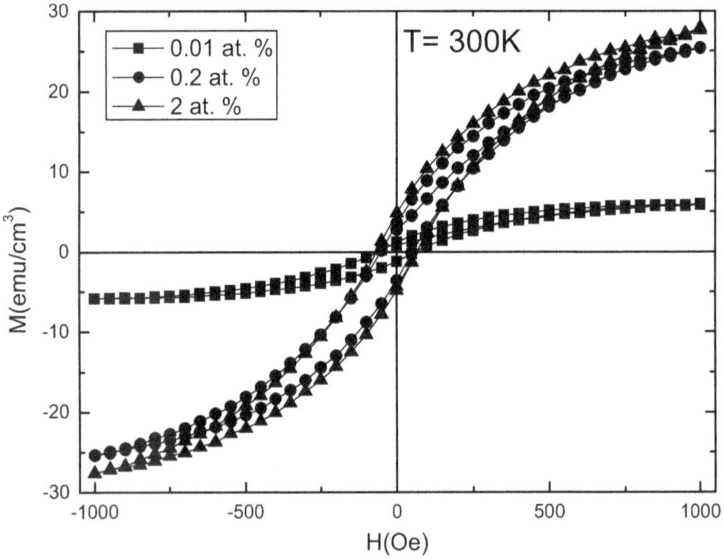

Fig. 19. Room temperature magnetization as a function of applied magnetic field for AlInN samples having different Mn concentrations.

A similar trend of variation in strain with dose was observed after second annealing except (a) strain immediately decreased after reaching to the saturation value and no plateau was observed (b) the samples implanted with doses $2x10^{16}$ cm^{-2} and $5x10^{16}$ cm^{-2} exhibited the strain higher in magnitude as compared to that measured after first annealing. It appears that the second annealing condition has provided sufficient thermal energy to the defects in the samples implanted with dose $2x10^{15}$ cm^{-2} so that the defect annihilation dominates and continues for higher doses. The reduction in XRD intensity (figures 9 and 10) may be either due to sputtering during implantation or decomposition during annealing. The reduction in Bragg reflections can also be explained using kinematics theory of scattering [44]. The change of the scattered (diffracted) X-rays is a consequence of the defects produced as a result of ion implantation.

The usual procedure to determine the fraction of the material removed is based on the comparison of the widths of the respective peaks in RBS before and after implantation [44]. The width of the peak is proportional to the number of atoms per unit area [52]. From RBS spectra measured after first annealing, it is clear that the width of Al peak is decreased preferentially on higher energy side while position of the peak remained the same. On the other hand the width of indium peak remained the same but its position shifted towards lower energy side. The behavior of both the peaks clearly indicates that a fraction of aluminum atoms has been sputtered off during implantation whereas indium atoms have been diffused inside. The probability of aluminum to be sputtered off is higher than that of indium because of lower indium concentration (especially on the surface) in our aluminum rich samples ($Al_{0.92}$ $In_{0.08}N$). The observed implantation induced in-diffusion of indium atoms is due to larger Rutherford cross-section because of larger atomic size as compared to aluminum atoms. Another reason for preferential in-diffusion of indium atoms is the lower bond strength of In-N in comparison to that of Al-N (7.7 eV and 11.5 eV respectively) [53]. The energy transferred by the interaction with ions forces the atoms to diffuse into the material. RBS spectra measured after second annealing indicated that indium and aluminum peaks are shifted towards higher energy which give a clue that annealing at 850 ^{o}C has provided sufficient energy to both Al and In atoms to move towards the surface. Outwards indium diffusion has been estimated from the shift of signal to higher energy side after second annealing [54]. Furthermore the decrease in yield as well as the peak widths for Al and In indicates the decrease in concentration of the respective atoms in the samples, which is probably due to the decomposition of the samples during annealing at 850 ^{o}C. These results indicate that annealing of AlInN at 850°C causes the significant decomposition of the samples. The AlInN films were grown at 820°C, so the decomposition is more likely to take place which is also indicated by our RBS measurements. Loss of nitrogen from AlInN during annealing at 800°C is already reported [55]. The preferential loss of nitrogen is accelerated during high temperature annealing that leads to the degradation of the near surface region [56]

Implantation has caused the change in shape of indium peak. A careful inspection of shape of indium signal (channel # 365 to 399) in RBS spectrum of as-grown samples (figure 17) indicates the presence of non-uniform indium distribution. Lower energy side of indium peak (channel # 365 to 376) indicates the presence of more abrupt gradient in comparison to that of higher energy side (channel # 376 to 399). This asymmetry of indium peak indicates the higher indium concentration in the near-interface region than in the near-surface region of as-grown samples. It is clear that implantation has modified the profile of indium peak. The RBS spectrum of indium peak of implanted samples depicts that (a) shape has become symmetrical and (b) peak position is shifted to lower energy side. These observations indicate the redistribution of indium atoms due to knock-in effects caused by impact of implanted Mn ions with indium atoms. The analysis of peak shape points to the presence of nearly equal gradient on lower energy side (channel # 356 to372) and on higher energy side (channel # 372 to 389). On the other hand, shift of peak towards lower energy side indicates the indium atoms have migrated toward interface. Both these observations points to a noticeable effect, that the implantation has caused the change in depth profile of indium in such a way that more indium is concentrated near the interface than in the near-surface region.

The presence of Mn signal (figure 17) after first annealing and suppression after second annealing can be explained by taking into account the atomic diffusion caused by high temperature annealing. Like Al and In atoms, Mn has also out-diffused and most probably have evaporated during high temperature annealing. Such evaporation of implanted atoms from GaN during annealing has been reported by Lorenz et al [57].

The presence of Ga peak in between the interface and indium peak can be explained by considering the bad quality of the interface resulting from the diffusion of Ga atoms into the AlInN layer during high temperature growth at 820 °C. The presence of Ga in the top layer near the interface may lead to the formation of a fuzzy interlayer of AlInGaN to shape the sample in the form of AlInN/AlInGaN/GaN structure. The suppression of Ga peak observed after first annealing is due to the in-diffusion of Ga atoms towards the GaN buffer during the energetic ion implantation. This may have reduced the interlayer thickness thereby improving quality of the interface. During second annealing out-diffusion of Ga atoms is expected similar to that Al and In that lead to increase in thickness of the fuzzy interlayer. The interface signal is now coming from the place up to which Ga atoms have diffused upwards. Therefore the observed shift of the interface towards higher energy side is the consequence of the upwards Ga diffusion. The different behavior of interface after both anneals is understandable by considering the annealing temperature below and above the growth temperature. It should be interesting to compare results of our magnetic measurements with that of Mn doped AlN. Saturation magnetization reported for single phase AlMnN with Mn concentration of 1 at. % is ~0.2 emu/cm^3 [58], which is much smaller as compared to the value observed for our samples even having lowest Mn concentration of 0.01at. %. The reason for this drastic difference in magnetic behavior can be explained by taking into account the lattice parameters which have significant effect on the magnetic properties. The lattice constants of our as-grown AlInN samples are c = 5.135 Å, a = 3.118 Å, which are larger as compared to that of AlN (c = 4.982 Å, a = 3.112 Å). Fermi level (E_F) of the AlInN having lattice constants larger than that of AlN will be lower in energy [59]. With the shift of E_F towards valence band, the strength of bonds reduces due to formation of vacancies. The reduction in bond strength is another explanation of ferromagnetism enhancement in Mn-AlInN system as compared to Mn-AlN. Decrease in bond strength due to indium incorporation into AlN will produce more vacancies, which makes AlInN comparatively a favorable environment for Mn to occupy the substitutional sites rather than making clusters. The tendency of self-compensation by Mn at interstitial sites is believed to be weak when substitutional Mn acceptor level shifts deeper into the band gap of the wide band gap materials [60] that also support the probability of Mn to occupy the substitutional sites. Our XRD measurements (figure 8) provide evidence that no second phase (like;

Ferromagnetic Mn_xN, Antiferromagnetic Mn_xN_y and ferromagnetic AlMn) are responsible for the observed magnetism in our implanted samples. It is reported that the formation of clusters takes place beyond 2% of Mn concentration [61] but Mn concentration in our samples is below this limit, therefore we expect the role of secondary phases, if any, is insignificant and ferromagnetic ordering in our samples is dominated by substitutional incorporation of Mn in the AlInN lattice.

The value of effective magnetic moment per Mn ion for our implanted samples is found extraordinarily large as compared to the Mn atomic moment of $5\mu_B$/ion, for two samples having lowest Mn concentration. The colossal magnetic moment in Gd implanted GaN has already been observed by Dhar et al [62] who have found extraordinary large values of effective magnetic moment per Gd atom (p_{eff}). The trend of dose dependence of M_s as observed in our samples, is similar to that of observed by Dhar et al for Gd implanted GaN. They developed a phenomenological model [63] to get a quantitative understanding of the range of spin polarization of GaN matrix by randomly positioned Gd atoms by assuming a rigid sphere of influence around every magnetic atom. The value of saturation magnetization for the Mn implanted samples based on their model is given by the following equation.

$$M_S = p_{M_n} N_{M_n} + p_o \tilde{v} N_o + p_1 N_o \sum_{n=2}^{N_{M_n}} n v_n$$

Where N_o is the concentration of AlInN matrix atoms, v is the volume of each sphere, $\tilde{v}$ is the volume fraction occupied by each sphere, i.e. $\tilde{v} = 1 - \exp(-v N_{M_n})$ and $\tilde{v}_n = \frac{(n N_{M_n})^n}{n!} e^{-v M_n}$ is the volume fraction of the regions contained within 'n' spheres. For the lower Mn concentrations, most of the spheres are assumed well separated and p_{eff} has its maximum value. Since $\tilde{v}$ is an increasing function of N_{Mn}, the value of M_s increases with increase in concentration of implanted ions and crosses the percolation threshold when $\tilde{v}$ approaches to unity for a certain concentration of ions. Finally, at very high Mn concentrations the entire matrix becomes polarized and first term in the equation starts to dominate over the rest of the terms and p_{eff} further decreases. Carrier mediated ferromagnetism does not appear to be applicable in the III-nitrides having transition metal related energy level very deep into the band gap [64]. Our implanted samples are expected to have low carrier concentration and resistive in nature like other transition metals implanted nitrides [65, 66]. This suggests that charge carriers are localized by the defects with random distribution of their localization centers in the lattice. It has been reported that ion implantation into nitrides and AlInN produces highly resistive regions and causes the decrease in carrier concentration [67]. We therefore suggest that ferromagnetic exchange interaction in our implanted samples is mediated by localized electrons that form bound magnetic polarons (BMP) whose overlap produces a spin-split Mn impurity band [12]. Radius of BMP increases logarithmically with decreasing temperature so that long range ferromagnetic ordering appears at low enough temperatures due to the overlapping of BMPs. Diluted magnetic semiconductors having random distribution of magnetic impurities are favorable to produce overlapping of BMPs. We calculated the radius of hydrogenic orbital of electron in BMP system for $Al_{0.92}In_{0.08}N$ using $a_B = \varepsilon^*(m_e/m^*) \times 0.53$ Å [ref. 12] and found its value to be 9.23 Å. Hydrogenic radius of electron in BMP system for AlN is reported to be 8.5 Å [68]. Large value of BMP radius increases the probability of BMP overlapping to mediate the ferromagnetic coupling, which is another indicator of enhanced ferromagnetism in implanted AlInN as compared to the AlN. The values calculated for inter Mn distance by assuming the uniform Mn distribution in the whole DMS layer are 19.6 nm, 2.58 nm and 1.18 nm for samples having Mn concentrations of about 0.01%, 0.2% and 2% respectively. Although these values are larger than the radius of BMP (0.92nm) but the values of inter Mn distances would be smaller, especially in the central region of DMS layer, if the true Gaussian distribution of implanted Mn is taken into account. Therefore, we are justified to assume that BMP is a possible mechanism for ferromagnetic ordering in Mn implanted AlInN samples.

6. Summary

The effects of neon and manganese ion implantation into MOCVD grown wurtzite AlInN have been studied in detail. Ne ions with doses $1x10^{14}$, $2x10^{15}$, $5x10^{15}$, $9x10^{15}$ /$cm^2$and Mn with doses $1x10^{14}$, $2x10^{15}$, $5x10^{15}$, $9x10^{15}$, $2x10^{16}$ and $5x10^{16}$ ions/cm^2 were implanted into the materials. After implantation, AlInN samples at 750°C and 850 ^{o}C using rapid thermal annealing processor. The structural, optical and magnetic characterization were performed with Rutherford backscattering spectroscopy (RBS), X-Ray diffraction (XRD), vibrating sample magnetometer (VSM) and superconducting quantum interference device (SQUID).

The XRD spectra of Ne and Mn implanted AlInN exhibits a systematic shift toward the lower angle side and a decrease in intensity of peak related to AlInN with respect to that of GaN with increase in dose up to $5x10^{15}$ cm^{-2}. The shift of the peak was found reversed for higher doses. Strain obtained from the shift of AlInN peak was found increased with dose until it reached a saturation value and after that it started decreasing with further increase in dose. RBS measurements indicated the atomic diffusion of In, Al, Ga and Mn in manganese implanted AlInN. The *in-* and *out-diffusion* of atoms has been observed after annealing at 750°C and 850°C respectively. A strong decomposition of AlInN samples has been noticed after annealing at 850°C. This points to the unsuitability of annealing of Mn implanted AlInN samples at temperature higher than the growth temperature of samples. Magnetic characterization carried out by SQUID magnetometer revealed the presence of well defined magnetic hysteresis at 300 K and a difference in FC and ZFC curves up to room temperature. These observations indicated the presence of room temperature ferromagnetic properties in Mn implanted AlInN samples which are prerequisites for DMS materials.

7. References

[1] Starosta K. Phys. Status Solidi a 68, K55 (1981).
[2] Kubota, Y. Kobayashi and K. Fujimoto, J. Appl. Phys. 66, 2984 (1989)
[3] C. Hums, J. Blasing, A. Dadgar, A. Diez, T. Hempel, J. Christen, A. Krost, K. Lorenz and E. Alves, Appl. Phys. Lett. 90, 022105 (2007)
[4] Tae Su Oh, Jong Ock Kim, Hyun Jeong, Yong Seok Lee, S Nagarajan, Kee Young Lim, Chang-Hee Hong and Eun-Kyung Suh, J. Phys. D: Appl. Phys. 41, 095402 (2008)
[5] Hadis Morkoc, Nitride semiconductors and devices, Springer (1999)
[6] O.Manasreh, III-nitride semiconductors: electrical, structural and defects properties, (2000).
[7] O. Ambacher. J. Phys. D: Appl. Phys. 31, 2563 (1998)
[8] J.-F. Carlin and M. Ilegems, Appl. Phys. Lett. 83, 668(2003)
[9] R. Butte et al, J. Phys. D: Appl. Phys. 40, 6328 (2008)

[10] S.J. Pearton et al, Physica B 340–342, 39 (2003)
[11] Pearton S J, Vartuli C B, Zolper J C, Yuan C and Stall R A, Appl. Phys. Lett. 67, 1435 (1995)
[12] J. M. D. Coey, M. Venkatesan and C. B. Fitzgerald, Nature mater. 4, 173 (2005)
[13] G Schmidt and L WMolenkamp, Semicond. Sci. Technol. 17, 310 (2002)
[14] Igor Zutic, Jaroslav Fabian and S. Das Sarma, REV. MOD. PHY. 76, 33 (2004)
[15] A. H. Macdonald, P. Schiffer and N. Samarth, nature materials. 4, 195 (2005).
[16] J. K. Furdyna. J. Appl. Phys. 64, R29 (1998)
[17] S. J. Pearton et al. J. Phys.:Condes. Matter. 16, R209 (2004)
[18] Dietl T, Ohno H, Matsukura F, Cibert J and Ferrand D, Science 287, 1019 (2000)
[19] S. Sonoda, S. Shimizu, T. Sasaki, Y. Yamamoto and H. Hori, J. Cryst. Growth. 237-239, 1358 (2002).
[20] K. Wang, R. W. Martin, E. Nogales, P. R. Edwards, K. P. O'Donnell, K. Lorenz, E. Alves and I. M. Watson, Appl. Phys. Lett. 89, 131912 (2006)
[21] R. W. Martin et al, phys. stat. sol. (c) 6, 1927 (2006)
[22] B. M. Paine, N.N. Hurvitz, and V. S. Speriosu, J. Appl. Phys. 61, 1335 (1987)
[23] K. Lorenz, N. Franco E. Alves, I. M. Watson, R.W. Martin, and K. P. O'Donnel, Phys. Rev. Lett. 97, 085501 (2006)
[24] S. J. Pearton, C. R. Abernathy, P. W. Wisk, W. S. Hobson, and F. Ren, Appl. Phys. Lett. 63, 1143 (1993)
[25] A. Buyanova, Mt. Wagner, W. M. Chen, B. Monemar, J. L. Lindstrom, H. Amano and I. Akasaki, Appl. Phys. Lett. 73, 2968 (1998)
[26] H. X. Jiang, S. X. Jin, J. Li, J. Shakya, and J. Y. Lin, Appl. Phys. Lett. 78, 1303 (2001)
[27] F. Briegleb and A. Geuther Ann. Chem. 123, 228 (1862)
[28] Tae Su Oh, Jong Ock Kim, Hyun Jeong, Yong Seok Lee, S Nagarajan, Kee Young Lim, Chang-Hee Hong and Eun-Kyung Suh, J. Phys. D: Appl. Phys. 41, 095402 (2008).
[29]V. Yu. Davydov, A. A. Klochikhin, V. V. Emtsev, D. A. Kurdyukov, S. V. Ivanov, V. A. Vekshin, F. Bechstedt, J. Furthmüller, J. Aderhold, J. Graul, A. V. Mudryi, H. Harma, A. Hashimoto, A. Yamamoto, and E. E. Haller, Phys. Status Solidi B 234, 787 (2002).
[30] Q. Guo and A. Yoshida, Jpn. J. Appl. Phys. 33, 2453 (1994)
[31] Nicolay S, Carlin J-F, Feltin E, Butt´e R, Mosca M, Tchernycheva M, Nevou L, Julien F H, Grandjean N and Ilegems M Appl. Phys. Lett. 87, 111106 (2005)
[32] Christmann G, Simeonov D, Butt´e R, Feltin E, Carlin J-F and Grandjean N Appl. Phys. Lett. 89, 261101 (2006)
[33] L. Vegard, Z. Phys. 5, 17 (1921)
[34] T. Seppanen, L. Hultman, J. Birch, M. Beckers and U. Kreissig, J. Appl. Phys. 101, 43519 (2007).
[35] M. J. Lukitsch, Y. V. Danylyuk, V. M. Naik, C. Huang, G. W. Auner, L. Rimai and R. Naik Appl. Phys. Lett. 79, 632 (2001)
[36] V S Harutyunyan, A P Aivazyan, E R Weber, Y Kim, Y Park and S G Subramanya, J. Phys. D: Appl. Phys. 34, A35–A39 (2001).
[37] D. Keith Bowen and Brian K. Tanner, X-Ray metrology in semiconductor manufacturing, Taylor & Francis group, (2006)
[38] L. T. Romano, C. G. Van de Walle, J. W. Ager III, W. Gotz and R. S. Kern, J. Appl. Phys. 87, 7745 (2000).
[39] S.O. Kucheyev, J.S. Williams, S.J. Pearton, Materials Science and Engineering Reports 33, 51(2001)
[40] Lindhard, Scharff and Schiottm Mat. Phys. Medd. Dan. Vid. Selsk. 33 (1963).
[41] www.SRIM.org

[42] E. Alves, U. Wahl, M.R. Correia, S. Pereira, B. De Vries, A. Vantomme, Nuclear Instruments and Methods in Physics Research B 206, 1042 (2003).
[43] W Liu, J F Wang, J J Zhu, D S Jiang and H Yang, J. Phys. D: Appl. Phys. 40, 5252 (2007).
[44] A. Seidel, S. Massing, B. Strehlau, and G. Linker, Phys. Rev. B, 38, 2273 (1988)
[45] A. Luukkainen, J. Keinonen and M. Erola, Phys. Rev. B. 32, 4814 (1985)
[46] C. A. Cima, H. Boudinov, J. P. de Souza, Yu. Suprun-Belevich and P. F. P. Fichtner, J. Appl. Phys. 88, 1771(2000)
[47] J. H. Evans, Nucl. Instrum. Methods Phys. Res. B. 196, 125 (2002).
[48] M. Chicoin, S. Roorda, R. A. Masut and P. Desjardins, J. Appl. Phys. 94, 6116 (2003)
[49] M. Luysberg et al., J. Appl. Phys. 92, 4290 (2002)
[50] V. A. Joshkin, C. A. Parker, S. M. Bedair, L. Y. Krasnobaev, J. J. Cuomo, R. F. Davis, A. Suvkhanov, Appl. Phys. Lett. 72, 2838 (1998)
[51] P. Partyka, R. S. Averback, D. V. Forbes, J. J. Coleman and P. Ehrhart, J. Appl. Phys. 83, 1265 (1998)
[52] W. K. Chu, J. W. Mayer, and M. A. Nicolet, Backscattering Spectroscopy (Academic, New York, 1978)
[53] A. Kerr. CRC Handbook of Chemistry and Physics (CRC Press, Boca Raton, FL, 1990)
[54] Y.Q. Wang, M. Curry, E. Tavenner, N. Dobson and R.E. Giedd, Nuclear Instruments and Methods in Physics Research B. 219-220, 798 (2004).
[55] J. Karpinski, J. Jun, and S. Porowski, J. Cryst. Growth. 66, 1(1984).
[56] S. O. Kucheyev, J. S. Williams, C. Jagadish, G. Li and S. J. Pearton, Appl. Phys. Lett.76, 3899 (2000).
[57] K. Lorenz, U. Wahl, E.Alves, T. Wojtowicz, P. Ruterana, S. Ruffenach, O. Briot, Superlattice and microstructures. 36, 737 (2004)
[58] R. M. Frazier, G. T. Thaler, B.P. Gila, J. Stapleton, M.E. Overberg, C.R. Abernathy, S.J. Pearton, F. Ren, And J.M. Zavada, J. Elect. Mater. 34, 365 (2005)
[59] Galanakis, P. H. Dederichs and N. Papanikolaou, Phys. Rev. B. 66, 134428 (2002).
[60] J. Masek, J. Kudrnovsky, F. Maca, Jairo Sinova, A. H. MacDonald, R. P. Campion, B. L. Gallagher, and T. Jungwirth, Phys. Rev. B. 75, 045202 (2007)
[61] Y. L. Soo, G. Kioseoglou, S. Kim, S. Huang, Y. H. Kao, S. Kuwabara, S. Owa, T. Kondo and H. Munekata: Appl. Phys. Lett. 79, 3926 (2001)
[62] S. Dhar, T. Kammermeier, A. Ney, L. Perez, K. H. Ploog, A. Melnikov and A. D. Wieck, Appl. Phys. Lett. 89, 062503 (2006)
[63] S. Dhar, L. Perez, O. Brandt, A. Trampert, K. H. Ploog, J. Keller and B. Beschoten, Phys. Rev. B. 72, 245203 (2005)
[64] Tobias Graf, Sebastian T. B. Goennenwein, and Martin S. Brandt, phys. stat. sol. 239, 277 (2003)
[65] A. Y. Polyakov, N. B. Smirnov, A. V. Govorkov, N. V. Pashkova, A. A. Shlensky, S. J. Pearton, M. E. Overberg, C. R. Abernathy, J. M. Zavada, R. G. Wilson, J. Appl. Phys. 93, 5388 (2003)
[66] R. M. Frazier, G. T. Thaler, C. R. Abenathy, S. J. Pearton, M. L. Nakarmi, k. b. Nam, J. Y. Lin, H. X. Jiang, J. Kelly, R. Rairigh, A. F. Hebard, J. M. Zavada and R. G. Wilson, J. Appl. Phys. 94, 4956 (2003)
[67] S. J. Pearton, Mater. Sci. Rep. 4, 313 (1990)
[68] Jun Zhang, S H Liou and D J Sellmyer, J. Phys.: Condens. Matter. 17, 3137 (2005).
[69] Abdul Majid, Akbar Ali, J. J. Zhu and Y. T. Wang, J Mater Sci: Mater Electron. 20:230 (2009)
[70] Abdul Majid, Akbar Ali, J J Zhu and Y T Wang, J. Phys. D: Appl. Phys. 41, 115404 (2008)
[71] Abdul Majid Rehana Sharif J.J. Zhu Akbar Ali, Appl. Phys. A 96, 979 (2009) and Abdul Majid, Rehana Sharif, Akbar Ali, and Jian Jun Zhu, Jpn. J. Appl. Phys. 48, 040202 (2009)

10

Nitridation of GaAs Surface by Low Energy Ion Implantation with *In Situ* Control of Chemical Composition

Valery Mikoushkin
Ioffe Institute, the Russian Academy of Sciences
Russia

1. Introduction

Nitridation of GaAs by low energy N_2^+ ion beams is an effective way for passivation and isolation of GaAs based structures (DeLouise, 1992, 1993; Pan et al., 1998; Hecht et al, 2001; Li et al., 2001; Kumar et al., 2009). The research activity in the field of ion induced GaAs nitridation accompanied by annealing for crystal structure restoration has been stimulated recently by the problem of perfect crystalline GaN film growth on GaAs substrate with using the implanted GaN layer as a buffer (Majlinger et al., 2008, 2009; Kumar et al., 2009). Low energy N_2^+ ion implantation enabled fabrication of GaN nanolayers even for device applications (Meškinis et al., 2004). Obvious advantages of the low energy implantation as a method for nitride nanolayer fabrication are relative simplicity, compatibility with other high vacuum techniques and possibility to control elemental and chemical composition of the nanolayer *in situ* by means of different kinds of electron spectroscopy. Qualitative chemical analysis of the implanted layers resulted in the conclusion that they are pure GaN films (DeLouise, 1992, 1993; Pan et al., 1998; Meškinis et al., 2004; Kumar et al., 2009) or the films mainly consisting of GaN (Hecht et al., 2001; Majlinger et al., 2008, 2009). Using x-ray photoelectron spectroscopy (XPS) with synchrotron radiation (SR) made it possible to reveal an essential contribution of the phase of dilute GaAsN alloy in addition to GaN (Majlinger et al., 2008, 2009; Mikoushkin et al., 2009) and to perform quantitative chemical analysis of the implanted layer with the help of the high resolution mode of this method (Mikoushkin et al., 2009). Efficient creation of the phase of GaAsN alloy has also been observed when GaAs was bombarded with a mixture of N_2^+ and Ar^+ ions (Gordeev et al., 2003; Mikoushkin et al., 2008).

The unique property of dilute $GaAs_{1-x}N_x$ alloy is that the band gap E_g sharply decreases with enhancement of the nitrogen concentration x even in the range of low concentrations x (Bi & Tu, 1997; Sik et al., 2001). The band gap becomes as narrow as $E_g \sim 1$ eV already at $x \sim 0.03$, which should be compared with $E_g \sim 1.44$ eV at $x = 0$. This makes dilute $GaAs_{1-x}N_x$ alloy to be one of the most promising materials for infrared applications (Henini, 2005). Unfortunately, higher nitrogen concentrations $x > 0.03$ in alloys are difficult to achieve by means of basic growth technologies (Henini, 2005). There have been only a few reports published since 1997 about more concentrated GaAsN alloys (Bi & Tu, 1997; Toivonen et al., 2000; Moody et al., 2002; Veal et al., 2004). The physical restriction in fabricating alloys with higher nitrogen concentrations may be in their instability at growth temperatures typically exceeding 550÷650°C (Mikoushkin et al., 2008). This instability results in decay of the alloy into thermodynamically more stable chemical phases: GaAsN → GaN and GaAs. Therefore the problem arises whether it is possible to fabricate nanofilms of $GaAs_{1-x}N_x$ alloys with higher nitrogen concentrations $x > 0.03$ at low temperatures by low energy implantation of GaAs. Based on the research done, the answer will be shown below to be positive. The implantation technique will be shown to enable fabrication of dilute $GaAs_{1-x}N_x$ alloys with nitrogen concentration $x \sim 0.09$ and, correspondingly, with the narrow band gap. The problem was solved with *in situ* control of chemical composition by high resolution XPS with synchrotron radiation and, surprisingly, by ordinary Auger Electron Spectroscopy (AES).

2. Nitridation of GaAs with high resolution synchrotron radiation based XPS diagnostics of chemical composition

Introduction. The classic research of De Louise seems to be the first study devoted to GaAs nitridation by low-energy N_2^+ ion implantation (DeLouise, 1992, 1993). It was performed under XPS control which is known to be one of the most powerful methods for elemental and chemical analysis (Briggs & Seah, 1983). Using this method with MgKα x-ray source the author showed that the implantation results in fabrication of nitrogen rich layers characterized by nitrogen concentration ranged from [N] ~ 15 at.% at energy E_i = 3.0 keV up to [N] ~ 30 at.% at energies E_i = 0.5 ÷ 1.2 keV. Creation of GaN layer was revealed by the analysis of Ga3d photoemission spectra. As a result of nitrogen implantation, the author observed a chemical shift of Ga3d photoelectron line due to additional contribution at higher binding energy which was attributed to the GaN phase. Analogous line shifts were observed later by other authors who came to similar conclusion about formation of GaN nanolayer on the GaAs surface in the implantation process. The only fact slightly disturbed the suggested interpretation: the photoemission signal contained contribution of GaAs besides the GaN. Presence of GaAs line in Ga3d spectrum was explained by the contribution of GaAs substrate which was possible to observe because the mean free path of Ga3d photoelectrons generated by MgKα/AlKα x-rays exceeded the thickness of nitrided layer. Therefore the GaAs signal was explained by the substrate contribution.

The geometry of the experiment described below according to Ref. (Mikoushkin et al., 2009) was made different. XPS chemical analysis of nitride film fabricated by implantation of N_2^+ ions with the energy E_i = 1.5 keV was performed at different experimental conditions compared to previous studies, namely at the conditions when the mean free path λ of photoelectrons generated by low energy synchrotron radiation is less than the thickness of the nitrided layer. Therefore photoelectrons of only nitrided layer were detected. Nevertheless contribution of a phase being chemically similar to the GaAs was observed apart from the phase of GaN. This contribution proved to be comparable with the content of the main phase. High energy resolution of the method made possible to show that the revealed phase is $GaAs_{1-x}N_x$ alloy and that the layer fabricated by implantation of nitrogen consists of the wide band gap matrix of GaN with the narrow band gap clusters of $GaAs_{1-x}N_x$ ($x \sim 0.05 \div 0.10$) alloy, which can be considered as a quantum dot system.

Experimental details. The experiment was carried out at the BESSY II storage ring in Berlin using monochromatic synchrotron radiation of the German-Russian beamline equipped with the plane-grating monochromator (PGM) and photoelectron spectrometer with hemispherical analyzer VG CLAM-4 (Fedoseenko et al., 2003). The total energy resolution of the method was better than 300 meV. Control of the elemental and chemical composition of sample was performed by measuring *in situ* core-level photoemission spectra: Ga3d, As3d, N1s. The photon energy scale of the monochromator was calibrated using Au $4f_{7/2}$ line of gold (E_B = 84.0 eV).

Commercial GaAs (100) n-type (n ~ 10^{18} cm^{-3}) wafer was taken as a sample. It was implanted at room temperature under high vacuum conditions by nitrogen ions N_2^+ of low energy E_i = 1500 eV in the preparation chamber of the electron spectrometer. The incidence angle was normal to the surface. The dose of the implanted ions Q ~ $3*10^{16}$ cm^3 was close to the dose sufficient for saturation of the nitrogen concentration in the near-surface area (DeLouise, 1992). The surface of the sample was preliminary cleaned by Ar^+ ions with the energy E_i = 1500 eV which was high enough to prevent considerable enrichment of the surface by the atoms of gallium. As a result of cleaning, the layer of native oxide and contaminations was completely removed from the surface, which was checked by measuring the O1s and C1s spectra. The thickness of the nitrided layer or the projected range of nitrogen ions was estimated by SRIM code (Ziegler et al., 1985) to be about d ~ 2.5 nm. The peculiarity of the conducted experiment was in relationship of the thicknesses of the nitrided layer and the near surface layer controlled by XPS which is determined by the mean free path of detected photoelectrons λ ~ 0.5 ÷ 1.0 nm. The last one was made less than the thickness of nitrided layer by diminishing the photoelectron kinetic energy down to E_e ~ 100 ÷ 126 eV using x-rays of low energies: hυ = 150 eV for measurement of Ga3d and As3d lines and hυ = 500 eV for N1s line. Therefore photoelectrons of only nitrided layer were registered.

Analysis of photoelectron specrta. The total concentration of the implanted nitrogen is one of the most important characteristics of the nitrided layer. It was estimated by the analysis of the N1s and Ga3d line intensities. To neglect the energy dependence of the spectrometer transmission, kinetic energies of these photoelectrons were made close to each other (E_e ~ 100 ÷ 126 eV). Mean free paths of these photoelectrons characterised the analysed layer proved to be equal at this condition too. Comparison of the intensities normalized to the corresponding photoemission cross sections σ(N1s, hv = 500 eV) and σ(Ga3d, hv = 150 eV) (Henke et al., 1993) gave the content of nitrogen [N] ~ 25 at.%. This value proved to be close to the values obtained in Ref. (DeLouise, 1992; Li et al., 2001). The embedded nitrogen can create different chemical compounds with matrix atoms besides GaN.

When colliding with the surface, molecular nitrogen ions dissociate, and chemically active atoms penetrate into the substrate to a projected range of about several nanometers before they completely lose their kinetic energy. At the end of this process, the nitrogen atoms can react with the matrix, substituting arsenic and resulting in formation of the following chemical phases: GaN, GaAsN, GaAs, As and AsN. High energy resolution SR-based XPS method enabled development of quantitative diagnostics of chemical composition for the nitrided layer with determining the concentrations of mentioned chemical phases. To solve this problem, photoelectron spectra of As3d, Ga3d and N1s measured for exceptionally nitrided layer without any substrate contribution should be considered. These spectra are shown in Fig. 1, 2 and 3.

Photoelectron spectrum of arsenic represented in Fig.1 confirms the qualitative model of nitrogen ion interaction with matrix, describing partial replacement of arsenic by nitrogen.

As3d spectrum on implanted GaAs surface (Fig.1b) shows some contribution of a phase of elementary arsenic As°. The spectrum confirms absence of AsN phase earlier reported for the layers nitrided by low energy implantation (DeLouise, 1992; Hecht et al., 2001; Li et al., 2001; Kumar et al., 2009). Finally, As3d spectra give new important information that the major part of arsenic remains in the nitrided layer and practically saves line position and corresponding chemical state. This fact implies creation of dilute alloy GaAsN with low nitrogen concentration which does not essentially influence the chemical state of arsenic atoms. This assumption was confirmed by the spectra of Ga3d and N1s.

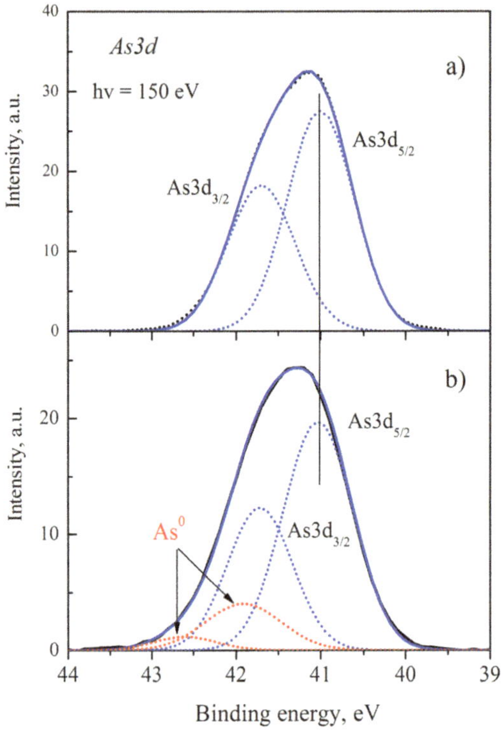

Fig. 1. As3d photoemission spectrum of GaAs (a) and GaAs implanted by nitrogen ions with energy E_i = 1500 eV (b). The last one is decomposed into different contributions.

Implanted nitrogen atoms substitute arsenic and create chemical bonds with gallium, which is seen in the Ga3d photoemission spectra shown in Fig.2. Curve 1 corresponds to Ga3d line of atomically clean surface of GaAs prepared by Ar^+ ion etching. Curve 2 is the spectrum of the surface implanted by nitrogen ions. Implantation shifts Ga3d spectrum to higher binding energies. Decomposition of the spectrum of the implanted surface into Gaussian contributions shows that the phase of GaN is dominating one, but a contribution of the substance being chemically similar to GaAs is observed in the nitrided layer too. This contribution is comparable with the content of the main phase. Before it will be shown that this phase is definitely the phase of $GaAs_{1-x}N_x$ alloy, some remarks about Ga3d and As3d spectra should be made.

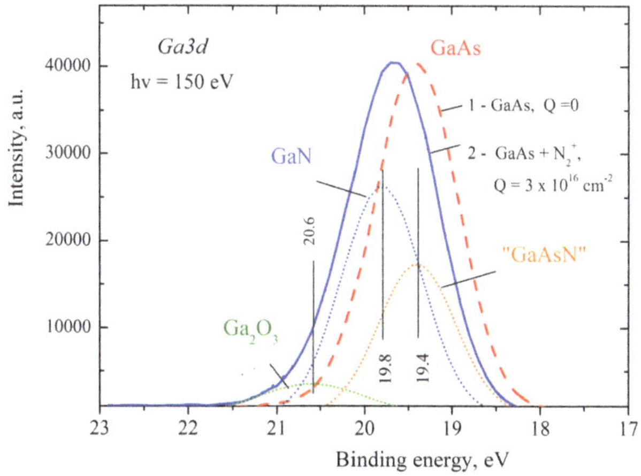

Fig. 2. Ga3d photoemission spectrum of GaAs (1) and GaAs implanted by nitrogen ions with energy E_i = 1500 eV (2). The last one is decomposed into different contributions.

Important characteristics of the gallium chemical state, the binding energies of Ga3d core electron, were obtained: E_B(Ga3d) = 19.4 eV and 19.8 eV for GaAs and GaN, correspondingly. These values proved to be in the frame of published data, though by 0.5 eV and 0.3 eV larger than those reported in Ref. (DeLouise, 1992; Li et al., 2001). A reason of this discrepancy may be in different positions of the Fermi level in samples being differently treated. The spectrum of implanted sample shows also a small contribution of gallium oxide Ga_2O_3. Origin of this contribution was in presence of residual oxygen in the gas system of the ion source which was not desirable and planned. But this small contamination not only did not essentially affect the spectra but it also was used for the additional analysis of the As3d photoemission spectrum (Fig. 1). This spectrum shows absence of As_2O_3 with binding energy E_B = 44.5 ÷ 46 eV (Briggs & Seah, 1983) despite the presence of residual oxygen in vacuum system, which illustrates higher resistivity of arsenic to oxidation and nitridation compared to gallium. Therefore the absence of AsN seems to be natural because of lower electronegativity of nitrogen. Stability of the As3d lines against nitridation mentioned above evidences also no influence of the embedded atoms and defects to the Fermi level position.

The phase GaAs observed in Ga and As spectra of nitrided layer was assumed to be $GaAs_{1-x}N_x$ ($x << 1$) dilute alloy which is chemically similar to the GaAs because of low nitrogen concentration x, and therefore it cannot be identified because of extremely small chemical shift of the corresponding lines. But this task was solved in N1s spectrum and the assumption was unambiguously confirmed. Fig. 3 shows N1s photoemission spectrum of the implanted sample (curve 1) with prominent line of the GaN phase. The binding energy of N1s core electron in GaN E_B(N1s) = 396.7 eV is in agreement with the previously published data. It coincides with that obtained in Ref. (DeLouise, 1992) and is less by 0.3 eV than the value given in Ref. (Zhu et al., 1992; Li et al., 2001). High total energy resolution of the beamline used and the endstation enabled revealing some asymmetry of N1s photoemission line related to additional contribution at higher binding energy where the line of GaAsN was expected. Indeed, the binding energy N1s in alloy Ga-As-N should be between the energies in N-Ga and N-O compounds. The asymmetry becomes stronger after bombardment of the film by Ar^+ ions (curve 2, Fig.3a), which was shown in our former works to enhance the contribution of the phase of $GaAs_{1-x}N_x$ alloy in nitrided layer due to mixing and homogenization of the material favoring to alloy creation (Gordeev et al., 2003; Mikoushkin et al., 2008). The difference of the spectra 2 and 1 (curve 3) shows growth of the discussed contribution at $E_B(N1s)$ = 397.3 eV as a result of Ar^+ ion bombardment of the film. Therefore, the conclusion was made that essential contribution of the phase of narrow band gap $GaAs_{1-x}N_x$ alloy was revealed in the film besides the dominating phase of the wide band gap GaN.

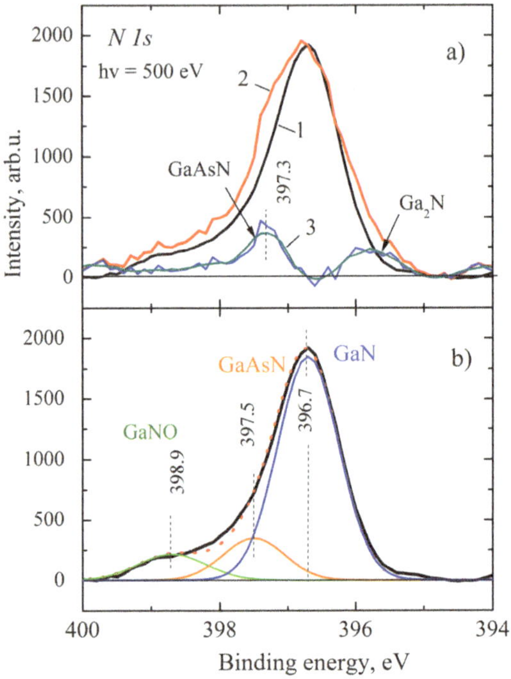

Fig. 3. N1s photoemission spectrum of GaAs implanted by nitrogen ions (curve 1, a) and its decomposition (b), spectrum of the implanted layer followed by Ar^+ ion bombardment (curve 2, a) and their difference (curve 3, a). The energy of implanting N_2^+ and bombarding Ar^+ ions is E_i = 1500 eV. The photon energy hv = 501.3 eV.

Fig. 3b demonstrates decomposition of the N1s spectrum of the nitrided layer corresponding to curve 1 in Fig.2a into contributions of different chemical phases. The decomposition gave the N1s binding energy $E_B(N1s)$ = 397.5 eV in $GaAs_{1-x}N_x$ alloy which turned out to be very close to the data of XPS measurements (~397.3 ÷ 397.8 and 397.6 eV) of single-phase dilute alloys $GaAs_{1-x}N_x$ with x = 0.04 and x = 0.01 grown by MBE in Ref. (Spruytte at al., 2001; Veal at al., 2004), correspondingly. Advantage of the present measurements is that the difference (0.8 eV) with the energy of GaN was obtained in one experiment and in one sample. The obtained information is necessary for diagnostics of the GaAs nitrides which is complicated by small energy difference of these chemical states and large spread in the binding energies published for GaN and other nitrides by different authors. Comparison of the binding energies obtained in this work for alloy in as prepared nitride layer (397.5 eV) and in the layer with lower nitrogen concentration after bombardment by Ar^+ ions (397.3 eV) points out to some natural dependence of the $GaAs_{1-x}N_x$ alloy binding energy on the nitrogen content: the higher content x, the lower energy E_B(N1s).

Decomposition of the N1s line gave also the portion of nitrogen in alloy ~ 10÷ 15 at.% in the conducted experiment, which provides at list 30 at.% of the alloy contribution. Taking into account the total concentration of nitrogen [N] ~ 25 at.%, and the ratio of GaN and $GaAs_{1-x}N_x$ phases (~ 0.6/0.4), which follows from the decomposition of Ga3d spectrum (Fig.2), the content of nitrogen in alloy $GaAs_{1-x}N_x$ was estimated to be x ~ 0.10. This value is high enough to decrease the band gap of the alloy below the value $\Delta = 1$ eV (Bi & Tu, 1997; Sik at al., 2001).

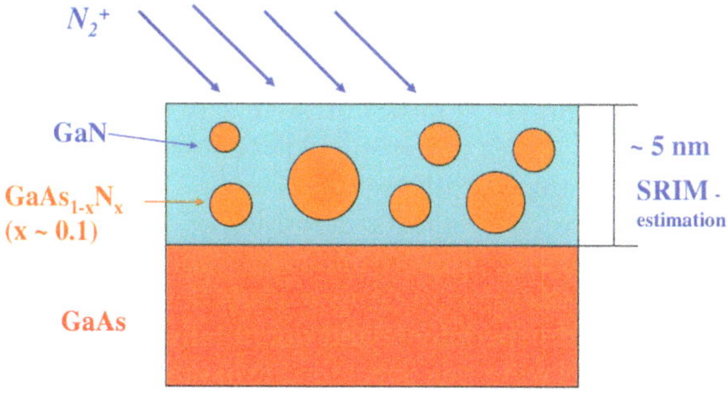

Fig. 4. Scheme of the nanostructure GaAsN/GaN formed on the GaAs surface as a result of implantation of low energy nitrogen ions N_2^+.

The revealed phase of $GaAs_{1-x}N_x$ alloy is generated on the atomic level due to segregation of atomic complexes into clusters. The characteristic size of the clusters cannot be more than the thickness of the nitrided layer. Thus we came to the conclusion that the fabricated nitrided nanofilm being a system of the narrow band gap clusters of $GaAs_{1-x}N_x$ (x ~ 0.10) alloy in the wide band gap matrix of GaN is characterized by the main attribute of a quantum dot system. Fig. 4 shows a scheme of the nitrided layer formed on the GaAs surface by implantation with low energy nitrogen ions N_2^+. The film thickness shown corresponds here to the length of nitrogen depth profile calculated by SRIM codes which is essentially longer than the SRIM projected range (see below). The implanted layer contains comparable contributions of two phases and represents a quantum dot like nanostructure GaAsN/GaN rather than an entire GaN film assumed in many former publications.

Conclusions. As a result of the conducted research, *in situ* chemical composition diagnostics of the GaAs nitrides has been developed using high resolution synchrotron radiation based photoelectron spectroscopy. Chemical phases of GaN, GaAsN and GaAs were distinguished by electron spectroscopy for the first time. Core level binding energies in $GaAs_{1-x}N_x$ (E_B = 397.5 eV) and in GaN (E_b = 396.7 eV) and their difference (ΔE_B = 0.8 eV) were measured in one experiment with high accuracy. The model of chemical composition and atomic structure of the nitrided layers fabricated by implantation with low energy N_2^+ ions into GaAs surface has been essentially improved. We revealed that the nitride nanolayer is not an entire GaN layer. It consists of the narrow band gap clusters of $GaAs_{1-x}N_x$ ($x \sim 0.10$) alloy in the wide band gap matrix of GaN, which makes the nanofilm similar to a quantum dot system.

3. Auger diagnostics of chemical composition of GaAs nitride nanolayers

Introduction. The research described above showed that low energy ion implantation makes it possible to fabricate nitrided layers with large contribution of the phase of dilute alloy $GaAs_{1-x}N_x$ ($x \sim 0.10$). Moreover, this alloy is characterized by high concentration x of nitrogen and by the narrow band gap desirable for infrared optoelectronic applications. High nitrogen concentration in alloy seems to be a result of specificity of implantation technology allowing fabrication at room temperature, which prevents the alloy decay. The aim of the next step of research was to increase the contribution of GaAsN alloy in the nitrided layer. To solve this problem, some measures were undertaken. For example, initial enrichment of the substrate surface by gallium and the total nitrogen concentration in implanted layer was reduced by some increasing the energies of etching and implanting ions.

Unfortunately synchrotron radiation based diagnostic methods are not easily available for routine experiments. Therefore, diagnostics of the GaAs nitrides on the basis of ordinary Auger Electron Spectroscopy (AES) was planned to be developed. The drawback of AES for chemical analysis is large natural width (against chemical shifts) and complicated shape of Auger lines. Fortunately, considerable relaxation shifts of Auger lines in nitride compounds make it possible to distinguish them from each other (Aksenov at al., 1998, 1999; Mikoushkin at al., 2008). Energies of Auger transitions in nitride compounds were determined in the conducted experiments and quantitative Auger diagnostics of chemical composition of nitrided nanolayers was developed (Mikoushkin, 2011).

Experimental details. The experiments were carried out in an electron spectrometer LHS-11 (Leybold AG) equipped with a hemispherical energy analyzer, in which the residual vacuum in the analytical chamber was $P < 2 \times 10^{-10}$ Torr. The native oxide removing, sample surface cleaning and nitrogen ion implantation were performed in a technological chamber ($P \sim 2 \times 10^{-9}$ Torr) equipped with a Penning type ion gun IQP-10/63 (Leybold AG) generating a beam with the diameter exceeding the sample target size. The object under study was the epitaxial film of GaAs (100) analogous to that grown in Ref. (Gordeev, 2003). The film was grown at Ioffe Institute at substrate temperature of 580 °C to a thickness of 100 nm on semi-insulating GaAs (100) substrate using Intevac Gen II MBE system (Zhukov, 2001). The crystalline structure was controlled by conventional High Energy Electron Diffraction system. The procedure of the surface cleaning with Ar^+ ions is known to unify the surface layer of GaAs and to result in formation of reproducible properties of the layer, characterized by amorphous structure and essential enrichment by gallium atoms. The energy of Ar^+ ions $E_i = 2500$ eV was taken to be high enough to prevent large enrichment of the surface by gallium and to eliminate conditions for efficient creation of GaN phase during nitridation. Then the film was implanted at room temperature by N_2^+ ions at normal incidence, with the dose of implanted ions $Q \sim 10^{17}$ cm^{-2} sufficient for saturation of the near-surface layer with nitrogen (DeLouise, 1992). All the operations were controlled by measuring the Auger spectra of the main elements, Ga*LMM*, As*LMM*, N*KVV*, and those of possible impurities, O*KVV*, C*KVV*. The Auger spectra were recorded relative to the Fermi level in the mode of constant relative energy resolution ($\Delta E/E$ = const). This mode enabled us to use the known coefficients for the "peak-to-peak" line intensities to obtain element composition of the material (Davis at al., 1976). Positions of Ga*LMM* and As*LMM* Auger lines (E_A = 1066.3 and 1224.7 eV, correspondingly) of the GaAs surface prepared coincided with those known for surfaces prepared by chemical etching and crystal cleavage (1066.2 and 1224.5 - 1225.0 eV, respectively) (Briggs & Seah, 1983) indicating negligible surface charging under the diagnostic beam. The energy of the implanted ions E_i = 2500 eV was taken also somewhat higher than typical energies used in former low energy implantation experiments. It provided relatively large thickness of the nitrided layer which was estimated by SRIM code as the projected range of nitrogen ions d ~ 3.6 nm (Ziegler at al., 1985). Thus, the mean free path of the detected electrons λ ~ 1.5 nm was markedly less than the thickness of the nitrided layer, and the experiment was performed under the conditions when the electrons ejected by only nitrided layer were collected without any contribution from the GaAs substrate. The choice of the relatively high energy of the implanted ions (E_i = 2500 eV) resulted also in twofold reduction of the total nitrogen concentration in the nitrided layer as compared to the maximum concentration achieved at energies below 1000 eV. The lack of nitrogen prevents creation of GaN and decay of GaAsN, which was assumed to help domination of GaAsN alloy in the nitrided layer aimed at in the research. A special experiment was performed by Electron Energy Loss Spectroscopy to confirm the information about chemical composition of the film by the analysis of plasmon spectra.

Analysis of Auger spectra. Formation of GaN in nitrided layers was confirmed in the previous studies of low energy implantation of nitrogen into GaAs by the chemical shift of Ga3d photoelectron line and the corresponding change of the Ga3d binding energy ΔE_{BE} = 0.7÷0.4 eV (e.g. Li et al., 2001; Mikoushkin at al., 2009). Fig. 5 shows a significantly larger shift ΔE_A = 3.5 eV of Ga*LMM* Auger line resulting from N_2^+ ion bombardment. This Auger line shift is caused by the large change in the extra-atomic relaxation energy related to the change in the response of the whole electron system to creation of a vacancy in core level. Therefore the shift ΔE_A = 2.4 eV of the As*LMM* Auger spectrum represented in Fig. 6 is of the same order despite the restricted involvement of arsenic into the creation of new chemical bonds accompanied by the insignificant chemical shift of photoelectron As3d line (Li at al., 2001; Kumar at al., 2009; Mikoushkin at al., 2009).

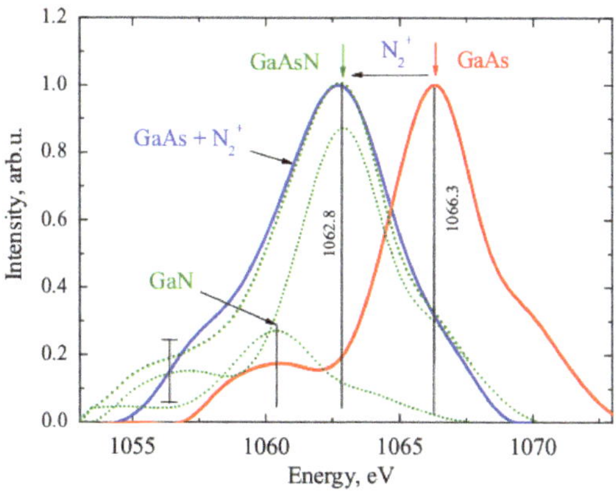

Fig. 5. Ga*LMM* Auger spectra of GaAs (right curve) and GaAs implanted by nitrogen ions with energy E_i = 2500 eV (left curve). The last one is decomposed into different contributions (dotted curves). The linear background is subtracted. The electron energy is measured relative to the Fermi level.

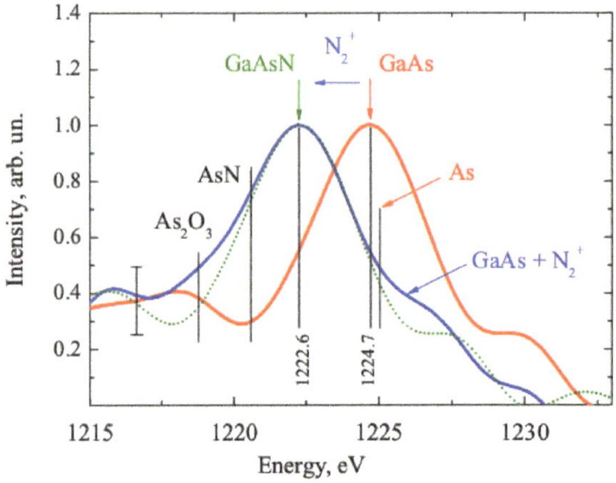

Fig. 6. As*LMM* Auger spectra of GaAs (right curve) and GaAs implanted by nitrogen ions with energy E_i = 2500 eV (left curve). The last one is mainly described by one contribution (dotted curve). Energy positions of Auger lines in AsN, As_2O_3 and As were taken in Ref. (Aksenov et al., 1999; Briggs & Seah, 1983).

Analysis of Ga*LMM* and As*LMM* spectra was performed by means of their decomposition into contributions corresponding to the chemical phases which should be expected in the nitrided layer: GaN, GaAsN, GaAs, AsN, As. Auger spectra of these contributions were approximated by the spectra of GaAs (elementary spectra) centered at the known energy positions or in the range of the values published in literature. Thus the shape and width of the elementary spectra were kept constant, whereas their relative intensity and energy position were varied for better agreement of the model with the experimental spectra. The assumption about the independence of the shape and width of the spectra from the chemical state seems to be reasonable since these Auger transitions involve only core levels. Fig. 6 shows that As*LMM* Auger spectrum of nitrided layer is mainly described by one line without essential contributions of GaAs, As and AsN phases. As was mentioned in previous part, absence of AsN was also reported in several studies of nitrided layers fabricated by low energy implantation (DeLouise, 1992; Hecht et al., 2001; Li et al., 2001; Kumar et al., 2009). Therefore the chemical state of arsenic at E_A = 1222.6 eV shown in Fig. 6 should be assigned to GaAsN. This conclusion can be made irrespective of the information about the position of the Auger line in AsN given in Ref. (Aksenov et al., 1998, 1999) and of the previous results cited. Indeed, AsN cannot be a dominant state of arsenic because the concentration of As atoms in the layer studied exceeds that of nitrogen atoms and exceeds manifold the concentration of nitrogen atoms unbound with Ga. Since the main state is not GaAs, it should be attributed to GaAsN (Mikoushkin, 2011).

Ga*LMM* spectrum also does not show any presence of GaAs, but it cannot be described by one line (Fig. 5). Decomposition of the spectrum shows additional contribution at lower kinetic energy E_A = 1060.5 eV besides the main line at E_A = 1062.8 eV. These lines were attributed to GaN and GaAsN phases, respectively, as the energy of Auger transition in GaN should be less than that in GaAsN because of larger core level chemical shift. The same sequence of Auger energies in these compounds follows from Ref. (Aksenov et al., 1998,

1999): E_A (GaN) = 1062.7 eV and E_A (GaAsN) = 1063.0 ÷ 1065.5 eV, though these values differ from the values obtained in this work. Thus, analysis of Ga*LMM* and As*LMM* Auger-spectra of the nitrided layer showed minor contributions of GaAs, As and AsN phases, confirmed formation of two nitrides observed earlier in Ref. (Majlinger et al., 2008, 2009; Mikoushkin et al., 2009) and revealed domination of the phase of GaAsN alloy in the layer studied.

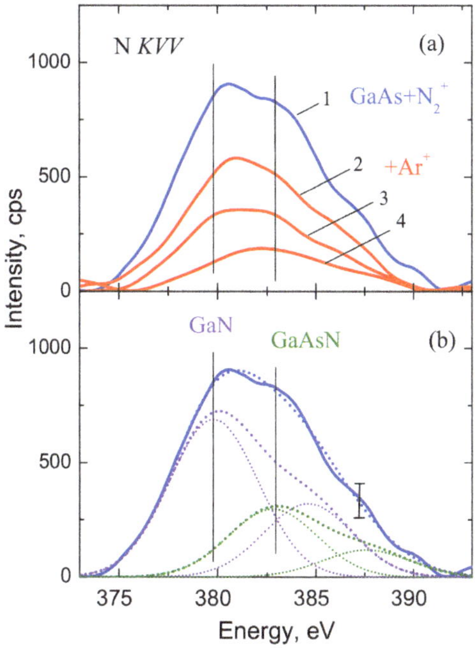

Fig. 7. a). N*KVV* Auger spectra of GaAs implanted by nitrogen ions with energy E_i = 2500 eV (1) and bombarded by Ar^+ ions with energy E_i = 2500 eV and increasing doses (2, 3, 4). b). N*KVV* Auger-spectrum of implanted GaAs (1) decomposed into different contributions (dotted lines). The linear background is subtracted.

GaN and GaAsN phases were identified unambiguously in N*KVV* spectra. Fig 7a shows N*KVV* Auger spectrum of GaAs after implantation (1). The spectrum demonstrates doublet structure in the region of the maximum which becomes even more reliably pronounced upon double differentiation and is indicative of the presence of two intense lines in the central part of each spectrum and two less intense components in the high energy wing. This shape is repeated in other N*KVV* spectra (2 - 4) measured in the course of etching the layer by Ar^+ ions. N*KVV* Auger spectra of nitrides are known to consist of two peaks of different intensity: the main and satellite ones (e. g., Aksenov et al., 1999). Therefore the experimental spectrum was modeled by two double peak elementary spectra with the main lines centered close to the above mentioned peculiarities in the central region of the experimental spectrum. The shape and width of the elementary spectrum were fixed again, while the intensity and position were varied. Fig. 7b gives an example of the spectrum decomposition, from which it can be seen that the sum of the elementary spectra (dotted curve) well describes the experimental spectrum (thick solid curve). Analysis of N*KVV* experimental spectrum shows two contributions with main peak positions at E_A = 379.8 ± 0.2 and 382.8 ± 0.2 eV. The first one coincides with the *NKVV* Auger energy of GaN and the second one turns up in the interval of Auger energies E_A = 381.5 ÷ 383 eV of disordered GaAsN phase generated in plasma nitridation of GaAs surface (Aksenov et al., 1998, 1999). Fig. 7b shows that the major part of nitrogen atoms is contained in GaN phase, which is in agreement with the results of the former studies of low energy ion nitridation of GaAs surface. The conducted analysis enabled us to determine the energies of *NKVV* Auger-transitions in GaN and GaAsN in one experiment with one sample, with the accuracy sufficient for reliable identification of these phases without referring to synchrotron radiation facilities.

Relative intensities of the elementary spectra in the decomposed *NKVV* spectrum give distribution of nitrogen atoms in GaN and GaAsN phases: $[N]_{GaN}$ = 0.7 and $[N]_{GaAsN}$ = 0.3. The concentration of chemically bound nitrogen atoms in the implanted layer was estimated to be [N] ~ 12 at %. Hence, the concentrations of GaN and GaAsN are: [GaN] ~ 12 * 0.7 *2 = 17 at.% and $[GaAs_{1-x}N_x]$ ~ 83 at.%, which is in qualitative agreement with the result of the less accurate decomposition of the Ga*LMM* spectrum represented in Fig.5: 25/75 at.% . The content of nitrogen in alloy is $[N]_{GaAsN}$ ~ 12 * 0.3 = 3.6 at %, hence x ~ 0.09. Thus, a nanofilm consisting mainly of dilute alloy $GaAs_{1-x}N_x$ with high concentration of nitrogen x ~ 0.09 was fabricated by low energy ion implantation.

Analysis of EELS spectra. A critical point in quantitative characterization of the chemical composition of the nitrided layer is the reliability of the conclusion made above about the negligible contribution of GaAs chemical phase. In contrast to the present study, an intensive signal of GaAs in photoelectron spectra was detected in former works, though it was explained by the contribution of GaAs substrate due to comparability of the mean free path of the photoelectrons with the thickness of nitrided layer. Experimental conditions of this work exclude the contribution of the substrate. But the phase of GaAs can be formed due to the low stability of alloy with high nitrogen concentration against external impacts such as ion bombardment: $GaAs_{1-x}N_x \rightarrow (1-x)GaAs + xGaN$ (Mikoushkin et al., 2008). Decay of alloy also enhances the content of the more stable chemical phase of GaN characterized by lower Gibbs energy. On the other hand, the fact of low GaAs concentration is the evidence of the reverse process of alloy creation. This process can be induced by secondary ion cascades generated by implanted ions. Ion cascades induce mixing atoms of the material, input of energy into the system and formation of the GaAsN phase. To confirm the possible role of ion cascades in forming nitride layer by low energy implantation, minor contribution of GaAs and other results of the conducted chemical analysis, an additional experiment with using Electron Energy Loss Spectroscopy (EELS) was performed. Fig. 8 shows EELS spectra of GaAs before and after implantation. The main peak of the spectra corresponds to the bulk plasmon. The plasmon energy of the implanted layer $\hbar\omega$ = 17.8 eV proved to be between the plasmon energies $\hbar\omega$ = 15.5 eV and $\hbar\omega$ = 19.5 eV (Sánchez et al, 2004) of GaAs and GaN, respectively. Decomposition of the implanted layer spectrum was done using GaAs spectrum as the elementary one for all constituents. The small contribution of transitions from Ga3d core level into unoccupied states revealed in GaN in Ref. (Sánchez et al, 2004) was added for better agreement of calculated and experimental spectra. The relative intensity of elementary spectrum in the decomposition is proportional to the corresponding fraction of the chemical phase. Fig. 8 shows contributions of GaAsN (main contribution), GaN (minor contribution) and GaAs (marginal contribution). Thus the conclusion about the domination of the GaAsN alloy in the implanted layer has been confirmed by the data of EELS experiment.

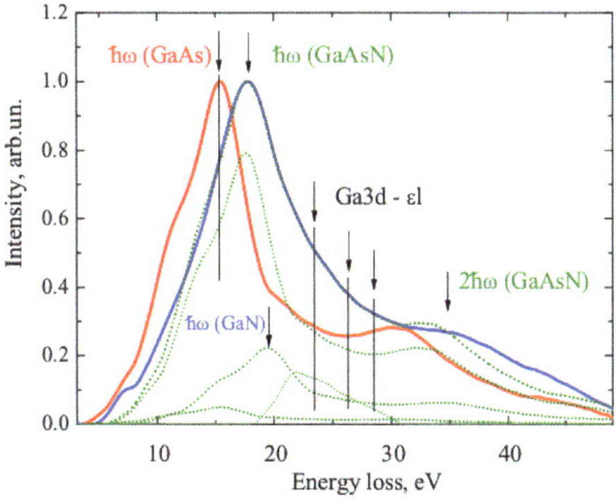

Fig. 8. EELS spectra of GaAs (left curve) and GaAs implanted by nitrogen ions with energy E_i = 2500 eV (right curve). The last one is decomposed into different contributions (dotted lines). The background of multiply scatted electrons is subtracted.

Conclusions. Thus, analysis of N*KVV* Auger spectrum of nitrided layer fabricated by low energy implantation revealed well resolved features associated with two compounds, GaN and GaAsN, characterized by small difference in core level chemical shifts. This made it possible to determine the energies of Auger transitions in these compounds in one experiment, with the accuracy being sufficient for reliable separation of corresponding contributions into experimental Auger line. Decomposition of N*KVV* Auger spectrum showed that GaN phase contains the major part of nitrogen atoms (~ 70%), which qualitatively agrees with former studies of implanted layers by XPS and is the evidence of reliability of the conducted Auger analysis. The experimental conditions, which ensured that the detected electrons did not contain signal of GaAs substrate, made it possible to determine the overall chemical composition of the nitrided layer. It proved to be a nanofilm with the thickness of about 4 nm, consisting mainly of dilute alloy $GaAs_{1-x}N_x$ with high concentration of nitrogen x ~ 0.09. High nitrogen concentration x implies formation of the narrow band gap (E_g < 1 eV) of the material. Domination of the nitrogen rich $GaAs_{1-x}N_x$ alloy in the nitrided layer was assumed to be provided by secondary ion cascades generated by implanted ions. The obtained results were confirmed by the analysis of EELS spectra of the nitrided layer. Thus it has been shown that the low energy implantation technique is a promising way for fabricating dilute $GaAs_{1-x}N_x$ alloys with high nitrogen concentration.

4. Formation of $GaAs_xN_{1-x}$ nanolayer on GaAs by manipulations with N_2^+ and Ar^+ ion beams

Introduction. The nitridation by low energy implantation of the epitaxial GaAs film described in previous section enabled obtaining the nitrided layer containing of about 80 at.% of dilute alloy $GaAs_{1-x}N_x$ with high nitrogen concentration x ~ 0.09. Surprisingly, the ordinary Auger electron spectroscopy proved to be an efficient tool for *in situ* diagnostics of chemical composition of GaAs nitrides. These two important results favored development of the implantation technology for fabrication of entire nanofilms or the nanofilms mainly consisting of dilute alloy $GaAs_{1-x}N_x$. The research devoted to this development resulted in enhancement of the $GaAs_{1-x}N_x$ alloy content up to 98 at.% due to chemical effect of "inert" Ar^+ ion beam on nitride nanolayers (Mikoushkin, 2010). This effect is caused in particular by the material homogenization due to intermixing of chemical phases by ion cascades efficiently generated by heavy argon ions. As a result of the nitrided layer modification, the major part of nitrogen goes from GaN to $GaAs_{1-x}N_x$.

Experimental details. The research was a continuation of the experiments described in previous section. Bombarding the sample surface with Ar^+ ions was performed directly in the analytical chamber using a scanning beam of the ionization type ion gun IQE-12/38 (Leybold AG). The ion energy was E_i = 2500 eV. The concentration of nitrogen in the implanted layer was [N] ~ 12 at.% and decreased to [N] ~ 2.5 at.% upon the bombardment with argon ions. The task of determining the chemical state of nitrogen in the implanted nitride layer was solved by the method of AES described above.

Analysis of depth profiles of the film chemical composition. Determining the thickness of fabricated nanofilms is of particular importance in their characterization. Rough estimation of the thickness considered as the projected range of ions was made by SRIM code (Ziegler at al., 1985). It is d ~ 3.6 nm for the nitrogen N_2^+ ions with energy E_i = 2500 eV. More accurate determination of the nitrided nanofilm thickness was made in this research by TRIM code calculations (Ziegler at al., 1985) of the dependence for the nitrogen concentration on the distance to the surface, or the nitrogen depth concentration profile. This depth profile normalized to the irradiating flux is shown in Fig. 9. The figure shows that real thickness of the implanted layer essentially exceeds the projected range d ~ 3.6 nm. The layer thickness considered as the full width at half maximum of the profile proved to be twice as much as the projected range: h ~ 8.5 nm.

Fig. 9 also shows the analogous depth profile for argon implanted into GaAs with energy E_i = 2500 eV which was calculated with the assumption that argon did not sputter the layer. Comparison of the profiles for argon and nitrogen results in conclusion important for the described experiment devoted to modification of the nitrided layer by Ar^+ ion bombardment: Ar^+ ions mix all atoms of the nitrided layer throughout the whole thickness. For further analysis of the modification process, one should take into account two processes induced by ion bombardment: (i) spattering the upper layer, and (ii) hammering atoms into deeper layers. These processes result in gradual diminishing the nitrogen concentration in the layer with retaining its thickness.

To analyse the chemical composition of the layer under modification, Auger spectra should be considered again. Some of the *NKVV* Auger electron spectra measured in this study were shown in Fig. 7a. Curve 1 corresponds to the spectrum of as prepared nitride layer. Curves 2 - 4 were obtained after the bombardment of the nitride layer by Ar^+ ions for 3, 7, and 12 min, respectively. A quantitative analysis of the revealed nitride phases was carried out by the spectra decomposition described above. The ratios of areas under the elementary spectra correspond to the content of nitrogen atoms in the revealed nitride phases. Decomposition of the as prepared nitrade film showed that only about 70% of nitrogen atoms enter into the GaN phase and about 30% of nitrogen is contained in the GaAsN phase.

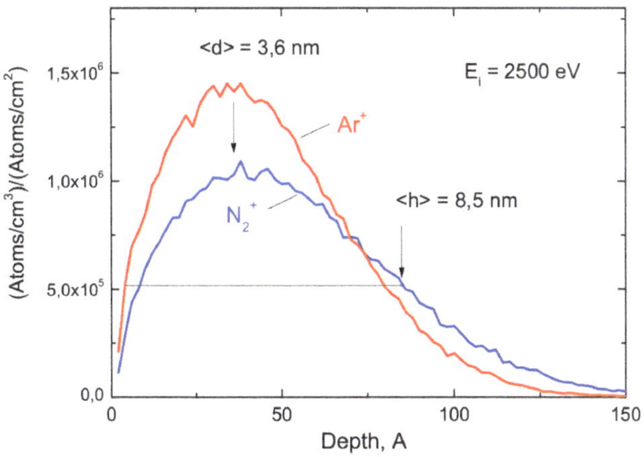

Fig. 9. Depth concentration profiles for nitrogen and argon implanted into GaAs with energy E_i = 2500 eV which were calculated by TRIM code (Ziegler at al., 1985).

As can be seen from Fig. 7a (curves 2 - 4), the bombardment of the nitride layer by Ar^+ ions leads to a redistribution of the intensities of the main nitride components, which is related to an increase in the fraction of the GaAsN phase. Fig. 10 shows plots of the relative content of nitrogen in the GaN and GaAsN phases versus duration of the sample bombardment by Ar^+ ions. In addition to the scale of irradiation duration, Fig. 10 also presents the scale of evaluated depth of the sputtered (etched-off) layer. This scale was calibrated to the TRIM calculated profile assuming that a decrease by half in the concentration of nitrogen in the

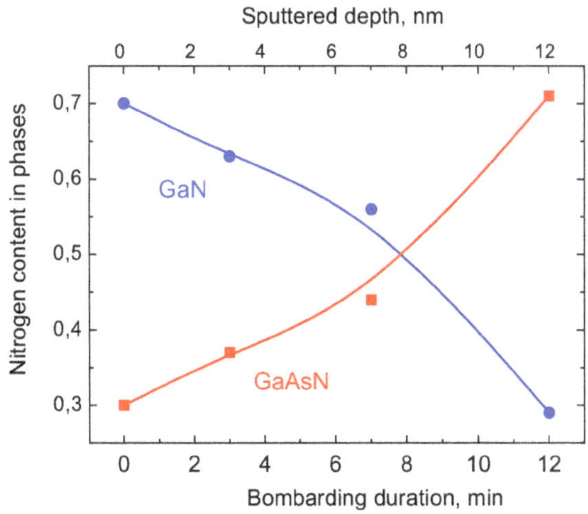

Fig. 10. Plots of the relative content of nitrogen in the GaN and $GaAs_{1-x}N_x$ phases versus duration of the sample bombardment by Ar^+ ions.

nitride layer approximately corresponded to the average depth ⟨h⟩ of nitrogen ion penetration. The data in Fig. 10 demonstrate the chemical effect of an argon ion beam on the nitride layer, which leads to a gradual disappearance of the GaN phase and a growth of the GaAsN phase. As a result of this treatment, the distribution of nitrogen between the two phases changes to the opposite.

Quantitative elemental and chemical composition of the nitrided layer and its variation in Ar^+ bombardment were determined in our experiment. Fig. 11 shows dose-dependencies of the Ga and As concentrations as well as the content of the main chemical phases GaN and GaAsN. Concentration of nitrogen is a difference between Ga and As concentrations. One can see that the initial nitrided layer contains ~83 at.% of GaAsN alloy. But the treatment of the layer by Ar^+ ion beam enhances this contribution above 98 at.%. The nitrogen content in $GaAs_{1-x}N_x$ alloy proved to be relatively high ranging from x ~ 0.09 to x ~ 0.04 in the course of Ar^+ bombardment.

The observed effect of the argon induced alloy formation can be explained by the intermixing of chemical phases (clusters of substances with different compositions) by ion cascades that are effectively generated by the heavy argon ions (Gordeev et al., 2003; Mikoushkin et al., 2008). In addition to the indicated chemical phases, this cascade process involves (i) residual (substituted) arsenic atoms not sputtered during the implantation and (ii) GaAs phase, the fraction of which increases as the nitride layer is sputtered by the argon ion beam and atoms of the GaAs substrate are involved in the process:

$$GaN + GaAsN + As + GaAs \rightarrow GaAsN. \quad (1)$$

The resulting solid solution is characterized by a greater relative content x of nitrogen and, hence, becomes unstable and exhibits partial decomposition (Mikoushkin et al., 2008):

$$GaAs_{1-x}N_x \rightarrow x\ GaN + (1-x)\ GaAs \quad (2)$$

The dynamic equilibrium of the processes (1) and (2) results in the distribution of nitrogen between the two nitride phases presented in Fig. 10. As the material is sputtered and the

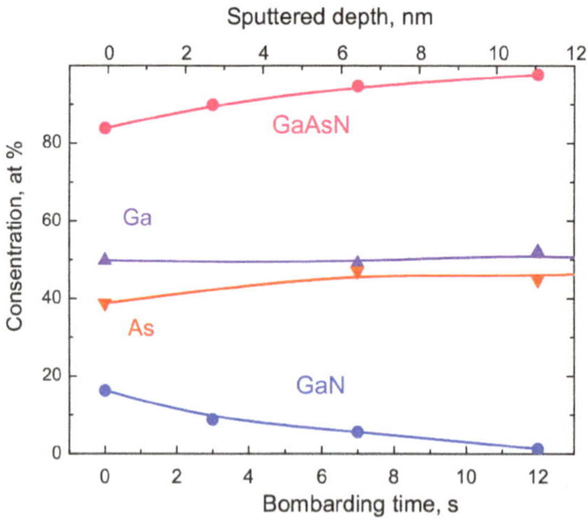

Fig. 11. Plots of the relative content of the main elements and chemical phases in the nitrided layer versus duration of the sample bombardment by Ar^+ ions.

nitrogen concentration [N] in the nitride layer decreases, the nitrogen concentration in $GaAs_{1-x}N_x$ alloy drops, this solution becomes stable, and its relative fraction increases.

Conclusions. Using the procedure of *in situ* Auger control, it has been established that only about 70% of nitrogen atoms of the nitrided layer prepared by implantation of N_2^+ ions with energy E_i = 2500 eV into an epitaxial layer of GaAs(100) is contained in the GaN phase. The remaining 30% enter into a GaAsN solid solution. It is established that the bombardment by accelerated argon ions produces a chemical effect on the nitride layer, which is related to a cascade mixing of the material. As a result, the nitrogen distribution between these competing phases changes to the opposite and the phase of GaAsN alloy becomes completely predominating. Finally, the 8 nm thick nitrided film consisting as far as 98 at.% of $GaAs_{1-x}N_x$ alloy with relatively high nitrogen concentration x ~ 0.04 has been fabricated by low energy implantation technique.

5. Conclusion

In situ quantitative chemical composition diagnostics of the nitrided nanolayers has been developed using high resolution synchrotron radiation based photoelectron spectroscopy and ordinary Auger electron spectroscopy. Chemical phases of GaN, GaAsN and GaAs were distinguished for the first time in nitrided layers formed on the GaAs surface by N_2^+ low energy ion implantation. Core level binding and Auger energies for these phases were measured in one experiment with high accuracy which made possible the developed diagnostics of the GaAs nitrides. We revealed that 5 nm thick nitride nanolayer created at high vacuum conditions by implantation of N_2^+ ions with energies below E_i = 1500 eV into GaAs(100) surface at room temperature is not an entire GaN layer. It consists of the narrow band gap clusters of $GaAs_{1-x}N_x$ alloy with high nitrogen concentration x ~ 0.10 in the wide band gap matrix of GaN. This system has a sign of a system of quantum dots. The implantation of N_2^+ ions at an energy of E_i = 2500 eV into an epitaxial layer of GaAs(100) was shown to form a nitride layer with a thickness of about 8 nm. Auger diagnostics for quantitative chemical analysis of the nitrided layers has been developed using the values of *NKVV* Auger energies in GaN and GaAsN chemical phases measured in one experiment, with the accuracy being sufficient for separating their contributions into the experimental spectrum. Using this diagnostics, it has been established that 70% of nitrogen atoms in the nitrided layer is contained in the GaN phase. The remaining 30% enter into a $GaAs_{1-x}N_x$ (x ~ 0.09) alloy. Nevertheless the alloy proved to be a dominant phase (> 80 at %) with inclusions of GaN clusters. It was established that the bombardment by accelerated argon ions produces a chemical effect which is related to a cascade mixing of the material. As a result of the nitride layer modification the distribution between these competing phases changes to the opposite and the domination of $GaAs_{1-x}N_x$ alloy becomes absolute (> 98 at %). Thus, the possibility of fabrication of the narrow band gap $GaAs_{1-x}N_x$ dilute alloy nanofilms with high nitrogen concentration x was demonstrated with the help of low energy ion implantation.

6. Acknowledgment

The research was supported by the Russian Foundation for Basic Research (RFBR № 08-08-12230-ofi), by the Ministry of education and science of Russia (Contract No P2431) and by the German- Russian Laboratory at BESSY II.

7. References

Aksenov, I.; Iwai, H.; Nakada, Y. & Okumura, H. (1998). Auger electron spectroscopy studies of nitridation of the GaAs (001) surface. , *Journal of Applied Physics*, Vol.84, No.6, pp. 3159-3166. ISSN 0021-8979

Aksenov, I.; Iwai, H.; Nakada, Y. & Okumura, H. (1999). Laser-induced interaction of ammonia with GaAs(100). I. Dissociation and nitridation, *Journal of Vacuum Science & Technology B*, Vol.17, No.4, pp. 1525-1539. ISSN 1071-1023

Bi, W.G. ; Tu, C.W. (1997). Bowing parameter of the band-gap energy of GaN_xAs_{1-x} , *Applied Physics Letters*, Vol.70, No.12, pp. 1608-1610. ISSN 0003-6951

Ziegler, J.F.; Biersack, J.P. & Littmark, U. (1985). *The Stopping and Range of Ions in Solids*, Ziegler, J.F. (Ed.), Pergamon Press, ISBN 008021603X , New York, USA

Briggs, D. & Seah, M.P. (1983). *Practical Surface Analysis by Auger and X-Ray Photoelectron Spectroscopy*, John Wiley & Sons, ISBN-10: 047126279X, Chichester/New York, USA

Davis, L.E.; MacDonald, N. C.; Palmberg, P.W.; Riach, G.E. & Weber, R.E. (1976). *Handbook of Auger Electron Spectroscopy, 2nd ed.*, Perkin-Elmer Corp., ISBN , Minnesota, USA

DeLouise, L.A. (1992). Reactive N_2^+ ion bombardment of GaAs(110): A method for GaN thin film growth. *Journal of Vacuum Science & Technology A*, Vol.10, No.4, pp. 1637-1641, ISSN 0734-2101

DeLouise, L.A. (1993). Nitridation ov GaAs(110) using energetic N^+ and N_2^+ ion beams. *Journal of Vacuum Science & Technology A*, Vol.11, No.3, pp. 609-614, ISSN 0734-2101

Fedoseenko, S.I.; Vyalikh, D.V.; Iossifov, I.E.; Follath, R.; Gorovikov, S.A.; Püttner, R.; Schmidt, J.-S.; Molodtsov, S.L.; Adamchuk, V.K.; Gudat, W. & Kaindl, G. (2003). *Nuclear Instruments & Methods in Physics Research A*, Vol. 505, No.3, p.p. 718-728. ISSN 0168-9002

Gordeev, Yu.S.; Bryzgalov, V.V.; Makarenko, B.N., Mikoushkin, V.M.; Konnikov, S.G.; Brunkov, P.N.; Ustinov, V.M. & Zhukov, A.E. (2003). Modification of GaAs by medium-energy N_2^+ ions. *Technical Physics*, Vol.48, No.7, pp. 885-888. ISSN 1063-7842

Hecht, J.-D.; F. Frost, Hirsch, D; Neumann, H.; Schindler, A.; Preobrajenski, A.B. & Chasse, T. (2001). Interstitiel nitrogen induced by low-enegry ion beam nitridation of AIII-BV semiconductor surfaces. *Journal of Applied Physics*, Vol. 90, No.13 , pp. 6066-6069. ISSN 0021-8979

Henini, M. (2005). *Dilute Nitride Semiconductors*, Elsevier Ltd., ISBN 0-08-044502-0, Amsterdam, Netherlands

Henke, B. L.; Gullikson, E. M. & Davis, J. C. (1993). *Atomic Data and Nuclear Data Tables*, Vol. 54, No.2, pp. 181. ISSN 0092-640X

Kumar, P.; Kumar, M.; Govind, S.; Mehta, B.R. & Shivaprasad, S.M. (2009). XPS investigation of ion beam induced conversion of GaAs(0 0 1) surface into GaN overlayer. *Applied Surface Science*, Vol.256, No.2, pp. 517-520. ISSN 0169-4332

Li, Y.G.; Wee, A.T.S.; Huan, C.H.A. & Zheng, J.C. (2001). Ion-induced nitridation of GaAs(100) surface. *Applied Surface Science*, Vol.174, No.3-4, pp. 275-282, ISSN 0169-4332

Majlinger, Z.; Bozanic, A.; Petravic, M.; Kim, K.-J.; Kim, B. & Yang, Y.-W. (2008). Interaction of low-energy nitrogen ions with GaAs surfaces. *Journal of Applied Physics*, Vol.104, No.6, pp. 063527-5. ISSN 0021-8979

Majlinger, Z.; Bozanic, A.; Petravic, M.; Kim, K.-J.; Kim, B. & Yang, Y.-W. (2009). NEXAFS and XPS study of GaN formation on ion-bombarded GaAs surfaces. *Vacuum*, Vol.84, No.1, pp. 41-44. ISSN 0042-207X

Meškinis, Š.; Šlapikas, K.; Puceta, M.; Tamulevičus, S. & Matukas, J. (2004). Effects of low-energy ion beam glancing angle nitridation on nGaAs surface and Co–nGaAs Schottky contact properties. *Vacuum*, Vol.77, No.1, pp. 79-86. ISSN 0042-207X

Mikoushkin, V.M.; Bryzgalov, V.V.; Gordeev, Yu.S. & Davidov, V.Yu. (2008). Formation and decomposition of nitrides under ion bombardment. *Bulletin of the Russian Academy of Sciences: Physics*, Vol.72, No.5, pp. 609-615. ISSN 1062-8738

Mikoushkin, V.M.; Bryzgalov, V.V.; Gordeev, Yu.S., Nikonov, S.Yu.; Solonitsina, A.P. & Brzhezinskaya, M.M. (2009). Chemical composition of GaAs-nitride nanolayers formed by implantation. *Physics Status Solidi C*, Vol.6, No.12, pp. 2655-2657. ISSN

Mikoushkin, V.M. (2010). Chemical Effect of Inert Argon Beam on Nitride Nanolayer Formed by Ion Implantation into GaAs Surface. *Technical Physics Letters*, Vol. 36, No.12, pp. 1136–1139. ISSN 1063-7850

Mikoushkin, V.M. (2011). Formation of $GaAs_{1-x}N_x$ nanofilm on GaAs by low energy N_2^+ implantation. *Applied Surface Science*, Vol.257, No.11, p.p. 4941-4944. ISSN 0169-4332

Moody, B.F.; Barletta, P.T.; El-Masry, N.A.; Roberts, J.C. ; Aumer, M.E. ; LeBoeuf, S.F. & Bedaira, S.M. (2002). Effect of H_2 on nitrogen incorporation in the metalorganic chemical vapor deposition of $GaAs_{1-y}N_y$ ($0<y<0.08$). *Applied Physics Letters*, Vol.80, No.14, pp. 2475-2477. ISSN 0003-6951

Pan, J.S.; Huan, C.H.A.; Wee, A.T.S.; Tan, H.S. & Tan, K.L. (1998). Auger electron spectroscopy and x-ray photoelectron spectroscopy analysis of angle of incidence effects of ion beam nitridation of GaAs. *Journal of Materials Research*, Vol.13, No.7, pp. 1799-1807, ISSN

Sánchez, A.M.; Gass, M.; Papworth, A.J.; Goodhew, P.J. & Ruterana, P. (2004). Physical Review B., Vol.70, No.3, pp. 035325-8. ISSN 1098-0121

Sik, J.; Schubert, M.; Leibiger, G.; Gottschalch, V. & Wagner, G. (2001). Band-gap energies, free carrier effects, and phonon modes in strained GaNAs/GaAs and GaNAs/InAs/GaAs superlattice heterostructures measured by spectroscopic ellipsometry. *Journal of Applied Physics*, Vol.89, No.1, pp. 294-305. ISSN 0021-8979

Spruytte, S.G.; Coldren, C.W.; Harris, J.S.; Wampler, W.; Krispin, P.; Ploog, K. & Larson, M.C. (2001). Incorporation of nitrogen in nitride-arsenides: Origin of improved luminescence efficiency after anneal, *Journal of Applied Physics*, Vol.89, No.8, p.p. 4401- 4406. ISSN 0021-8979

Toivonen, J.; Hakkarainen, T.; Sopanen, M. & Lipsanen, H. (2000). High nitrogen composition GaAsN by atmospheric pressure metalorganic vapor-phase epitaxy. *Journal of Crystal Growth*, Vol.221, No.1-4, pp. 456-460. ISSN 0022-0248

Veal, T.D.; Mahboob, I.; Piper, L.F.J.; McConville, C.F. & Hopkinson, M. (2004). Core-level photoemission spectroscopy of nitrogen bonding in GaN_xAs_{1-x} alloys. *Applied Physics Letters*, Vol. 85, No.9, pp. 1550-1552. ISSN 0003-6951

Zhu, X.-Y.; Wolf, M.; Huett, T.; White, J.M. (1992). Laser-induced interaction of ammonia with GaAs(100). I. Dissociation and nitridation, *Journal of Chemical Physics*, Vol.97, No.8, p.p. 5856-5867. ISSN 0021-9606

Zhukov, A.E.; Semenova, E.S.; Ustinov, V.M. & Weber, E.R. (2001). GaAsN-on-GaAs MBE Using a DC Plasma Source. *Technical Physics*, Vol.46, No.10, pp. 1265-1269. ISSN 1063-7842

11

Implantation Damage Formation in GaN and ZnO

Katharina Lorenz[1] and Elke Wendler[2]
[1]*Instituto Tecnológico e Nuclear, Instituto Superior Técnico, Universidade Técnica de Lisboa, Sacavém & Centro de Física Nuclear da Universidade de Lisboa, Lisboa*
[2]*Institut für Festkörperphysik, Friedrich-Schiller-Universität Jena, Jena*
[1]*Portugal*
[2]*Germany*

1. Introduction

GaN and ZnO and related group-III-N and group-II-O compounds are wide bandgap semiconductors suitable for the production of optoelectronic as well as high temperature, high frequency and high power electronic devices. While GaN based light emitting diodes and lasers are already realized on an industrial scale, ZnO still suffers from fundamental problems, in particular, the lack of efficient and reproducible p-type doping. Nevertheless, the two materials are considered competitors on the way to novel (opto)electronic applications (Yao & Hong, 2009).

1.1 Basic properties of GaN and ZnO

GaN and ZnO show several common characteristics in structural and optical properties. Besides the wide and direct band gap (3.437 eV in ZnO and 3.510 eV in GaN at low temperature) both materials show a similar valence band structure, and both crystallize preferentially in the wurtzite lattice structure. Fig. 1 illustrates the wurtzite structure that consists of two hexagonal lattices that are shifted relative to each other along the c-axis characterized by the so called u-parameter; one lattice is occupied by the cations (Ga/Zn) and the other by the anions (N/O). Some basic parameters of the two materials are summarized in Table 1. The interest in ZnO is increasing mainly due to its high exciton binding energy allowing efficient short wavelength emitters working at room temperature and due to the availability of inexpensive bulk crystals for homoepitaxial growth.

Another common feature of both semiconductors is their strong resistance to radiation damage (Kucheyev et al., 2004). The latter is an important characteristic for device processing steps involving ion implantation, for example for doping or isolation and for the use of electronic circuits in radiation environments, such as in space or nuclear reactors.

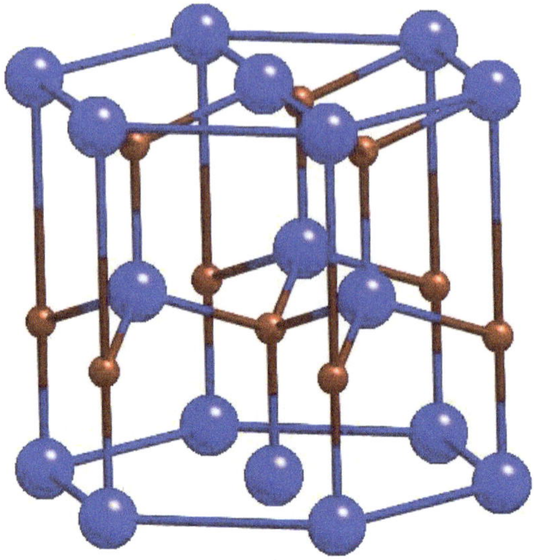

Fig. 1. Schematic of the wurtzite lattice consisting of two simple hexagonal lattices occupied by the cations (Ga/Zn) and the anions (N/O).

	GaN	ZnO
a lattice parameter (Å)[a]	3.189	3.250
c lattice parameter (Å) [a]	5.185	5.204
c/a ratio (ideal 8/3 = 1.633) [a]	1.626	1.601
u-parameter (ideal 3/8 = 0.375) [a]	0.376	0.382
Density (g/cm^3)	6.15[b]	5.675[d]
Band gap (eV) [a]	3.510	3.437
Exciton binding energy (eV) [a]	23.44	60
Melting point (K) [b]	2773	2248
Phillips Ionicity[c]	0.5	0.616

Table 1. Crystal parameters and properties of GaN and ZnO taken from [a](Yao & Hong, 2009), [b](Ioffe Institute, St. Petersburg, http://www.ioffe.rssi.ru/SVA/NSM/), [c](Adachi, 2005), [d](Klingshirn et al., 2010)

1.2 Implantation in GaN and ZnO

While ion implantation is routinely applied in silicon industry for electrical doping or implant isolation, this technique is not very much developed for devices based on wide band gap semiconductors. Nevertheless, prototypes of GaN-based field effect transistors and LEDs employing processing by ion implantation were reported (Pearton et al., 2006) and ion implantation was for example successfully used for optical doping of nitrides using rare earth ions (see e.g. Lorenz et al., 2010a; and references therein). Also magnetic doping for spintronics research was realized for both GaN and ZnO using ion implantation of transition metals or rare earth ions (Pearton et al. 2003, 2004, Brandt et al., 2010). Furthermore many groups investigate the damage formation in ZnO and GaN upon ion implantation and its effects on the materials' properties (Kucheyev et al., 2001,2004; Ronning et al., 2001; Pearton et al., 2006; and references therein).

The main effect of medium energy ions (several hundred keV range) on materials is caused by energy transfer through atomic collisions (nuclear energy loss). However, other processes active during the implantation are influencing the damage retained after the implantation such as the diffusion of defects (strongly dependent on the implantation temperature) as well as electronic excitations due to the electronic energy loss. The latter can for example change the charge state of defects and influence their diffusivities. Typically, samples are implanted and measured *ex situ* after the implantation to describe the retained lattice damage. An important technique to study implantation damage is Transmission Electron Microscopy (TEM) allowing a direct visualization of defects and the determination of their microscopic nature. Other techniques such as Rutherford backscattering/channelling (RBS/C), X-ray diffraction, Raman scattering, optical spectroscopy or electrical measurements, among others, determine the effect of radiation defects on structural, optical or electrical properties of the materials.

RBS/C allows the determination of defect concentrations with depth resolution and was applied by many groups to investigate implantation damage build-up in GaN and ZnO (e.g. Alves et al., 2003; Jiang et al., 2000; Kucheyev et al., 2001, 2003, 2004; Lorenz et al., 2005, 2009; Wendler et al., 2003). All these reports agree that both GaN and ZnO exhibit strong dynamic annealing during implantation for a vast variety of ion species, energies and implantation temperatures. Dynamic annealing denotes the crystal recovery during the implantation by the recombination of interstitials and vacancies due to close pair recombination as well as defect migration over longer ranges. These effects are of course strongly dependent on the implantation temperature and ionization processes caused by the irradiation. Such processes can be evidenced in RBS/C measurements since they lead to a non-linear increase of defect concentration with ion fluence. Very high amorphisation thresholds are reported for both ZnO and GaN due to this efficient dynamic annealing. In fact, amorphisation of ZnO was only observed for high fluence Si implantation attributed to strong chemical effects (Kucheyev et al., 2003), while heavy ion Au implantation at 77 K (Kucheyev et al., 2003) and Ar implantation at 15 K (Lorenz et al., 2005) did not render ZnO amorphous for fluences as high as 4×10^{16} at./cm^2. GaN can be rendered amorphous at room temperature and below for high fluence implantation (e.g. 2×10^{16} Ar/cm^2 at 300 keV and 15 K. Wendler et al., 2003) but the exact processes are still under debate (Ruterana et al., 2011). Like in ZnO the implanted species have strong effects on the amorphisation behaviour (Kucheyev et al., 2004). In general ZnO was found to be more resistant to radiation damage. One reason for this may be the higher ionicity of the bonds (see Table 1) which was shown to enhance the radiation hardness in semiconductors (see e.g. Trachenko, 2004; Wendler, 2011).

Besides defect recombination, mobility of point defects during the dynamic annealing processes also favours the formation of extended defects such as point defect clusters, stacking faults or dislocation loops. Indeed, TEM investigations for GaN (Wang et al., 2002; Gloux et al., 2006) and ZnO (Perillat-Merceroz et al., 2011) showed the prominence of these defect types in both materials for room temperature implantations.

In this work we review studies on the implantation damage build-up at 15 K in GaN and ZnO measured by RBS/C (Wendler et al., 2003; Lorenz et al., 2005) and discuss similarities and differences of both materials. At this low temperature thermal diffusion of atoms and defects is minimized. However, defect movement and rearrangement of atoms are still possible, eventually enhanced by lattice strain and electronic excitation processes within ion tracks. Furthermore, the mobility of point defects at temperatures well below room temperature will be demonstrated. Finally, we will discuss the recent advances of optical doping of ZnO and GaN by rare earth implantation.

2. Damage build-up for implantation at 15 K and low temperature annealing

2.1 Experimental details

Ion implantations were performed at 15 K in a special target chamber at the Institut für Festkörperphysik in Jena, where *in situ* RBS/C measurements allow monitoring the defect accumulation without changing the target temperature (Breeger et al., 2001). A schematic of the set-up is shown in Fig. 2. The target chamber is connected to both the 400 keV ion implanter and the 3 MV Tandetron accelerator used to produce the He^{+} beam for the RBS/C

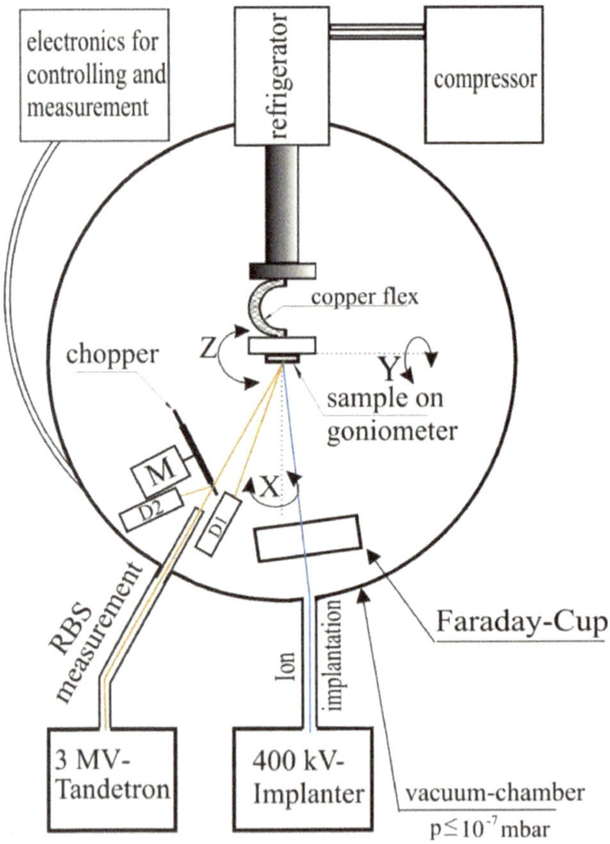

Fig. 2. Schematic representation of the two-beam chamber used for the low temperature *in situ* RBS/C measurements (Breeger et al., 2001).

measurements. In each implantation series a sample is implanted step-wise and after each implantation step a simple rotation of the sample is required to bring it in position for the RBS/C measurement. The sample is mounted on a four axis goniometer and cooled by a closed-cycle He refrigerator. A heating system allows tuning the temperature between 15 K and room temperature. The implantations were carried out with the surface normal tilted by 7° off the ion beam direction to prevent channelling effects. The RBS/C analysis was performed with 1.4 MeV He^+ ions and a Si surface barrier detector at 170°.

Commercial (Eagle-Picher) O-face (0001) ZnO single crystals were implanted with N, Ar, and Er ions to fluences from 1×10^{11} to 7×10^{16} at./cm^2 using energies of 80, 200, and 380 keV, respectively. GaN layers grown, either by Metal Organic Chemical Vapour Deposition (MOCVD) or by Hydride Vapour Phase Epitaxy (HVPE), on (0001) sapphire substrates were implanted with 300 keV Ar ions to fluences from 3×10^{12} to 4×10^{16} at./cm^2 as well as with 150 keV O, 780 keV Te, 800 keV Xe or 380 keV Au ions.

2.2 Damage build-up in ZnO

Figure 3 shows RBS/C spectra for Ar implanted ZnO (Wendler et al., 2009). The aligned spectrum of the as-grown sample is also plotted for comparison. The backscattering yields of the aligned spectra increase progressively with the implantation fluence due to the build-up of implantation damage, but even for the highest fluence of 7×10^{16} Ar/cm^2 the yield remains well below the random backscattering level. As a measure for the produced damage concentration, the difference in minimum yield $\Delta\chi_{min}$ is determined, where $\Delta\chi_{min}$ is given by $\Delta\chi_{min} = (Y_{al} - Y_{al}^{virgin})/Y_{ra}$. Y_{al}^{virgin} and Y_{al} are the RBS yields of the aligned spectra for the unimplanted and implanted sample, respectively, and Y_{ra} is the maximum yield measured in random direction. To calculate the damage level the yields were integrated within windows comprising the entire implanted layer leaving out the surface peak, as indicated in Fig. 3.

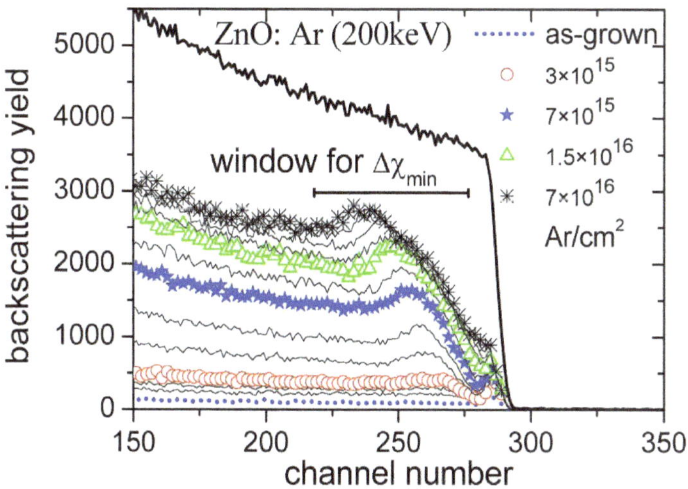

Fig. 3. RBS/C spectra for ZnO implanted with Ar to different fluences compared to the as-grown sample. The window used for the determination of $\Delta\chi_{min}$ is indicated.

Figure 4a shows $\Delta\chi_{min}$ as a function of the ion fluence N_I for the three implanted ion species (Lorenz et al., 2005). The curves do not show a simple linear increase of damage with the fluence but two plateaus can be distinguished, where, within wide fluence ranges, the damage level increases only slightly. Such behaviour is common in materials with strong dynamic annealing effects (Kucheyev et al., 2004).

Four stages of damage production can be distinguished, labelled stage I to IV in Fig. 4, and which are interpreted as follows: In stage I, an almost linear increase of the lattice damage with ion fluence is observed due to the formation of point defects in well separated collision cascades. In stage II the damage saturates; this is attributed to the overlapping of the damage cascades of single ions, allowing for an increased recombination of point defects and the rearrangement of atoms to different defect structures. In stage III the damage increases steeply indicating a slowing down of recombination processes and an enhanced formation of stable defect structures. Stage IV shows a second plateau, being still considerably below the amorphisation level. The last stage was only reached in the case of Ar implantation.

The lines in Fig. 4 are calculated using a simplified version of the Hecking model (Hecking et al., 1986). From the fit of this model to the data, in particular from the first low fluence regime (stage I), the displacement energy for Zn could be estimated to be 65 eV (Lorenz et al., 2005, Wendler, 2011). Although this value is only an upper limit since in-cascade recombination of point defects can occur even at 4 K, the obtained value agrees well with those found by other authors (see Zinkle & Kinoshita, 1997 and references therein).

Simulations with the Monte Carlo program SRIM 2008 (Ziegler et al., 2008) were performed to determine the number of Zn displacements per ion and unit depth at the maximum of the damage profile, N_{displ}, using values for the displacement energies of Zn and O of 65 and 50 eV, respectively. Subsequently, the number of displacements per Zn atom, n_{dpa}, was calculated for each fluence and ion using $n_{dpa}=N_{displ}\cdot N_I/N_0$, where N_I is the ion fluence and $N_0=4.1475\times10^{22}$ Zn/cm^3 the atomic density of Zn in ZnO. This procedure allows the direct comparison of damage build-up curves for different ions or energies. In Fig. 4b the damage build-up curves are re-plotted as a function of n_{dpa}. In this representation the curves for different ions almost overlap. The slight differences observed may be due to the different nature of defects created by ions with largely different masses. The collision cascades are denser in the case of Er implantation, favouring the formation of extended defects and defect clusters, which may be the reason why the steep damage increase in stage III sets in at lower n_{dpa} values than it does for the lighter ions.

Fig. 5 shows selected aligned spectra for N and Er implantation for two different fluences. The fluences were chosen to give approximately the same values of $\Delta\chi_{min}$ for both Er and N implantation and the defect levels DL1 and DL2 corresponding to the selected spectra are marked in Fig. 4 (b). As expected, the fluence to produce a certain damage level is lower for the heavier Er ion. The shape of the aligned spectra is very different for N and Er implantation even for samples showing similar $\Delta\chi_{min}$ values. For N implantation, for both defect levels the expected approximately Gaussian defect distribution is clearly seen from the increased backscattering yield within the implanted region (channel 250 and higher) while the backscattering yield decreases again for deeper layers (below channel 250).

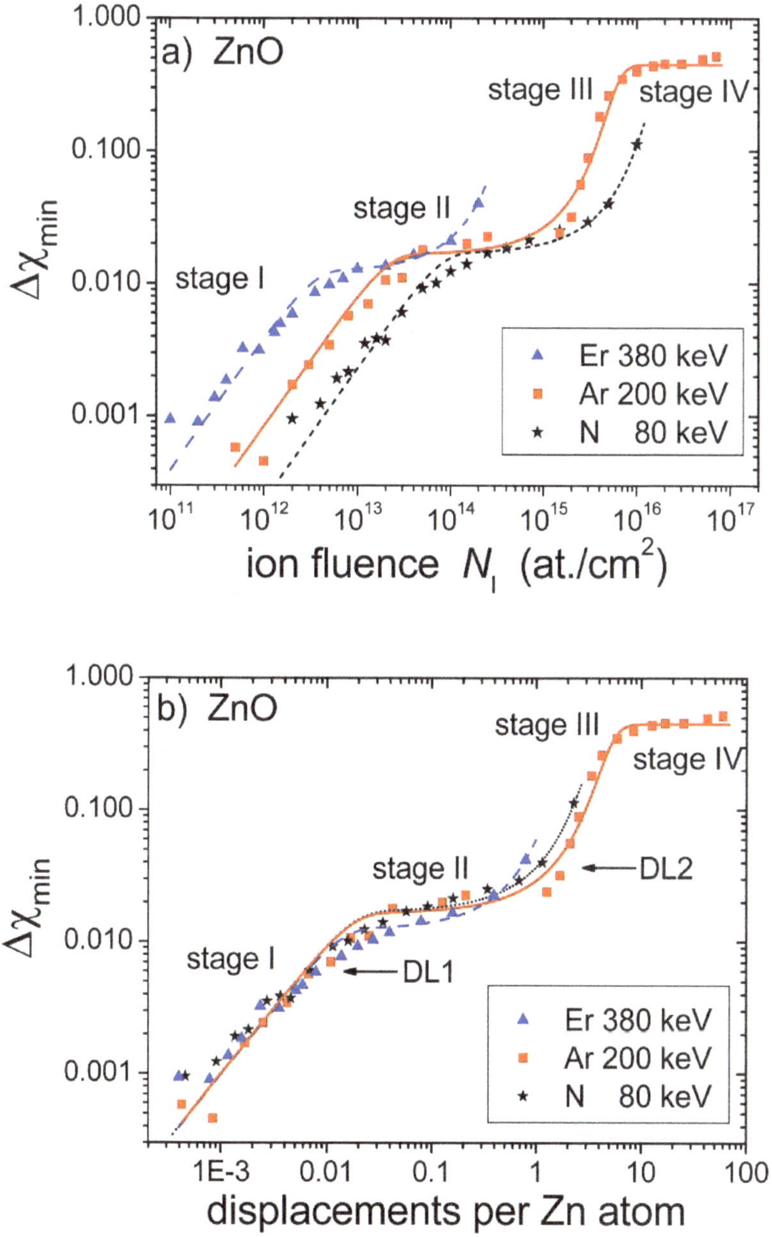

Fig. 4. Difference in minimum yield $\Delta\chi_{min}$ in ZnO as a function of the ion fluence (a) and as a function of the number of displacements per Zn atom, n_{dpa} (b) for 80 keV N, 200 keV Ar and 380 keV Er implantation into ZnO single crystals. The plotted curves are fits based on the defect interaction and amorphisation model of Hecking (Hecking et al., 1986).

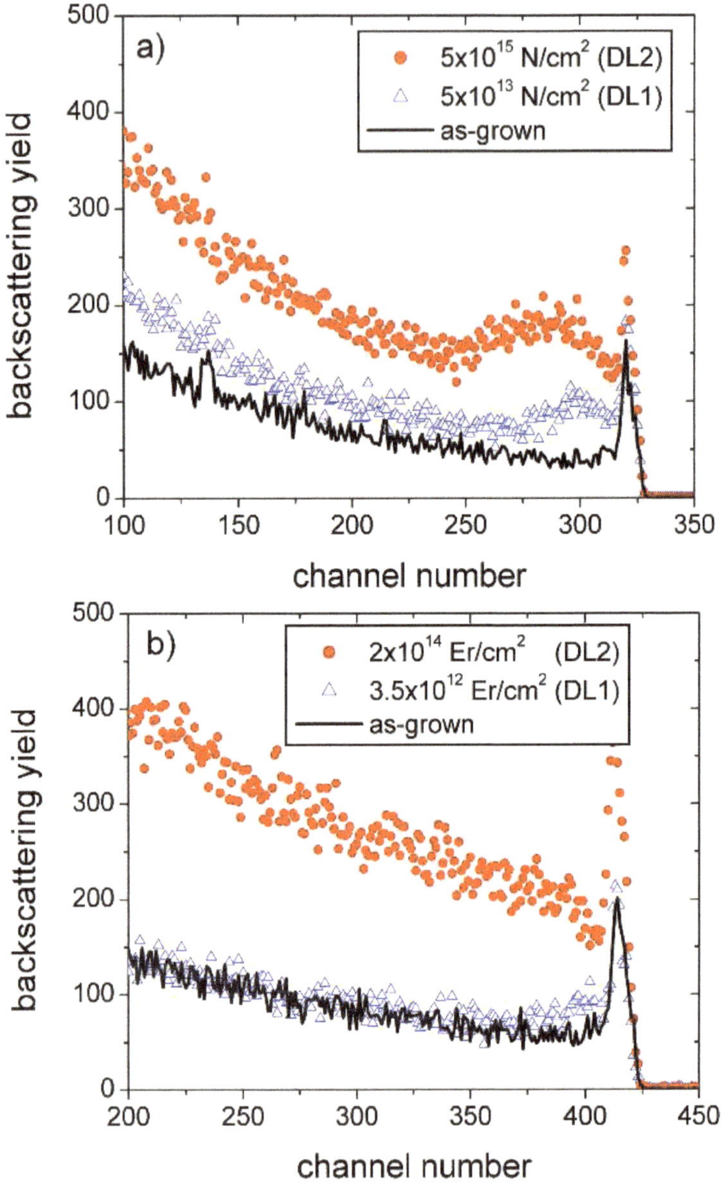

Fig. 5. <0001> aligned RBS/C spectra for ZnO implanted with N (a) and Er (b) to different fluences corresponding to defect levels DL1 and DL2 as marked in Fig. 4b compared to the spectrum of the as-grown sample.

Contrary, for Er implantation to DL2 such a decrease of $\Delta\chi_{min}$ is not observed for depths beyond the implanted region (below channel number 350). It was shown previously that point defects and amorphous zones preferentially lead to direct backscattering at displaced atoms while stacking faults and dislocation loops rather contribute to the increasing minimum yield by dechannelling of the analyzing beam (see e.g. Feldmann, 1982; Wendler et al., 2009). Therefore, both the high dechannelling rates (Fig. 5b) and the early on-set of stage III in the damage build-up curve (Fig. 4b) for Er suggest that heavy ions facilitate the formation of extended defects like dislocation loops.

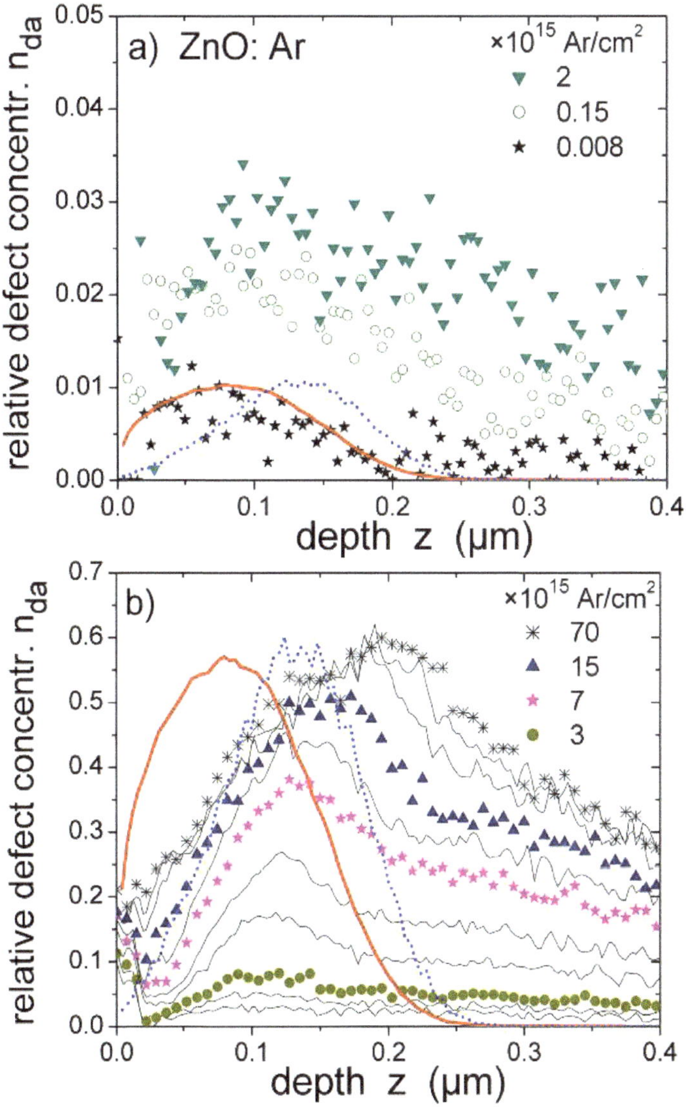

Fig. 6. Defect profiles $n_{da}(z)$ for 200 keV Ar implanted ZnO for various ion fluences. For clarity in (b) only some fluence values are given in the legend. The calculated distributions of primary displacements (red solid line) and implanted ions (blue dashed line) derived from SRIM simulations are given in arbitrary units for comparison.

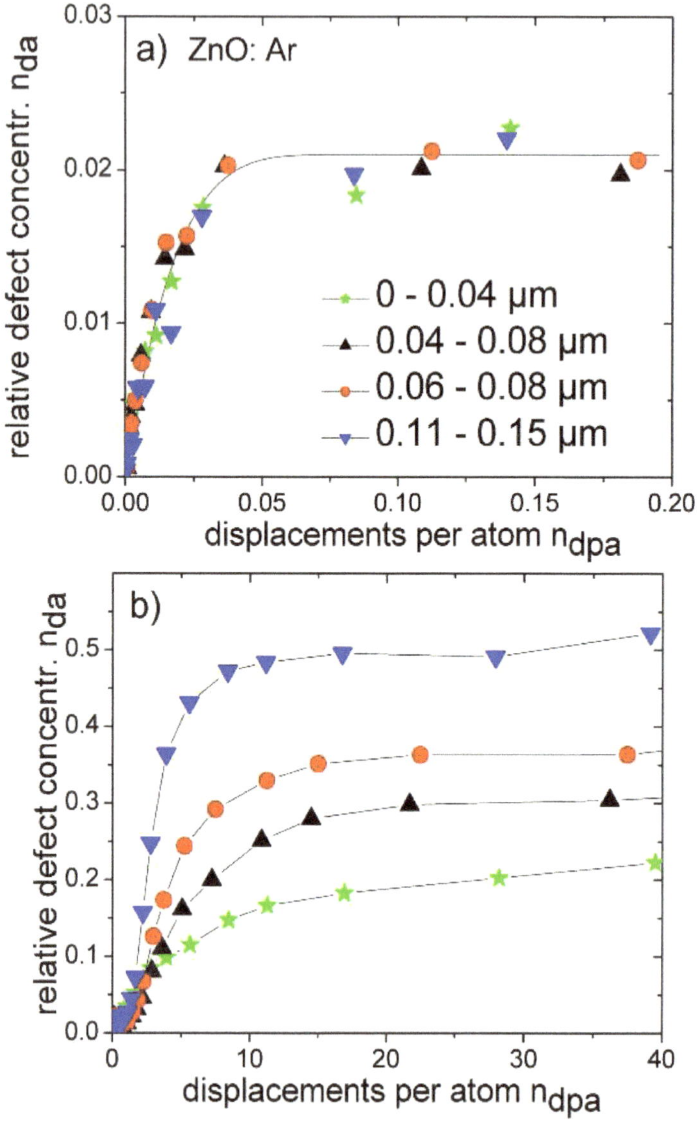

Fig. 7. Relative defect concentration n_{da} at different depth regions versus the corresponding number of displacements per lattice atom n_{dpa} for 200 keV Ar implanted ZnO. Part (a) represents stages I and II and part (b) stages III and IV. The maximum nuclear energy deposition is at 0.06-0.10 µm and the maximum ion concentration at 0.11-0.15 µm (see Fig. 6).

The occurrence of extended defects becomes especially obvious when analyzing the RBS/C spectra in more detail, which is done for Ar implanted ZnO (Wendler et al., 2009). Figure 6 plots the relative concentration of randomly displaced lattice atoms n_{da} versus the depth z (referred to as defect profiles in the following text). The calculated distribution of primary displacements and the Argon profile, both derived from SRIM simulations, are also included for comparison. The profiles were calculated with the computer code DICADA (Gärtner, 2005) in order to take into account dechannelling. The calculations assume the occurrence of uncorrelated displaced lattice atoms which follow when point defects, clusters of point defects or amorphous zones exist within the implanted layer. These defect types will contribute to the backscattering yield primarily by direct backscattering. Dislocations and stacking faults, on the other hand, will induce strong dechannelling of the analysing beam while the direct backscattering yield is relatively low (Feldman et al., 1982). Behind the implanted layer (below ~0.3 μm), n_{da} should reach zero if the correct type of defects is assumed in the calculation. Within the experimental error this is only the case for very small ion fluences ranging within stage I (see profile for 8×10^{12} Ar/cm^2 in Fig. 6 (a)). This supports the interpretation that stage I is mainly correlated with the production of point defects. Within stage II strong defect recombination results in a nearly constant defect concentration (see Fig. 4). Furthermore, the profiles in Fig. 6 (a) for 1.5×10^{14} and 2×10^{15} Ar/cm^2 (both fluences belong to stage II) suggest that there is an extension of the defect distribution towards deeper regions and the defect concentration appears to deviate from zero below the implanted layers. These findings indicate that, within stage II, point defects do not only recombine but extended defects start to form like previously observed for room temperature implantation in ZnO (Perillat-Merceroz et al., 2011). The processes within stages I and II are solely determined by the energy deposition into collision processes. This can be seen from Fig. 4 (b) which compares results for various ion species and shows a perfect overlap of the damage build-up curves and also from Fig. 7 (a) which compares the defect concentration for Ar implanted layers at various depths. The curves overlap for stages I and II.

For stages III and IV, on the other hand, Fig. 4 (b) shows that the curves $\Delta\chi_{min}$ versus n_{dpa} start to split for the various ion species suggesting that the implanted ion species themselves become important for the damage formation. The same result can be deduced from the profiles of the Ar implanted ZnO layers in Fig. 6 (b). The dominant rise of defect concentration within stage III starts at a depth close to that of maximum displacements (see profile for 3×10^{15} in Fig. 6 (b)) but then shifts progressively to larger depths near the implanted Ar ions (see profile for 7×10^{15} Ar/cm^2 in Fig. 6 (b)). From Fig. 6 (b) it is obvious that for ion fluences within stages III and IV, the calculated profiles do not reach zero and long artificial tails extend far beyond the implanted layer (the maximum range of primary displacements or ions is about 0.3 μm). This suggests that the dechannelling of the analyzing ions within the implanted layer is not correctly treated when assuming only uncorrelated displaced lattice atoms. Therefore the implanted layers must contain extended defects. Most probably the damage consists of a mixture of point defect clusters, stacking faults and dislocation loops. The occurrence of secondary effects in the damage formation, being not solely determined by the nuclear energy deposition for stages III and IV, is also seen in Fig. 7 (b), which shows the defect concentration versus the number of displacements per lattice atom at various depths. In contrast to the results in Fig. 7 (a) no uniform dependence is found.

2.3 Damage build-up in GaN

In Fig. 8 a set of RBS/C spectra is plotted for Ar implanted GaN (Wendler et al., 2003). For low and medium fluences the spectra look similar to those of ZnO and also reveal high dechannelling rates. GaN shows a more pronounced surface damage peak than ZnO. Strong

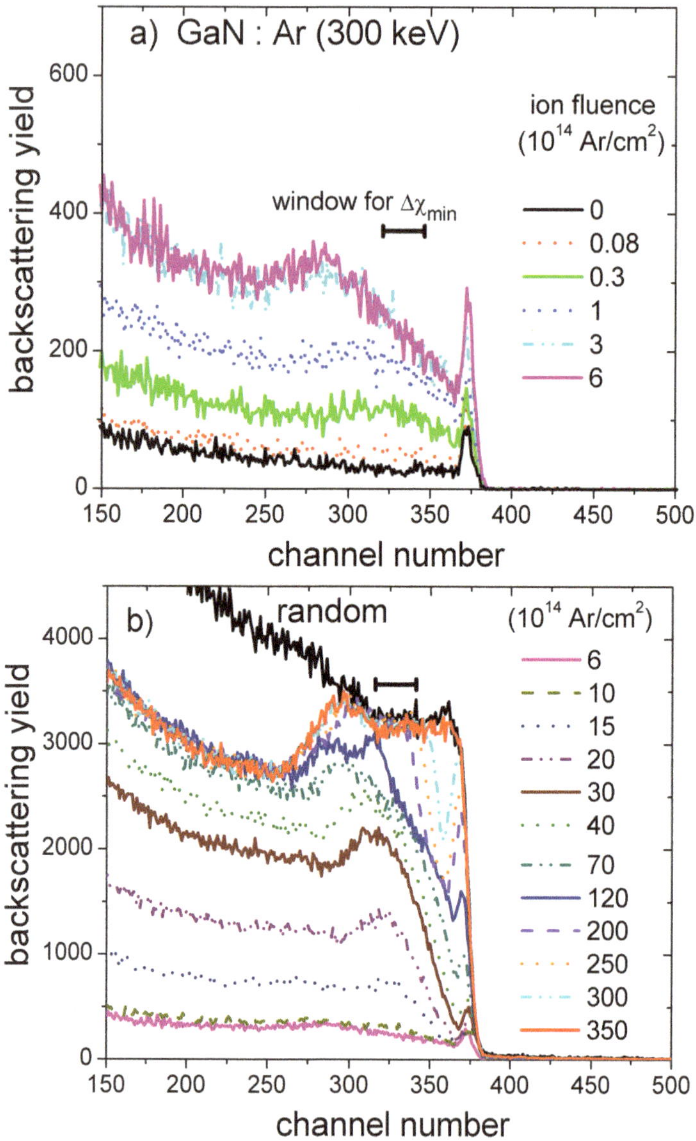

Fig. 8. <0001> aligned RBS/C spectra for GaN implanted with 300 keV Ar to different fluences: a) low ion fluences compared to the as-grown sample and b) high fluences with random spectrum as reference. The window used for the determination of $\Delta\chi_{min}$ is indicated.

surface damage in GaN is also often seen for room temperature implantation usually attributed to the surface acting as a sink for migrating point defects. For fluences between 1.2 and 2×10^{16} Ar/cm^2 the random level is reached (around channel no. 320) which is commonly taken as an indication for amorphisation. Amorphisation processes in GaN are complex and not well understood. For room temperature implantation it was believed to start at the surface (Kucheyev et al., 2004). However, recent studies indicate that this highly damaged surface layer often consists of randomly oriented nanocrystals rather than an amorphous phase (Ruterana et al., 2011). At 15 K we assume that real amorphisation occurs since dynamic annealing is slowed down, however, from RBS/C measurements alone it is not possible to distinguish amorphous and poly/nanocrystalline material.

To see this effect more clearly in the damage build-up curves, the window for the determination of the $\Delta\chi_{min}$ does not contain the whole implanted layer (see Fig. 8). Fig. 9 (a) shows the damage build-up curves as a function of the fluence for different ion species. In order to compare the results for the various implanted ion species, in Fig. 9 (b) the ion fluence was converted into displacements per Ga atom. Herein a displacement energy of 41eV for Ga and O was taken in the SRIM calculations, with the value for Ga being determined in the same way as for ZnO by fitting the Hecking model to the damage build-up curve (Wendler et al., 2004). This value of the displacement energy is in good agreement with molecular dynamics calculations yielding an average value of 45eV for Ga and 109 eV for N (Nord et al., 2003). The curves in Fig. 9 exhibit five stages of damage formation with the first four being similar as discussed above for ZnO and stage V being the amorphisation which occurs for all ion species (Wendler et al., 2003).

Figure 9 suggests that stage I occurs at given numbers of n_{dpa} independent of the ion species. The first saturation value in stage II is similar for all ion species except for Au ions which may be connected with the fact that in this case the thickness of the implanted layer is only one third of that for the other ion species. Stages III, IV and V shift to larger n_{dpa} values with decreasing mass of the implanted ions, pointing to the role of the implanted ions themselves in these cases. More details about the defect evolution can be seen in the defect profiles plotted in Fig. 10 for Ar implanted GaN, which were calculated as described above for ZnO. Again, simulations for the profiles of primary displacements and Ar are included for comparison. Within stage II there is not only a recombination of defects causing the very weak increase of the defect concentration with rising ion fluence as seen in Fig. 9, but the defect profiles extend into larger depth (see Fig. 10 (a) and (b); stage II for Ar ranges from 1×10^{14} Ar/cm^2 to about 1×10^{15} Ar/cm^2). This indicates the occurrence of some defect mobility even at 15 K and suggests that the formation of more complex defect structures starts already within stage II. The formation of dislocation loops and stacking faults is typically observed in GaN implanted at room temperature (e.g. Kucheyev et al., 2001; Gloux et al., 2006). Their formation may be driven by the large strains induced into the implanted layers by defects and implanted ions which cause an expansion of the *c* lattice parameter as frequently observed by X-ray diffraction (Liu et al., 1997). At the onset of stage III, defect accumulation occurs preferentially at the depth of maximum nuclear energy deposition and within stage IV the defects again extend into larger depth (see Fig. 10 b) and c)). For all ion species studied, amorphous seeds nucleate within a rather narrow layer at a depth below the surface but which is neither that of maximum nuclear energy deposition nor that of maximum ion concentration (see Fig. 10 d)). A further increase of the ion fluence renders the implanted layer amorphous similar to a process known for silicon called ion beam induced interfacial amorphisation (IBIIA) (Priolo & Rimini, 1990). As explained above, within stage III and IV a complex defect structure of point defect clusters and extended defects is to be expected with the extended defects causing the artificial tails in the defect profiles at depths beyond 0.35 μm. The transformation into the amorphous state is also suggested by the successive disappearance of these tails in the defect profiles with increasing ion fluence (see Fig. 10 d)).

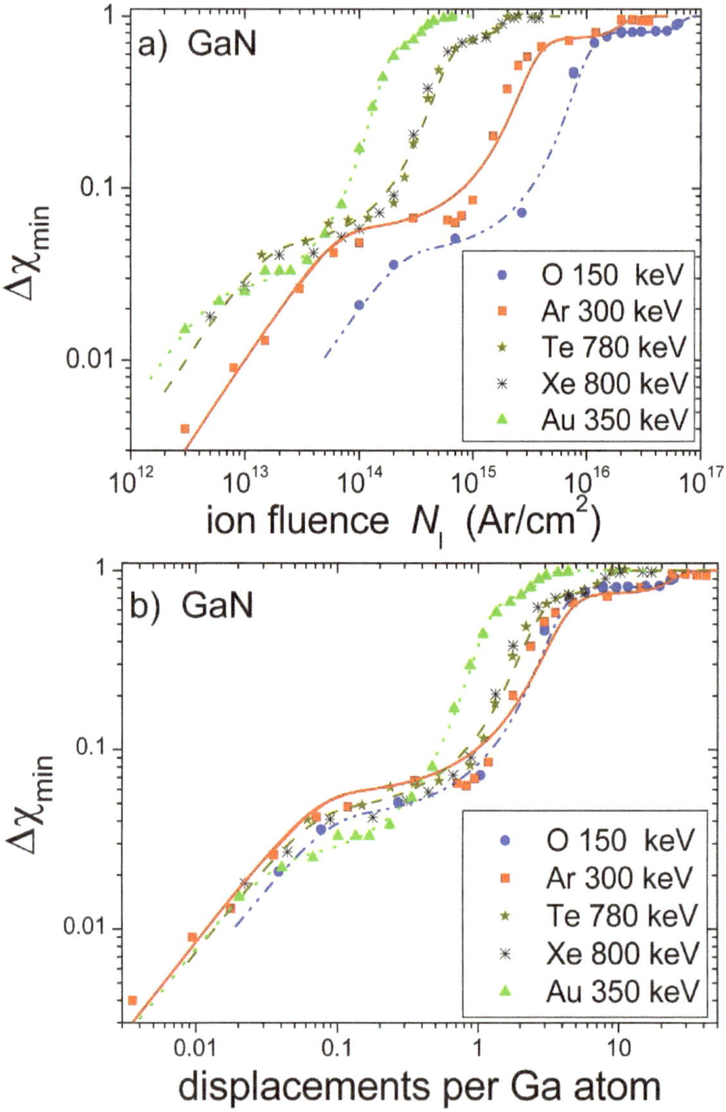

Fig. 9. Difference in minimum yield $\Delta\chi_{min}$ as a function of the fluence (a) and of the number of displacements per Ga atom, n_{dpa} (b) for 150 keV O, 300 keV Ar, 780 keV Te, 800 keV Xe and 350 keV Au implantation into GaN.

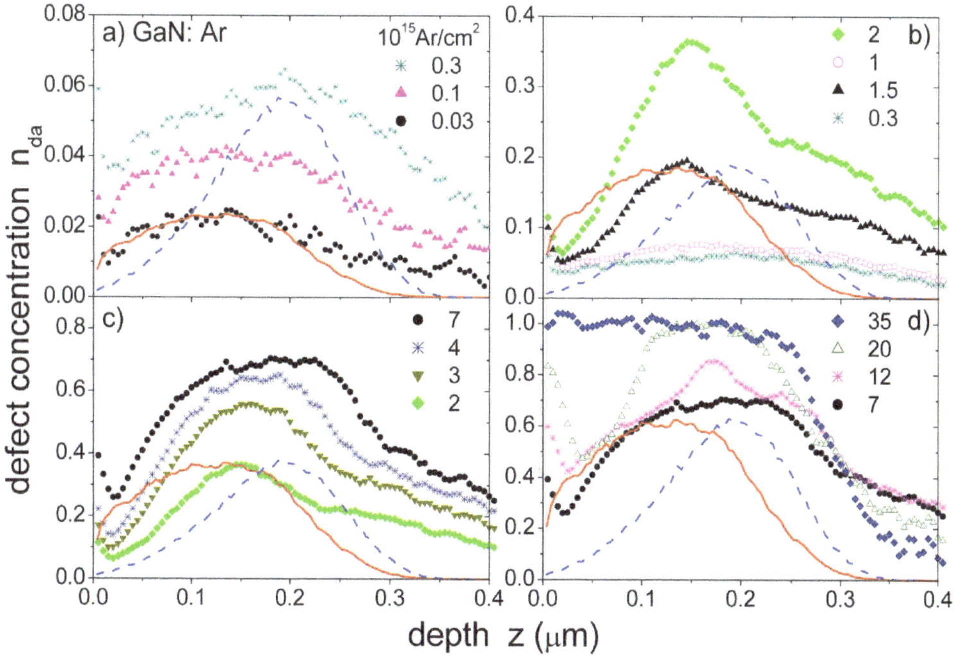

Fig. 10. Defect profiles $n_{da}(z)$ for 300 keV Ar implanted GaN for various ion fluences (in 10^{15} Ar/cm^2). The calculated distributions of primary displacements (solid lines) and implanted ions (dashed lines) are given in arbitrary units for comparison. Notice the different scales of the ordinate.

2.4 Comparison of damage build-up in ZnO and GaN

Both ZnO and GaN were shown above to exhibit strong resistance to implantation damage even at 15 K. Fig. 11 compares directly the damage build-up curves for 15 K Ar implantation in ZnO and GaN. As already mentioned, GaN exhibits a very similar damage accumulation behaviour to ZnO for the first four stages. In contrast to ZnO, in GaN the damage level reaches the random level for the highest fluences (stage V), showing that the channelling effect is completely inhibited and indicating the amorphisation of the material for fluences around 2×10^{16} Ar/cm^2 (n_{dpa}~24) and higher. In ZnO, on the other hand, the damage level stays well below the random level up to the highest fluence of 7×10^{16} Ar/cm^2 (n_{dpa}~59).

For low fluences in stage I (up to n_{dpa}~0.02) very similar damage levels are observed for GaN and ZnO. This is obviously connected with the fact that the masses and displacement energies of the two materials are comparable. In this stage the defect profiles agree well with the distribution of primary displacements calculated with the SRIM code. This suggests that the nuclear energy loss is the primary factor influencing the damage accumulation.

The second stage is interpreted as the fluence regime in which individual collision cascades start to overlap allowing recombination of point defects and the formation of new defect structures. The damage profiles expand towards deeper regions of the sample for both materials, suggesting that other factors besides the nuclear energy deposition start to influence the damage build-up. The broadened defect distributions and the high dechannelling yields point to a migration of point defects and a formation of extended defects. Since the experiments were carried out at 15 K the defect mobility is most likely induced by athermal processes. High lattice strain upon irradiation has been observed for both materials (Liu et al. 1997; Perillat-Merceroz et al., 2011) and may be one origin for the enhanced defect mobility. Electronic excitations possibly also influence the damage formation. Despite these similarities between the behaviour of GaN and ZnO, in this stage the damage level within the plateau is considerably higher for GaN (n_{da}~0.05) than for ZnO (n_{da}~0.02). This indicates a lower diffusivity of point defects and therefore lower recombination rates or less efficient rearrangement of interstitials to extended defects in GaN than in ZnO.

In stage III, both materials show a strong increase of implantation damage which occurs at higher n_{dpa} values for ZnO than for GaN. The damage profiles exhibit both a pronounced peak typical for direct backscattering from point defect clusters arising in the implanted zone and strong artificial backscattering tails for the deeper, unimplanted regions of the sample suggesting the formation of stacking faults and dislocation loops. Indeed, for the case of room temperature implantation into GaN the strongly increasing damage level was attributed to both displaced atoms in point defect clusters as well as extended defects (Pagowska et al., 2011). However, in GaN the direct backscattering peak arises at the depth of maximum nuclear energy deposition while for ZnO the maximum of the defect peak also starts in this region but then shifts deeper towards the end of range of the implanted Ar ions. In stage IV the maximum defect level remains almost constant, being

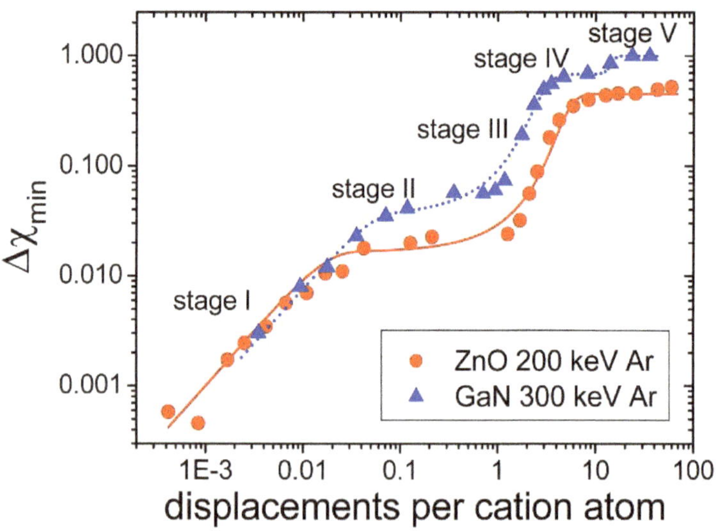

Fig. 11. Difference in minimum yield $\Delta\chi_{min}$ as a function of the number of displacements per cation atom (Ga for GaN and Zn for ZnO) for 200 keV Ar implantation into ZnO and 300 keV Ar implantation into GaN. The plotted curves are fits based on the defect interaction and amorphisation model of Hecking (Hecking et al., 1986).

lower in ZnO than in GaN, showing again its better radiation resistance. In GaN we observe a widening of the defect profile towards the surface and the bulk. For ZnO the shape of the defect profile remains similar to that for stage III and the defect distribution only widens slightly towards deeper regions. The different shapes of the defect profiles for GaN and ZnO in stages III and IV indicate a different microscopic structure within the implanted layers in the two materials.

Finally amorphisation is only reached for GaN (stage V). Amorphous seeds nucleate at a depth with both high damage and high impurity concentration and grow during further irradiation. Contrary, ZnO could not be rendered amorphous even for the highest fluence of 7×10^{16} Ar/cm^2.

2.5 Low temperature annealing of implantation damage

In order to investigate the mobility of defects (in particular Zn and Ga interstitials), annealing experiments were performed at temperatures between 15 K and 250 K (Lorenz et al., 2005, 2011). Samples were first implanted at 15 K to a certain fluence. Then isochronal annealing steps were performed up to 250 K by heating the sample up to a certain temperature for 10 min and then cooling it back to 15 K in order to perform the RBS/C measurement.

Fig. 12 shows $\Delta\chi_{min}$ values as a function of the annealing temperature after 15 K implantation of 2×10^{12} Er/cm^2 into ZnO and 1.5×10^{13} Ar/cm^2 into GaN, corresponding to stage I for which mainly point defects are to be expected. Both materials show partial recovery of the lattice at low temperatures. In the case of ZnO the defect removal starts between 80 and 130 K and for GaN the first annealing step at 100 K already reveals a

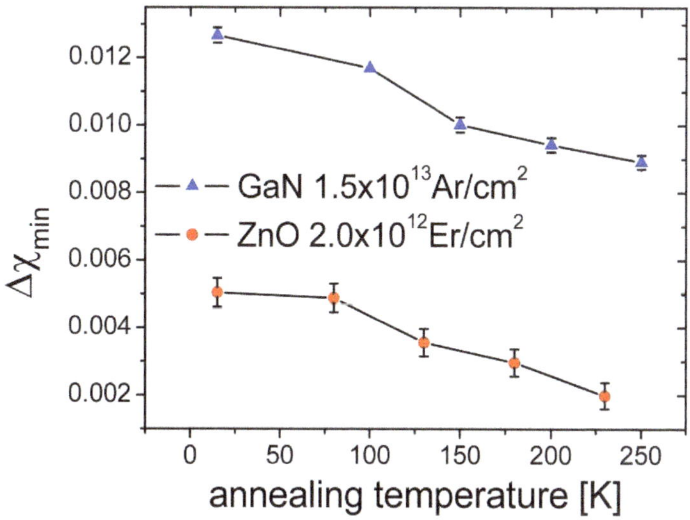

Fig. 12. Difference in minimum yield $\Delta\chi_{min}$ as a function of the annealing temperature after 15 K implantation of 2×10^{12} Er/cm^2 into ZnO and 1.5×10^{13} Ar/cm^2 into GaN. $\Delta\chi_{min}$ was derived using the same window for GaN as shown in Fig. 8 and a window comprising the entire implanted region for ZnO.

decrease of $\Delta\chi_{min}$. The total lattice recovery in the measured temperature range was seen to be better in ZnO. The damage level in ZnO decreases by 60% while that in GaN only decreases by 30%. However, the initial damage level in GaN was higher which may also influence the subsequent annealing results.

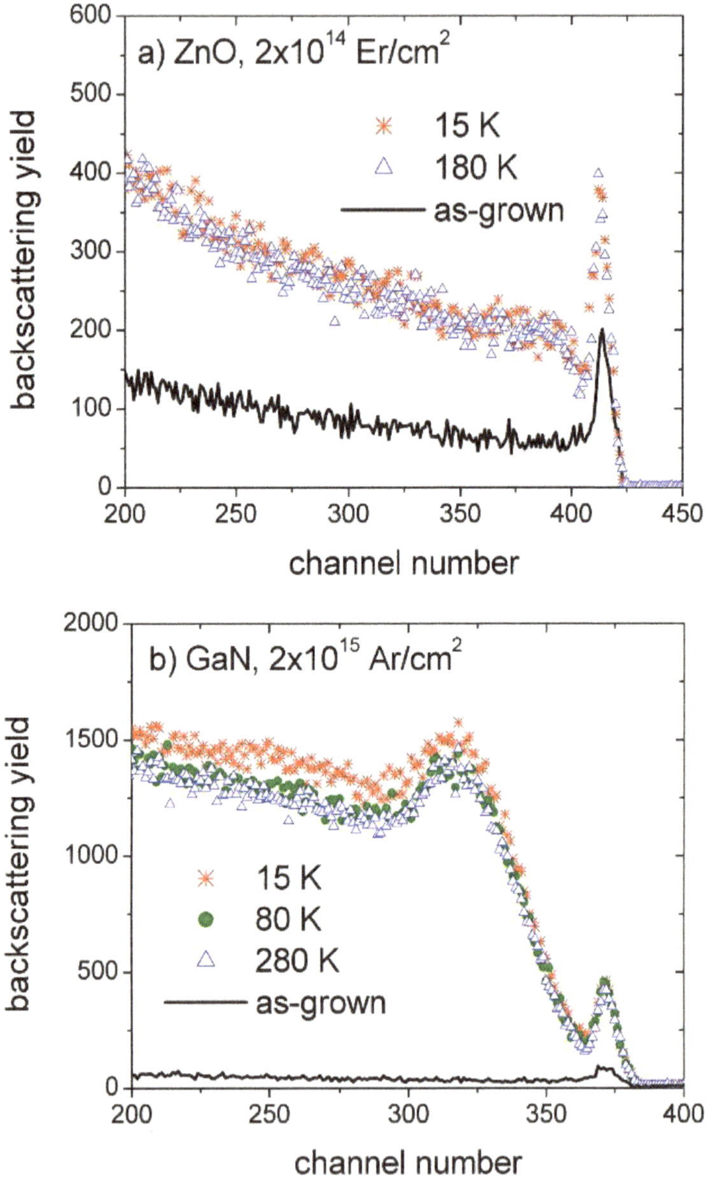

Fig. 13. RBS/C aligned spectra for different annealing temperatures after 15 K implantation of 2×10^{14} Er/cm^2 into ZnO (a) and 2×10^{15} Ar/cm^2 into GaN (b).

To investigate the low temperature defect annealing for higher damage levels ZnO was implanted to a fluence of 2×10^{14} Er/cm^2 and GaN to a fluence of 2×10^{15} Ar/cm^2, both fluences corresponding to stage III (see Figs. 4 and 9). RBS/C aligned spectra after the implantation and different annealing steps are shown in Fig. 13. In the case of ZnO no defect removal is observed up to 180 K, the highest temperature applied. In the case of GaN a very slight recovery of the crystal occurs after annealing at 80 K but increasing the temperature further has no effect on the spectra. The annealing in GaN occurs mainly in a deep region beyond the main defect peak.

We see a clear lattice recovery for fluences corresponding to stage I where mainly point defects are expected. This finding indicates that point defects in both materials are mobile at temperatures as low as 100 K. In ZnO annealing of point defects created by electron irradiation was seen to occur in several steps starting at 110 K in good agreement with our observations (Gorelkinskii and Watkins, 2004). Also in the case of GaN indications were found for a long range motion of an intrinsic defect starting at about 100 K (Chow et al., 2000).

In contrast to the high mobility of point defects in GaN and ZnO, extended defects are known to be thermally stable. A significant reduction of stacking faults in GaN created during implantation was only observed for temperatures well above 1000 °C (Wojtowicz et al., 2006). Similar results are found for ZnO where in particular basal loops and linear dislocations appeared to be stable after 900 °C annealing (Perillat-Merceroz et al., 2011). The fact that almost no lattice recovery takes place below room temperature (see Fig. 13) supports our previous assumption that in stages III and IV a complex network of extended defects forms already during implantation at 15 K. The slight annealing in the case of GaN can be attributed to the annealing of remaining mobile point defects similar to the effects seen for the lower fluence (Fig. 12).

3. Rare earth doping and annealing of GaN and ZnO

3.1 Rare earth doped wide bandgap semiconductors

Rare earth (RE) ions incorporated into transparent compounds often prevail in the trivalent RE^{3+} state and show characteristic light emission properties with wavelengths from infrared to ultraviolet, covering the entire visible spectrum. These emission lines originate from transitions within the *4f* electron shell which is only partially filled while it is surrounded by filled *5s* and *5p* orbitals. This shielding is the reason why RE emissions are relatively insensitive to the surrounding matrix (although it is the crystal field of the host matrix that relaxes the selection rules allowing optical transitions which are forbidden in the free ion). Therefore RE^{3+} emission lines are very narrow, show similar wavelengths in different materials and the wavelength is very stable in a wide temperature range. The most prominent example exploiting RE emissions is probably the Nd:YAG laser.

Since Favennec (Favennec et al., 1989) has shown that RE^{3+} emission quenching with temperature is decreasing with increasing energy gap of the semiconductor host, strong research efforts have been undertaken in RE doping of wide bandgap semiconductors like III-nitrides and II-oxides. The wide bandgap of GaN and ZnO furthermore allows light emission in the visible. In fact, GaN was successfully doped with RE ions both by ion implantation or in situ doping techniques and red (Eu, Pr), green (Er, Tb), and blue (Tm) light emission has been achieved (Steckl et al., 2002). Recently the first current injection LED, operating at low voltages, based on GaN:Eu was realized for doping during epitaxial growth (Nishikawa et al., 2009). Results on RE doping of GaN by ion implantation have been reviewed in (Lorenz et al., 2010a). Optical activation of RE ions in ZnO was found to be more challenging. In both lattices the large RE ions will be more easily incorporated on the cation site. While RE^{3+} in GaN is isoelectronic to Ga, in the case of ZnO some defects or impurities are necessary for charge compensation. RE emission has been reported from polycrystalline pellets and nanomaterials but doping single crystals was less successful (Jadwisienczak et al., 2002, Alves et al., 2003).

Directly after implantation GaN and ZnO samples are typically optically inactive. Implantation defects act as nonradiative recombination paths for excitons and quench both the native luminescence of the host and the RE emissions. Thermal annealing is necessary to remove defects and optically activate the rare earth ions. An empirical rule states that efficient removal of implantation damage should take place during annealing at two thirds of a material's melting temperature (in Kelvin) i.e. ~1600 °C for GaN and ~1200 °C for ZnO. However, at such high temperatures the more volatile N and O start to diffuse out of the samples and induce dissociation of the material. For example GaN was shown to start to dissociate at temperatures as low as 800 °C (Bartels et al., 1999). Also in ZnO outdiffusion of O occurred at temperatures as low as 500 °C if no oxidizing annealing atmosphere was chosen (Pearton et al., 2004).

In GaN, the effect of thermal annealing on RE emission intensity and spectral shape has been intensively studied. Best results were obtained for annealing well above 1000 °C using special measures to protect the surface such as thin AlN capping layers (Lorenz et al., 2004) or extremely high nitrogen pressures (Roqan et al., 2010). Although Europium luminescence intensity is seen to rise strongly with the annealing temperature, implantation defects are thermally very stable and full recovery of structural defects could only be achieved for low implantation fluences. To reach technologically interesting Eu concentrations of > 0.1 at.% high implantation fluences are necessary and it was shown that in such heavily damaged samples residual damage remains even after annealing at temperatures as high as 1450 °C (Lorenz et al., 2010a). Nevertheless, high quality films were produced with emission properties similar to *in situ* doped films used in GaN:Eu LEDs (Lorenz et al., 2010b).

Implantation damage in ZnO after high fluence RE implantation was shown to be fully recovered for 1000 °C annealing in air (Alves et al., 2003), i.e. remarkable 200 °C below the temperature expected for efficient defect removal. Despite this promising result, optical activation of the RE was not achieved. RBS/C shows that the RE diffuses towards the surface during the annealing where it accumulates leading to a second peak in the RBS spectrum (Alves et al., 2003). Furthermore, during the 30 min annealing at 1050 °C about 50% of Er ions were lost (Alves et al., 2003). These results indicate a very low solubility of RE in ZnO. In fact, optical activation of RE implanted single crystalline ZnO was only shown for annealing temperatures below 300 °C (Jadwisienczak et al., 2002), insufficient to effectively remove implantation damage. In some polycrystalline or nanostructured materials, on the other hand, RE emission was observed probably arising from RE residing in grain boundaries or secondary phases (Monteiro et al., 2006).

This chapter does not aim to give an extensive review on RE doping of GaN and ZnO, which would be beyond the scope of this book. In the following we will discuss some RBS/C results in GaN and ZnO in a comparative manner in order to shed some light on the processes mentioned above leading to efficient RE emission from GaN while RE doping of ZnO remains challenging.

3.2 Experimental details

GaN and ZnO thin films grown by metal organic vapour phase epitaxy (MOCVD) on sapphire substrates were implanted with 160 or 300 keV Eu ions at room temperature. The fluences (1-2×10^{15} Eu/cm^2) were chosen to introduce technologically relevant concentrations of Eu while keeping the implantation damage low. Annealing of the samples was performed at 1000 °C in N_2 atmosphere either for 2 min in a rapid thermal annealing (RTA) equipment or for 20 min in a conventional tube furnace. RBS/C random and aligned spectra as well as full angular scans in the <0001> axis and one additional axis were acquired using 2 MeV He^+ particles.

3.3 Experimental results

Fig. 14 presents RBS/C spectra for 1.3×10^{15} Eu/cm^2 implantation into ZnO (a) and GaN (b) at room temperature and 300 keV. The aligned spectra of the as-grown samples are also shown for comparison. After implantation the minimum yield within the implanted area increases from 4% to 34% for ZnO while for GaN it increases from 2% to 54%. Therefore, like for low temperature implantation, also at room temperature the damage level for a given implantation fluence is lower in the case of ZnO. The surface damage peak is clearly seen for GaN and is absent for ZnO. After annealing GaN for 20 min and ZnO for 2 min at 1000 °C and in nitrogen atmosphere the minimum yields are comparable (~15%). It was shown that for similar implantation conditions, annealing ZnO for 20 min in air could restore the minimum yield of the as-grown sample (Miranda et al., 2011). However, for such treatment the RE is seen to diffuse out of the sample. Therefore in this study RTA was used in an attempt to suppress diffusion. In fact, the Eu profile after annealing did not change as it was shown in (Miranda et al., 2011) but no optical activation of Eu was reached while the GaN sample shown in Fig. 14 b revealed the typical red Eu emission at ~620 nm (Lorenz et al., 2009).

In order to analyse this behaviour in more detail, full angular scans were performed for Eu and Zn/Ga allowing the determination of the lattice site location of Eu in the ZnO and GaN host matrix. Fig. 15 shows these scans for the <0001> axis as well as for an additional axis. Directly after the implantation the <0001> scans for Eu and both Ga and Zn overlap completely suggesting that Eu is located on substitutional lattice sites in both hosts.

In the case of GaN, the minimum yield for Ga decreases after RTA at 1000 °C due to the lattice recovery while the scan for Eu remains basically unchanged. This shows that annealing does not promote the further incorporation of Eu on substitutional sites. Nevertheless, the substitutional fraction remains high after annealing.

It is worth to note that the Eu scan in the tilted $<10\bar{1}1>$ axis shows a slight narrowing compared to the Ga scan indicating that Eu occupies near-substitutional Ga-sites in GaN with a slight displacement along the c-axis (Monteiro et al., 2001). The high substitutional fraction is a prerequisite for efficient RE emission in GaN, in fact, a correlation between substitutional fraction and Eu luminescence intensity was found for implanted samples (Lorenz et al., 2009).

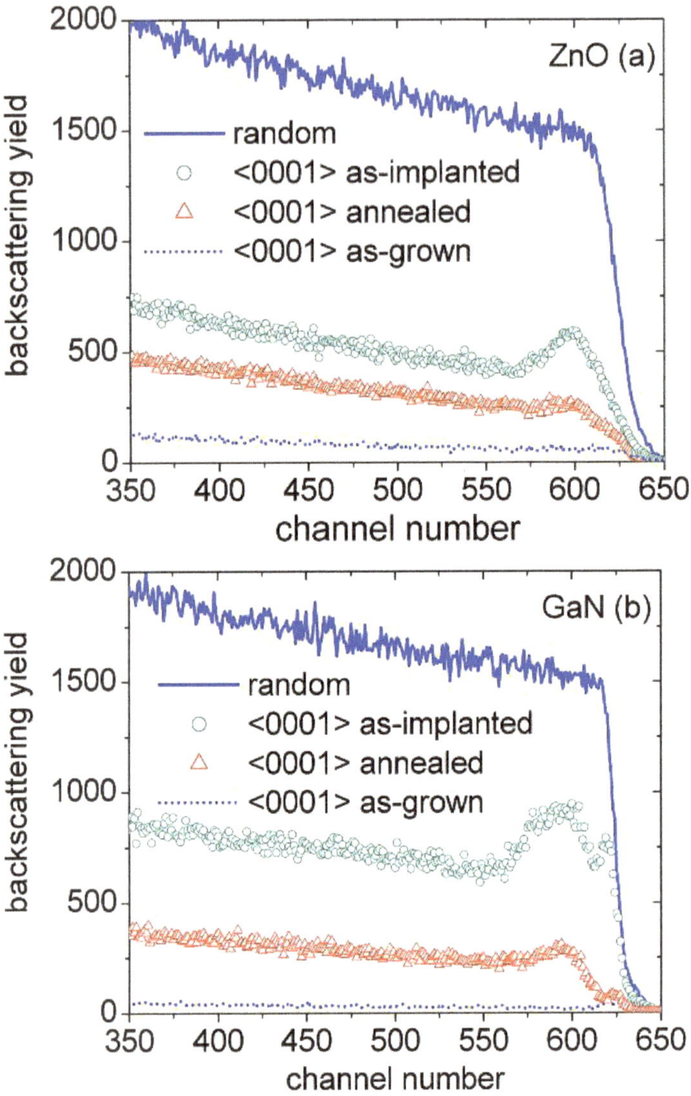

Fig. 14. RBS/C spectra for ZnO (a) and GaN (b) implanted with 1.3×10^{15} Eu/cm^2 with 300 keV at room temperature. The aligned spectra are shown for the as-implanted samples and after annealing at 1000 °C in N_2 (2 min for ZnO and 20 min for GaN).

For ZnO a different behaviour is observed. RTA successfully removes implantation damage but the minimum yield of Eu is increasing (Fig. 15a). The scans after annealing reveal a very low substitutional fraction of only 13% while the remaining Eu is incorporated on random sites in the ZnO lattice. So although the RTA treatment suppresses long distance diffusion of Eu, it drives the ions out of the substitutional into random lattice sites. A decreasing substitutional fraction with increasing annealing temperature has also been observed using the emission channelling technique for considerably lower implantation fluences than used in this study (Wahl et al., 2003).

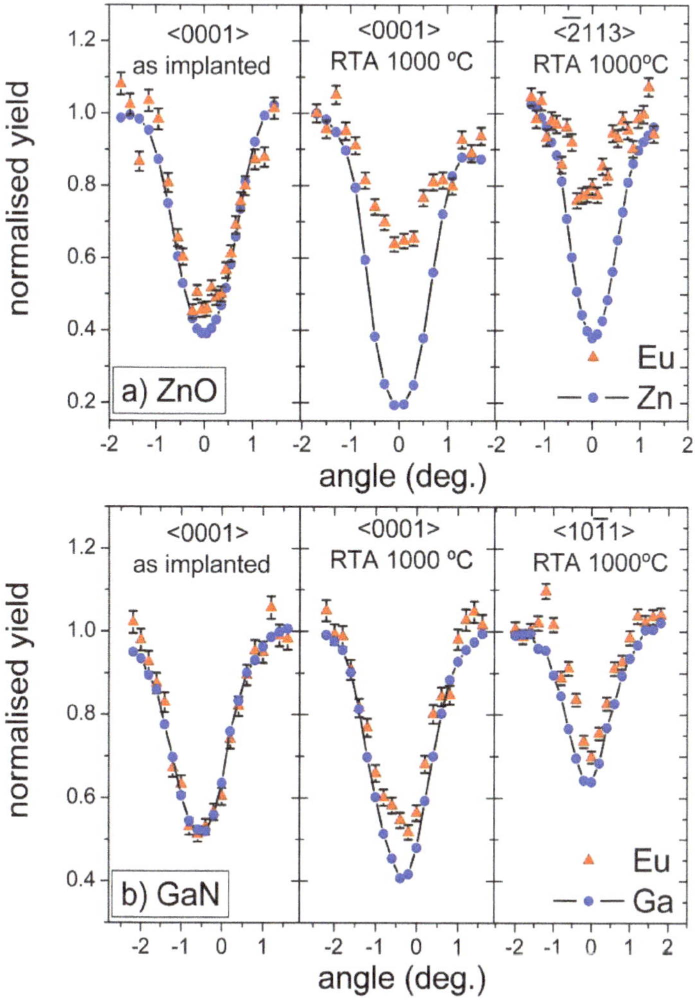

Fig. 15. a) Full angular RBS/C scans in the <0001> and the < $\bar{2}113$ > axes for ZnO implanted to 1.3×10^{15} Eu/cm^2 with 300 keV at room temperature. b) Full angular RBS/C scans in the <0001> and the < $10\bar{1}1$ > axes for GaN implanted to 1.5×10^{15} Eu/cm^2 with 160 keV at room temperature. Both samples were annealed for 2 min in N_2 atmosphere in a rapid thermal annealing (RTA) equipment.

The exact reasons for the difficulty to achieve optical activation of RE ions implanted into ZnO are still under debate. Radiative transitions may be suppressed due to the microstructure in the direct vicinity of the RE ions which may possess an unsuited crystal field or promote non-radiative recombination. Optical activation of implanted RE ions in

ZnO was reported for annealing at 150 °C but the emission gradually vanished with increasing annealing temperature (Jadwisienczak et al., 2002). These results together with the structural results discussed before indicate that the observed change of lattice site of the RE ions during annealing leads to the suppression of RE luminescence.

4. Summary and conclusions

The radiation damage formation upon low temperature ion implantation has been compared for GaN and ZnO. It was shown that both materials exhibit strong dynamic annealing effects even at 15 K. The damage build-up with fluence was found to proceed in a similar way for GaN and ZnO for low and intermediate fluences. Several stages could be distinguished in the damage formation processes. For low ion fluences an almost linear increase of damage with the ion fluence is observed attributed to the formation of point defects in isolated collision cascades (stage I). When the damaged regions produced by single ions start to overlap, vacancies and interstitials can recombine resulting in a plateau-like slow increase of the damage level (stage II). Furthermore our results point to a long range migration and the formation of extended defects within stage II driven by athermal processes. For higher fluences more stable defect complexes form (stage III) and the damage level increases steeply before reaching a second plateau (stage IV). It can be assumed that within stages III and IV a complex network of extended defects (dislocation loops and stacking faults) and point defect clusters forms, similar as observed for room temperature implantation. For high fluences (stage V) GaN amorphises. Contrary, stage V was not reached in ZnO, i.e. it was not rendered amorphous up to the highest fluence applied.

Annealing studies were performed in the temperature range from 15 K to room temperature. For low fluences (stage I) implantation defects are shown to be mobile at temperatures well below room temperature in both semiconductors. However, for higher fluences (stage III) nearly no defect recovery is observed supporting the assumption that thermally stable extended defects have formed even at 15 K.

In conclusion, GaN and ZnO are proved to be radiation resistant even during implantation at very low temperatures. However, the effective defect migration processes favour the formation of extended defects at higher fluences. These defect types are thermally very stable posing a challenge for post-implant annealing which is necessary for implant activation in devices.

Finally, the implantation technique was applied to dope both materials with optically active rare earth (RE) ions. Despite the better radiation properties of ZnO, optical activation of RE could not be achieved which is attributed to the fact that annealing causes a site change of the RE from substitutional to random interstitial sites. For many annealing treatments out-diffusion of the RE is observed due to a low solubility limit. Annealing procedures beyond thermal equilibrium such as laser annealing or taking advantage of the high electronic energy loss by swift heavy ion irradiation may circumvent the problem of site stability of the implanted RE atoms. Further investigations in this direction would be desirable. GaN:RE, on the other hand, shows bright RE emission in the three primary colours. However, the incomplete recovery of implantation damage still poses problems mainly caused by the out-diffusion of N at high temperatures. Above mentioned advanced annealing techniques are worth to be studied as a means to avoid N out-diffusion. Also multi-step implantation/annealing processes may improve the structural and optical characteristics of implanted GaN. Here fluences should be kept low to avoid the formation of extended defects (preferably corresponding to stage I) and after each implantation step the sample is annealed.

5. Acknowledgements

Funding by FCT Portugal (Ciência 2007, PTDC/CTM/100756/2008) and the German-Portuguese bilateral collaboration program (DAAD-FCT project number 50750957) is gratefully acknowledged.

6. References

Adachi S. (2005). *Properties of Group IV, III/V and II/VI Semiconductors*, Wiley, ISBN 0-470-09032-4, Chichester

Alves E., Rita E., Wahl U., Correia J. G., Monteiro T., Soares J., & Boemare C. (2003). Lattice site location and optical activity of Er implanted ZnO. *Nucl. Instrum. Meth. Phys. Res. B*, Vol. 206, (May 2003), pp. 1047-1051, ISSN 0168-583X

Bartels J., Freitag K., Marques J. G., Soares J. C., & Vianden R. (1999). Incorporation of the transition metal Hf into GaN. *Hyp. Int.*, Vol. 120-121, No. 1-8, (August 1999), pp. 397-402, ISSN 0304-3843

Brandt O., Dhar S., Pérez L., & Sapega V. (2010). High-Temperature Ferromagnetism in the Super-Dilute Magnetic Semiconductor GaN:Gd, In *Rare Earth Doped III-Nitrides for Optoelectronic and Spintronic Applications*, Topics in Applied Physics, Volume 124/2010, O'Donnell K. P. & Dierolf V., pp. 309-342, Springer, ISBN 978-90-481-2876-1, Dordrecht

Breeger B., Wendler E., Trippensee W., Schubert Ch., & Wesch W. (2001). Two-beam irradiation chamber for in situ ion-implantation and RBS at temperatures from 15 K to 300 K. *Nucl. Instrum. Meth. Phys. Res. B*, Vol. 174, No. 1-2, (March 2001), pp. 199-204, ISSN 0168-583X

Chow K. H., Watkins G. D., Usui A. & Mizuta M. (2000). Detection of Interstitial Ga in GaN. *Phys. Rev. Lett.*, Vol. 85, (September 2000), pp. 2761-2764, ISSN 0031-9007

Favennec P.N., L'Haridon H., Salvi M., Moutonnet D., & Le Guillou Y. (1989). Luminescence of Erbium Implanted in Various Semiconductors: IV, III-V and II-VI Materials, *Electronics Lett.*, Vol. 25, No. 11, (May 1989), pp. 718-719, ISSN 0013-5194

Feldman L. C., Mayer J. W., & Picraux S. T. (1982). *Material Analysis by Ion Channeling*, Academic Press, ISBN 0-12-252680-5, New York

Gärtner K. (2005). Modified master equation approach of axial dechannelling in perfect compound crystals. *Nucl. Instrum. Meth. Phys. Res. B*, Vol. 227, No. 4, (October 2004), pp. 522-530, ISSN 0168-583X

Gloux F., Wojtowicz T., Ruterana P., Lorenz K., & Alves E. (2006). Investigation of the crystallographic damage formed in GaN by medium range energy rare earth ion implantation. *J. Appl. Phys.*, Vol. 100, No. 7, (October 2006) p. 073520, ISSN 0021-8979

Gorelkinskii Yu. V. & Watkins G. D. (2004). Defects produced in ZnO by 2.5-MeV electron irradiation at 4.2 K: Study by optical detection of electron paramagnetic resonance. *Phys. Rev. B*, Vol. 69, (March 2004), p. 115212, ISSN 1098-0121

Hecking N., Heidemann K. F., & TeKaat E. (1986). Model of temperature dependent defect interaction and amorphization in crystalline silicon during ion irradiation. *Nucl. Instrum. Meth. Phys. Res. B*, Vol. 15, No. 1-6, (April 1986), pp. 760-764, ISSN 0168-583X

Ioffe Institute, St. Petersburg, http://www.ioffe.rssi.ru/SVA/NSM/ (last accessed August 2011)

Jadwisienczak W. M., Lozykowski H. J., Xu, A., & Patel B. (2002). Visible emission from ZnO doped with rare-earth ions. *Journal of Electronic Materials*, Vol. 31, No. 7, (July 2002), pp. 776-784, ISSN 0361-5235

Jiang W., Weber W. J., & Thevuthasan S. (2000). In situ ion channeling study of gallium disorder and gold profiles in Au-implanted GaN. *J. Appl. Phys.*, Vol. 87, No. 11, (June 2000), pp. 7671-7678, ISSN 0021-8979

Klingshirn C. F., Meyer B. K., Waag A., Hoffmann A., & Geurts J. (2010). *Zinc Oxide from Fundamental Properties towards Novel Applications*, Springer ISBN 978-3-642-10576-0, Heidelberg Dordrecht London New York

Kucheyev S. O., Williams J. S., & Pearton S. J. (2001). Ion implantation into GaN. *Mater. Sci. Eng. R*, Vol. 33, No. 2-3, (May 2001), pp. 51-107, ISSN 0927-796X

Kucheyev S. O., Williams J. S., Jagadish C., Zou J., Evans C., Nelson A. J., & Hamza A. V. (2003). Ion-beam-produced structural defects in ZnO. *Phys. Rev. B*, Vol. 67, (March 2003), p. 094115, ISSN 1098-0121

Kucheyev S. O., Williams J. S., & Jagadish C. (2004). Ion-beam-defect processes in group-III nitrides and ZnO. *Vacuum*, Vol. 73, No. 1, (March 2004), pp. 93–104, ISSN 0042-207X

Liu C., Mensching B., Volz K., Rauschenbach B. (1997). Lattice expansion of Ca and Ar ion implanted GaN. *Appl. Phys. Lett.*, Vol. 71, (October 1997), pp. 2313, ISSN 0003-6951

Lorenz K., Wahl U., Alves E., Dalmasso S., Martin R. W., O'Donnell K. P., Ruffenach S., & Briot O. (2004). High temperature annealing and optical activation of Eu implanted GaN. *Appl. Phys. Lett.*, Vol. 85, No. 14, (October 2004), pp. 2712-2714, ISSN 0003-6951

Lorenz K., Alves E., Wendler E., Bilani O., Wesch W., & Hayes M. (2005). Damage formation and annealing at low temperatures in ion implanted ZnO. *Appl. Phys. Lett.*, Vol. 87, No. 18, (November 2005), p. 191904, ISSN 0003-6951

Lorenz K., Barradas N. P., Alves E., Roqan I. S., Nogales E., Martin R. W., O'Donnell K. P., Gloux F., & Ruterana P. (2009). Structural and optical characterization of Eu-implanted GaN. *J. Phys. D: Appl. Phys.*, Vol. 42, No. 16, (August 2009), pp. 165103, ISSN 0022-3727

Lorenz K., Alves E., Gloux F., & Ruterana P. (2010a). RE Implantation and Annealing of III-Nitrides. In *Rare Earth Doped III-Nitrides for Optoelectronic and Spintronic Applications*, Topics in Applied Physics, Volume 124/2010, O'Donnell K. P. & Dierolf V.,pp. 25-54, Springer, ISBN 978-90-481-2876-1, Dordrecht

Lorenz K., Alves E., Roqan I. S., O'Donnell K. P., Nishikawa A., Fujiwara Y., & Boćkowski M. (2010b). Lattice site location of optical centers in GaN:Eu light emitting diode material grown by organometallic vapor phase epitaxy, *Appl. Phys. Lett.*, Vol 97, No. 11, (September 2010), p. 111911, ISSN 0003-6951

Lorenz K., Peres M., Franco N., Marques J. G., Miranda S. M. C., Magalhães S., Monteiro T., Wesch W., Alves E., & Wendler E. (2011). Radiation damage formation and annealing in GaN and ZnO., *Proceedings of Oxide-based Materials and Devices II, Proc.*

of SPIE Vol. 7940, p. 79400O, ISBN: 978-0-8194-8502-1, San Francisco, USA, January 2011

Miranda S. M. C., Peres M., Monteiro T., Alves E., Sun H. D., Geruschke T., Vianden R., Lorenz K. (2011). Rapid Thermal Annealing of Rare Earth Implanted ZnO Epitaxial Layers. *Optical Materials*, Vol. 33, No. 7, (May 2011), pp. 1139–1142, ISSN 0925-3467

Monteiro T., Boemare C., Soares M. J., Sa-Ferreira R. A., Carlos L. D., Lorenz K., Vianden R., & Alves E. (2001). Photoluminescence and lattice location of Eu and Pr implanted GaN samples, *Physica B*, Vol. 308-310, (December 2001), pp. 22-25, ISSN 0921-4526

Monteiro T., Soares M. J., Neves A., Pereira S., Correia M. R., Peres M., Alves E., Rogers D., Teherani F., Munoz-SanJose V., Trindade T., & Pereira A. (2006). Optical active centres in ZnO samples. *J. Non-Cryst. Solids*, Vol. 352, No. 9-20, (June 2006), pp. 1453–1456, ISSN 0022-3093

Nishikawa A., Kawasaki T., Furukawa N., Terai Y., & Fujiwara Y. (2009). *Appl. Phys. Express*, Vol. 2, No. 7, (July 2009), p. 071004, ISSN 1882-0778

Nord J., Nordlund K., Keinonen J., Albe K. (2003). Molecular dynamics study of defect formation in GaN cascades. *Nucl. Instrum. Meth. Phys. Res. B*, Vol. 202, (November 2003), pp. 93-99, ISSN 0168-583X

Pagowska K., Ratajczak R., Stonert A., Turos A., Nowicki L., Sathish N., Józwika P., & A. Muecklich (2011). RBS/Channeling and TEM Study of Damage Buildup in Ion Bombarded GaN. Acta Physica Polonica A, Vol. 120, No. 1, (July 2011), pp. 153-155, ISSN 0587-4246

Pearton S. J., Abernathy C. R., Overberg M. E., Thaler G. T., Norton D. P., Theodoropoulou N., Hebard A. F., Park Y. D. Ren F., Kim J., & Boatner L. A. (2003). Wide band gap ferromagnetic semiconductors and oxides. *J. Appl. Phys.*, Vol. 93, No. 1, (January 2003), pp. 1-13, ISSN 0021-8979

Pearton S. J., Norton D. P., Ip K., Heo Y. W., & Steiner T. (2004). Recent advances in processing of ZnO. *J. Vac. Sci. Technol. B*, Vol. 22, No.3, (May/June 2004) , pp. 932-948, ISSN 0734-211X

Pearton S. J., Abernathy C. R., & Ren F. (2006). *Gallium Nitride Processing for Electronics, Sensors and Spintronics*, Springer, ISBN 1-85233-935-7, London

Perillat-Merceroz G., Gergaud P., Marotel P., Brochen S., Jouneau P.-H., & Feuillet G. (2011). Formation and annealing of dislocation loops induced by nitrogen implantation of ZnO. *J. Appl. Phys.*, Vol. 109, No. 2, (January 2011) p. 023513, ISSN 0021-8979

Priolo F. & Rimini E. (1990). Ion-beam-induced epitaxial crystallization and amorphisation in silicon. *Mat. Sci. Rep.* Vol. 5, No. 6, (June 1990), pp. 319-379, ISSN 0927-796X

Ronning C., Carlson E. P., & Davis R. F. (2001). Ion implantation into gallium nitride, *Physics Reports-Review Section of Physics Letters*, Vol. 351, No. 5, (September 2001), pp. 349-385, ISSN 0370-1573

Roqan I. S., O'Donnell K. P., Martin R. W., Edwards P. R., Song S. F., Vantomme A., Lorenz K., Alves E., & Boćkowski M. (2010). Identification of the prime optical center in GaN:Eu^{3+}. *Phys. Rev. B*, Vol. 81, No. 8, (February 2010), p. 085209, ISSN 1098-0121

Ruterana P., Lacroix B., & Lorenz K. (2011). A mechanism for damage formation in GaN during rare earth ion implantation at medium range energy and room temperature. *J. Appl. Phys.*, Vol. 109, No. 1, (January 2011), p. 013506, ISSN 0021-8979

Steckl A. J., Heikenfeld J. C., Lee D. S., Garter M. J., Baker C. C., Wang Y., & Jones R. (2002). Rare-Earth-Doped GaN: Growth, Properties, and Fabrication of Electroluminescent

Devices. *IEEE J. Sel. Top. Quantum Electron,* Vol. 8, No. 4, (July/August 2002), pp. 749-766, ISSN 1077-260X

Trachenko K. (2004). Understanding resistance to amorphization by radiation damage. *J. Phys.: Condens. Matter,* Vol. 16 (December 2004), pp. R1491–R1515, ISSN 0953-8984

Wahl U., Rita E., Correia J. G., Alves E., & Araujo, J. P. (2003). Implantation site of rare earths in single-crystalline ZnO. *Appl. Phys. Lett.,* Vol. 82, (February 2003), pp. 1173-1175, ISSN 0003-6951

Wang C. M., Jiang W., Weber W. J., & Thomas L. E. (2002). Defect clustering in GaN irradiated with O^+ ions. *J. Mater Res.,* Vol. 17, No. 11, (November 2002), pp. 2945-2952, ISSN 0884-2914

Wendler E., Kamarou A., Alves E., Gärtner K., & Wesch W. (2003). Three-step amorphisation process in ion-implanted GaN at 15 K. *Nucl. Instrum. Meth. Phys. Res. B,* Vol. 206, (May 2003), pp. 1028-1032, ISSN 0168-583X

Wendler E., Wesch W., Alves E., & Kamarou A. (2004). Comparative study of radiation damage in GaN and InGaN by 400 keV Au implantation. *Nucl. Instrum. Meth. Phys. Res. B,* Vol. 218, (June 2004), pp. 36-41, ISSN 0168-583X

Wendler E., Bilani O., Gärtner K., Wesch W., Hayes M., Auret F. D., Lorenz K., & Alves E. (2009). Radiation damage in ZnO ion implanted at 15 K. *Nucl. Instrum. Meth. Phys. Res. B,* Vol. 267, No. 16, (August 2009), pp. 2708–2711, ISSN 0168-583X

Wendler E. (2011). In-Situ RBS Channelling Studies of Ion Implanted Semiconductors and Insulators. *AIP Conf. Proc.,* Vol. 1336, pp. 621-625, *Proceedings of Twenty-First International Conference Application of Accelerators in Research and Industry,* ISBN: 978-0-7354-0891-3, Fort Worth, Texas, USA, August 2010

Wojtowicz T., Gloux F., Ruterana P., Lorenz K., Alves E. (2006). TEM investigation of Tm implanted GaN, the influence of high temperature annealing. *Optical Materials,* Vol. 28, (May 2006), pp. 738-741, ISSN 0925-3467

Yao, T. & Hong S.-K. (Eds.). (2009). *Oxide and Nitride Semiconductors Processing, Properties, and Applications,* Springer, ISBN 978-3-540-88846-8, Berlin Heidelberg.

Ziegler J. F., Biersack J. P., Ziegler M. D. (2008). *SRIM - The Stopping and Range of Ions in Matter,* SRIM Co., ISBN 0-9654207-1-X, Chester, http://www.srim.org/ (last accessed August 2011)

Zinkle S. J. & Kinoshita C. (1997). Defect production in ceramics. *J. Nucl. Mater.,* Vol. 251, (October 1997), pp. 200-217, ISSN 0022-3115

Section 4

Advances in Processing

12

Optical Waveguides Fabricated by Ion Implantation/Irradiation: A Review

Ovidio Peña-Rodríguez[1,2], José Olivares[1,2],
Mercedes Carrascosa[3], Ángel García-Cabañes[3],
Antonio Rivera[4] and Fernando Agulló-López[1]
[1]*Centro de Microanálisis de Materiales (CMAM), Universidad Autónoma de Madrid (UAM), Cantoblanco, Madrid*
[2]*Instituto de Óptica, Consejo Superior de Investigaciones Científicas (IO-CSIC), Madrid*
[3]*Departamento de Física de Materiales C-IV, Universidad Autónoma de Madrid, Madrid*
[4]*Instituto de Fusión Nuclear, Universidad Politécnica de Madrid, Madrid*
Spain

1. Introduction

Optical waveguides are key elements of many photonic devices; for this reason many materials and methods have been intensively studied to fabricate them. Ion implantation provides a general and flexible method to achieve this goal, with several advantages over the alternative techniques (Townsend *et al.*, 1994). Most optical waveguides and integrated optical devices are manufactured using low-energy light ions (H and He), taking advantage of the effects caused by nuclear damage. However, this is achieved at the expense of using very high fluences (10^{16}-10^{17} cm^{-2}), which reduces the practical utility of the method (Olivares *et al.*, 2007c).

The study of irradiations with heavier ions (A > 12) and higher energies (4-100 MeV), where the electronic stopping power dominates over the nuclear, has increased recently, as a way to overcome this limitation (Olivares *et al.*, 2005a). The characteristics associated with these processes differ significantly from those applied to nuclear collisions. In particular, the complete amorphization of the lattice can only be achieved when the electronic stopping power is above a certain threshold (Olivares *et al.*, 2005a). In this respect, it has been shown that a controlled damage can be generated by selecting the type of ion and its mass, energy and fluence, obtaining a micro-processing of crystals with a degree of accuracy, flexibility and efficiency well beyond the current state of the art. The optical waveguides produced with this method are both an end in itself (Caballero *et al.*, 2005; Olivares *et al.*, 2005b, 2007c; García-Navarro *et al.*, 2006; Manzano *et al.*, 2010) and a mean that allows the precise study of several fundamental aspects of the generation and accumulation of electronic damage (Agulló-López *et al.*, 2006; García-Navarro *et al.*, 2007; Rivera *et al.*, 2010a; b; García *et al.*, 2011).

For example, by irradiating $LiNbO_3$ with moderate ion fluences ($2x10^{14}$ cm^{-2}) of F were obtained waveguides with nonlinear properties comparable to those of other $LiNbO_3$ guides

(Bentini *et al.*, 2004), having high confinement profiles clearly advantageous over other types of guides and losses around 1 dB/cm. Waveguides have also been achieved by using the ions O, Si and Mg (Hu *et al.*, 2001; Bentini *et al.*, 2002), and in the KGW crystal (García-Navarro *et al.*, 2006), indicating the potential generality of the method. Of particular interest are the waveguides obtained by irradiating in the regime of high electronic stopping power (~5-6 keV/nm) and ultra-low fluence ($2x10^{12}$ cm^{-2}) (Olivares *et al.*, 2007c), because of the short irradiation time required.

In this book chapter, we intend to review the current state of the art on the fabrication of optical waveguides in insulating materials by means of ion implantation. The first overviews of this topic were performed by Townsend and colleagues in 1987 (Townsend, 1987) and 1994 (Townsend *et al.*, 1994), summarizing the results obtained up to these dates. Later, in a follow-up review paper, Chen and co-workers (Chen *et al.*, 2007) summarized the progresses between 1994 and 2006. Moreover, there have been some focused reviews, presenting the progresses on the 2D waveguide production in insulating materials by ion implantation (Chen, 2008), waveguides fabricated on lithium niobate ($LiNbO_3$) by the irradiation with swift heavy ions (SHI) (Olivares *et al.*, 2007c) and the production of photonic guiding structures on $LiNbO_3$ by energetic ion beams (Chen, 2009b). However, despite these precedents, there have been in recent years a large number of achievements in these areas that make necessary and updated review summarizing them. In this work we have focused on the results reported between 2007 and 2011, putting a particular emphasis on differentiate the waveguides fabricated by ion implantation (i.e., the traditional ones, where nuclear collisions are the dominant process) from those made by ion irradiation (i.e., where the electronic energy loss is dominant) because, in our experience, they are frequently confused.

We have divided this chapter in four sections, where the first one is this introduction. The basic methods used for the fabrication and characterization of ion implanted optical waveguides are described in Section 2. Section 3 summarizes the principal materials where this technique has been used to generate optical waveguides. Finally, the main applications of the fabricated waveguides are exposed in Section 5.

2. Experimental methods

In this section we will describe the basic methods used for the fabrication and characterization of ion implanted optical waveguides. First, we will describe some important aspects of ion implantation, then analyze the subtle but important differences between ion implantation and ion irradiation and finally discuss the thermal treatments and refractive index reconstruction methods.

2.1 Modification of materials using ion beams

Bombardment of a material with ion beams constitutes a very efficient and controllable way to modify the properties of the material in the near-surface region. This effect provides an useful technique to fabricate optical waveguides and other photonic devices requiring that at least one of the dimensions scales with the light wavelength. There are two main contributions to the ion-beam modification (damage) process; one of them, usually defined as *irradiation damage*, is associated to the interaction of the incoming ions with the material;

i.e., to the energy deposition. In turn, two different energy deposition processes can be considered, either through elastic atomic collisions or through electronic excitation processes. The two processes are, respectively, characterized by the nuclear S_n and electronic S_e stopping powers, defined as the energy loss per unit depth of the ion trajectory.

The other contribution to the modification of the material has to do with the incorporation of those bombarding ions into the structure of the material, i.e. *ion implantation.* Ion implantation occurs mostly near the end of the ion trajectory when the ion energy has decreased below ~10 keV/amu, where the nuclear stopping power is dominant. In many of the earlier works dealing with the preparation of optical waveguides using ion beams these two effects were strongly intermixed. However, more recently, a strong research activity is being devoted to the irradiation with ion beams of ions of medium mass and high energy that results in the separation of the region of strong electronic damage from that associated to nuclear damage and implantation.

2.1.1 Passage of ions through matter: Stopping powers

We will briefly review here the energy loss mechanisms and the main processes describing the energy transfer from the incoming ion to the material, including the possible formation of point defects. This is a specialized topic in nuclear physics that has been extensively reviewed. As we said before, elastic collisions between the incoming ions and the atoms of the material and electronic excitation processes associated to ion-electron interactions (atoms are not affected) are the two main mechanisms for energy loss. Their relative importance depends on the type of ion and input energy. Both losses mechanisms are described by the so-called *stopping powers* or more properly *stopping forces* (either nuclear or electronic) that are defined as the energy loss per unit length of the trajectory. They are dependent on the ion charge and mass as well as on the ion energy (velocity). As a general rule, electronic stopping power is dominant at high input energies, whereas nuclear processes take the lead

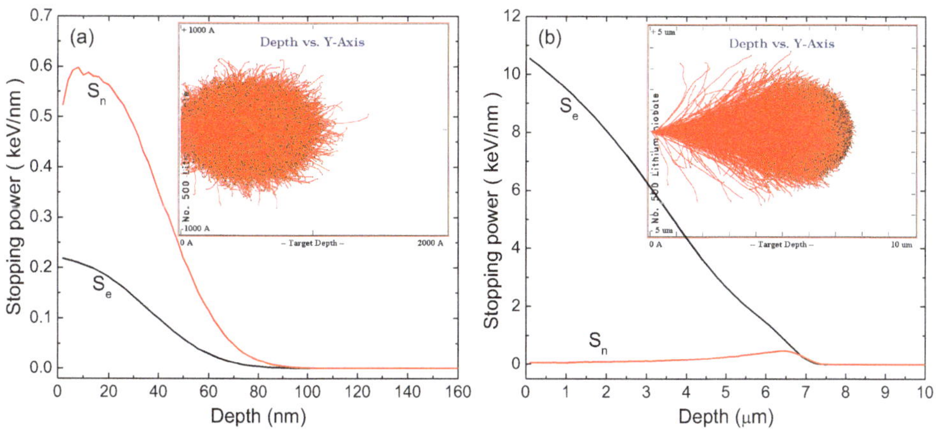

Fig. 1. Simulated stopping powers and ion distributions for bromine implanted on $LiNbO_3$ at 100 keV (a) and 40 MeV (b). The insets represent the lateral straggling of the implanted ions. Note the different depth scales. Calculations were performed using SRIM 2008 (Ziegler, 1985, 2008).

at low energies (< 10 keV/amu). The stopping power curves as a function of penetration depth are illustrated in Figure 1. Since ion-electron interactions involve low momentum transfers, the ion trajectories are straight at high energies and become zigzagging in the nuclear collision regime, where momentum transfer is relevant. The main parameters of the trajectory, *projected range and straggling*, are indicated on the figure. The bombarding ions are finally implanted in the material once they have slow down sufficiently. The transversal definition of such trajectory reaches the nanometric scale in the electronic regime and is of micrometric size at lower energies. These features are quite relevant when using ion beams for patterning of materials. A complicating feature when trying to determine the stopping powers and compare to experiments is the change in the charge kept by the ion during its path in the material. In other words, the charge associated to the ion is a function of ion velocity and it is continuously readjusting during its motion. The point is that electrons whose orbital velocity is lower than the ion velocity are stripped from the ion.

2.1.2 Nuclear collision processes and displacement damage

Elastic nuclear collisions are described within a classical, often non-relativistic, framework. The nuclear stopping power is linear with energy up to a maximum value at an energy E_1 and then decreases according to the law, $S_n \propto \ln E / E$. To describe energy transfer in this regime a rather simple but useful approach can be used. It considers binary elastic collisions and that the material is amorphous (random atomic locations). Let us consider an incident ion of charge Z and mass M_1 impinging with a kinetic energy E_0 on an atom at rest and mass M_2, both energies in the laboratory frame, Figure 2. The cross-section for transfer of a kinetic energy T to the target atom is often described by the Rutherford law:

$$\sigma(T) = \frac{\pi b^2}{4} T_M \frac{1}{T^2}, \tag{1}$$

b being the distance of closest approach in a head-on collision and $T_M = 4E_0 M/(M_1 + M_2)^2$ the maximum transferred energy.

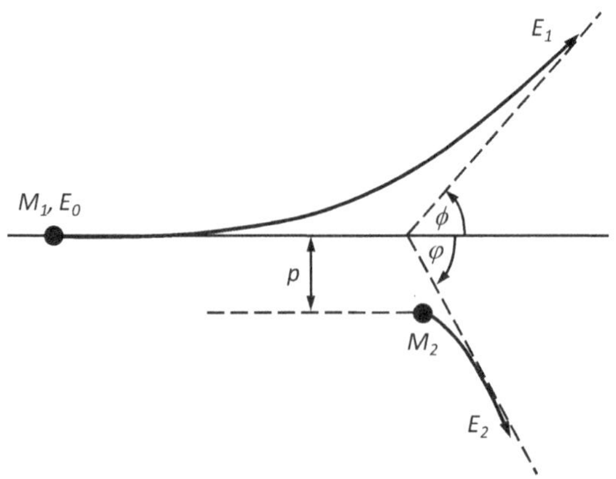

Fig. 2. Schematic representation of a classical two-body atomic collision in the laboratory frame.

In fact, the Rutherford formula only applies for high enough incident energies. For lower energies a screened Coulomb potential applies and the formula has to be adequately corrected. In the limit of very small energies one must use a hard-sphere potential leading to an energy-independent cross-section. If the transferred energy, T, overcomes a certain value E_D for displacement energy the knocked atom is ejected from its lattice position creating a vacancy and an interstitial; i.e., *a Frenkel pair*. This is the primary damage process in the nuclear regime. In the Rutherford approximation the displacement cross-section is:

$$\sigma_d = \int_E^E \sigma(T)dT = \frac{\pi b^2}{4}\frac{T_M}{E_d}, \tag{2}$$

which is proportional to the incident energy. To calculate the overall defect concentration produced by the ion bombardment one has to take into account the secondary displacements caused by each primary displaced atom. The number ν of secondary displacements is obviously 1 for $T > E_D$, whereas for $T >> E_D$ is given by $\nu = \frac{1}{2}T/E_D$. Then, one can obtain for $E >> E_D$ the following approximate expression for the average number of secondary displacements per a primary displaced atom is:

$$\overline{\nu}(T) = \frac{1}{\sigma_D}\int_0^{T(\max)} \sigma(T)\nu(T)dT = \frac{1}{2}\frac{\overline{T}}{E_D} = \frac{1}{2}\ln\frac{T_{\max}}{E_D}. \tag{3}$$

Taking into account the low range of the displaced atoms within the crystal lattice all of them will be quite close to each other forming a region of high disorder (*displacement spike*).

2.1.2.1 Calculation methods

Efficient codes are now available to compute, both, the energy loss and the amount of displaced atoms during ion propagation. Stopping and range of ions in matter (SRIM) (Ziegler, 1985, 2008), the most widely used, is based on the binary collision approximation, a Monte Carlo simulation method, with a random selection of the impact parameter of the next colliding ion. It allows the calculation of various quantities such as the vacancy concentration, sputtering rate, ionization and phonon production in the target material, energy deposition rate (divided in the nuclear and electron losses) and the three-dimensional distribution of the ions in the solid and its parameters (penetration depth, lateral spread – i.e., the straggling , etc.). Unfortunately it also has some limitations; for instance, the cumulative damage produced by the electronic energy deposition cannot be simulated with this software.

More recently, molecular dynamics (MD) methods have been developed to deal with the problem of damage in a more rigorous way. The evolution of the system (solid) is calculated once the interatomic potentials are known (usually from *ab initio* calculations). One can calculate the net force acting on any particle of the system from the gradient of the potential and then solve the classical motion equations for all the atoms in the system. The result is that the trajectory of every atom is obtained at any given time. Actually, the atomic interactions are described by the Quantum Mechanics; however, the classical approach turns out to be very appropriate to describe many effects at a much lower computational cost than the Quantum formalism.

Molecular dynamics is not limited by the binary collision approach and hence provide a more realistic view of the defects formed upon different conditions. For instance, by means of molecular dynamics methods one can study defects produced through elastic collisions, thermal treatments or mechanical stress. As an example, Figure 3 shows fission tracks produced in zircon (Moreira *et al.*, 2010). The amorphous structure of the tracks becomes evident with molecular dynamics calculations. An important aspect of molecular dynamics is that it makes possible to follow not only the formation of defect cascades upon irradiation but their subsequent evolution. For example, by means of molecular dynamics calculations (Trachenko *et al.*, 2006) studied the resistance to amorphization of different ceramics (SiO_2, GeO_2, TiO_2, Al_2O_3 and MgO) bombarded with U ions of 40 keV (to simulate U recoil in alpha decay). The authors concluded that the resistance to amorphization is primarily governed by the cascade relaxation process, being very efficient in MgO or Al_2O_3 (most defects annihilate) whereas it is inefficient for SiO_2 and GeO_2 (as a result significant damage is accumulated).

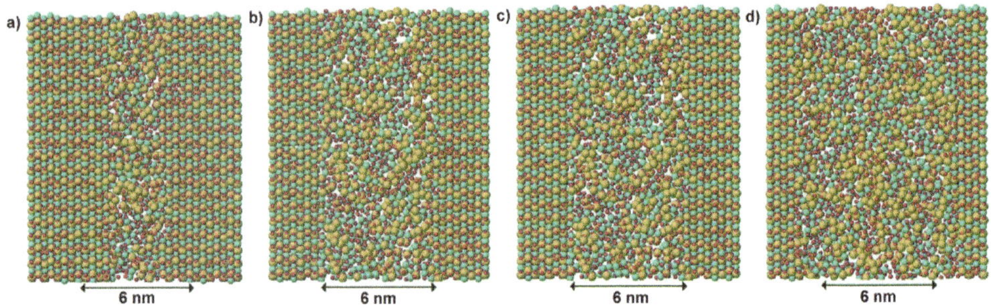

Fig. 3. Projection of atom positions in a slice through the center of fission tracks from the final configuration for Zircon with the following dE/dx values: (a) 3.90; (b) 5.94; (c) 7.65; and (d) 12.75 keV nm^{-1}. Blue circles are Zr, yellow ones are Si and red ones are O (Moreira *et al.*, 2010).

The application of Molecular Dynamics to swift heavy ion irradiation of ceramics requires more care. The reason is that in this case most of the ion energy is transferred to the target by electronic excitation mechanisms. At the moment defect production in the electronic excitation regime is described by phenomenological models presented later in this chapter. In some cases Molecular Dynamics simulations have been applied to swift heavy ion irradiation of ceramics making use of an electron-phonon coupling mechanism for the energy transfer from electrons to the lattice. For example, it was shown (Pakarinen *et al.*, 2010) how the high energy density induced by swift heavy ions in a nanometric-size track leads to density variations in both amorphous and crystalline materials. The changes induced in ZnO by swift heavy ion passage, as well as the subsequent lattice relaxation, are shown in Figure 4. This type of simulations does not consider possible mechanisms of defect formation via non-radiative decay mechanisms of electronic states (e.g., trapped excitons). When these mechanisms play a role they must be considered because the energy stored in these states is similar to that finally transferred to the lattice via electron-phonon interaction. The point here is that including such mechanisms is beyond the classical approach.

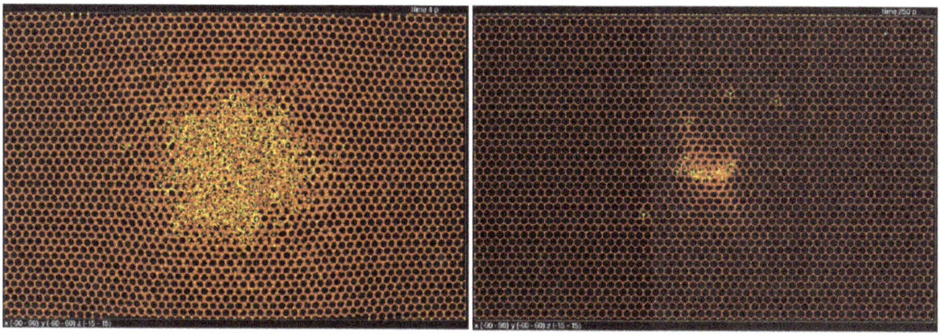

Fig. 4. Recrystallization in a ZnO track at 13 keV/nm stopping power, at 4 ps after the ion passage (left) and after 250 ps cooling (right) (Pakarinen *et al.*, 2010).

An appropriate description of these mechanisms requires in general a Quantum Mechanical formulation difficult to include in classical Molecular Dynamics. Some efforts are being carried out to include electronic effects in Molecular Dynamics simulations in a computationally cost-effective manner (Duffy & Rutherford, 2009). Additional work in this line may seriously contribute to describe defect production in the electronic excitation regime. Note that the topic of swift ion irradiation applied to waveguide fabrication constitutes an important part of this chapter.

2.1.3 Processes triggered by electronic excitation: Energy loss

The analysis of the electronic stopping power and its implication on the generation of lattice disorder is more complicated and requires a mechano-quantical formalism. However, a number of theoretical, mostly phenomenological, approaches are available. As for the energy loss processes, the ion velocity is crucial to determine its charge and, therefore, its electronic interactions. One should distinguish two main limiting cases depending on whether the ion velocity is lower or higher than the electron velocity in the atomic orbitals.

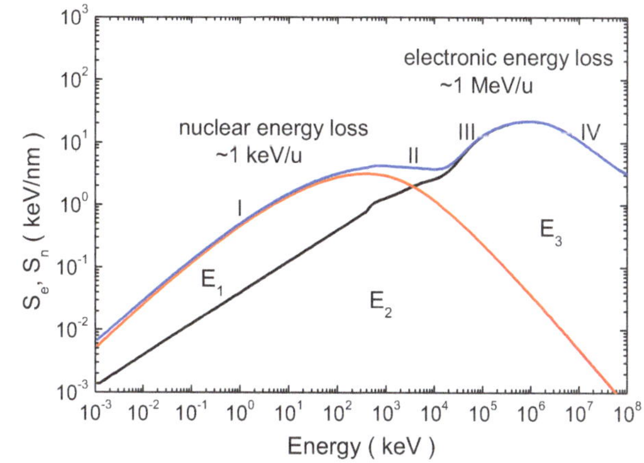

Fig. 5. Relation of nuclear and electronic energy loss as a function of the ion energy for Au ions on SiO_2. Calculations were performed using SRIM 2008 (Ziegler, 1985, 2008).

In a first case, corresponding to low ion energies (velocities) the Lindhard and Firsov theories (formation of a quasi-molecule) apply yielding a quadratic dependence on energy, $S_e \propto E^{1/2}$, up to an energy E_3 where a maximum stopping is reached (Bragg peak). This corresponds to stage III of the curve in Figure 5. In the second case, the Bethe-Bloch models based on a pure Coulomb interaction between a fully stripped ion and the electron are adequate predicting a dependence, $S_e \propto \ln E / E$. This is the behavior observed for stage IV of the curve. One notices that the electronic stopping is clearly dominant except for low enough energies up to E_2.

2.1.4 Damage caused by electronic excitation: Experimental observations

2.1.4.1 Amorphous tracks

The most clearly observable effect of ion-beam damage in the electronic stopping regime for many oxide, semiconductor and even metallic materials, is the formation of heavily-defective and even amorphous linear tracks around the trajectory of single ion impacts (Itoh *et al.*, 2009). This fact has been known since middle seventies and has been successfully applied to charge-particles dosimetry and age determination of rocks and minerals. There is strong evidence that the generation of tracks is not related to nuclear collision processes and it is only possible for electronic stopping powers above a certain threshold value. The threshold depends on material, ion type and even slightly on ion velocity. A list of thresholds is given in Table 1. A number of techniques have been successfully used to observe and characterize those tracks, including transmission electron microscopy (TEM), atomic force microscopy (AFM), Rutherford backscattering in the channeling condition (RBS/C), Mossbauer spectroscopy, etc. As an illustrative example we show in Figure 6 a high-resolution TEM picture of a track in $LiNbO_3$. Another interesting effect associated to the presence of tracks is the swelling of the sample surface that forms a hillock of nanometric height and radius at the ion impact point (Canut *et al.*, 1996). The effect is, indeed, related to the elastic strains associated to the linear amorphous inclusion inside the unirradiated material.

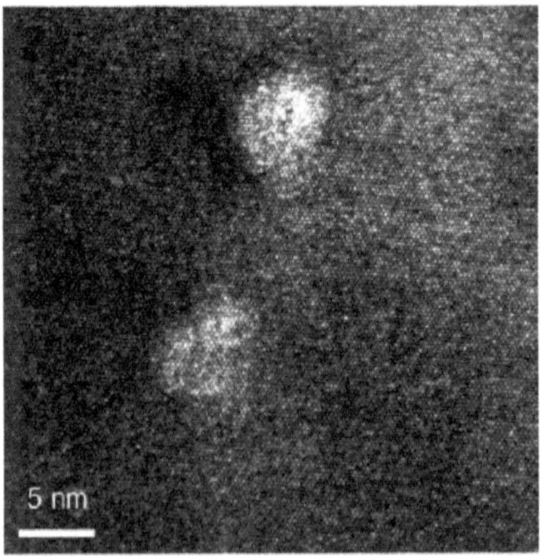

Fig. 6. High-resolution TEM micrograph of a $LiNbO_3$ sample irradiated with Br 45 MeV ions at a fluence of 3×10^{10} cm^{-2} after annealing in air for 30 min at 275°C, showing amorphous tracks of about 2.5 nm (Crespillo *et al.*, 2011).

Threshold range (keV/nm)	Target (threshold, keV/nm)	Reference
0–5	SiO_2 (2)	(Meftah *et al.*, 1994)
	a-Ge (3)	(Furuno *et al.*, 1996)
	SiO_2 (3.5 #)	(Arnoldbik *et al.*, 2003)
	LiF (4)	(Trautmann *et al.*, 2000)
	$BaFe_{12}O_{19}$ (4)	(Meftah *et al.*, 1994)
	$Pd_{80}Si_{20}$ (amorphous)	(Klaumünzer *et al.*, 1986)
	Mica (5)	(Toulemonde *et al.*, 1994)
	CaF_2 (5)	(Boccanfuso *et al.*, 2002)
5–10	$Y_3Fe_5O_{12}$ (6.5)	(Meftah *et al.*, 1994)
	$LiNbO_3$ (7)	(Meftah *et al.*, 1994)
	SiO_2 (7 #)	(Khalfaoui *et al.*, 2003)
	$MgAl_2O_4$ (7.5)	(Zinkle & Skuratov, 1998)
	$Y_3Al_5O_{12}$ (7.5)	(Meftah *et al.*, 1994)
10–15	InP (14*)	(Kamarou *et al.*, 2008)
15–20	Si_3N_4 (15)	(Zinkle *et al.*, 2002)
	MgO (15.8 #)	(Skuratov *et al.*, 2003)
	$Y_3Fe_5O_{12}$ (16 #)	(Meftah *et al.*, 1994)
	a-Si (17)	(Furuno *et al.*, 1996)
	GeS_2 (18)	(Vetter *et al.*, 1994)
	U_3Si (19)	(Hou & Klaumünzer, 2003)
20–25	InP (20)	(Gaiduk *et al.*, 2000)
	MgO (20)	(Beranger *et al.*, 1996)
	Al_2O_3 (21)	(Canut *et al.*, 1995)
25–30	Al_2O_3 (25 #)	(Skuratov *et al.*, 2003)
	GaSb (28)	(Szenes *et al.*, 2002)
	InSb (28)	(Szenes *et al.*, 2002)
	InAs (28)	(Szenes *et al.*, 2002)
	UO_2 (29)	(Matzke *et al.*, 2000)
30–35	Zr (30)	(Dunlop & Lesueur, 1993)
	Bi (31, 31*)	(Wang *et al.*, 1996)
	GaAs (31*)	(Kamarou *et al.*, 2008)
	Ge* (33)	(Kamarou *et al.*, 2008)
	GaN (<34)	(Kucheyev, 2004)
	SiC (>34)	(Zinkle *et al.*, 2002)
	AlN (>34)	(Zinkle *et al.*, 2002)
	$Si_{0.5}Ge_{0.5}$ (34)	(Gaiduk *et al.*, 2002)
35–40	Si (37*)	(Kamarou *et al.*, 2008)
	Co (37)	(Dunlop & Lesueur, 1993)
	Ge (38)	(Colder *et al.*, 2001)
	GaAs (38*)	(Colder *et al.*, 2001)
	Fe (40)	(Dunlop *et al.*, 1994)
Above 40	Ge (42)	(Komarov, 2003)
	Si (46*)	(Dunlop *et al.*, 1998)

Table 1. Thresholds (keV/nm) reported from various sources [reproduced from (Itoh *et al.*, 2009)]. Values for C_{60} projectiles are marked *, and those where surface data are given are marked #. Values in the 7.5-14 keV/nm range are rare, so group I and group II are separated. Group I all have low bandgaps and/or do not show exciton self-trapping.

Above threshold the radius of the tracks increases monotonically with the incident stopping power (Szenes, 1995). Even for amorphous materials like silica, tracks associated to definite structural changes (compaction) that induce a change in density are induced by ion bombardment. The detailed structure of track is still a matter of debate, particularly, near the stopping power threshold. Several reports indicate that near threshold tracks may be discontinuous. The subject is related with the problem of defect formation to be described below.

2.1.4.2 Amorphous layers

For stopping powers above the threshold value the random density of tracks grows until they overlap and constitute a homogeneous defective (amorphous) layer. Depending on the stopping power curve one may form the layer at the surface or buried into the material. The two situations are clearly illustrated in Figure 7 for Si at 7.5 MeV and F at 22 MeV. In the first case the maximum stopping power occurs at the sample surface whereas in the second case it lies around 4 microns buried below the surface. The formation of these layers can be monitored by RBS/C and optical techniques like Raman spectroscopy, ellipsometry and dark mode propagation. Due to the same reasons mentioned above for the occurrence of hillocks at the track emergence sites the amorphous layers show a clear swelling in relation to the unirradiated regions of the material.

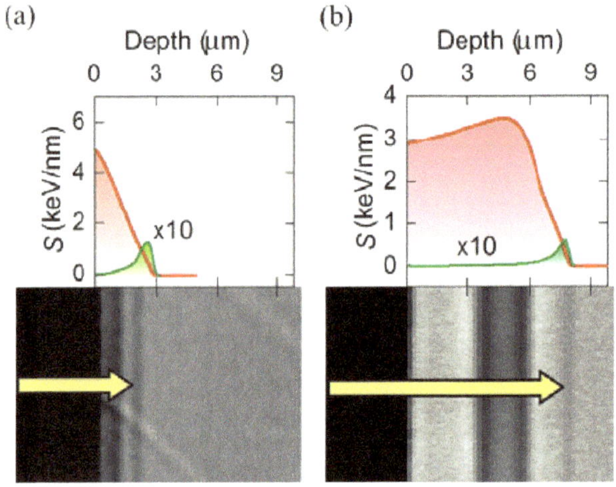

Fig. 7. Micrographs of $LiNbO_3$ samples irradiated at a fluence of 2×10^{14} cm^{-2} with (a) Si ions at 7.5 MeV and (b) F ions at 22 MeV. The corresponding nuclear Sn (green curve) and electronic Se (red curve) stopping powers as a function of depth are drawn above the graphs using the same depth scale for comparison (Rivera *et al.*, 2009).

2.1.5 Mechanisms for electronic damage: Theoretical models

To understand damage production by electronic excitation processes one needs a detailed analysis of the sequence of processes triggered by the initial electronic excitation produced during the passage of a fast (swift) ion. For these ions the major direct cause of energy loss is the Coulomb electrostatic interaction with the electrons of the material, which become free and behave as ballistic electrons. Due to kinematical reasons these so-called delta electrons mostly move in a direction perpendicular to the ion trajectory. A few femtoseconds after the fast passage of an energetic ion, the electronic system of the material is in a high excitation state whereas the ionic system remains essentially unperturbed. Then, the excited electrons interact among themselves and consequently the electronic system undergoes a rapid thermalization process and is generated a hot electron plasma in rough thermal equilibrium. When one reaches the time scale of lattice vibrations (10^{-14}-10^{-13} s) electron-phonon interaction enters into play and the energy is transferred from the electronic to the ionic system until a common temperature is reached (*electron-lattice relaxation*). From here on, both subsystems cool down under thermal equilibrium until a final stage is reached at room temperature. This process is very complicated and provides diverse routes for energy storage. As we will show later, thermal energy may be sufficient to break bonds and displace atoms. In other words, defect generation can be one of the outputs of this electron-lattice relaxation stage. Moreover, bound electron-hole states can be formed and relaxed either by light emission or non-radiatively. This is the most difficult and controversial part of the story.

Each of the above steps in the process give rise to certain radial energy distributions, that are briefly discussed below. Ignoring the possible damage events, one can focus on the energy transfer process that can be described by two coupled equations. They simply describe the temperature evolution of the two subsystems.

2.1.5.1 Radial profile of energy delivered by δ electrons

Most authors dealing with this topic follow the MonteCarlo calculations by Waligórski *et al.* (Waligórski *et al.*, 1986). They calculated the dose or energy invested by the delta electrons per unit radial distance. They proposed an analytical expression that adequately fits the MonteCarlo results. This distribution of energy is the starting point for the calculations of energy transfer to the lattice.

2.1.5.2 Electron-lattice relaxation

In the standard approach one assumes a certain effective temperature for both electrons and lattice after the passage of the bombarding ion. Then in the time scale of 10^{-14}-10^{-13} s the electron-phonon interaction between the electrons and lattice atoms sets in and leads to energy transfer between the two subsystems. Moreover, one should consider the heat transport (energy diffusion) for each charged species. The scheme is simple but not free of problems. Aside from the difficulties associated to the definition of temperature in such short time scales, the mathematical formulation presents some uncertainties, mostly related to the meaning and reliability of the involved physical parameters. Due to the straight trajectory of ions one can use cylindrical coordinates and write the two coupled equations in the form (Toulemonde *et al.*, 1992, 2006):

$$C_e(T_e)\frac{\partial T_e}{\partial t} = \frac{1}{r}\frac{\partial}{\partial r}\left[rK_e(T_e)\frac{\partial T_e}{\partial r}\right] - g(T_e - T_a) + A(r,t)$$
$$, \qquad (4)$$
$$C_a(T_a)\frac{\partial T_a}{\partial t} = \frac{1}{r}\frac{\partial}{\partial r}\left[rK_a(T_a)\frac{\partial T_a}{\partial r}\right] + g(T_e - T_a)$$

where *T*, *C*, and *K* denote, respectively, the temperatures, the specific heats and thermal conductivities of electrons (index *e*) and lattice (index *a*). *g* is the electron-phonon coupling for the material. *A(r,t)* stands for the energy deposition profile caused by the ion impact. According to Waligórski *et al.* (Waligórski *et al.*, 1986) it can be expressed as:

$$A(r,t) = b\, S_e \exp\left\{-(t-t_0)^2 / 2s^2\right\} F(r)\,, \tag{5}$$

where b is an energy normalization factor and t_o is the delay time needed for the electrons to reach equilibrium (~10^{-15} s). *F(r)* is the radial distribution of the generated delta electrons.

The key point of the model is to propose meaningful and reliable values for the parameters appearing in the above equations. Solving equation (4) yields the time and radial evolution of the electron and lattice temperatures, the details can be consulted in the appropriate references. Maximum temperatures T_e and T_a are reached in around 10^{-13} s and then a rapid cooling to near RT takes place in about 10^{-11} s. Note that the cooling rates are in the range of 10^{15} s^{-1}; i.e. much higher than can be usually achieved in a laboratory. In principle, the overall energy invested in the material is not modified by the operative transfer and diffusion processes involved in eq. (4). During and after cooling the electrons and holes forming the excitation spike may experience recombination and trapping processes, including self-trapping or generation of localized excitons at given lattice sites.

2.2 Ion implantation versus ion irradiation

Recently, a strong research activity is being devoted to the fabrication of optical waveguides with ions of medium mass and high energy (swift heavy ions or SHI), where the amorphous region is produced by electronic excitation processes when the electronic stopping power is above a certain threshold value (Agulló-López *et al.*, 2005). This regime allows the separation of the region of electronic excitation from that associated to nuclear damage and implantation (Figure 8, hence the name *irradiation*). The waveguides obtained in this way are different than the conventional ones, where the optical barrier is constituted by the strongly damaged (quasi-amorphous) region at the end of the ion range, whose refractive index is lower than that of the pristine crystal (Davis *et al.*, 1996). The damage in this latter case is produced by the elastic nuclear collisions experienced by low energy (< 10 keV/amu), light- or medium-mass ions near their stopping. Obviously, this region is heavily overlapped with that corresponding to ion implantation that also contributes to the lattice disorder.

Waveguides fabricated by SHI irradiation were first reported by Olivares *et al.* (Olivares *et al.*, 2005b) and they have a number of distinctive advantages over those prepared by standard ion implantation. In particular, the amorphized optical barrier limiting the waveguiding layer is mainly produced by electronic excitation. Its location and width can be suitably controlled by the irradiation fluence and it can be well separated from the deeper implantation region (Figure 8) (Olivares *et al.*, 2007a). Moreover, the refractive index contrast reaches higher values than for conventional implanted guides and the boundary between waveguiding and barrier layers is sharper. Due to all these advantages, there has recently been a considerable increase in the number of research works pushing the use of electronic excitation as a novel means to fabricate optical waveguides and other integrated optics devices (Majkić *et al.*, 2008; Caballero-Calero *et al.*, 2009b; Ren *et al.*, 2010a; Dong *et al.*, 2011a). Unfortunately, both types of waveguides are frequently confused in the current literature (i.e., both are designated as ion-implanted waveguides). This confusion is understandable to some degree because obviously both effects, irradiation and implantation, always occur in any irradiation experiment either with light ions or swift heavy ions. However, we believe that only in the cases when the waveguides have been produced by electronic damage alone one should talk of *SHI waveguides* or *ion-irradiated waveguides with optical barrier of electronic origin* whereas for all other cases we are in the presence of *ion-implantation waveguides*.

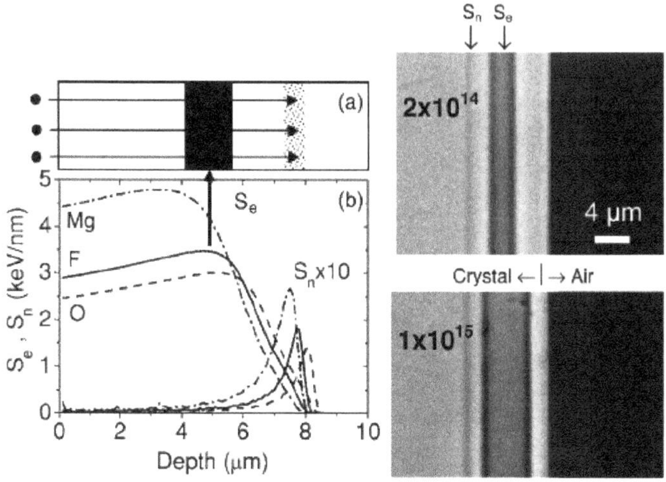

Fig. 8. (a) Illustrative scheme of the generated buried amorphous layer (black strip) near the maximum of the electronic stopping curves. The dotted strip indicates the region of nuclear damage and implantation. (b) Plot of the electronic and nuclear stopping power curves for F at 22 MeV (continuous lines), O at 20 MeV (dashed lines), and Mg at 28 MeV (dashed-dotted lines). Optical microphotographs of a polished Y-cut cross section for samples irradiated with F at fluences of (c) 2×10^{14} and (d) 1×10^{15} cm^{-2}. The depths of the maximum electronic stopping power (Se) and of the nuclear stopping power (Sn) are indicated with arrows. Note also a faint line corresponding to the region of nuclear damage (end of ion range) (Olivares *et al.*, 2007a).

2.3 Annealing

During ion implantation, all the energy deposited by electronic excitation and nuclear collisions on the lattice of the target material creates color centers and alters the original structure, which results in an unavoidable increase of the waveguide losses. Consequently, a thermal annealing is necessary to improve the optical performance of all the ion-implanted waveguides fabricated by this technique (Townsend *et al.*, 1994); with it, the unwanted defects within the guide region are removed, improving the electro-optic or non-linear optic properties. The most commonly used methods of annealing include conventional thermal treatments using furnaces (Gumennik *et al.*, 2005), rapid thermal annealing (Fleuster *et al.*, 1994) and laser annealing (Townsend & Olivares, 1997). In the first case, thermal treatments are performed in conventional furnaces at 200-500 ^{0}C using times that go from a few tens of minutes to several hours, depending on the substrate properties. Rapid thermal annealing, on the other hand, can annihilate the induced defects and avoid, at the same time, the unwanted diffusion of the implanted species (Fleuster *et al.*, 1994). Finally, laser annealing may be confined to the surface layer by using strongly absorbed radiation, resulting in local heating which can anneal intrinsic defects, allow lattice restructuring to remove stress and induce crystalline regrowth or solid phase epitaxy (Townsend & Olivares, 1997).

2.4 Determination of the refractive index profiles

Typical refractive index profiles in the waveguides are several-micrometers thick and, consequently, their direct measurement is quite complicated. For this reason, they are usually determined using a two-step method where, firstly, the effective refractive index (N_{eff}) is measured and then the refractive index profile is calculated using a reconstruction technique. Measurement of N_{eff}, is commonly achieved by a prism coupling method such as the dark modes (m-lines) (Ulrich & Torge, 1973) whereas the preferred reconstruction techniques are the inversed Wentzel-Kramer-Brillouin (iWKB) method (Chiang, 1985), and the reflectivity calculation method (RCM) (Chandler & Lama, 1986).

2.4.1 Prism coupling technique: Dark modes (m-lines)

The optical characterization of waveguides is frequently carried out using the prism coupling technique for the measurement of the effective indices of the modes. This measure allows the determination of the refractive index profile of the waveguide. In this method the prism is placed over the waveguide, pressing up until the thickness of the layer of air between it and the guide is decreased and the evanescent wave escaping the prism penetrates within the optical waveguide. Only the light coupled at certain discrete angles that satisfy the transverse resonant condition will be guided by the optical waveguide (Ulrich & Torge, 1973). Light is propagated in the waveguide only if the effective index in the prism is equal to the effective index of a propagating mode. With this condition and using simple trigonometric expressions it is easy to obtain the expressions (6) from which one can extract the effective indices as a function of the coupling angle measured experimentally $\phi_{exp} = \phi_1 + \phi_2$, if the prism angles, α_1 and α_2, are known (Figure 9).

$$N_{eff} = n_p \cdot \sin\left[\alpha_{1,2} - \arcsin\left(\sin\left(\alpha_{1,2} - \phi_{1,2}\right)/n_p\right)\right]. \quad (6)$$

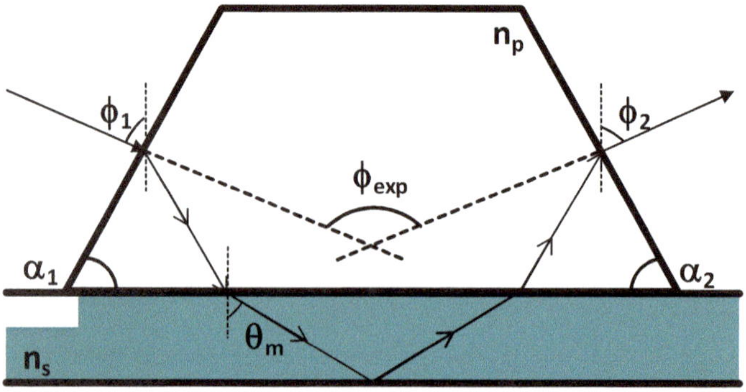

Fig. 9. Schematic representation of the prism coupling technique.

In the dark modes technique a previously expanded beam is focused on the base of the prism (at the point where the optical contact occurred), as seen in Figure 10. This will span a range of coupling angles simultaneously. The light reflected at the base of the prism is then collected on a screen as shown in Figure 10, and then a dark lines pattern appear, corresponding to the angles for which the light is coupled into the waveguide.

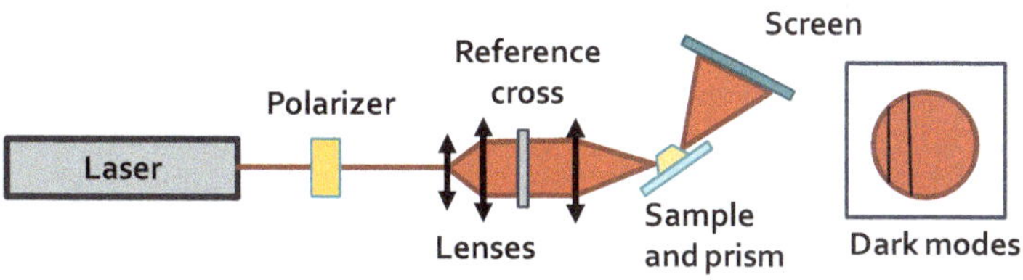

Fig. 10. Schematic of the experimental setup used to measure dark modes.

2.4.2 Inverse Wentzel-Kramer-Brillouin approximation (iWKB)

Once the effective indices of the guide are known, there is a range of refractive index profiles that are mathematically compatible with the same set of values for the effective rate. Normally the largest gap occurs in the surface. It is, therefore, necessary to make certain approximations in order to obtain the refractive index profiles. Fortunately, by considering only physically possible profiles, the uncertainty is reduced to less than ~20% of the difference between the effective indexes of the first two modes (m = 0, 1) (Hertel & Menzler, 1987). Moreover, there are usually indications that can give us a clue as to what type of profile is expected for each case. For instance, by knowing the mechanism of production of the waveguide, one can predict the approximate shape of the profile, if it will be rough or smooth, etc. In this section we will discuss one of the main approaches used for the determination of the index profile of the waveguide from the effective refractive indices: the inverse Wentzel-Kramer-Brillouin approximation (iWKB).

The Wentzel-Kramers-Brillouin (WKB) approximation is a mathematical method used to solve the wave equation of planar waveguides for TE modes (Gedeon, 1974). There is a complementary method, called inverse WKB (iWKB) (Chiang, 1985), giving the refractive index profile through effective rates, based on this approach, which applies to gradual $n_g = n_g(x)$ profiles which meet the condition that the refractive index variation is small over distances comparable to the wavelength. This method uses the following expression:

$$k \int_0^{x(m)} \sqrt{n_g^2 - N_m^2}\, dx = m\,\pi + \phi_r + \phi_s\,, \tag{7}$$

where N_m is the effective index of mode m, and $x(m)$ is its point of return, i.e., $n(x(m)) = N_m$. The phase shift at the interface guide-substrate is $\pi/4$ for all modes, and at the interface air-guide it is:

$$\phi_a = \arctan\left[r_a \left(\frac{N_m^2 - 1}{n_0^2 - N_m^2} \right)^{1/2} \right], \tag{8}$$

where n_0 is the refractive index at the surface and ϕ_a is equal to 1 for TE modes and $(n_0/n_{air})^2$ for TM modes. The method is based on the construction of a continuous function of effective index N_m, adjusting the measured values of effective index to a polynomial by means of the least squares (Chiang, 1985). The main advantage of this method is that the profile can be as soft as desired, requiring only the change of the number of intervals.

3. Materials

Precisely one of the great advantages of ion implantation for the production of optical waveguides is that, by using the adequate ion and energy, you can modify the properties of virtually any material. So, it is not surprising the wide range of materials, ranging from crystals to polymers and glasses, where this technique has been used to generate waveguides. Chen and colleagues (Chen *et al.*, 2007) summarized in a recent review the state of the art on the materials and ions used to fabricate optical waveguides up to 2007. However, there has been a lot of activity in recent years on this field. For instance, the traditional ion implantation (i.e., where the waveguides are produced by means of the nuclear collisions) has been used to fabricate waveguides in various materials, such as SnP_2S_6 (Guarino *et al.*, 2006), KGW (Merchant *et al.*, 2006a; Chen *et al.*, 2008e), KLTN (Gumennik *et al.*, 2005; Ilan *et al.*, 2008), ZnO (Ming *et al.*, 2011), $ZnWO_4$ (Zhao *et al.*, 2010a), Nd:YAG (Chen *et al.*, 2009; Tan & Chen, 2010b), Nd:$YLiF_4$ (Tan *et al.*, 2007b), Nd:LGS (Ren *et al.*, 2010c), Nd:YGG (Zhao *et al.*, 2011a), Nd:BSO (Liu *et al.*, 2011), Nd:CBN (Tan *et al.*, 2009b), DAST (Jazbinsek *et al.*, 2008) etc.) and Er^{3+}-doped telluride glass (Berneschi *et al.*, 2007b), achieving unprecedented control over the refractive index changes of the substrate. Moreover, irradiation with swift heavy ions have been also used to produce waveguides in a wide number of materials, like $LiNbO_3$ (Olivares *et al.*, 2005b, 2007b; c, 2009; Chen, 2009b; Dong *et al.*, 2011a), KGW (García-Navarro *et al.*, 2006), KLTN (Ilan *et al.*, 2006), BGO (Chen *et al.*, 2011), Nd:YAG (Ren *et al.*, 2010a, 2011a), Nd:GdCOB (Ren *et al.*, 2011b), a-SiO_2 (Manzano *et al.*, 2010), c-SiO_2 (Manzano-Santamaría *et al.*, 2012) and chalcogenide glasses (Qiu & Narusawa, 2011a), allowing a fast fabrication of high-quality waveguides. In this section we briefly review the fabrication and properties of ion implanted waveguides in various materials, emphasizing the most recent results.

3.1 Non-linear and photorefractive crystals

3.1.1 Lithium niobate ($LiNbO_3$ or LN)

Lithium niobate is one of the favorite crystals for the fabrication of optical waveguides (Arizmendi, 2004). This interest comes from the combination of properties and characteristics of this material that make possible to tailor its behavior for useful devices. One of its main advantages is that it can be grown in quite large crystals by the use of relatively easy techniques. Moreover, the control of its intrinsic point defects as well as impurities offers a wide range of variation of responses. Consequently, most works related to the fabrication of waveguides by ion implantation (and, indeed, by any technique) have been performed on this material. In particular, light ions (i.e., where the nuclear regime plays an important role) have been used for many years to produce optical waveguides in lithium niobate (Hamelin *et al.*, 1994). On the other hand, the effects of the irradiation with swift heavy ions (i.e., where the electronic stopping power is dominant) have been studied for some time (Canut *et al.*, 1996); however, only recently they started being used to fabricate waveguides (Olivares *et al.*, 2005b).

It is generally accepted that the implantation of $LiNbO_3$ with high fluences of light ions (~10^{16} cm^{-2}) will create an optical barrier with decreased refractive indices at the end of the ion track by nuclear energy deposition, which produces a waveguide structure between the air and the barrier. However, for moderate fluences the extraordinary refractive index (n_e) experiences positive changes in the waveguide region (Chandler *et al.*, 1989; Rams *et al.*, 2000). Fabrication of nontunneling waveguides by means of the implantation with low-dose heavy ions is based on this very same effect, which increases n_e (Bentini *et al.*, 2002; Vázquez *et al.*, 2003; Olivares *et al.*, 2005b). Jiang *et al.* (Jiang *et al.*, 2007) have developed a simple model to explain the observed changes in the refractive-index. Based on this, they concluded that damage in the crystal lattice can be responsible for the refractive-index changes. This conclusion is reinforced by the results reported by Vitova *et al.* (Vitova *et al.*, 2009), who found by means of a x-ray absorption near edge structure (XANES) study that the displacements of Nb and Li atoms, as well as Li and O vacancies, are likely to cause structural disorder, leading to changes in the refractive index of LN and a reduced birefringence.

Among the light ions, H and He have been the two more frequently used for waveguide fabrication in recent years. For example, Vincent *et al.* (Vincent *et al.*, 2005) have reported the fabrication of optical waveguides in 10 μm-periodically poled lithium niobate (PPLN) crystals by implantating He^+. The obtained waveguides supported both TE and TM guided modes; by annealing at 200 ^{0}C during two hours they were able to decrease losses down to 1 dB/cm. Moreover, second order quasi phase matching (QPM) was used to achieve frequency doubling within the obtained waveguides. Peithmann *et al.* (Peithmann *et al.*, 2006b), on the other hand, developed a new method to fabricate embedded, polarization sensitive channel waveguides in $LiNbO_3$ by means of 40 MeV ^3He ions. For this, they used the shadow projected by a tungsten blade with a thickness of 25 μm and a width of 1 mm, placed in front of the ion beam. This procedure was repeated for two perpendicular faces, obtaining a region inside the crystal which had not being exposed to the ion beam. They found changes of the order of $\Delta n \approx 3\times10^{-3}$, which are stable at room temperature but can be erased by annealing the crystals at 500 ^{0}C. The same group performed a detailed study on the changes of the ordinary and extraordinary refractive indexes of lithium niobate crystals irradiated with 41 MeV ^3He ions (Peithmann *et al.*, 2006a). They found that small fluences yielded refractive index changes about 5×10^{-4} while values around 3×10^{-3} were reached for the highest fluences. These index modulations were stable up to 100°C but could be erased thermally using temperatures close to 500°C. Tan and Chen, on the other hand, developed a method to form optical ridge waveguides in $LiNbO_3$ crystals using combination of proton implantation and selective wet etching that preserved the thermo-optic features (Tan & Chen, 2010a). The measured modal field was found to be well confined in the ridge waveguide region; moreover, propagation losses as low as ~0.9 dB/cm were obtained, after applying a thermal annealing at 400 °C.

Meanwhile, Kumar *et al.* (Kumar *et al.*, 2007) studied the influence of proton implantation on the formation of optical waveguides in lithium niobate. They implanted optically polished titanium-doped congruent lithium niobate single crystals with 120 keV protons at fluences in the range 10^{15}-10^{17} cm^{-2}. Nanoscale three-dimensional defect clusters as well as variations of O-H bond stretching vibrations as a function of fluences were observed on the implanted samples. Finally, Tan *et al.* (Tan *et al.*, 2008c) reported the formation of reconfigurable optical channel waveguides on iron-doped lithium niobate crystals (Figure 11). In this two-step process, planar waveguides were first formed using 500 keV protons and a moderate post-implantation annealing. Afterwards, the sample surface was illuminated with well-defined stripe patterns of green light using lithographic masks, forming two-dimensional waveguides in the overlap region on top of the crystal, due to the photovoltaic effect in lithium niobate that results in negative index changes in illuminated regions. They argue that the obtained structures may be used for reconfigurable optical interconnections and splitters.

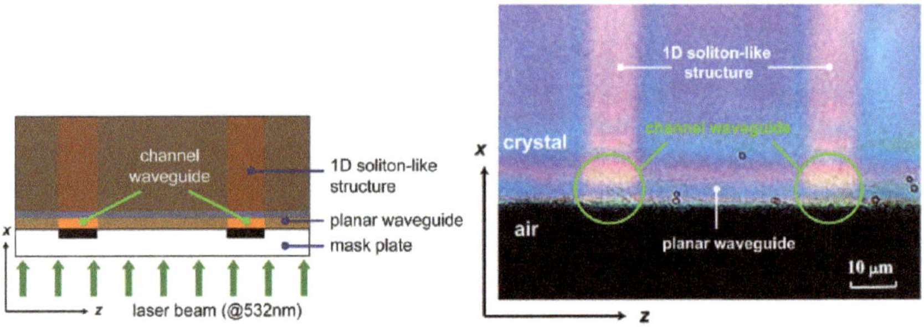

Fig. 11. (left) Schematic plot of the experimental setup for cannel waveguide formation and (right) superposition of the refractive index profile of planar layer and light-imprinted (1D dark soliton-like) stripes, imaged from the sample's entrance facet when illuminated by polarized light (Tan *et al.*, 2008c).

On the other hand, high energy implantation of medium-light mass ions, followed by a suitable annealing, produces a damage profile with a low defect density in the first region below the surface and a higher density defect distribution at the end of the ion ranges, which is an effective method for fabrication of buried waveguides (Bentini *et al.*, 2002). This is because two different kinds of damage are produced in this case: (a) in the near-surface region the damage is mainly produced by ionization and (b) at the end-of-range of the incoming ions the damage comes from the nuclear collision cascades. Most of the damage produced in the first layer can be eliminated by means of a low-temperature annealing whereas the damage of the buried layer requires higher annealing temperature (Bentini *et al.*, 2004). Examples of this approach are the fabrication of planar and channel waveguides on $LiNbO_3$ using low-dose, few-MeV lithium (Raj & Kumar, 2010), carbon (Sugliani *et al.*, 2010), fluorine (Jia *et al.*, 2006) and oxygen (Wang *et al.*, 2006b; Liu *et al.*, 2007, 2008a; b; Song *et al.*, 2008; Tan *et al.*, 2008a; d; Wang & Lu, 2009; Zhao *et al.*, 2011b) ions.

Yet another kind of kinetics has been exploited to fabricate optical waveguides by means of swift heavy ions (Olivares *et al.*, 2005b). In this regime, lattice amorphization is induced (Toulemonde *et al.*, 1994; Canut & Ramos, 1998) along the trajectories of impinging ions whenever the electronic stopping power S_e is above a certain threshold $S_{e,th}$. Then, a homogeneous amorphous layer is generated when the irradiation fluence assures full overlapping of these nanometer-sized tracks. Olivares and coworkers (Caballero *et al.*, 2005; Olivares *et al.*, 2005a; b) pioneered this method and demonstrated its ability to generate high-confinement, step-like optical waveguides in $LiNbO_3$ irradiated with silicon or fluorine ions at fluences in the range 10^{13}-10^{15} cm^{-2}. These results were later confirmed by Wang *et al.* (Wang *et al.*, 2006a), who obtained and characterized planar waveguides on z-cut stoichiometric lithium niobate crystals irradiated with 3.6 MeV Si^{2+} ions at doses of 3-10×10^{13}

cm^{-2}. They also found that the lattice damage could be considerably reduced or even almost removed by means of a thermal treatment. This technique is also suitable to produce 2D waveguides. For instance, Majkić *et al.* (Majkić *et al.*, 2008) used a two-step method to produce and characterize optical microring resonators (Figure 12) and optical channel waveguides by implanting 2.5×10^{14} cm^{-2} of F ions on $LiNbO_3$. After producing planar waveguides by ion implantation, the planar structuring was achieved by laser lithography masking and Ar^+ sputtering. Moreover, they reported the first ring resonators with 80-μm radius fabricated by this method, having an extinction ratio of 14 dB, a free spectral range of 2.0 nm and a finesse of 4. Likewise, Zhao *et al.* (Zhao *et al.*, 2010b) fabricated ridge waveguides on lithium niobate using implantation of O^+ ions combined with Ar ion beam etching.

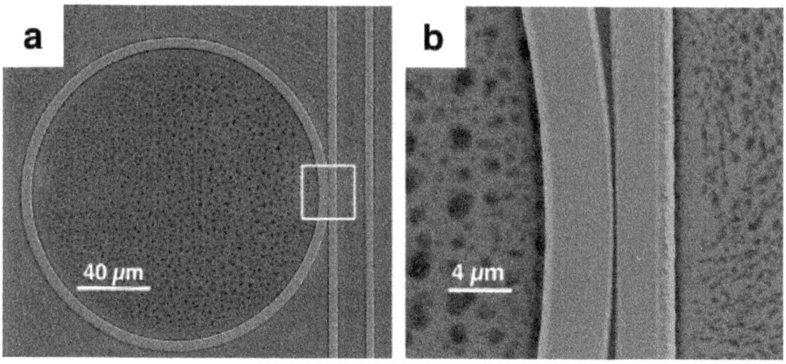

Fig. 12. Scanning electron micrograph of a microring resonator and a bus waveguide, structured in $LiNbO_3$. (a) The whole ring and the bus waveguide. Ring radius is 80 μm, ridge height is 1.2 μm. (b) Enlarged coupling region, the gap size is ~0.2 μm (Majkić *et al.*, 2008).

Quasi-phase matching (QPM) in periodically poled $LiNbO_3$ (PPLN) is a widely used technique for efficient frequency conversion (Batchko *et al.*, 1999). Recently Caballero-Calero *et al.* (Caballero-Calero *et al.*, 2009a; b) have investigated the feasibility of using swift-heavy-ion (SHI) irradiation for the generation of periodically poled structures in optical waveguides. They described two successful ways of producing PPLN: by irradiating bulk PPLN samples with F ions at 22 MeV or by electric periodic poling of previously fabricated swift-ion-irradiated waveguides. Optical waveguide structures were obtained in both cases at the end of the trajectory, combined with a PPLN structure. Good nonlinear and electrooptical properties and high optical confinement makes them good candidates for nonlinear optical devices.

Olivares *et al.* (Olivares *et al.*, 2007b) also demonstrated a novel method to produce optical waveguides on lithium niobate using irradiations with high-energy medium-mass ions (i.e., swift heavy ions) at ultralow fluences (10^{12}-10^{13} cm^{-2}). In this regime amorphous nanotracks are created by every single impact leading to an effective nanostructured medium that behaves as an optical waveguide where light propagates transversally to the ion tracks. This method implied a reduction of four orders of magnitude with respect to a typical He implantation. The obtained waveguides presented losses of ~5-10 dB/cm and significant second-harmonic generation and electro-optic responses (>50% bulk). In a further refinement, Ruiz *et al.* (Ruiz *et al.*, 2007) irradiated proton-exchanged $LiNbO_3$ planar optical waveguides with very low fluences (5×10^{10}-5×10^{12} cm^{-2}) of Cl at 30 MeV. They reported large modifications in the refractive index profiles that resulted in an improved optical performance. This study opened the possibility of fine tuning the refractive index by a suitable control of the fluence. Moreover, Villarroel *et al.* (Villarroel *et al.*, 2009) investigated the photorefractive behavior of this type of optical waveguides. They determined an electro-optic coefficient r_{33} = 18.1±0.5 pm/V for these guides and also studied their recording and light-induced and dark erasure of holographic gratings, as well as the optical beam degradation in single-beam configuration, finding that their damage thresholds are of the same order but a factor of 2–3 greater than that of α-phase guides commonly used in nonlinear applications.

More exotic applications of the irradiation with swift heavy ions at ultralow fluences, like the fabrication of thick optical waveguides have also been reported (Figure 13) (Olivares *et al.*, 2009). In this case Kr and Xe ions with energies ~10 MeV/amu were used to produce waveguides with a thickness of 40-50 micrometers. The obtained waveguides exhibited refractive index jumps up to 0.04 (λ = 633 nm) and propagation of ordinary and extraordinary modes with low losses (~3 dB/cm) and a high nonlinear optical response.

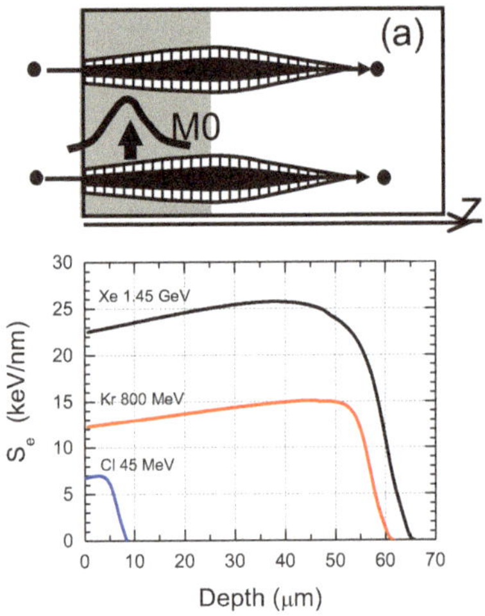

Fig. 13. (a) Schematic depth morphology of the tracks, showing the amorphous core (black) and the surrounding damage halo (dashed) generated after the passage of the ion, which stops approximately where S_n is maximum. A schematic light profile (M0) illustrates waveguiding behavior that is generated in the top effective medium layer (emphasized with gray color). (b) Electronic, Se, stopping power curves for Cl at 45 MeV, Kr at 800 MeV and Xe at 1.45 GeV in $LiNbO_3$ calculated with SRIM2003.

3.1.2 β-Barium borate (β-BaB_2O_4 or BBO)

β-Barium borate is a promising non-linear crystal, which combines wide transparency range, large non-linear coefficients, wide thermal acceptance bandwidth and high damage threshold (Chen *et al.*, 1990). This material is particularly interesting for applications in the ultraviolet region. Ion implantation is one of the better methods available for producing optical waveguides on BBO. He^+, Cu^{2+} and Ni^{2+} ions have been used for this task in recent years. For example, Degl'Innocenti and collaborators (Degl'Innocenti *et al.*, 2006a; b) have reported the fabrication of ridge optical waveguides on top of β-BaB_2O_4 (BBO) crystals (Figure 14). First, BBO crystals were implanted with He^+ ions to form planar optical waveguides and subsequently ridge waveguides were micromachined using femtosecond laser ablation (Degl'Innocenti *et al.*, 2006b). Moreover, they also produced the same type of waveguides by means of He^+ implantation, photolithography masking, and plasma etching (Degl'Innocenti *et al.*, 2006a). Finally, they demonstrated the potential of these waveguides for the second harmonic generation of continuous-wave deep-UV laser light, exploiting the nonlinear optical properties of BBO crystal (Degl'Innocenti *et al.*, 2008; Poberaj *et al.*, 2009). Likewise, Jia and coworkers fabricated non-leaky planar waveguides on BBO by implanting medium doses (~10^{14} cm^{-2}) of 3 MeV Cu^{2+} (Jia *et al.*, 2008b) and Ni^{2+} (Jia *et al.*, 2008d) ions.

Fig. 14. (a) SEM top picture of a ridge defined by two adjacent grooves ablated on the BBO surface; (b) 4.3 μm height ridge with a reduced sidewall roughness after successive Ar^+ ion sputtering (Degl'Innocenti *et al.*, 2006b).

3.1.3 Potassium gadolinium tungstate ($KGd(WO_4)_2$ or KGW)

Potassium gadolinium tungstate is a very attractive material for the fabrication of active optical components due to its high Raman cross-section (Major *et al.*, 2005) and its suitability as a rare-earth host material, with strong emission lines at 1.067 and 1.35 μm (Basiev *et al.*, 2003), which is also suitable for efficient high-power diode pumping around 800 nm (Graf, 1995). García-Navarro *et al.* (García-Navarro *et al.*, 2006) fabricated waveguides in this material using irradiations with low fluences (10^{13}-10^{14} cm^{-2}) of F and O at 25 MeV and C at 14 MeV. The energy of the ions was selected with the criterion of placing the maximum of the electronic stopping power a few microns inside the crystal. Consequently, buried isotropic layers (amorphous-like) were formed. These layers had a low refractive index value of 1.87, in contrast to the three crystalline values which are above 2.0. Subsequently Merchant et al. (Merchant *et al.*, 2006b, 2009) characterized the obtained waveguides by optical techniques, finding that the Raman properties were preserved in the guiding region of the waveguide when irradiated with carbon ions but were considerably reduced when using oxygen and fluorine ions.

3.1.4 Potassium titanium oxide phosphate ($KTiOPO_4$ or KTP)

Potassium titanyl phosphate (KTP) possesses large non-linear optical and electro-optical coefficients (in particular second-harmonics generation SHG is very effective), high optical damage threshold, high thermal stability and transparency over a large wavelength range. High-quality commercial lasers in visible spectral regimes have been realized in this material via frequency doubling of infrared lasers (Bierlein & Arweiler, 1986). Fabrication of waveguides on KTP have been reported using proton exchange (Roelofs *et al.*, 1993), pulsed laser deposition (PLD) (Wang *et al.*, 1998), and ion implantation (Bindner *et al.*, 2001; Schrempel *et al.*, 2004; Chen *et al.*, 2008d; Wang *et al.*, 2008; Tan *et al.*, 2008b). Ion implantation stands among these techniques because it allows an accurate control of the waveguides refractive index profiles by selecting the species, energies and fluences of the incident ions. Fabrication of waveguides on KTP by ion implantation have been reported using H^+ (Tan *et al.*, 2008b), He^+ (Schrempel *et al.*, 2004; Chen *et al.*, 2008d; Dong *et al.*, 2011b), C^{3+} (Wang *et al.*, 2008), N^+, B^+ (Wesch *et al.*, 2001) and Li^+ (Schrempel *et al.*, 2002).

Chen *et al.* (Chen *et al.*, 2008d) reported the fabrication of trapezoidal-shaped optical channel waveguides on $KTiOPO_4$ crystal fabricated by He^+-ion implantation using photoresist masks with wedged-shaped cross sections (Figure 15). First, they formed a photoresist-mask with thickness of 5 μm, containing a series of strongly wedged smooth stripes with a period

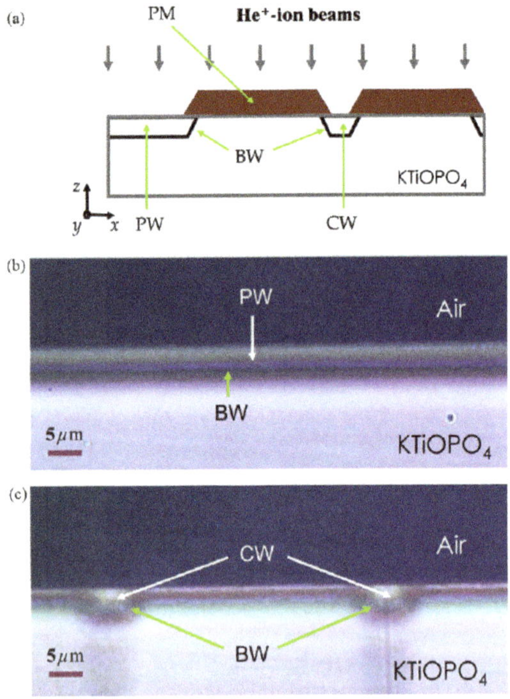

Fig. 15. Schematic of the waveguide fabrication processes in $KTiOPO_4$ (a), microscopic photographs of the cross sections of planar (b), and channel waveguides with trapezoidal shapes (c). PM, photoresist mask; BW, barrier walls; PW, planar waveguide; CW, channel waveguide. The sample crystal axes were marked (Chen *et al.*, 2008d).

of 50 μm. Afterwards, the ion implantation was performed, resulting on the channel waveguides (Figure 15c). Some sections of the samples remained unmasked, in order to also obtain planar waveguides (Figure 15b). The very same method was used by Tan *et al.* (Tan *et al.*, 2008b) and Wang *et al.* (Wang *et al.*, 2008) to fabricate channel and planar waveguides using protons and C^{3+}, respectively. These waveguides were characterized optically (Dong *et al.*, 2011b), finding that the structural and nonlinear properties of the KTP lattice have been only modified at the end of ions' trajectory, which makes these structures good candidates for integrated laser frequency conversion components.

3.1.5 Other non-linear and photorefractive crystals

In addition to the described materials, ion implantation has been used to fabricate optical waveguides in a large number of other non-linear and photorefractive crystals. Table 2 summarizes some of the crystals that have been reported recently, as well as the irradiation conditions and types of guides produced.

Crystal	Ion	Energy (MeV)	Fluence (cm^{-2})	Type of waveguide	Reference
LiB_3O_5	He^+	2.8	5.0×10^{16}	planar	(Wang *et al.*, 2002)
$Ca_4GdO(BO_3)_3$	He^+	1.8	1.5×10^{16}	channel	(Vincent *et al.*, 2003)
$Sn_2P_2S_6$	He^+	2.0	$(0.5\text{-}2.0)\times10^{15}$	planar	(Guarino *et al.*, 2006)
$Ca_xBa_{1-x}Nb_2O_6$	He^+	2.8	1.0×10^{16}	planar	(Fu *et al.*, 2007)
$Sn_2P_2S_6$	He^+	2.0	2.5×10^{14}	planar	(Mosimann *et al.*, 2009)
$ZnWO_4$	He^+	0.5	1.0×10^{16}	planar	(Zhao *et al.*, 2010a)
$Bi_{12}TiO_{20}$	He^+	0.55	1.0×10^{16}	planar	(Zhao *et al.*, 2010d)
$K_{1(-)}T_{0.68}N_{0.32}O_3$:Li	^{12}C	30.0, 40.0	$(0.47\text{-}1.86)\times10^{15}$	slab	(Ilan *et al.*, 2006)
$Sr_xBa_{1-x}Nb_2O_6$	C^{3+}	6.0	1.0×10^{14}	ridge	(Tan *et al.*, 2007c)
$RbTiOPO_4$	C^{3+}	6.0	$(0.5\text{-}1.0)\times10^{14}$	planar	(Wang *et al.*, 2009)
$LiTaO_3$	C^{3+}	6.0	$(0.3\text{-}1.0)\times10^{15}$	planar	(Gang, 2010)
$ZnWO_4$	C^+	5.0	1.0×10^{15}	planar	(Zhao *et al.*, 2010a)
CaF_2	N^+	3.5	5.0×10^{12}-8.0×10^{16}	planar	(Banyasz *et al.*, 2010)
$KTiOAsO_4$	O^+	3.0	5.0×10^{12}-1.0×10^{14}	planar	(Jiang *et al.*, 2006b)
$Bi_{12}TiO_{20}$	O^{2+}	4.5	6.0×10^{14}	planar	(Zhao *et al.*, 2010d)
ZnO	O^+	2.0	1.0×10^{15}	planar	(Ming *et al.*, 2011)
ZnO	O^+	4.0	$(0.5\text{-}2.0)\times10^{15}$	planar	(Ming *et al.*, 2011)
ZnO	O^+	6.0	1.0×10^{15}	planar	(Ming *et al.*, 2011)
LiB_3O_5	Cu^+	6.0	1.0×10^{15}	planar	(Jiang *et al.*, 2006a)
$Ba_{11}Sr(BO_3)_8$	Ag^{13+}	120.0	5.0×10^{11}	planar	(Ishwar Bhat *et al.*, 2002)

Table 2. Data reported on the fabrication of optical waveguides by ion implantation in some non-linear and photorefractive crystals.

3.2 Laser crystals

A laser consists of a resonant cavity containing an amplifying medium. For solid state lasers the cavity is usually obtained by fabricating a doped crystal into a Fabry-Perot configuration, with the excited impurity atoms providing the amplifying medium. For this reason, rare earth ions such as Nd^{3+}, Er^{3+} and Yb^{3+} embedded in a variety of host crystals are frequently encountered in today solid state lasers (Kiss & Pressley, 1966; Kaminskii, 1996). Moreover, the fabrication of optical waveguides on laser crystals is a matter of paramount importance, as can be deduced by the abundant literature available on this topic (Pollnau & Romanyuk, 2007).

3.2.1 Yttrium aluminum garnet ($Y_3Al_5O_{12}$ or YAG)

Yttrium aluminum garnet (YAG) is commonly used as a host material in various solid-state lasers (Kiss & Pressley, 1966; Fricke, 1970). Rare earth elements such as neodymium and erbium can be doped into YAG as active laser ions, yielding Nd:YAG and Er:YAG lasers, respectively (Saiki *et al.*, 2006). Its high chemical stability complicates the formation of optical waveguides in this crystal using metal diffusion or ion exchange techniques, leaving ion implantation as one of the few viable methods for this task. H^+, C^{3+}, N^{3+}, O^{3+} and Ar^{4+} are some of the ions that have been used to produce waveguides on YAG. For example, Vázquez *et al.* (Vázquez *et al.*, 2006) reported the fabrication of optical waveguides in Nd:YAG crystals by proton implantation. They characterized optically these waveguides, finding typical optical barrier profiles, good laser emission characteristics at 1064 nm and high stability in the continuous-wave regime. The same ion was used by Szachowicz *et al.* (Szachowicz *et al.*, 2008) to study if the presence of Nd^{3+} ions favors blue luminescence at 486 nm on Nd,Tm:YAG codoped single crystal waveguides. For this, they fabricated channel waveguides by multiple implantation of H^+, using various irradiation angles and doses. Their main conclusion was that that the Nd^{3+} codoping considerably favors the Tm^{3+} blue emission due to an efficient energy transfer upconversion process. Multi-energy proton implantation was also used by Tan and Chen (Tan & Chen, 2010b) to fabricate optical channel waveguides in Nd:YAG laser ceramics. They found important changes in the Nd spectroscopic properties, such as line broadening and shifting as well as a luminescence quenching, which were attributed to the strong structural modifications induced by the high-dose proton implantation.

Chen and co-workers attained optical channel waveguides in Nd/Ce codoped YAG laser crystal (Figure 16) using mask-assisted carbon ion implantation (Kong *et al.*, 2009) and in Nd:YAG crystals implanted with low doses of oxygen (Kong *et al.*, 2008; Chen *et al.*, 2009). For the first system, they found that the fluorescence properties of both Ce and Nd ions (including the energy transfer between them) were not significantly affected by the waveguide formation processing. Afterwards, Tan *et al.* (Tan *et al.*, 2010) showed generation of continuous wave lasers at a wavelength of ~1064 nm in Nd:YAG implanted with C^{3+} at room temperature. Similar results were obtained for N^{3+}-implanted Nd:YAG crystals by Ren *et al.* (Ren *et al.*, 2011a). Finally, Ren and collaborators (Ren *et al.*, 2010a) took advantage of the ion-induced electronic damage to fabricate a Nd:YAG planar waveguide laser by means of ultra-low-fluence (2×10^{12} cm^{-2}) irradiation with 60 MeV Ar^{4+} ions. The obtained buried waveguiding was associated with an increased refractive index layer as a consequence of the electronic damage. Moreover, continuous-wave laser oscillations at 1064.2 nm were observed from the waveguide under 808 nm optical excitation, with the absorbed pump power at threshold and laser slope efficiency close to 26 mW and 5.9%, respectively.

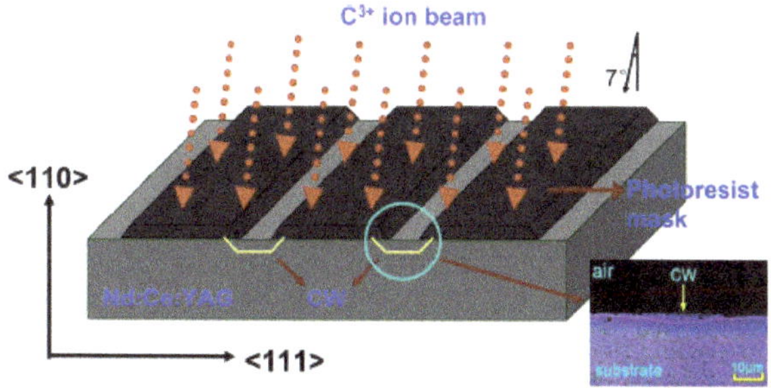

Fig. 16. Schematic plot of the channel waveguide fabrication process in Nd:Ce:YAG substrate by C^{3+} ion implantation. The inset shows the microscopic photograph of the cross section of the Nd:Ce:YAG channel waveguide sample (Kong *et al.*, 2009).

3.2.2 Yttrium vanadate (YVO_4)

Yttrium vanadate (YVO_4) is a promising solid-state laser material because of its stability and good quality. The excellent fit of the Nd^{3+} dopant in the crystal lattice gives it some added advantages such as a low threshold and a large stimulated emission cross-section at the lasing wavelength (Zeller & Peuser, 2000). Finally, Nd:YVO_4 is a highly birefringent crystal that has a broader halfwidth and a higher absorption coefficient than those of Nd:YAG. For those reasons, Nd:YVO4 has been proposed as a better candidate than Nd:YAG for low-power lasers (Yang & Huang, 2000). Successful waveguide formation has been reported in the last five years for this crystal using H^+, He^+, C^{2+}, O^+ and Cu^{2+} ions. Sánchez-Morales *et al.* (Sánchez-Morales *et al.*, 2007) reported the formation of planar waveguides by proton and helium implantation in Nd:YVO_4. Wide barriers were generated in this work using double and triple implants with energies between 0.4 MeV and 1.75 MeV and fluences around 10^{16} cm^{-2}. Moreover, a double waveguide was generated with a triple implantation. The same group also fabricated of waveguides on Nd:YVO_4 crystals by means of a C^{2+} irradiation, using fluences as low as 5×10^{14} cm^{-2} (Sánchez-Morales *et al.*, 2006; Vázquez *et al.*, 2006). They found that carbon implantation produced a considerable refractive index variation in the guiding region, increasing the ordinary index and reducing the extraordinary index. Emission bands coming from the $^4F_{3/2}$ level were not altered by the ion beam process, suggesting the potential use of carbon implanted waveguides for laser integrated devices. Likewise, other groups produced planar (Jia *et al.*, 2008a) and channel (Chen *et al.*, 2006a) waveguides using 3 MeV O^+ ions as well as 3 MeV Cu^{2+} ions (Jia *et al.*, 2011) at intermediate fluences (10^{13}-10^{15} cm^{-2}).

3.2.3 Rare-earth-doped lithium niobate (X:$LiNbO_3$)

The main advantages of lithium niobate have been discussed above; however, an additional benefit of this material arises when it is doped with rare earth elements (particularly Nd^{3+} and Er^{3+}). In this case, it is possible to fabricate new monolithic laser or amplifier components combining the medium's gain and the large range of integrated optical functions already demonstrated including coupled-cavity, single-frequency, tunable, Q-switched, mode-locked, or frequency-doubled lasers and amplified integrated optical circuits with no insertion losses (Lallier, 1992). H^+, C^+, O^+ and Ar^{4+} are some of the ions that have been used recently to fabricate optical waveguides on rare-earth-doped LN laser crystals. Dong and co-workers produced optical channel waveguides in Nd^{3+}:MgO:$LiNbO_3$

crystals by using 500 keV protons (fluence of 6×10^{16} cm^{-2}) and a stripe photoresist mask (Dong *et al.*, 2008). With this procedure the emission intensity of Nd^{3+} ions was only slightly modified with respect to the bulk crystal. Likewise, 3 MeV carbon (7.5×10^{14} cm^{-2}) (Chen *et al.*, 2008b) and oxygen (6×10^{14} cm^{-2}) (Chen *et al.*, 2008c) ions have also been successfully used to produce guides on this material. The fluorescence image of the proton-implanted waveguides has been found to be completely different from that obtained by medium-mass ion implantation, revealing that different waveguide formation mechanisms are taking place in each case (Jaque & Chen, 2009). Optical waveguides can also be obtained on doped near-stoichiometric lithium niobate (SLN) by applying low-doses of medium-mass ions. For instance, 7.5×10^{14} cm^{-2} of 3 MeV C^+ ions (Chen, 2009a) and 6×10^{14} cm^{-2} of 3 MeV O^+ ions (Chen *et al.*, 2008a) were used to produce optical waveguides on Nd:MgO:SLN and Nd:SLN crystals, respectively. Ultra-low fluences (2×10^{12} cm^{-2}) of 60 MeV Ar^{4+} ions achieved the same result on the latter material (Dong *et al.*, 2011a). Finally, embedded channel waveguides were reported for the first time on a *z*-cut erbium-doped lithium niobate (Er:$LiNbO_3$) substrate implanted with 1.65×10^{15} cm^{-2} of 3.9 MeV C^{3+} ions (Sher *et al.*, 2011).

3.2.4 Other laser crystals

In addition, optical waveguides have been also fabricated on a wide range of laser crystals that are not described individually due to the space constrains. The specific crystals and dopants, as well as the irradiation conditions and types of guides produced are summarized on Table 3.

Dopant:Crystal	Ion	Energy (MeV)	Fluence (cm^{-2})	Type of waveguide	Reference
Er:$YAlO_3$	H^+, He^+	1-1.5	$(1\text{-}4)\times10^{16}$	channel and planar	(Szachowicz *et al.*, 2006)
Nd:GGG	H^+	0.46|0.48|0.5	$(2\|2\|4)\times10^{16}$	planar	(Ren *et al.*, 2010d)
Nd:LGS	H^+	0.5	9.6×10^{16}	channel	(Ren *et al.*, 2010c)
Ce: KNSBN	He^+	2-2.2	$(0.9\text{-}1.5)\times10^{16}$	channel	(Chen *et al.*, 2006b)
Nd:KGW	He^+	1.9|2.0|2.1	$(3.24\|3.24\|5.4)\times10^{15}$	planar	(Chen *et al.*, 2008e)
Nd:CBN	He^+	2.6|2.7|2.8	$(1.3\|1.3\|2.2)\times10^{15}$	channel	(Tan *et al.*, 2009b)
Nd:SGG	He^+	0.5	$(1\text{-}3)\times10^{16}$	planar	(Guo *et al.*, 2010)
Nd:BSO	He^+	0.5	$(1\text{-}3)\times10^{16}$	planar	(Liu *et al.*, 2011)
Nd:KGW	C^{3+}	6.0	1.0×10^{15}	planar	(Chen *et al.*, 2008e)
Nd:CNGG	C^{3+}	6.0	$(1\text{-}8)\times10^{14}$	planar	(Wang & Yu, 2010)
Nd:GGG	C^{3+}	6.0	6.0×10^{14}	planar	(Ren *et al.*, 2010d)
Nd:BSO	C^{3+}	6.0	1.0×10^{15}	planar	(Liu *et al.*, 2011)
Nd:GdCOB	C^{5+}	17.0	2.0×10^{14}	planar	(Ren *et al.*, 2011b)
Yb:YCOB	O^{3+}	3, 6	$(0.5\text{-}20)\times10^{14}$	planar	(Jiao *et al.*, 2007)
Nd:$LuVO_4$	O^{3+}	2.4, 3.0, 3.6	$(1.4\text{-}3.1)\times10^{14}$	planar	(Jia *et al.*, 2008c)

Table 3. Data reported on the fabrication of optical waveguides by ion implantation in some laser crystals.

3.3 Glasses

Various optical glass materials are excellent hosts of laser dopants; for this reason rare-earth doped laser glasses are now commercially available. The most common method for producing waveguides in glass is ion exchange; however, ion implantation is also frequently used for this task (Townsend *et al.*, 1994). In recent years have been reported the fabrication of optical waveguides on a wide range of glasses (silicates, tellurites, phosphates and chalcogenides) doped with various elements (Nd, Er, Yb and Ti) and using several ions (H^+, He^+, O^+, $N^{+,3+}$, F^{3+}, Cl^{6+} and Ar^{4+}). For example, high chemical stability coupled with a large solubility for rare-earth ions makes silicate glasses excellent candidates for the fabrication of lasers and waveguide amplifiers. Wang *et al.* (Wang *et al.*, 2007) reported the fabrication of low-loss planar and stripe waveguides in a Nd^{3+}-doped silicate glass by implanting 6 MeV oxygen ions at a fluence of 10^{15} cm^{-2}. It was found that the ion beam processing created a refractive index enhanced region as well as an optical barrier. The same group also reported the fabrication of ridge optical waveguides in Er^{3+}/Yb^{3+} co-doped phosphate glasses by 2.8 MeV He^+ implantation and a subsequent Ar^+ ion beam etching (at 500 eV) (Tan *et al.*, 2007a). Optical characterization showed that an optical barrier-type waveguide had been formed. These results are relevant due to the importance of this glass in telecommunication applications.

On the other hand, tellurite glasses are promising materials for the fabrication of active integrated optical circuits due to their high rare-earth solubility, large stimulated emission cross sections and broad emission bandwidth around the 1.55 micron wavelength. Fabrication of channel (Berneschi *et al.*, 2007a; Khánh *et al.*, 2009) and slab (Berneschi *et al.*, 2011) waveguides on this glass (doped with Er^{3+}) using 1.5 MeV N^+ ions (fluences in the range 5×10^{12} to 8×10^{16} cm^{-2}) have been reported by Berneschi and collaborators. Optical characterization revealed that the implanted layer exhibited a decrease of the refractive index with respect to the virgin bulk glass, while the region comprised between the sample surface and the end of the ion track acted as an optical guiding structure. It was also found that a post-implantation annealing process resulted in a reduction of the barrier region.

Equally important are chalcogenide glasses, and in particular gallium lanthanum sulphide (GLS). These novel materials exhibit promising properties, such as a wide transmission window, large Kerr nonlinear coefficient, low toxicity, excellent thermal and mechanical stabilities, and high solubility for rare-earth dopants (Zakery & Elliott, 2003). Qiu and coworkers have reported recently the fabrication of planar optical waveguides on this material using both, light (Qiu & Narusawa, 2010, 2011b) and swift heavy ions (Qiu & Narusawa, 2011a; Qiu *et al.*, 2011). Firstly, they used 350 keV protons at a fluence of 10^{15} cm^{-2}; finding an optical barrier-type waveguide, as expected when the nuclear damage is dominant. In the second case, they employed 20 MeV N^{3+} and 60 MeV Ar^{4+} ions at intermediate (1.5×10^{14} cm^{-2}) and ultra-low (2×10^{12} cm^{-2}) fluences, respectively. The Ar^{4+}-implanted waveguides exhibited an step-like distribution, while the N^{3+}-implanted ones showed a "well" region with increased refractive index near the surface of the glass coming from the electronic energy deposition and a "barrier" layer with decreased refractive index inside the glass, due to the nuclear energy deposition. Similarly, Manzano *et al.* (Manzano *et al.*, 2010) reported the preparation of planar optical waveguides in silica (a-SiO_2) by means of swift heavy-ion irradiation (F^{3+} at 5 MeV, Cl^{6+} at 20 MeV), confirming that for SHI the refractive index enhancement is strongly correlated with the electronic stopping power. The importance of this glass stems from its availability, low optical absorption, near-zero thermal expansion, low dielectric losses and good chemical inertness that makes it one of the most widely used material for photonics applications, including commercial fibers for information technologies (Devine *et al.*, 2000).

4. Applications

Obviously applications are the ultimate goal of most research efforts. Despite being a relatively recent field, where new results are generated at a very high rate, the waveguides fabricated by ion implantation has been already used for many applications such as beam splitting/coupling, frequency doubling, two-wave mixing, photorefractive solitons and waveguide lasers. In this Section we make a brief overview of the most recent developments in this area.

4.1 Beam splitters/couplers

Beam splitters and couplers, and particularly simple Y-junctions, are key components for waveguide devices such as integrated interferometers. Some recent works have reported the fabrication of these elements by means of ion implantation, using different ions and materials. Hadjichristov and Stefanov (Hadjichristov & Stefanov, 2010) reported the fabrication of a beam splitter/coupler at the telecommunication wavelength of 1.55 micrometers by (50 KeV) silicon implantation of PMMA. In turn, Bentini *et al.* (Bentini *et al.*, 2007) and Zhao et al. (Zhao *et al.*, 2010c) used $LiNbO_3$ substrates. The first group fabricated a Mach-Zehnder interferometer by using high energy ion implantation of carbon ions on $LiNbO_3$ whereas Zhao and collaborators obtained power splitters in $LiNbO_3$ using multi-energy O ion implantation with standard lithography. The design consisted of a cascade Y-branch waveguide structure that distributed equally the input power between four output waveguides. Finally, two interesting recent works have reported the formation of reconfigurable channel waveguides and beam splitters (Figure 17), combining ion implantation with selective light illumination of iron doped $LiNbO_3$ crystals (Tan *et al.*, 2008c; d).

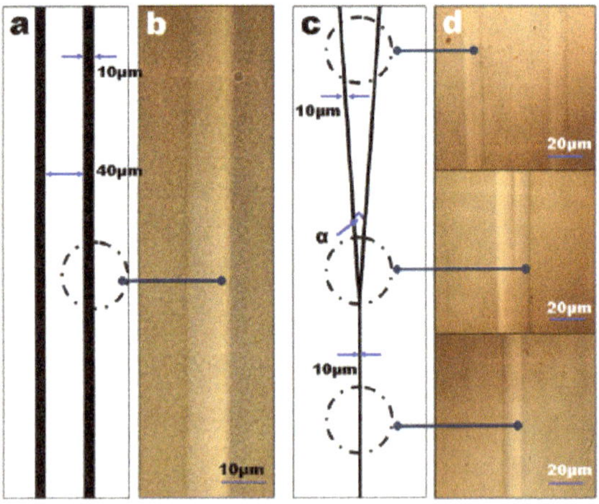

Fig. 17. Lithographic mask patterns of (a) straight stripe and (c) Y-branch waveguides (tan(α) = 0.01); and microscopy photographs of the top view of (b) straight channel and (d) Y-branch waveguide, imaged by using polarized light reflection (Tan *et al.*, 2008c).

4.2 Second harmonic generation

Probably ion implantation is one of the best available processes to obtain optical waveguides with applications in second harmonic generation (SHG). This is because the nonlinear susceptibility $\chi^{(2)}$ is preserved or only slightly reduced in the guiding layer (Dong *et al.*, 2011b). However, a practical device has not been developed yet as it has been, for instance, in $LiNbO_3$ where optical frequency doubling devices were probed for Ti-indiffused (Sohler *et al.*, 2008) and proton exchanged (Langrock *et al.*, 2006) waveguides. The main reason is the high cost of obtaining a waveguide with low optical losses, high optical damage threshold, good phase-matching (PM) or quasi-phase-matching (QPM) conditions, etc. Therefore, the capability of ion implanted waveguides for SGH devices strongly depends of the starting material and the reduction of fabrication costs.

In the last years, much effort has been made to improve the optical properties of materials. In periodically poled $LiNbO_3$ (PPLN), Vincent *et al.* (Vincent *et al.*, 2005) have found conversion efficiencies up to $6.5\times10^{-3}\%\ W^{-1}$ using second order QPM in low losses waveguides (>1 dB/cm) obtained by several implantation of He^+ with energy in the range of 1.5 to 1.7 MeV. When PPLN is doped with 6 mol. % of Zn, to increase the optical damage resistance, green light at first order QPM has been generated with an efficiency of $10^{-2}\ W^{-1}$ in similar waveguides (Vincent *et al.*, 2007). SHI waveguides are a promising candidate to SHG integrated devices in $LiNbO_3$. It has been demonstrated that its high step-like refractive index change is accompanied with low decreasing of nonlinear coefficients (Olivares *et al.*, 2007a). Furthermore, PPLN structures can be made both before and after waveguide fabrication (Caballero-Calero *et al.*, 2009a; b) for efficient QPM. To all of this, it must be added an important reduction in time and fabrication costs with respect to the process of implantation with light ions.

Important progresses have been made as well in the development of ultraviolet integrated optics devices based on barium-borate (BBO) crystals (Poberaj *et al.*, 2009). For instance, continuous-wave UV laser light at 278 nm and 266 nm by optical frequency doubling of visible light have been reported in β-BaB_2O_4 waveguides produced by He^+ ions implantation (Degl'Innocenti *et al.*, 2006a, 2008). Finally, optical planar waveguides produced in the new nonlinear $Ca_4GdO(BO_3)_3$ crystal by ionic implantation have also shown SHG significant conversion efficiency that can be increased after applying a thermal treatment (Boudrioua *et al.*, 2005). Planar waveguides fabricated in Nd^{3+}-doped crystals by 17 MeV carbon ion irradiation have been reported recently (Ren *et al.*, 2011b). They exhibit high optical conversion efficiency of frequency doubling for CW ($0.48\%\ W^{-1}$) and pulsed ($6.8\%\ W^{-1}$) laser beams.

4.3 Photorefractive applications

A number of photorefractive applications have been implemented in ion implanted waveguides. Two relevant examples are photorefractive solitons and two-wave mixing (TWM) applications.

4.3.1 Two-wave mixing

Two-wave mixing in photorefractive crystals is a non linear process that allows energy transfer between the involved beams and, consequently, optical amplification. This process was also demonstrated in photorefractive waveguides that in some cases were prepared by ion implantation. Two materials have been initially used to prepare the waveguides: $BaTiO_3$ (Youden *et al.*, 1992; Dazzi *et al.*, 1999; Mathey *et al.*, 2001a; b) and $KNbO_3$ (Brülisauer *et al.*, 1995, 1996). The main parameters of the ion implantation process and two-wave mixing operation reported in those works have been summarized by Chen *et al.* (Chen *et al.*, 2007).

Later on, TWM was proved in iron-doped congruent $LiNbO_3$ waveguides fabricated by O^{3+} ion implantation obtaining gain coefficients similar to that of bulk crystals (Tan *et al.*, 2008a). Waveguides have been also fabricated on near-stoichiometric Fe:$LiNbO_3$ substrates by proton implantation (Peng *et al.*, 2011). High gain coefficients of 15 cm^{-1} have been obtained in these TWM experiments. Finally, Mossiman *et al.* (Mosimann *et al.*, 2009) have demonstrated TWM in He^+ implanted waveguides fabricated on a promising material for infrared operation: the ferroelectric semiconductor ($Sn_2P_2S_6$). In this case a gain of 2.5 cm^{-1} has been obtained at the telecom wavelength of 1.55 μm.

4.3.2 Photorefractive solitons

Spatial solitons, i.e. non diffracting beams, have very interesting properties to implement photonic applications such as all optical switches and routers. Photorefractive spatial solitons have attracted considerable interest because they can be generated with low light power (~mW) (del Re *et al.*, 2006). As for other photorefractive applications, first experiments were demonstrated in bulk crystals (del Re *et al.*, 2006) but later on it was also proved on waveguide configuration. Photorefractive bright spatial solitons were observed in SBN waveguides produced by He^+ implantation in the visible and infrared regions (Kip *et al.*, 1998; Wesner *et al.*, 2001). More recently, dark photovoltaic spatial solitons have been also observed in a planar waveguide produced by ion implantation of protons in cupper doped $LiNbO_3$ crystal (Kruglov *et al.*, 2008). Finally, gap solitons have been observed very recently in lithium niobate combining a waveguide array fabricated by proton implantation with light-induced refractive index modification via the photorefractive effect (Tan *et al.*, 2009a).

4.4 Waveguide lasers

Rare earth (Nd, Er, etc.) and transition metal (Ti, Cr, etc.) doped crystals are widely used as gain media for solid state lasers. Conservation of the luminescence properties of the irradiated material, along with a low pump threshold, are very important characteristics for ion-implanted waveguide lasers. The former point was addressed in Section 3.2, where we described how it has been demonstrated that, for various crystals, the luminescence is preserved after the waveguide formation. In this section we will describe the most recent examples of working laser prototypes on ion-implanted optical waveguides.

Yao *et al.* (Yao *et al.*, 2010) reported a waveguide laser in Nd:YAG crystals, produced by proton beam writing, with an output power of 60 mW at 1064 nm. This system exhibited the additional advantage of having highly symmetric guided modes (Figure 18), which would facilitate its connection with the optical fibers. For the first ion implanted Nd:YAG ceramic planar waveguide, a 1064 nm laser was generated with a 19.5 mW threshold and 11% slope efficiency (Tan *et al.*, 2010). Moreover, swift-ion-irradiated Nd:YAG waveguide laser systems have similar performances (laser slope efficiencies of 16% and pump power thresholds of 38.3 mW) than the ion implanted samples (Ren *et al.*, 2010a, 2011a).

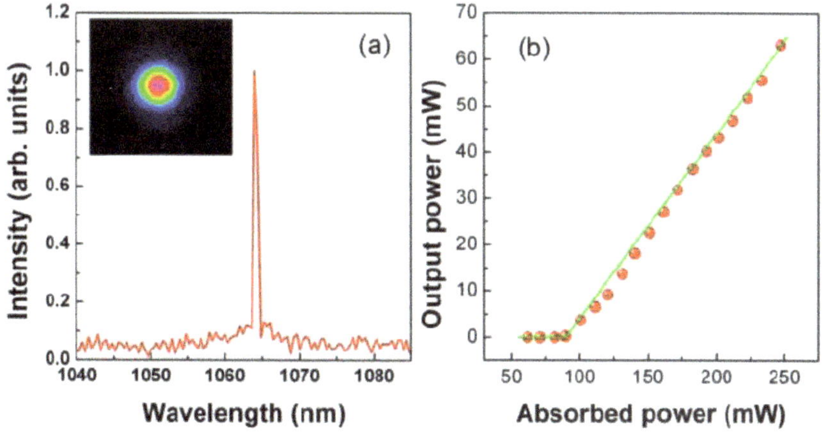

Fig. 18. (a) Laser oscillation spectra from the waveguide produced by 1 MeV PBW at fluence of 10^{16} cm^{-2}, showing a keen-edged peak at 1064.2 nm with a FWHM of 0.75 nm. The mode image of the waveguide laser is shown as inset. (b) The measured output laser power as a function of the absorbed pump power (balls) from the waveguide. The green solid line shows the linear fit of the experimental data (Yao *et al.*, 2010).

Moreover, the use of He^+ ion implantation, combined with Ar^+ ion etching, have been demonstrated to fabricate ridge waveguides in Cr:$LiSrAlF_6$ (Majkić *et al.*, 2007). They found that the fluorescence emissions from the waveguides are similar to those from the bulk material, indicating that the Cr^{3+} ions are not altered by the ion implantation process. The output PL power of this system was 13 μW at 165 mW pump power. Unfortunately, their application as a tunable integrated light source is somewhat limited by its high loss (~10 dB/cm). This limitation is not present in the buried Ti:sapphire channel waveguides produced by proton implantation (Grivas *et al.*, 2006; Pollnau *et al.*, 2007), whose very low loss (~1dB/cm) makes them a better choice. In this case laser operation (~780 nm) was achieved at room temperature, with an absorbed pump power threshold of 230 mW; the maximum output power and the slope efficiency were 17 mW and 3%, respectively.

A planar waveguide laser have been also demonstrated on H^+-implanted Nd:GGG crystals, reaching a slope efficiency of 30% and 22 mW of output power (Ren *et al.*, 2010b). Similarly, channel (Sánchez-Morales *et al.*, 2008) and planar (Jia *et al.*, 2007) waveguide lasers have been produced on Nd:YVO_4 using, respectively, 7 MeV carbon and 180 MeV argon ions. In the former system laser oscillations were generated at 1064 nm, with a pump threshold power of 45 mW and a slope efficiency of 30%. The SHI waveguide laser, on the other hand, emitted at 1067 nm with a slope efficiency of 8.5%.

5. Concluding remarks

Ion implantation is a well-established technique to modify in a controllable way all kinds of materials i.e. dielectrics, semiconductors and metals. In this review we have focused on the effects of ion beams on the optical properties of dielectric materials that can be used to fabricate optical waveguides and other related photonic devices. A large span of possibilities exists, depending on ion type and energy. Although a number of reviews and monographs are available on this subject, we have concentrated on recent developments (e.g. the production of waveguides with optical barrier by electronic excitation) and novel crystal hosts with a promising potential for laser and nonlinear optical applications. Special attention/emphasis has been paid as well to the novel processing capability due to the high electronic excitation ("electronic damage") provided by high energy (> 0.1 MeV/amu) heavy ions and/or swift heavy ions (> 1 MeV/amu). This field is still alive, both from the scientific and technological sides, and exciting achievements are likely to show up in the coming years. Particularly attractive is the connection/synergy between the (lateral) nano-structuring with the deep structuring capability (microns thick) of swift heavy ions.

6. References

Agulló-López, F., García, G. & Olivares, J. (2005). Lattice preamorphization by ion irradiation: Fluence dependence of the electronic stopping power threshold for amorphization. *J. Appl. Phys.* Vol.97, No.9, pp.093514, ISSN 0021-8979.

Agulló-López, F., Mendez, A., García, G., Olivares, J. & Cabrera, J. M. (2006). Synergy between thermal spike and exciton decay mechanisms for ion damage and amorphization by electronic excitation. *Phys. Rev. B* Vol.74, No.17, pp.174109, ISSN 1550-235X.

Arizmendi, L. (2004). Photonic applications of lithium niobate crystals. *Phys. Stat. Sol. A* Vol.201, No.2, pp.253-283, ISSN 1521-396X.

Arnoldbik, W. M., Tomozeiu, N. & Habraken, F. H. P. M. (2003). Electronic sputtering of thin SiO_2 films by MeV heavy ions. *Nucl. Instrum. Meth. B* Vol.203, pp.151-157, ISSN 0168-583X.

Banyasz, I., Berneschi, S., Lohner, T., Fried, M., Petrik, P., Khanh, N. Q., Zolnai, Z., Watterich, A., Bettinelli, M., Brenci, M., Nunzi-Conti, G., Pelli, S., Righini, G. C. & Speghini, A. (2010). Characterisation of slab waveguides, fabricated in CaF_2 and Er-doped tungsten-tellurite glass by MeV energy N^+ ion implantation, using spectroscopic ellipsometry and m-line spectroscopy. *Proceedings of Silicon Photonics and Photonic Integrated Circuits II*, Brussels, Belgium, 2010. p 77190G. Brussels, Belgium: SPIE.

Basiev, T. T., Osiko, V. V., Prokhorov, A. M. & Dianov, E. M. (2003). Crystalline and fiber Raman lasers. In: Sorokina, I. T. & Vodopyanov, K. L. (Eds.) *Solid-State Mid-Infrared Laser Sources*. pp 359-408. Berlin, Heidelberg: Springer Berlin Heidelberg., ISBN 978-3-540-00621-3.

Batchko, R. G., Shur, V. Y., Fejer, M. M. & Byer, R. L. (1999). Backswitch poling in lithium niobate for high-fidelity domain patterning and efficient blue light generation. *Appl. Phys. Lett.* Vol.75, No.12, pp.1673-1675, ISSN 0003-6951.

Bentini, G. G., Bianconi, M., Cerutti, A., Chiarini, M., Pennestrì, G., Sada, C., Argiolas, N., Bazzan, M. & Mazzoldi, P. (2007). Integrated Mach-Zehnder micro-interferometer on $LiNbO_3$. *Opt. Laser Technol.* Vol.45, No.3, pp.368-372, ISSN 0143-8166.

Bentini, G. G., Bianconi, M., Correra, L., Chiarini, M., Mazzoldi, P., Sada, C., Argiolas, N., Bazzan, M. & Guzzi, R. (2004). Damage effects produced in the near-surface region of x-cut $LiNbO_3$ by low dose, high energy implantation of nitrogen, oxygen, and fluorine ions. *J. Appl. Phys.* Vol.96, No.1, pp.242-247, ISSN 0021-8979.

Bentini, G. G., Bianconi, M., Chiarini, M., Correra, L., Sada, C., Mazzoldi, P., Argiolas, N., Bazzan, M. & Guzzi, R. (2002). Effect of low dose high energy O^{3+} implantation on refractive index and linear electro-optic properties in X-cut $LiNbO_3$: Planar optical waveguide formation and characterization. *J. Appl. Phys.* Vol.92, No.11, pp.6477-6483, ISSN 0021-8979.

Beranger, M., Brenier, R., Canut, B., Ramos, S. M. M., Thevenard, P., Balanzat, E. & Toulemonde, M. (1996). Defect creation induced by GeV ions in MgO containing Na precipitates. *Nucl. Instrum. Meth. B* Vol.112, No.1-4, pp.112-115, ISSN 0168-583X.

Berneschi, S., Brenci, M., Nunzi Conti, G., Pelli, S., Bettinelli, M., Speghini, A., Bányász, I., Fried, M., Khanh, N. Q., Lohner, T., Petrik, P., Watterich, A. & Zolnai, Z. (2011). Slab optical waveguides in Er^{3+}-doped tellurite glass by N^+ ion implantation at 1.5 MeV. *Opt. Eng.* Vol.50, No.7, pp.071110, ISSN 0091-3286.

Berneschi, S., Brenci, M., Nunzi Conti, G., Pelli, S., Righini, G. C., Bányász, I., Watterich, A., Khanh, N. Q., Fried, M. & Pászti, F. (2007a). Channel waveguides fabrication in Er^{3+}-doped tellurite glass by ion beam irradiation. *Proceedings of Integrated Optics: Devices, Materials, and Technologies XI*, San Jose, CA, USA, 2007. p 647509. San Jose, CA, USA: SPIE.

Berneschi, S., Nunzi Conti, G., Bányász, I., Watterich, A., Khanh, N. Q., Fried, M., Pászti, F., Brenci, M., Pelli, S. & Righini, G. C. (2007b). Ion beam irradiated channel waveguides in Er^{3+}-doped tellurite glass. *Appl. Phys. Lett.* Vol.90, No.12, pp.121136, ISSN 0003-6951.

Bierlein, J. D. & Arweiler, C. B. (1986). Electro-optic and dielectric properties of $KTiOPO_4$. *Appl. Phys. Lett.* Vol.49, No.15, pp.917-919, ISSN 0003-6951.

Bindner, P., Boudrioua, A., Loulergue, J. C. & Moretti, P. (2001). Formation of planar optical waveguides in potassium titanyl phosphate by double implantation of protons. *Appl. Phys. Lett.* Vol.79, No.16, pp.2558-2560, ISSN 0003-6951.

Boccanfuso, M., Benyagoub, A., Schwartz, K., Trautmann, C. & Toulemonde, M. (2002). Study of the damage produced in CaF_2 by swift heavy ion irradiation. *Nucl. Instrum. Meth. B* Vol.191, No.1-4, pp.301-305, ISSN 0168-583X.

Boudrioua, A., Vincent, B., Kremer, R., Moretti, P., Tascu, S. & Aka, G. (2005). Linear and nonlinear optical properties of implanted $Ca_4GdO(BO_3)_3$ planar waveguides. *J. Opt. Soc. Am. B* Vol.22, No.10, pp.2192-2199, ISSN 1520-8540.

Brülisauer, S., Fluck, D. & Günter, P. (1995). High gain two-wave mixing in H^+-implanted photorefractive Fe:$KNbO_3$ planar waveguides. *Electron. Lett.* Vol.31, No.4, pp.312-313, ISSN 0013-5194.

Brülisauer, S., Fluck, D., Günter, P., Beckers, L. & Buchal, C. (1996). Photorefractive effect in proton-implanted Fe-doped $KNbO_3$ waveguides at telecommunication wavelengths. *J. Opt. Soc. Am. B* Vol.13, No.11, pp.2544-2548, ISSN 1520-8540.

Caballero, O., García-Cabañes, A., Carnicero, J., Carrascosa, M., Agulló-López, F., Olivares, J., García-Navarro, A. & García, G. (2005). Non-linear and photorefractive characterisation of highly confined $LiNbO_3$ waveguides prepared by high-energy, low-fluence ion irradiation. In: Zhang, G. (Ed.) *Proceedings of Photorefractive Effects, Materials, and Devices,* Juli 19 2005. p 179. Optical Society of America.

Caballero-Calero, O., García-Cabañes, A., Carrascosa, M., Agulló-López, F., Villarroel, J., Crespillo, M. & Olivares, J. (2009a). Periodic poling of optical waveguides produced by swift-heavy-ion irradiation in $LiNbO_3$. *Appl. Phys. B* Vol.95, No.3, pp.435-439, ISSN 1432-0649.

Caballero-Calero, O., García-Cabañes, A., Carrascosa, M., Bermúdez, V., Crespillo, M. & Olivares, J. (2009b). Fabrication of periodically poled swift ion-irradiation waveguides in $LiNbO_3$. *Ferroelectrics* Vol.390, No.1, pp.29-35, ISSN 1563-5112.

Canut, B., Benyagoub, A., Marest, G., Meftah, A., Moncoffre, N., Ramos, S. M. M., Studer, F., Thevenard, P. & Toulemonde, M. (1995). Swift-uranium-ion-induced damage in sapphire. *Phys. Rev. B* Vol.51, No.18, pp.12194-12201, ISSN 1550-235X.

Canut, B. & Ramos, S. M. M. (1998). The concept of effective electronic stopping power for modelling the damage cross-section in refractory oxides irradiated by GeV ions or MeV clusters. *Rad. Eff. Def. Solids* Vol.145, No.1-2, pp.1-27, ISSN 1042-0150.

Canut, B., Ramos, S. M. M., Brenier, R., Thevenard, P., Loubet, J. L. & Toulemonde, M. (1996). Surface modifications of $LiNbO_3$ single crystals induced by swift heavy ions. *Nucl. Instrum. Meth. B* Vol.107, No.1-4, pp.194-198, ISSN 0168-583X.

Colder, A., Marty, O., Canut, B., Levalois, M., Marie, P., Portier, X., Ramos, S. M. M. & Toulemonde, M. (2001). Latent track formation in germanium irradiated with 20, 30 and 40 MeV fullerenes in the electronic regime. *Nucl. Instrum. Meth. B* Vol.174, No.4, pp.491-498, ISSN 0168-583X.

Crespillo, M. L., Caballero-Calero, O., Joco, V., Rivera, A., Herrero, P., Olivares, J. & Agulló-López, F. (2011). Recrystallization of amorphous nanotracks and uniform layers generated by swift-ion-beam irradiation in lithium niobate. *Appl. Phys. A* Vol.104, No.4, pp.1143-1152, ISSN 1432-0630.

Chandler, P. J. & Lama, F. L. (1986). A new approach to the determination of planar waveguide profiles by means of a non-stationary mode index calculation. *Opt. Acta* Vol.33, No.2, pp.127-143, ISSN 0030-3909.

Chandler, P. J., Zhang, L., Cabrera, J. M. & Townsend, P. D. (1989). "Missing modes" in ion-implanted $LiNbO_3$ waveguides. *Appl. Phys. Lett.* Vol.54, No.14, pp.1287-1289, ISSN 0003-6951.

Chen, C., Wu, Y. & Li, R. (1990). The development of new NLO crystals in the borate series. *J. Cryst. Growth* Vol.99, No.1-4, pp.790-798, ISSN 0022-0248.

Chen, F. (2008). Construction of two-dimensional waveguides in insulating optical materials by means of ion beam implantation for photonic applications: Fabrication methods and research progress. *Crit. Rev. Sol. Stat. Mat. Sci.* Vol.33, No.3-4, pp.165-182, ISSN 1547-6561.

Chen, F. (2009a). Carbon-ion-implanted Nd, MgO-codoped near-stoichiometric lithium-niobate optical waveguides. *J. Korean Phys. Soc.* Vol.55, pp.2701-2704, ISSN 0374-4884.

Chen, F. (2009b). Photonic guiding structures in lithium niobate crystals produced by energetic ion beams. *J. Appl. Phys.* Vol.106, No.8, pp.081101, ISSN 0021-8979.

Chen, F., Jaque, D., Tan, Y., Yao, S. & Liu, H. (2008a). Low-dose ion implanted active waveguides in Nd^{3+} doped near-stoichiometric lithium niobate: Promising candidates for near infrared integrated laser. *Phys. Stat. Sol.-Rapid Res. Lett.* Vol.2, No.3, pp.141-143, ISSN 1862-6270.

Chen, F., Tan, Y. & Jaque, D. (2009). Ion-implanted optical channel waveguides in neodymium-doped yttrium aluminum garnet transparent ceramics for integrated laser generation. *Opt. Lett.* Vol.34, No.1, pp.28-30, ISSN 1539-4794.

Chen, F., Tan, Y., Jaque, D., Wang, L., Wang, X.-L. & Wang, K.-M. (2008b). Active waveguide in Nd^{3+}:MgO:$LiNbO_3$ crystal produced by low-dose carbon ion implantation. *Appl. Phys. Lett.* Vol.92, No.2, pp.021110, ISSN 0003-6951.

Chen, F., Tan, Y. & Ródenas, A. (2008c). Ion implanted optical channel waveguides in Er^{3+}/MgO co-doped near stoichiometric $LiNbO_3$: A new candidate for active integrated photonic devices operating at 1.5 μm. *Opt. Express* Vol.16, No.20, pp.16209-16214, ISSN 1094-4087.

Chen, F., Tan, Y., Wang, L., Hou, D.-C. & Lu, Q.-M. (2008d). Optical channel waveguides with trapezoidal-shaped cross sections in $KTiOPO_4$ crystal fabricated by ion implantation. *Appl. Surf. Sci.* Vol.254, No.6, pp.1822-1824, ISSN 0169-4332.

Chen, F., Tan, Y., Wang, L., Wang, X.-L., Wang, K.-M. & Lu, Q.-M. (2008e). Diverse mechanism of refractive index modification in neodymium-doped $KGd(WO_4)_2$ crystal induced by MeV He^+ or C^{3+} ion implantation for waveguide construction. *J. Appl. Phys.* Vol.103, No.8, pp.083123, ISSN 0021-8979.

Chen, F., Wang, L., Jiang, Y., Wang, X.-L., Wang, K.-M., Fu, G., Lu, Q.-M., Rüter, C. E. & Kip, D. (2006a). Optical channel waveguides in Nd:YVO_4 crystal produced by O^+ ion implantation. *Appl. Phys. Lett.* Vol.88, No.7, pp.071123, ISSN 0003-6951.

Chen, F., Wang, L., Wang, X.-L., Wang, K.-M. & Lu, Q.-M. (2006b). Channel waveguide array in Ce-doped potassium sodium strontium barium niobate crystal fabricated by He^+ ion implantation. *Appl. Phys. Lett.* Vol.89, No.19, pp.191102, ISSN 0003-6951.

Chen, F., Wang, X.-L. & Wang, K.-M. (2007). Development of ion-implanted optical waveguides in optical materials: A review. *Opt. Mater.* Vol.29, No.11, pp.1523-1542, ISSN 0925-3467.

Chen, F., Yang, J., Zhang, C., Lu, Q., Akhmadaliev, S. & Zhou, S. (2011). Planar optical waveguides in $Bi_4Ge_3O_{12}$ crystal fabricated by swift heavy-ion irradiation. *Appl. Opt.* [online], Vol.In press, ISSN 1539-4522. Available from: http://www.opticsinfobase.org/abstract.cfm?msid=150631. [Accessed 2011-09-12].

Chiang, K. S. (1985). Construction of refractive-index profiles of planar dielectric waveguides from the distribution of effective indexes. *J. Lightwave Technol.* Vol.3, No.2, pp.385-391, ISSN 0733-8724.

Davis, G. M., Zhang, L., Chandler, P. J. & Townsend, P. D. (1996). Planar and channel waveguide fabrication in LiB_3O_5 using MeV He^+ ion implantation. *J. Appl. Phys.* Vol.79, pp.2863-2867, ISSN 0021-8979.

Dazzi, A., Mathey, P., Lompré, P., Jullien, P., Moretti, P. & Rytz, D. (1999). High performance of two-wave mixing in a $BaTiO_3$ waveguide realized by He^+ implantation. *J. Opt. Soc. Am. B* Vol.16, No.11, pp.1915-1920, ISSN 1520-8540.

Degl'Innocenti, R., Guarino, A., Poberaj, G. & Günter, P. (2006a). Second harmonic generation of continuous wave ultraviolet light and production of β-BaB_2O_4 optical waveguides. *Appl. Phys. Lett.* Vol.89, No.4, pp.041103, ISSN 0003-6951.

Degl'Innocenti, R., Majkić, A., Sulser, F., Mutter, L., Poberaj, G. & Günter, P. (2008). UV second harmonic generation at 266 nm in He^+ implanted β-BaB_2O_4 optical waveguides. *Opt. Express* Vol.16, No.15, pp.11660-11669, ISSN 1094-4087.

Degl'Innocenti, R., Reidt, S., Guarino, A., Rezzonico, D., Poberaj, G. & Günter, P. (2006b). Micromachining of ridge optical waveguides on top of He^{+}-implanted β-BaB_2O_4 crystals by femtosecond laser ablation. *J. Appl. Phys.* Vol.100, No.11, pp.113121, ISSN 0021-8979.

Devine, R. A. B., Duraud, J.-P. & Dooryhée, E. (2000). *Structure and imperfections in amorphous and crystalline silicon dioxide.* 1. ed. John Wiley & Sons Ltd., ISBN 0471975362.

Dong, N., Chen, F., Jaque, D., Benayas, A., Qiu, F. & Narusawa, T. (2011a). Characterization of active waveguides fabricated by ultralow-fluence swift heavy ion irradiation in lithium niobate crystals. *J. Phys. D: Appl. Phys.* Vol.44, No.10, pp.105103, ISSN 1361-6463.

Dong, N., Jaque, D., Chen, F. & Lu, Q. (2011b). Second harmonic and Raman imaging of He^{+} implanted $KTiOPO_4$ waveguides. *Opt. Express* Vol.19, No.15, pp.13934-13939, ISSN 1094-4087.

Dong, N.-N., Chen, F., Tan, Y. & Kong, Y.-X. (2008). Proton-implanted optical channel waveguides in Nd^{3+}:MgO:$LiNbO_3$: Fabrication, guiding properties, and luminescence investigation. *Appl. Phys. B* Vol.94, No.2, pp.283-287, ISSN 1432-0649.

Duffy, D. M. & Rutherford, A. M. (2009). Including electronic effects in damage cascade simulations. *J. Nucl. Mater.* Vol.386-388, pp.19-21, ISSN 0022-3115.

Dunlop, A., Jaskierowicz, G. & Della-Negra, S. (1998). Latent track formation in silicon irradiated by 30 MeV fullerenes. *Nucl. Instrum. Meth. B* Vol.146, No.1-4, pp.302-308, ISSN 0168-583X.

Dunlop, A. & Lesueur, D. (1993). Damage creation via electronic excitations in metallic targets part I: Experimental results. *Rad. Eff. Def. Solids* Vol.126, No.12, pp.123-128, ISSN 1042-0150.

Dunlop, A., Lesueur, D., Legrand, P., Dammak, H. & Dural, J. (1994). Effects induced by high electronic excitations in pure metals: A detailed study in iron. *Nucl. Instrum. Meth. B* Vol.90, No.1-4, pp.330-338, ISSN 0168-583X.

Fleuster, M., Buchal, C., Snoeks, E. & Polman, A. (1994). Rapid thermal annealing of MeV erbium implanted $LiNbO_3$ single crystals for optical doping. *Appl. Phys. Lett.* Vol.65, No.2, pp.225-227, ISSN 0003-6951.

Fricke, W. C. (1970). Fundamental mode YAG:Nd laser analysis. *Appl. Opt.* Vol.9, No.9, pp.2045-2052, ISSN 2155-3165.

Fu, G., Wang, K.-M., Wang, X.-L., Zhang, H.-J., Xu, X.-G., Song, H.-L. & Ma, H.-J. (2007). Planar waveguides in calcium barium niobate fabricated by MeV He ion implantation. *Appl. Phys. B* Vol.87, No.2, pp.289-292, ISSN 1432-0649.

Furuno, S., Otsu, H., Hojou, K. & Izui, K. (1996). Tracks of high energy heavy ions in solids. *Nucl. Instrum. Meth. B* Vol.107, No.1-4, pp.223-226, ISSN 0168-583X.

Gaiduk, P. ., Komarov, F. . & Wesch, W. (2000). Damage evolution in crystalline InP during irradiation with swift Xe ions. *Nucl. Instrum. Meth. B* Vol.164-165, pp.377-383, ISSN 0168-583X.

Gaiduk, P. I., Larsen, A. N., Trautmann, C. & Toulemonde, M. (2002). Discontinuous tracks in arsenic-doped crystalline $Si_{0.5}Ge_{0.5}$ alloy layers. *Phys. Rev. B* Vol.66, No.4, pp.045316, ISSN 1550-235X.

Gang, F. (2010). Optical waveguide formed in $LiTaO_3$ crystal by MeV C^{3+} ion implantation. *J. Korean Phys. Soc.* Vol.56, No.4, pp.1364-1368, ISSN 0374-4884.

García, G., Rivera, A., Crespillo, M. L., Gordillo, N., Olivares, J. & Agulló-López, F. (2011). Amorphization kinetics under swift heavy ion irradiation: A cumulative overlapping-track approach. *Nucl. Instrum. Meth. B* Vol.269, No.4, pp.492-497, ISSN 0168-583X.

García-Navarro, A., Agulló-López, F., Bianconi, M., Olivares, J. & García, G. (2007). Kinetics of ion-beam damage in lithium niobate. *J. Appl. Phys.* Vol.101, No.8, pp.083506, ISSN 0021-8979.

García-Navarro, A., Olivares, J., García, G., Agulló-López, F., García-Blanco, S., Merchant, C. & Aitchison, J. S. (2006). Fabrication of optical waveguides in KGW by swift heavy ion beam irradiation. *Nucl. Instrum. Meth. B* Vol.249, No.1-2, pp.177-180, ISSN 0168-583X.

Gedeon, A. (1974). Comparison between rigorous theory and WKB-analysis of modes in graded-index waveguides. *Opt. Commun.* Vol.12, No.3, pp.329-332, ISSN 0030-4018.

Graf, T. (1995). Lasing properties of diode-laser-pumped Nd:KGW. *Opt. Eng.* Vol.34, No.8, pp.2349-2352, ISSN 0091-3286.

Grivas, C., Shepherd, D. P., Eason, R. W., Laversenne, L., Moretti, P., Borca, C. N. & Pollnau, M. (2006). Room-temperature continuous-wave operation of Ti:sapphire buried channel-waveguide lasers fabricated via proton implantation. *Opt. Lett.* Vol.31, No.23, pp.3450-3452, ISSN 1539-4794.

Guarino, A., Jazbinsek, M., Herzog, C., Degl'Innocenti, R., Poberaj, G. & Günter, P. (2006). Optical waveguides in $Sn_2P_2S_6$ by low fluence MeV He^+ ion implantation. *Opt. Express* Vol.14, No.6, pp.2344-2358, ISSN 1094-4087.

Gumennik, A., Agranat, A. J., Shachar, I. & Hass, M. (2005). Thermal stability of a slab waveguide implemented by α particles implantation in potassium lithium tantalate niobate. *Appl. Phys. Lett.* Vol.87, No.25, pp.251917, ISSN 0003-6951.

Guo, S.-S., Liu, T., Zhao, J.-H., Guan, J. & Wang, X.-L. (2010). Planar waveguides in Nd:SGG crystal formed by He ion implantation. *Appl. Opt.* Vol.49, No.31, pp.6039-6042, ISSN 1539-4522.

Hadjichristov, G. B. & Stefanov, I. L. (2010). Ion-implanted polymethyl methacrylate beam splitter/coupler for 1.55 μm applications. *Appl. Opt.* Vol.49, No.10, pp.1876-1879, ISSN 2155-3165.

Hamelin, N., Lifante, G., Chandler, P. J., Townsend, P. D., Pityana, S. & Mccaffery, A. J. (1994). Second harmonic generation in ion implanted lithium niobate planar waveguides. *J. Mod. Opt.* Vol.41, No.7, pp.1339-1348, ISSN 1362-3044.

Hertel, P. & Menzler, H. P. (1987). Improved inverse WKB procedure to reconstruct refractive index profiles of dielectric planar waveguides. *Appl. Phys. B* Vol.44, No.2, pp.75-80, ISSN 1432-0649.

Hou, M. D. & Klaumünzer, S. (2003). Heavy ion-induced deformation of radiation-amorphized U_3Si. *Nucl. Instrum. Meth. B* Vol.209, pp.149-153, ISSN 0168-583X.

Hu, H., Lu, F., Chen, F., Shi, B.-R., Wang, K.-M. & Shen, D.-Y. (2001). Monomode optical waveguide in lithium niobate formed by MeV Si^+ ion implantation. *J. Appl. Phys.* Vol.89, No.9, pp.5224-5226, ISSN 0021-8979.

Ilan, H., Gumennik, A., Fathei, R., Agranat, A. J., Shachar, I. & Hass, M. (2006). Submerged waveguide constructed by the implantation of ^{12}C ions in electro-optic crystals. *Appl. Phys. Lett.* Vol.89, No.24, pp.241130, ISSN 0003-6951.

Ilan, H., Gumennik, A., Perepelitsa, G., Israel, A. & Agranat, A. J. (2008). Construction of an optical wire imprinted in potassium lithium tantalate niobate by He^+ implantation. *Appl. Phys. Lett.* Vol.92, No.19, pp.191101, ISSN 0003-6951.

Ishwar Bhat, S., Mohan Rao, P., Ganesh Bhat, A. P. & Avasthi, D. K. (2002). Irradiation effects on the optical properties of a new NLO mixed borate crystal. *Surf. Coat. Technol.* Vol.158-159, pp.725-728, ISSN 0257-8972.

Itoh, N., Duffy, D. M., Khakshouri, S. & Stoneham, A. M. (2009). Making tracks: Electronic excitation roles in forming swift heavy ion tracks. *J. Phys.: Condens. Matter* Vol.21, No.47, pp.474205, ISSN 0953-8984.

Jaque, D. & Chen, F. (2009). High resolution fluorescence imaging of damage regions in H^+ ion implanted Nd:MgO:$LiNbO_3$ channel waveguides. *Appl. Phys. Lett.* Vol.94, No.1, pp.011109, ISSN 0003-6951.

Jazbinsek, M., Mutter, L. & Gunter, P. (2008). Photonic applications with the organic nonlinear optical crystal DAST. *IEEE J. Sel. Top. Quant. Electron.* Vol.14, No.5, pp.1298-1311, ISSN 1077-260X.

Jia, C.-L., Jiang, Y., Wang, L., Wang, X.-L., Wang, K.-M. & Zhang, H.-J. (2008a). Characterization of optical waveguides in $YbVO_4$ crystals formed by 3.0 MeV oxygen ion implantation. *Appl. Opt.* Vol.47, No.8, pp.1117-1121, ISSN 1539-4522.

Jia, C.-L., Jiang, Y., Wang, X.-L., Chen, F., Wang, L., Jiao, Y., Wang, K.-M., Lu, F., Shen, D.-Y., Ma, H.-J. & Nie, R. (2006). Formation of waveguides in $LiNbO_3$ by 6.0 MeV F^{3+} implantation. *J. Appl. Phys.* Vol.100, No.3, pp.033505, ISSN 0021-8979.

Jia, C.-L., Wang, X.-L., Wang, K.-M., Jiang, Y. & Wang, L. (2008b). Characterization of optical waveguides in β-BaB_2O_4 crystals formed by 3.0-MeV Cu^{2+}-ion implantation. *Appl. Phys. B* Vol.91, No.1, pp.139-143, ISSN 1432-0649.

Jia, C.-L., Wang, X.-L., Wang, K.-M., Ma, H.-J. & Nie, R. (2007). Characterization of optical waveguide in Nd:$GdVO_4$ by triple-energy oxygen ion implantation. *Appl. Surf. Sci.* Vol.253, No.24, pp.9311-9314, ISSN 0169-4332.

Jia, C.-L., Wang, X.-L., Wang, K.-M. & Zhang, H.-J. (2008c). Characterization of optical waveguide in Nd:$LuVO_4$ crystals by triple-energy oxygen ion implantation. *Physica B* Vol.403, No.4, pp.679-683, ISSN 0921-4526.

Jia, C.-L., Wang, X.-L., Wang, L., Wang, K.-M., Ma, H.-J. & Nie, R. (2008d). Characterization of optical waveguides in β-BaB_2O_4 crystals formed by 3.0 MeV Ni^{2+} ions implantation. *Appl. Surf. Sci.* Vol.254, No.16, pp.5095-5099, ISSN 0169-4332.

Jia, C.-L., Zhang, T. & Wang, L. (2011). Structural and optical properties of waveguides in $YbVO_4$ crystals formed by 3.0 MeV Cu^{2+}-ions implantation. *Opt. Laser Technol.* Vol.43, No.7, pp.1138-1142, ISSN 0030-3992.

Jiang, Y., Jia, C.-L., Wang, L., Wang, X.-L., Chen, F., Wang, K.-M., Lu, Q.-M., Ma, H.-J. & Shen, D.-Y. (2006a). The optical properties of planar waveguides in LiB_3O_5 crystals formed by Cu^+ implantation. *Appl. Surf. Sci.* Vol.253, No.5, pp.2674-2677, ISSN 0169-4332.

Jiang, Y., Wang, K.-M., Wang, X.-L., Chen, F., Jia, C.-L., Wang, L., Jiao, Y. & Lu, F. (2007). Model of refractive-index changes in lithium niobate waveguides fabricated by ion implantation. *Phys. Rev. B* Vol.75, No.19, pp.195101, ISSN 1550-235X.

Jiang, Y., Wang, K.-M., Wang, X.-L., Jia, C.-L., Wang, L., Jiao, Y., Lu, Q.-M., Ma, H.-J., Nie, R. & Shen, D.-Y. (2006b). Planar optical waveguide in potassium titanyl arsenate formed by oxygen ion implantation at low doses. *Appl. Phys. Lett.* Vol.88, No.1, pp.011114, ISSN 0003-6951.

Jiao, Y., Chen, F., Wang, X.-L., Wang, K.-M., Wang, L., Wang, L.-L., Zhang, H.-J., Lu, Q.-M., Ma, H.-J. & Nie, R. (2007). Fabrication of waveguides in Yb:YCOB crystal by MeV oxygen ion implantation. *Appl. Surf. Sci.* Vol.253, No.18, pp.7360-7364, ISSN 0169-4332.

Kamarou, A., Wesch, W., Wendler, E., Undisz, A. & Rettenmayr, M. (2008). Radiation damage formation in InP, InSb, GaAs, GaP, Ge, and Si due to fast ions. *Phys. Rev. B* Vol.78, No.5, pp.054111, ISSN 1550-235X.

Kaminskii, A. (1996). *Crystalline lasers: Physical processes and operating schemes.* 1. ed. CRC-Press., ISBN 0849337208.

Khalfaoui, N., Beuve, M., Bouffard, S., Caron, M., Rothard, H., Schlutig, S., Stoquert, J. P. & Toulemonde, M. (2003). Neutral atom sputtering yields from $Gd_3Ga_5O_{12}$ and $Y_3Fe_5O_{12}$ garnets: Observation of a velocity effect. *Nucl. Instrum. Meth. B* Vol.209, pp.304-309, ISSN 0168-583X.

Khánh, N. Q., Berneschi, S., Bányász, I., Brenci, M., Fried, M., Nunzi Conti, G., Pászti, F., Pelli, S., Righini, G. C. & Watterich, A. (2009). Fabrication of channel waveguides in Er^{3+}-doped tellurite glass via N^+ ion implantation. *Nucl. Instrum. Meth. B* Vol.267, No.12-13, pp.2327-2330, ISSN 0168-583X.

Kip, D., Wesner, M., Shandarov, V. & Moretti, P. (1998). Observation of bright spatial photorefractive solitons in a planar strontium barium niobate waveguide. *Opt. Lett.* Vol.23, No.12, pp.921-923, ISSN 1539-4794.

Kiss, Z. J. & Pressley, R. J. (1966). Crystalline solid lasers. *Proc. IEEE* Vol.54, No.10, pp.1236-1248, ISSN 0018-9219.

Klaumünzer, S., Hou, M.-dong & Schumacher, G. (1986). Coulomb explosions in a metallic glass due to the passage of fast heavy ions? *Phys. Rev. Lett.* Vol.57, No.7, pp.850-853, ISSN 1079-7114.

Komarov, F. F. (2003). Defect and track formation in solids irradiated by superhigh-energy ions. *Phys.-Usp.* Vol.46, No.12, pp.1253-1282, ISSN 1468-4780.

Kong, Y.-X., Chen, F., Jaque, D., Lu, Q.-M. & Ma, H.-J. (2009). Optical channel waveguide in Nd/Ce codoped YAG laser crystal produced by carbon ion implantation. *Appl. Opt.* Vol.48, No.23, pp.4514-4518, ISSN 1539-4522.

Kong, Y.-X., Chen, F., Jaque, D., Tan, Y., Dong, N.-N., Lu, Q.-M. & Ma, H.-J. (2008). Low-dose O^{3+} ion-implanted active optical planar waveguides in Nd:YAG crystals: Guiding properties and micro-luminescence characterization. *J. Phys. D: Appl. Phys.* Vol.41, No.17, pp.175112, ISSN 1361-6463.

Kruglov, V. G., Shandarov, V. M., Tan, Y., Chen, F. & Kip, D. (2008). Dark photovoltaic spatial solitons in a planar waveguide obtained by proton implantation in lithium niobate. *Quantum Electron.* Vol.38, No.11, pp.1045-1047, ISSN 1468-4799.

Kucheyev, S. O. (2004). Lattice damage produced in GaN by swift heavy ions. *J. Appl. Phys.* Vol.95, No.10, pp.5360-5365, ISSN 0021-8979.

Kumar, P., Moorthy Babu, S., Ganesamoorthy, S., Karnal, A. K. & Kanjilal, D. (2007). Influence of swift ions and proton implantation on the formation of optical waveguides in lithium niobate. *J. Appl. Phys.* Vol.102, No.8, pp.084905, ISSN 0021-8979.

Lallier, E. (1992). Rare-earth-doped glass and $LiNbO_3$ waveguide lasers and optical amplifiers. *Appl. Opt.* Vol.31, No.25, pp.5276-5282, ISSN 2155-3165.

Langrock, C., Kumar, S., McGeehan, J. E., Willner, A. E. & Fejer, M. M. (2006). All-optical signal processing using $\chi^{(2)}$ nonlinearities in guided-wave devices. *J. Lightwave Technol.* Vol.24, No.7, pp.2579-2592, ISSN 0733-8724.

Liu, H. P., Lu, F., Liu, X. Z., Zhang, R. F., Song, Q., Wang, X. L., Ma, X. J., Yang, T. L., Lv, Y. B. & Li, Y. H. (2008a). Reconstruction of refractive index profiles of 3 MeV O^{2+} ion-implanted MgO-doped $LiNbO_3$ using wet etching and ellipsometry. *J. Phys. D: Appl. Phys.* Vol.41, No.6, pp.065302, ISSN 1361-6463.

Liu, T., Guo, S.-S., Zhao, J.-H., Guan, J. & Wang, X.-L. (2011). Planar optical waveguides in Nd:BSO crystals fabricated by He and C ion implantation. *Opt. Mater.* Vol.33, No.3, pp.385-388, ISSN 0925-3467.

Liu, X., Lu, F., Chen, F., Tan, Y., Zhang, R., Liu, H., Wang, L. & Wang, L. (2008b). Reconstruction of extraordinary refractive index profiles of optical planar waveguides with single or double modes fabricated by O^{2+} ion implantation into lithium niobate. *Opt. Commun.* Vol.281, No.6, pp.1529-1533, ISSN 0030-4018.

Liu, X., Lu, F., Chen, F., Zhang, R., Liu, H., Wang, L., Fu, G. & Wang, H. (2007). Reconstruction of extraordinary refractive index profile of O^{2+} ion-implanted $LiNbO_3$ single-mode channel waveguide based on beam propagation method and image processing. *Opt. Commun.* Vol.274, No.1, pp.80-84, ISSN 0030-4018.

Majkić, A., Koechlin, M., Poberaj, G. & Günter, P. (2008). Optical microring resonators in fluorine-implanted lithium niobate. *Opt. Express* Vol.16, No.12, pp.8769-8779, ISSN 1094-4087.

Majkić, A., Poberaj, G., Degl'Innocenti, R., Döbeli, M. & Günter, P. (2007). Cr:$LiSrAlF_6$ channel waveguides as broadband fluorescence sources. *Appl. Phys. B* Vol.88, No.2, pp.205-209, ISSN 1432-0649.

Major, A., Aitchison, J. S., Smith, P. W. E., Langford, N. & Ferguson, A. I. (2005). Efficient Raman shifting of high-energy picosecond pulses into the eye-safe 1.5-μm spectral region by use of a $KGd(WO_4)_2$ crystal. *Opt. Lett.* Vol.30, No.4, pp.421-423, ISSN 1539-4794.

Manzano, J., Olivares, J., Agulló-López, F., Crespillo, M. L., Moroño, A. & Hodgson, E. (2010). Optical waveguides obtained by swift-ion irradiation on silica (a-SiO_2). *Nucl. Instrum. Meth. B* Vol.268, No.19, pp.3147-3150, ISSN 0168-583X.

Manzano-Santamaría, J., Olivares, J., Rivera, A. & Agulló-López, F. (2012). Electronic damage in quartz (c-SiO_2) by MeV ion irradiations: Potentiality for optical waveguiding applications. *Nucl. Instrum. Meth. B* Vol.272, pp.271-274, ISSN 0168-583X, doi:10.1016/j.nimb.2011.01.081.

Mathey, P., Dazzi, A., Jullien, P., Rytz, D. & Moretti, P. (2001a). Guiding properties and nonlinear wave mixing at 854 nm in a rhodium-doped $BaTiO_3$ waveguide implanted with He^+ ions. *J. Opt. Soc. Am. B* Vol.18, No.3, pp.344-347, ISSN 1520-8540.

Mathey, P., Dazzi, A., Jullien, P., Rytz, D. & Moretti, P. (2001b). Two-wave mixing at 854 nm in $BaTiO_3$:Rh planar waveguide implanted with He^+. *Opt. Mater.* Vol.18, No.1, pp.69-71, ISSN 0925-3467.

Matzke, H., Lucuta, P. G. & Wiss, T. (2000). Swift heavy ion and fission damage effects in UO_2. *Nucl. Instrum. Meth. B* Vol.166-167, pp.920-926, ISSN 0168-583X.

Meftah, A., Brisard, F., Costantini, J. M., Dooryhee, E., Hage-Ali, M., Hervieu, M., Stoquert, J. P., Studer, F. & Toulemonde, M. (1994). Track formation in SiO_2 quartz and the thermal-spike mechanism. *Phys. Rev. B* Vol.49, No.18, pp.12457-12463, ISSN 1550-235X.

Merchant, C. A., Aitchison, J. S., García-Blanco, S., Hnatovsky, C., Taylor, R. S., Agulló-Rueda, F., Kellock, A. J. & Baglin, J. E. E. (2006a). Direct observation of waveguide formation in $KGd(WO_4)_2$ by low dose H^+ ion implantation. *Appl. Phys. Lett.* Vol.89, No.11, pp.111116, ISSN 0003-6951.

Merchant, C. A., Scrutton, P., García-Blanco, S., Helmy, A. S. & Aitchison, J. S. (2006b). Raman characterization of $KGd(WO_4)_2$ waveguides formed by swift heavy-ion irradiation. *Proceedings of 19th Annual Meeting of the IEEE Lasers and Electro-Optics Society, 2006. LEOS 2006*, Oktober 2006. pp 320-321. IEEE.

Merchant, C. A., Scrutton, P., García-Blanco, S., Hnatovsky, C., Taylor, R. S., García-Navarro, A., García, G., Agulló-López, F., Olivares, J., Helmy, A. S. & Aitchison, J. S. (2009). High-resolution refractive index and micro-Raman spectroscopy of planar waveguides in $KGd(WO_4)_2$ formed by swift heavy ion irradiation. *IEEE J. Quant. Elect.* Vol.45, No.4, pp.373-379, ISSN 0018-9197.

Ming, X., Lu, F., Yin, J., Chen, M., Zhang, S., Liu, X., Qin, Z. & Ma, Y. (2011). Optical confinement achieved in ZnO crystal by O^+ ions implantation: Analysis of waveguide formation and properties. *Opt. Express* Vol.19, No.8, pp.7139-7146, ISSN 1094-4087.

Moreira, P. A. F. P., Devanathan, R. & Weber, W. J. (2010). Atomistic simulation of track formation by energetic recoils in zircon. *J. Phys.: Condens. Matter* Vol.22, No.39, pp.395008.

Mosimann, R., Juvalta, F., Jazbinsek, M., Günter, P. & Grabar, A. A. (2009). Photorefractive waveguides in He^+ implanted pure and Te-doped $Sn_2P_2S_6$. *J. Opt. Soc. Am. B* Vol.26, No.3, pp.444-449, ISSN 1520-8540.

Olivares, J., Crespillo, M. L., Caballero-Calero, O., Ynsa, M. D., García-Cabañes, A., Toulemonde, M., Trautmann, C. & Agulló-López, F. (2009). Thick optical waveguides in lithium niobate induced by swift heavy ions (~10 MeV/amu) at ultralow fluences. *Opt. Express* Vol.17, No.26, pp.24175-24182, ISSN 1094-4087.

Olivares, J., García, G., Agulló-López, F., Agulló-Rueda, F., Kling, A. & Soares, J. C. (2005a). Generation of amorphous surface layers in $LiNbO_3$ by ion-beam irradiation: thresholding and boundary propagation. *Appl. Phys. A* Vol.81, No.7, pp.1465-1469, ISSN 0947-8396.

Olivares, J., García, G., García-Navarro, A., Agulló-López, F., Caballero, O. & García-Cabañes, A. (2005b). Generation of high-confinement step-like optical waveguides in $LiNbO_3$ by swift heavy ion-beam irradiation. *Appl. Phys. Lett.* Vol.86, No.18, pp.183501, ISSN 0003-6951.

Olivares, J., García-Navarro, A., García, G., Agulló-López, F., Agulló-Rueda, F., García-Cabañes, A. & Carrascosa, M. (2007a). Buried amorphous layers by electronic excitation in ion-beam irradiated lithium niobate: Structure and kinetics. *J. Appl. Phys.* Vol.101, No.3, pp.033512, ISSN 0021-8979.

Olivares, J., García-Navarro, A., García, G., Méndez, A., Agulló-López, F., García-Cabañes, A., Carrascosa, M. & Caballero, O. (2007b). Nonlinear optical waveguides generated in lithium niobate by swift-ion irradiation at ultralow fluences. *Opt. Lett.* Vol.32, No.17, pp.2587-2589, ISSN 1539-4794.

Olivares, J., García-Navarro, A., Méndez, A., Agulló-López, F., García, G., García-Cabañes, A. & Carrascosa, M. (2007c). Novel optical waveguides by in-depth controlled electronic damage with swift ions. *Nucl. Instrum. Meth. B* Vol.257, No.1-2, pp.765-770, ISSN 0168-583X.

Pakarinen, O. H., Djurabekova, F. & Nordlund, K. (2010). Density evolution in formation of swift heavy ion tracks in insulators. *Nucl. Instrum. Meth. B* Vol.268, No.19, pp.3163-3166, ISSN 0168-583X.

Peithmann, K., Zamani-Meymian, M.-R., Haaks, M., Maier, K., Andreas, B. & Breunig, I. (2006a). Refractive index changes in lithium niobate crystals by high-energy particle radiation. *J. Opt. Soc. Am. B* Vol.23, No.10, pp.2107-2112, ISSN 1520-8540.

Peithmann, K., Zamani-Meymian, M.-R., Haaks, M., Maier, K., Andreas, B., Buse, K. & Modrow, H. (2006b). Fabrication of embedded waveguides in lithium-niobate crystals by radiation damage. *Appl. Phys. B* Vol.82, No.3, pp.419-422, ISSN 1432-0649.

Peng, B., Chen, F., Tan, Y. & Kip, D. (2011). Two-wave mixing of ion-implanted photorefractive waveguides in near-stoichiometric Fe:$LiNbO_3$ crystals. *Opt. Mater.* Vol.33, No.6, pp.773-776, ISSN 0925-3467.

Poberaj, G., Degl'Innocenti, R., Medrano, C. & Günter, P. (2009). UV integrated optics devices based on beta-barium borate. *Opt. Mater.* Vol.31, No.7, pp.1049-1053, ISSN 0925-3467.

Pollnau, M., Grivas, C., Laversenne, L., Wilkinson, J. S., Eason, R. W. & Shepherd, D. P. (2007). Ti:Sapphire waveguide lasers. *Laser Phys. Lett.* Vol.4, No.8, pp.560-571, ISSN 1612-202X.

Pollnau, M. & Romanyuk, Y. E. (2007). Optical waveguides in laser crystals. *Comptes Rendus Physique* Vol.8, No.2, pp.123-137, ISSN 1631-0705.

Qiu, F. & Narusawa, T. (2010). Proton-implanted planar waveguide in gallium lanthanum sulphide glass. *Jpn. J. Appl. Phys.* Vol.49, pp.092503, ISSN 1347-4065.

Qiu, F. & Narusawa, T. (2011a). Application of swift and heavy ion implantation to the formation of chalcogenide glass optical waveguides. *Opt. Mater.* Vol.33, No.3, pp.527-530, ISSN 0925-3467.

Qiu, F. & Narusawa, T. (2011b). Ion-implanted Ti-doped chalcogenide glass waveguide as a candidate for tunable lasers. *J. Opt. Soc. Am. B* Vol.28, No.6, pp.1490-1492, ISSN 1520-8540.

Qiu, F., Narusawa, T. & Zheng, J. (2011). Swift and heavy ion implanted chalcogenide laser glass waveguides and their different refractive index distributions. *Appl. Opt.* Vol.50, No.5, pp.733-737, ISSN 1539-4522.

Raj, S. G. & Kumar, G. R. (2010). Swift ion beam irradiation on lithium niobate single crystals. *Rad. Eff. Def. Solids* Vol.165, pp.370-379, ISSN 1042-0150.

Rams, J., Olivares, J., Chandler, P. J. & Townsend, P. D. (2000). Mode gaps in the refractive index properties of low-dose ion-implanted $LiNbO_3$ waveguides. *J. Appl. Phys.* Vol.87, No.7, pp.3199.

del Re, E., Segev, M., Christodoulides, D., Crosignani, B. & Salamo, G. (2006). Photorefractive solitons. In: Günter, P. & Huignard, J. P. (Eds.) *Photorefractive materials and their applications 1: Basic effects*. 1. ed., pp 317-367. Springer., ISBN 038725191X.

Ren, Y., Dong, N., Chen, F., Benayas, A., Jaque, D., Qiu, F. & Narusawa, T. (2010a). Swift heavy-ion irradiated active waveguides in Nd:YAG crystals: Fabrication and laser generation. *Opt. Lett.* Vol.35, No.19, pp.3276-3278, ISSN 1539-4794.

Ren, Y., Dong, N., Chen, F. & Jaque, D. (2011a). Swift nitrogen ion irradiated waveguide lasers in Nd:YAG crystal. *Opt. Express* Vol.19, No.6, pp.5522-5527, ISSN 1094-4087.

Ren, Y., Dong, N., Tan, Y., Guan, J., Chen, F. & Lu, Q. (2010b). Continuous wave laser generation in proton implanted Nd:GGG planar waveguides. *J. Lightwave Technol.* Vol.28, No.24, pp.3578-3581, ISSN 0733-8724.

Ren, Y., Jia, Y., Chen, F., Lu, Q., Akhmadaliev, S. & Zhou, S. (2011b). Second harmonic generation of swift carbon ion irradiated Nd:GdCOB waveguides. *Opt. Express* Vol.19, No.13, pp.12490-12495, ISSN 1094-4087.

Ren, Y., Tan, Y., Chen, F., Jaque, D., Zhang, H., Wang, J. & Lu, Q. (2010c). Optical channel waveguides in Nd:LGS laser crystals produced by proton implantation. *Opt. Express* Vol.18, No.15, pp.16258-16263, ISSN 1094-4087.

Ren, Y.-Y., Chen, F., Lu, Q.-M. & Ma, H.-J. (2010d). Optical waveguides in Nd:GGG crystals produced by H^+ or C^{3+} ion implantation. *Appl. Opt.* Vol.49, No.11, pp.2085-2089, ISSN 1539-4522.

Rivera, A., Crespillo, M. L., Olivares, J., García, G. & Agulló-López, F. (2010a). Effect of defect accumulation on ion-beam damage morphology by electronic excitation in lithium niobate: A Monte Carlo approach. *Nucl. Instrum. Meth. B* Vol.268, No.13, pp.2249-2256, ISSN 0168-583X.

Rivera, A., Crespillo, M. L., Olivares, J., Sanz, R., Jensen, J. & Agulló-López, F. (2010b). On the exciton model for ion-beam damage: The example of TiO_2. *Nucl. Instrum. Meth. B* Vol.268, No.19, pp.3122-3126, ISSN 0168-583X.

Rivera, A., Olivares, J., García, G., Cabrera, J. M., Agulló-Rueda, F. & Agulló-López, F. (2009). Giant enhancement of material damage associated to electronic excitation during ion irradiation: The case of $LiNbO_3$. *Phys. Stat. Sol. A* Vol.206, No.6, pp.1109-1116, ISSN 1862-6300.

Roelofs, M. G., Ferretti, A. & Bierlein, J. D. (1993). Proton- and ammonium-exchanged optical waveguides in $KTiOPO_4$. *J. Appl. Phys.* Vol.73, No.8, pp.3608-3613, ISSN 0021-8979.

Ruiz, T., Méndez, A., Carrascosa, M., Carnicero, J., García-Cabañes, A., Olivares, J., Agulló-López, F., García-Navarro, A. & García, G. (2007). Tailoring of refractive index profiles in $LiNbO_3$ optical waveguides by low-fluence swift-ion irradiation. *J. Phys. D: Appl. Phys.* Vol.40, pp.4454-4459, ISSN 1361-6463.

Saiki, T., Imasaki, K., Motokoshi, S., Yamanaka, C., Fujita, H., Nakatsuka, M. & Izawa, Y. (2006). Disk-type Nd/Cr:YAG ceramic lasers pumped by arc-metal-halide-lamp. *Opt. Commun.* Vol.268, No.1, pp.155-159, ISSN 0030-4018.

Sánchez-Morales, M. E., Vázquez, G. V., Márquez, H., Rickards, J. & Trejo-Luna, R. (2006). Optical waveguides formed in $Nd:YVO_4$ crystals by C^{2+} implantation. *J. Mod. Opt.* Vol.53, pp.539-545, ISSN 1362-3044.

Sánchez-Morales, M. E., Vázquez, G. V., Mejía, E. B., Márquez, H., Rickards, J. & Trejo-Luna, R. (2008). Laser emission in $Nd:YVO_4$ channel waveguides at 1064 nm. *Appl. Phys. B* Vol.94, No.2, pp.215-219, ISSN 1432-0649.

Sánchez-Morales, M. E., Vázquez, G. V., Moretti, P. & Márquez, H. (2007). Optical waveguides in Nd:YVO_4 crystals by multi-implants with protons and helium ions. *Opt. Mater.* Vol.29, No.7, pp.840-844, ISSN 0925-3467.

Schrempel, F., Höche, T., Ruske, J.-P., Grusemann, U. & Wesch, W. (2002). Depth dependence of radiation damage in Li^+-implanted $KTiOPO_4$. *Nucl. Instrum. Meth. B* Vol.191, No.1-4, pp.202-207, ISSN 0168-583X.

Schrempel, F., Opfermann, T., Ruske, J.-P., Grusemann, U. & Wesch, W. (2004). Properties of buried waveguides produced by He-irradiation in KTP and Rb:KTP. *Nucl. Instrum. Meth. B* Vol.218, pp.209-216, ISSN 0168-583X.

Sher, S. M., Pintus, P., Di Pasquale, F., Bianconi, M., Montanari, G. B., De Nicola, P., Sugliani, S. & Prati, G. (2011). Design of 980 nm-pumped waveguide laser for continuous wave operation in ion implanted Er:$LiNbO_3$. *IEEE J. Quant. Elect.* Vol.47, No.4, pp.526-533, ISSN 0018-9197.

Skuratov, V. A., Zinkle, S. J., Efimov, A. E. & Havancsak, K. (2003). Swift heavy ion-induced modification of Al_2O_3 and MgO surfaces. *Nucl. Instrum. Meth. B* Vol.203, pp.136-140, ISSN 0168-583X.

Sohler, W., Hu, H., Ricken, R., Quiring, V., Vannahme, C., Herrmann, H., Büchter, D., Reza, S., Grundkötter, W., Orlov, S., Suche, H., Nouroozi, R. & Min, Y. (2008). Integrated optical devices in lithium niobate. *Opt. Photon. News* Vol.19, No.1, pp.24-31, ISSN 1047-6938.

Song, Q., Lu, F., Ma, X., Liu, H., Liu, X., Zhang, R. & Wang, X. (2008). MgO:$LiNbO_3$ planar waveguide formed by MeV O^{2+} implantation and its annealing characteristics. *Laser Phys.* Vol.18, pp.815-818, ISSN 1555-6611.

Sugliani, S., Bianconi, M., Bentini, G. G., Chiarini, M., De Nicola, P., Montanari, G. B., Menin, A., Malacarne, A. & Potì, L. (2010). Refractive index tailoring in congruent Lithium Niobate by ion implantation. *Nucl. Instrum. Meth. B* Vol.268, No.19, pp.2911-2914, ISSN 0168-583X.

Szachowicz, M., Joubert, M.-F., Moretti, P., Couchaud, M., Ferrand, B., Borca, C. & Boudrioua, A. (2008). H^+ implanted channel waveguides in buried epitaxial crystalline YAG:Nd,Tm layers and infrared-to-blue upconversion characterization. *J. Appl. Phys.* Vol.104, No.11, pp.113104, ISSN 0021-8979.

Szachowicz, M., Tascu, S., Joubert, M.-F., Moretti, P. & Nikl, M. (2006). Realization and infrared to green upconversion luminescence in Er^{3+}:$YAlO_3$ ion-implanted optical waveguides. *Opt. Mater.* Vol.28, No.1-2, pp.162-166, ISSN 0925-3467.

Szenes, G. (1995). General features of latent track formation in magnetic insulators irradiated with swift heavy ions. *Phys. Rev. B* Vol.51, No.13, pp.8026-8029, ISSN 1550-235X.

Szenes, G., Horváth, Z. E., Pécz, B., Pászti, F. & Tóth, L. (2002). Tracks induced by swift heavy ions in semiconductors. *Phys. Rev. B* Vol.65, No.4, pp.045206, ISSN 1550-235X.

Tan, Y. & Chen, F. (2010a). Optical ridge waveguides preserving the thermo-optic features in $LiNbO_3$ crystals fabricated by combination of proton implantation and selective wet etching. *Opt. Express* Vol.18, No.11, pp.11444-11449, ISSN 1094-4087.

Tan, Y. & Chen, F. (2010b). Proton-implanted optical channel waveguides in Nd:YAG laser ceramics. *J. Phys. D: Appl. Phys.* Vol.43, No.7, pp.075105, ISSN 1361-6463.

Tan, Y., Chen, F., Beličev, P. P., Stepić, M., Maluckov, A., Rüter, C. E. & Kip, D. (2009a). Gap solitons in defocusing lithium niobate binary waveguide arrays fabricated by proton implantation and selective light illumination. *Appl. Phys. B* Vol.95, No.3, pp.531-535, ISSN 1432-0649.

Tan, Y., Chen, F., Hu, L.-L., Xing, P.-F., Chen, Y.-X., Wang, X.-L. & Wang, K.-M. (2007a). Ridge optical waveguide in an Er^{3+}/Yb^{3+} co-doped phosphate glass produced by He^{+} ion implantation combined with Ar^{+} ion beam etching. *J. Phys. D: Appl. Phys.* Vol.40, No.21, pp.6545-6548, ISSN 1361-6463.

Tan, Y., Chen, F., Jaque, D., Gao, W.-L., Zhang, H.-J., Solé, J. G. & Ma, H.-J. (2009b). Ion-implanted optical-stripe waveguides in neodymium-doped calcium barium niobate crystals. *Opt. Lett.* Vol.34, No.9, pp.1438-1440, ISSN 1539-4794.

Tan, Y., Chen, F. & Kip, D. (2008a). Photorefractive properties of optical waveguides in Fe:$LiNbO_3$ crystals produced by O^{3+} ion implantation. *Appl. Phys. B* Vol.94, No.3, pp.467-471, ISSN 1432-0649.

Tan, Y., Chen, F., Wang, L. & Jiao, Y. (2007b). Carbon ion-implanted optical waveguides in Nd:$YLiF_4$ crystal: Refractive index profiles and thermal stability. *Nucl. Instrum. Meth. B* Vol.260, No.2, pp.567-570, ISSN 0168-583X.

Tan, Y., Chen, F., Wang, L., Wang, X.-L., Wang, K.-M. & Lu, Q.-M. (2008b). Optical channel waveguides in $KTiOPO_4$ crystal produced by proton implantation. *J. Lightwave Technol.* Vol.26, No.10, pp.1304-1308, ISSN 0733-8724.

Tan, Y., Chen, F., Wang, X.-L., Wang, L., Shandarov, V. M. & Kip, D. (2008c). Formation of reconfigurable optical channel waveguides and beam splitters on top of proton-implanted lithium niobate crystals using spatial dark soliton-like structures. *J. Phys. D: Appl. Phys.* Vol.41, No.10, pp.102001, ISSN 1361-6463.

Tan, Y., Chen, F. & Zhang, H.-J. (2007c). Optical ridge waveguides in SBN crystal produced by low-dose carbon ion implantation followed by a sputter etching technique. *Opt. Express* Vol.15, No.25, pp.16696-16701, ISSN 1094-4087.

Tan, Y., Feng Chen, Stepic, M., Shandarov, V. & Kip, D. (2008d). Reconfigurable optical channel waveguides in lithium niobate crystals produced by combination of low-dose O^{3+} ion implantation and selective white light illumination. *Opt. Express* Vol.16, No.14, pp.10465-10470, ISSN 1094-4087.

Tan, Y., Zhang, C., Chen, F., Liu, F.-Q., Jaque, D. & Lu, Q.-M. (2010). Room-temperature continuous wave laser oscillations in Nd:YAG ceramic waveguides produced by carbon ion implantation. *Appl. Phys. B* Vol.103, No.4, pp.837-840, ISSN 1432-0649.

Toulemonde, M., Assmann, W., Dufour, C., Meftah, A., Studer, F. & Trautmann, C. (2006). Experimental phenomena and thermal spike model description of ion tracks in amorphisable inorganic insulators. In: Sigmund, P. (Ed.) *Ion Beam Science: Solved and Unsolved Problems*. pp 263-291. Copenhagen: Royal Danish Academy of Sciences and Letters., ISBN 87-7304-330-3.

Toulemonde, M., Bouffard, S. & Studer, F. (1994). Swift heavy ions in insulating and conducting oxides: Tracks and physical properties. *Nucl. Instrum. Meth. B* Vol.91, No.1-4, pp.108-123, ISSN 0168-583X.

Toulemonde, M., Dufour, C. & Paumier, E. (1992). Transient thermal process after a high-energy heavy-ion irradiation of amorphous metals and semiconductors. *Phys. Rev. B* Vol.46, No.22, pp.14362-14369, ISSN 1550-235X.

Townsend, P. D. (1987). Optical effects of ion implantation. *Rep. Prog. Phys.* Vol.50, No.5, pp.501-558, ISSN 1361-6633.

Townsend, P. D., Chandler, P. J. & Zhang, L. (1994). *Optical effects of ion implantation.* Cambridge, UK: Cambridge University Press., ISBN 0521394309.

Townsend, P. & Olivares, J. (1997). Laser processing of insulator surfaces. *Appl. Surf. Sci.* Vol.109-110, pp.275-282, ISSN 0169-4332.

Trachenko, K., Dove, M. T., Artacho, E., Todorov, I. T. & Smith, W. (2006). Atomistic simulations of resistance to amorphization by radiation damage. *Phys. Rev. B* Vol.73, No.17, pp.174207, ISSN 1550-235X.

Trautmann, C., Toulemonde, M., Costantini, J. M., Grob, J. J. & Schwartz, K. (2000). Swelling effects in lithium fluoride induced by swift heavy ions. *Phys. Rev. B* Vol.62, No.1, pp.13-16, ISSN 1550-235X.

Ulrich, R. & Torge, R. (1973). Measurement of thin film parameters with a prism coupler. *Appl. Opt.* Vol.12, No.12, pp.2901-2908, ISSN 2155-3165.

Vázquez, G., Rickards, J., Lifante, G., Domenech, M. & Cantelar, E. (2003). Low dose carbon implanted waveguides in Nd:YAG. *Opt. Express* Vol.11, No.11, pp.1291-1296, ISSN 1094-4087.

Vázquez, G. V., Sánchez-Morales, M. E., Flores-Romero, E., Márquez, H., Rickards, J., Trejo-Luna, R. & Moretti, P. (2006). Study of optical waveguides in Nd:YAG and Nd:YVO_4 crystals. *Proceedings of Fifth Symposium Optics in Industry*, 2006. p 604609. SPIE.

Vetter, J., Scholz, R. & Angert, N. (1994). Investigation of latent tracks from heavy ions in GeS crystals by high resolution TEM. *Nucl. Instrum. Meth. B* Vol.91, No.1-4, pp.129-133, ISSN 0168-583X.

Villarroel, J., Carrascosa, M., García-Cabañes, A., Caballero-Calero, O., Crespillo, M. & Olivares, J. (2009). Photorefractive response and optical damage of $LiNbO_3$ optical waveguides produced by swift heavy ion irradiation. *Appl. Phys. B* Vol.95, pp.429-433, ISSN 1432-0649.

Vincent, B., Boudrioua, A., Kremer, R. & Moretti, P. (2005). Second harmonic generation in helium-implanted periodically poled lithium niobate planar waveguides. *Opt. Commun.* Vol.247, No.4-6, pp.461-469, ISSN 0030-4018.

Vincent, B., Boudrioua, A., Loulergue, J. C., Moretti, P., Tascu, S., Jacquier, B., Aka, G. & Vivien, D. (2003). Channel waveguides in $Ca_4GdO(BO_3)_3$ fabricated by He^+ implantation for blue-light generation. *Opt. Lett.* Vol.28, No.12, pp.1025-1027, ISSN 1539-4794.

Vincent, B., Kremer, R., Boudrioua, A., Moretti, P., Zhang, Y.-C., Hsu, C.-C. & Peng, L.-H. (2007). Green light generation in a periodically poled Zn-doped $LiNbO_3$ planar waveguide fabricated by He^+ implantation. *Appl. Phys. B* Vol.89, No.2-3, pp.235-239, ISSN 1432-0649.

Vitova, T., Zamani-Meymian, M. R., Peithmann, K., Maier, K. & Hormes, J. (2009). Atomic displacement and disorder in $LiNbO_3$ single crystal caused by high-energy ^3He ion irradiation: An x-ray absorption spectroscopy study. *J. Phys.: Condens. Matter* Vol.21, No.40, pp.495401, ISSN 0953-8984.

Waligórski, M. P. R., Hamm, R. N. & Katz, R. (1986). The radial distribution of dose around the path of a heavy ion in liquid water. *Nucl. Tracks Radiat. Meas.* Vol.11, No.6, pp.309-319, ISSN 1359-0189.

Wang, K.-M., Hu, H., Lu, F., Chen, F., Shi, B.-R., Ma, C.-Q., Lu, Q.-M. & Shen, D.-Y. (2002). Refractive index profiles of LiB_3O_5 waveguides formed by MeV He ion irradiation. *J. Appl. Phys.* Vol.92, No.7, pp.3551-3553, ISSN 0021-8979.

Wang, K.-M., Shi, B.-R., Cue, N., Zhu, Y.-Y., Xiao, R.-F., Lu, F., Li, W. & Liu, Y.-G. (1998). Waveguide laser film in erbium-doped $KTiOPO_4$ by pulsed laser deposition. *Applied Physics Letters* Vol.73, No.8, pp.1020-1022, ISSN 0003-6951.

Wang, L., Chen, F., Wang, X.-L., Wang, K.-M., Jiao, Y., Wang, L.-L., Li, X.-S., Lu, Q.-M., Ma, H.-J. & Nie, R. (2007). Low-loss planar and stripe waveguides in Nd^{3+}-doped silicate glass produced by oxygen-ion implantation. *J. Appl. Phys.* Vol.101, No.5, pp.053112, ISSN 0021-8979.

Wang, L., Chen, F., Wang, X.-L., Wang, L.-L., Wang, K.-M., Gao, L., Ma, H.-J. & Nie, R. (2006a). Si^{2+} ion implanted into stoichiometric lithium niobate crystals: Waveguide characterization and lattice disorder analysis. *Nucl. Instrum. Meth. B* Vol.251, No.1, pp.104-108, ISSN 0168-583X.

Wang, L. & Lu, Q.-M. (2009). Formation and characterization of a near-stoichiometric $LiNbO_3$ waveguide by MeV oxygen implantation. *Appl. Opt.* Vol.48, No.14, pp.2619-2624, ISSN 1539-4522.

Wang, L.-L., Wang, K.-M., Lu, Q.-M. & Ma, H.-J. (2008). Enhanced refractive index well-confined planar and channel waveguides in $KTiOPO_4$ produced by MeV C^{3+} ion implantation with low dose. *Appl. Phys. B* Vol.94, No.2, pp.295-299, ISSN 1432-0649.

Wang, L.-L., Wang, L., Wang, K.-M., Lu, Q.-M. & Ma, H.-J. (2009). Annealing effect on mono-mode refractive index enhanced $RbTiOPO_4$ waveguides formed by ion implantation. *Opt. Express* Vol.17, No.7, pp.5069-5074, ISSN 1094-4087.

Wang, L.-L. & Yu, Y.-G. (2010). Characterization of laser waveguides in Nd:CNGG crystals formed by low fluence carbon ion implantation. *Appl. Surf. Sci.* Vol.256, No.8, pp.2616-2619, ISSN 0169-4332.

Wang, X.-L., Chen, F., Wang, L. & Jiao, Y. (2006b). Channel waveguides of $LiNbO_3$ crystals fabricated by low-dose oxygen ion implantation. *J. Appl. Phys.* Vol.100, No.5, pp.056106, ISSN 0021-8979.

Wang, Z. G., Dufour, C., Paumier, E. & Toulemonde, M. (1996). Defects in metals induced by nuclear collisions and their modifications by swift heavy ion irradiations. *Nucl. Instrum. Meth. B* Vol.115, No.1-4, pp.577-580, ISSN 0168-583X.

Wesch, W., Opfermann, T., Schrempel, F. & Höche, T. (2001). Track formation in $KTiOPO_4$ by MeV implantation of light ions. *Nucl. Instrum. Meth. B* Vol.175-177, pp.88-92, ISSN 0168-583X.

Wesner, M., Herden, C., Kip, D., Krätzig, E. & Moretti, P. (2001). Photorefractive steady state solitons up to telecommunication wavelengths in planar SBN waveguides. *Opt. Commun.* Vol.188, No.1-4, pp.69-76, ISSN 0030-4018.

Yang, P. K. & Huang, J. Y. (2000). An inexpensive diode-pumped mode-locked Nd:YVO_4 laser for nonlinear optical microscopy. *Opt. Commun.* Vol.173, No.1-6, pp.315-321, ISSN 0030-4018.

Yao, Y., Tan, Y., Dong, N., Chen, F. & Bettiol, A. A. (2010). Continuous wave Nd:YAG channel waveguide laser produced by focused proton beam writing. *Opt. Express* Vol.18, No.24, pp.24516-24521, ISSN 1094-4087.

Youden, K. E., James, S. W., Eason, R. W., Chandler, P. J., Zhang, L. & Townsend, P. D. (1992). Photorefractive planar waveguides in $BaTiO_3$ fabricated by ion-beam implantation. *Opt. Lett.* Vol.17, No.21, pp.1509-1511, ISSN 1539-4794.

Zakery, A. & Elliott, S. . (2003). Optical properties and applications of chalcogenide glasses: A review. *J. Non-Cryst. Sol.* Vol.330, No.1-3, pp.1-12, ISSN 0022-3093.

Zeller, P. & Peuser, P. (2000). Efficient, multiwatt, continuous-wave laser operation on the $^4F_{(3/2)}$-$^4I_{(9/2)}$ transitions of Nd:YVO_4 and Nd:YAG. *Opt. Lett.* Vol.25, No.1, pp.34-36, ISSN 1539-4794.

Zhao, J.-H., Huang, Q., Liu, P. & Wang, X.-L. (2011a). An He-implanted optical planar waveguide in an Nd:YGG laser crystal preserving fluorescence properties. *Appl. Surf. Sci.* Vol.257, No.16, pp.7310-7313, ISSN 0169-4332.

Zhao, J.-H., Huang, Q., Wang, L., Fu, G., Qin, X.-F., Liu, P., Guo, S.-S., Liu, T. & Wang, X.-L. (2011b). The properties of ion-implanted $LiNbO_3$ waveguides measured by the RBS and ion beam etching stripping methods. *Opt. Mater.* Vol.33, No.8, pp.1357-1361, ISSN 0925-3467.

Zhao, J.-H., Liu, T., Guo, S.-S., Guan, J. & Wang, X.-L. (2010a). Optical properties of planar waveguides on $ZnWO_4$ formed by carbon and helium ion implantation and effects of annealing. *Opt. Express* Vol.18, No.18, pp.18989-18996, ISSN 1094-4087.

Zhao, J.-H., Liu, X.-H., Huang, Q., Liu, P. & Wang, X.-L. (2010b). Lithium niobate ridge waveguides fabricated by ion implantation followed by ion beam etching. *J. Lightwave Technol.* Vol.28, No.13, pp.1913-1916, ISSN 0733-8724.

Zhao, J.-H., Wang, X.-L. & Chen, F. (2010c). 1×4-Branch waveguide power splitters in lithium niobate by means of multi-energy O ion implantation. *Opt. Mater.* Vol.32, No.11, pp.1441-1445, ISSN 0925-3467.

Zhao, J.-H., Wang, X.-L., Fu, G., Liu, X.-H., Huang, Q. & Liu, P. (2010d). The fabrication of planar waveguides on $Bi_{12}TiO_{20}$ crystals by oxygen and helium ion implantation. *Nucl. Instrum. Meth. B* Vol.268, No.22, pp.3434-3437, ISSN 0168-583X.

Ziegler, J. *SRIM - The Stopping and Range of Ions in Matter: http://www.srim.org/.* [online] (2008). Available from: http://www.srim.org/. [Accessed 2009-09-16].

Ziegler, J. F. (1985). *The stopping and range of ions in solids.* Pergamon Pr., ISBN 008021603X.

Zinkle, S. J. & Skuratov, V. A. (1998). Track formation and dislocation loop interaction in spinel irradiated with swift heavy ions. *Nucl. Instrum. Meth. B* Vol.141, No.1-4, pp.737-746, ISSN 0168-583X.

Zinkle, S. J., Skuratov, V. A. & Hoelzer, D. T. (2002). On the conflicting roles of ionizing radiation in ceramics. *Nucl. Instrum. Meth. B* Vol.191, No.1-4, pp.758-766, ISSN 0168-583X.

13

Ga^{3+} Focused Ion Beam for Piezo Electric Nano Structuration Fabrication

D. Rémiens[1], D. Deresmes[2], D. Troadec[2] and J. Costecalde[1]
[1]*Institut d'Electronique de Microélectronique et Nanotechnologies (IEMN)*
[2]*Université de Valenciennes et du Hainaut Cambrésis (UVHC)*
France

1. Introduction

Piezoelectric nanostructures are currently of the potential interest for the development of MEMS / NEMS (Micro / Nano-Electro-Mechanical Systems). Their main applications in the biomedical domains (micro / nano bio actuators and sensors), in automobile industry (actuators and sensors) and in aeronautic industry (health control monitoring) are based on $PbZr_{0.54}Ti_{0.46}O_3$ (PZT) recognized as leading material for these piezoelectric micro/nano devices.[1, 2, 3, 4] The fabrication of piezoelectric nanostructures on pure PZT (or "doped- PZT") films is frequently performed by means of Reactive Ion Etching (RIE or ICP (Inductively Coupled Plasma) -RIE with fluorine and/or chlorine gases), Reactive Ion Beam Etching (RIBE), Focused Ion Beam (FIB) Ga^{3+} etching and electron beam direct writing. [5, 6, 7, 8]In the present study we focus our results on Ga^{3+} FIB etching of PZT nano structures. Typically, the obtained results show that it is possible to fabricate nanostructures with very small lateral size (50 nm on PZT films) when the films are etching in amorphous state followed by a post annealing treatment at the temperature which correspond to the perovskite formation temperature; the piezoelectric properties are similar to those obtained on un etched PZT films. Any degradation is observed. For crystallized films, the situation is completely different: no ferroelectric properties are observed after etching and the piezoelectric-response is strongly degraded. A post annealing treatment in oxygen results in a partial recovery of the ferroelectric properties. The main goal of the present study is to evaluate the damage induced by FIB etching of PZT films: implantation of Ga^{3+} and film amorphisation. The damaged layer that appears on the films' surface was thoroughly characterized in terms of the composition and charge implantation but a more detailed characterization seems to be necessary. To address the quality of the nanostructures, the measurement of local electromechanical activity via piezoelectric hysteresis loops acquisition using Piezoelectric response Force Microscopy (PFM) is certainly one of the best options. The conclusion of this work is the development of a suitable process of nano-structures manufacture without introducing piezoelectric property degradation.

2. Description of the structures and experimental set up

2.1 Structure

In this work, we have studied two types of $Si/SiO_2/TiO_x/Pt/PZT/Pt$ structures, namely PZT films being either in amorphous or crystallized state before etching [9]. The films in both states are deposited by radio frequency magnetron sputtering without substrate heating and therefore the films were amorphous. The film composition is fixed at the morphotropic phase boundary (Zr/Ti = 54/46). A post annealing was used to crystallize the PZT film in the Perovskite phase; the annealing temperature was 625°C (conventional annealing) in air during 1/2 hour [10]. The film thickness varied between 100 and 300 nm depending on the island diameter, i.e., for a small island diameter the film was thinner to reduce the etching time. Figure 1 presents a typical scheme of the PZT nanostructure (nano island) that we fabricate with the FIB system.

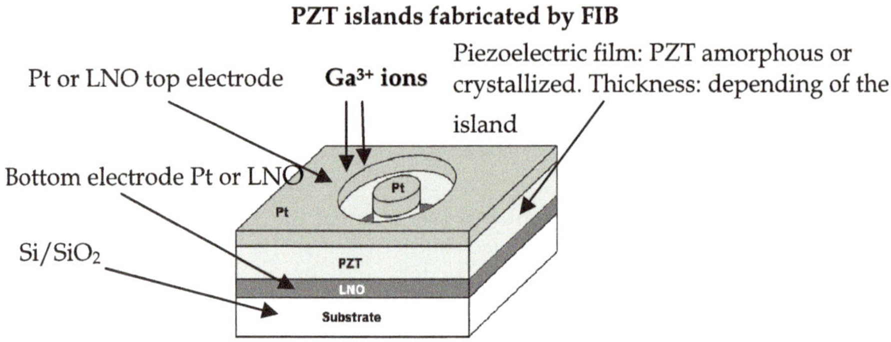

Fig. 1. PZT nano structures (nano island) fabricated by FIB etching on crystallized sample (conventional etching) and on amorphous sample (post annealing after etching).

2.2 FIB system

A focused beam of Ga^{3+} ions (STRATA DB 235 – FEI) was used to pattern the PZT film, Figure 2 shows our FIB system. The system was equipped with both the electron and ion beams.

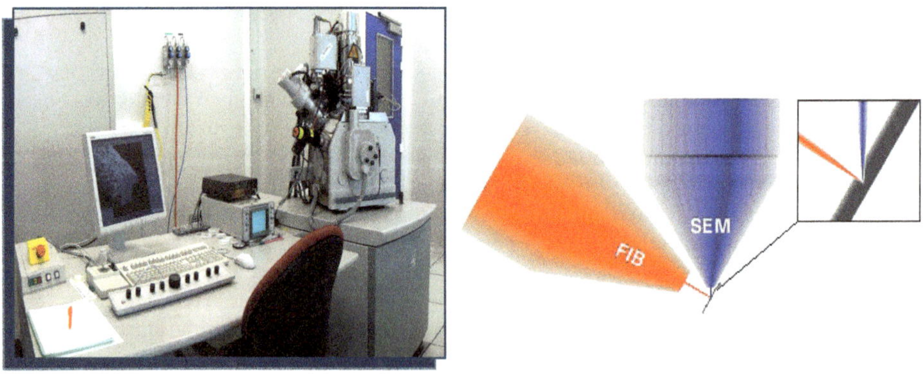

Fig. 2. FIB Machine: FEI dual STRATA 235 – SEM coupling with Ga^{3+} ion beam

The advantage of this system is the possibility to monitor the etching process in real time. The acceleration voltage was 30 kV and ion current was in the range 100 – 10pA depending on the island diameter. We have studied the relation between the current density of the ion beam and its diameter. The ion beam was at the 90° angle to the etched surface.

2.3 Piezoelectric characterization at nano-scale

At nano-scale, the Atomic Force Microscopy (AFM) is currently used [11-17]. Several AFM techniques, namely, Piezoelectric-response Force Microscopy (PFM) (and its spectroscopic tool which enables the measurement of local piezoelectric hysteresis loops) and Kelvin Force Microscopy (KFM) have been developed and these techniques were used in this work to characterize the piezoelectric nanostructures. For such investigations, a commercial AFM (Multimode, Nanoscope IIIA, Veeco) operating under environmental conditions was used; a scheme of the system is given Figure 3. The local piezoelectric hysteresis loops were obtained by plotting the amplitude of piezoelectric vibration A as a function of the DC voltage in so-called "in-field" mode in order to provide information about electromechanical activity. The frequency and the amplitude of the driving AC voltage were f = 2 kHz and Vac = 1.5 V, respectively. The applied DC voltage was gradually swept between -10 V and +10 V within the period of 200 s. To minimize the electrostatic force effect, Pt/Ir coated Si tips/cantilevers with relatively high spring constant of about 45 N.m-1 were used[16]. PFM domain imaging was also performed using the same cantilevers with the spring constant of 3.8 $N.m^{-1}$ (the frequency and the amplitude of the driving AC voltage were f = 8 kHz and Vac = 1.5 V, respectively). KFM is a non contact AFM-based technique based on the electrostatic interaction between the tip and the surface. Surface potential maps are easily detected due to work function variation as a result of surface charges, electric dipoles or absorption layer

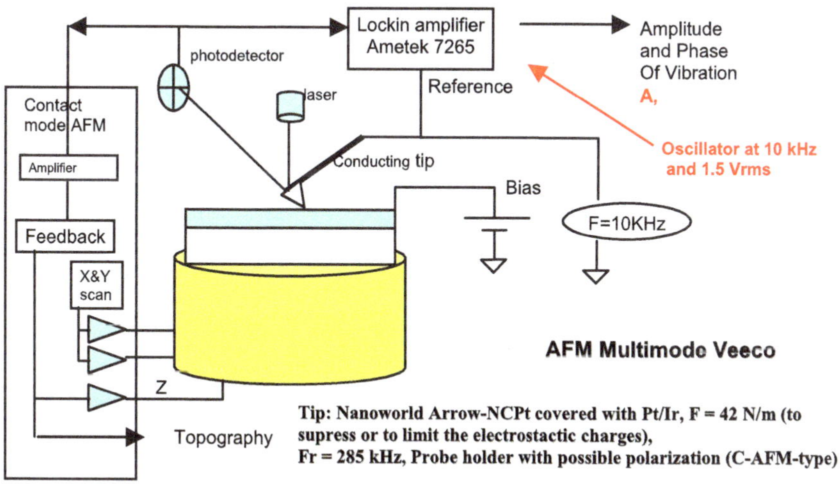

Fig. 3. AFM Multimode Veeco Tip: Nanoworld Arrow-NCPt covered with Pt/Ir, F = 42 N/m (to suppress or to limit the electrostatic charges), Fr = 285 kHz, Probe holder with possible polarization (C-AFM-type)

2.4 Results and discussion

Before the fabrication of Pt/PZT/Pt islands, we have evaluated the degradation induce by the Ga^{3+} ion beam on the crystallized PZT films deposited on the entire substrate. A large zone 7 x 7 μm^2 was etched at different doses (Figure 4); the Ga^{3+} dose is in relation with number of path of the beam. More precisely, the number of ion beam paths was varied between 1 and 20. According to [18], the relation between the number of paths and the Ga^{3+} ion dose is the following: ion dose = 4^{13} for 1 path and it become 3^{15} for 20 paths

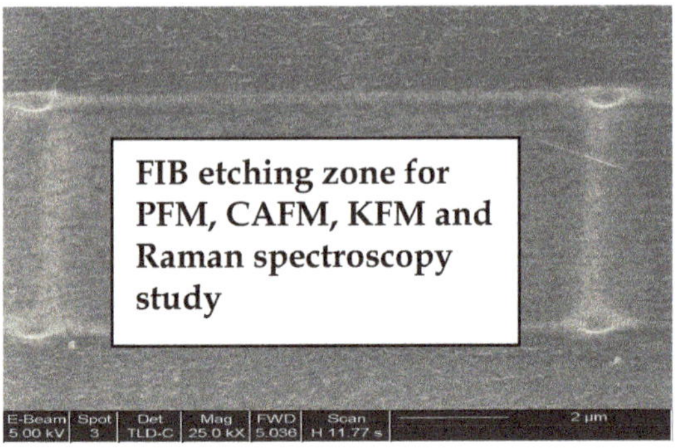

Fig. 4. Scanning Electron Microscopy (SEM) image of an etched zone by Focused Ion Beam (FIB). The dimension of the zone is 7 µm x 7 µm. Piezo-response Force Microscopy (PFM) and Kelvin Force Microscopy (KFM) images are made in this zone.

Firstly, we have analyzed macroscopically the defects induced in the material by Raman Spectroscopy. We have compared the responses of different samples: (a) the un-etched crystallized PZT film, (b) after hardest amorphous FIB etching process (Ga ion dose value =3^{15}), c) after lightest conventional FIB etching process (dose value =4^{13}), (d) after hardest conventional FIB etching process (dose value =3^{15}). The results are presented Figure 5.

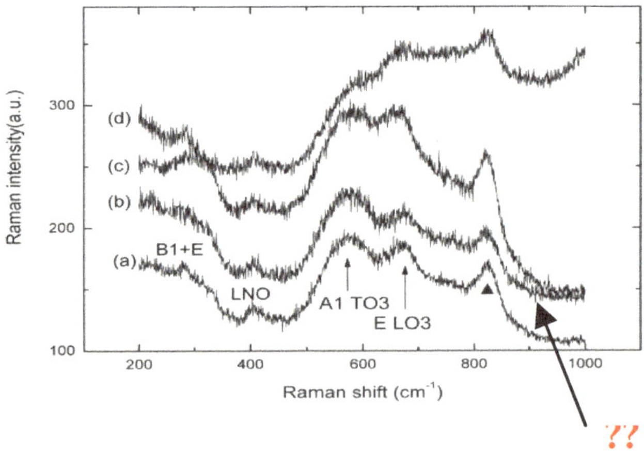

Fig. 5. Raman Spectroscopy on PZT by conventional FIB etching process. The broadband centered at 280, 572 and 676cm^{1}➔ modes in the single PZT crystals

For the hardest conventional etching case, the Raman spectra recorded in the region exposed to the FIB shows an important modification in comparison of single PZT crystals [19, 20]: we observe a sharp increase in the intensity of the Raman peak at 676cm^{-1} which corresponds to the (E(LO3) mode). This result is certainly connected with the distortion of the tetragonal symmetry of the PZT crystalline lattice due to the ion implantation and the modification of the stress state [21, 22]. The broadening of Raman scattering in 550-900 cm^{-1} region could be attributed to the existence of the amorphous material which is usually expected during the conventional FIB etching process. [23, 24, 25] In conclusion, whatever the FIB etching process, the peak position of Raman shift does not show any change, so the crystalline structure still exhibits perovskite phase.

We have compared the piezoelectric response of three samples (Figure 6): un-etched crystallized PZT film (denoted as sample (a)), amorphous PZT film etched before crystallization (sample (b)) and PZT film etched after crystallization (sample (c)). In crystallized PZT films (sample (c)) we observed strong piezo-response degradation whatever is the ion dose during etching. In fact, no piezoelectric response was obtained even after 1st path. At the same time no ferroelectric domains could be imaged.

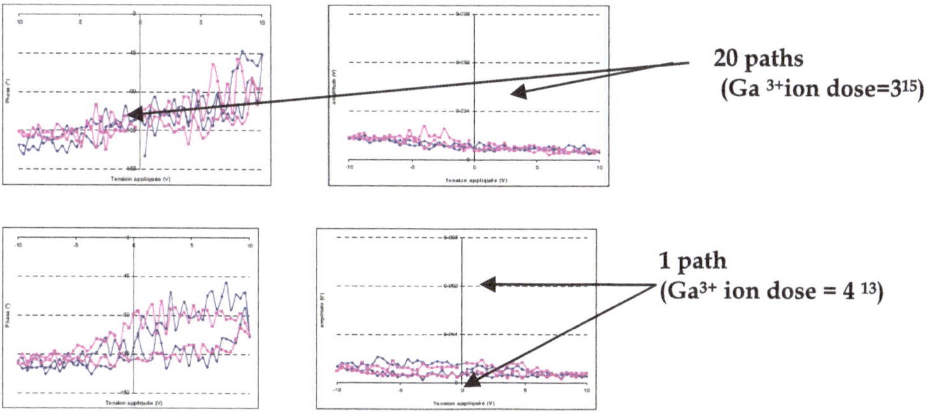

❖ Results: PFM measurements of FIB-etched PZT square patterns show that there is no piezoelectric response inside the square. The results suggest that there exist a damaged layer on the surface of the square, which can affect the piezoelectric performance of PZT

Fig. 6. PFM signal (amplitude) on the etched zone (conventional process: PZT crystallized) as a function of the number of paths.

The degradations induce by the FIB are not limited to the etched region; we have observed that even outside the exposed zone, some defects appears. Figure 7 shows this phenomenon, in this example we use the conventional FIB process. The PFM signal is restore at 100 μm outside the etched zone, the explanation can be in relation with the spatial extension of the ions beam. It is well known that the ions beam has a Gaussian form and as a consequence the zone outside the etched region is also in contact with the Ga^{3+} ions beam.

We tried to rejuvenate at least partially, the piezoelectric activity using a post annealing treatment (as it is currently proposed in the literature, e.g. for PZT films etched by RIBE[22]), but irrespective of the conditions (temperature, atmosphere) we were unable to restore the initial properties of the films. These results suggest that a damaged layer exists on the surface of the film which affects strongly the piezoelectric performance of PZT films. These damages could be: amorphisation of film surface, modification of the film composition, etc. In order to understand more the phenomena we analyse the film surface by KFM. The results are presented Figure 8. We show the evolution of the KFM signal as a function of the number of paths, the changing of colour of the etched zone is in relation with the number of paths; when the etched zone is completely degraded, the zone becomes black. Some effects can explain these modifications: lead loss, gallium implantation; some authors have observed the substitution of Ga in the place of Pb to form: GZT (Ga,Zr,Ti) at the film surface.[26]

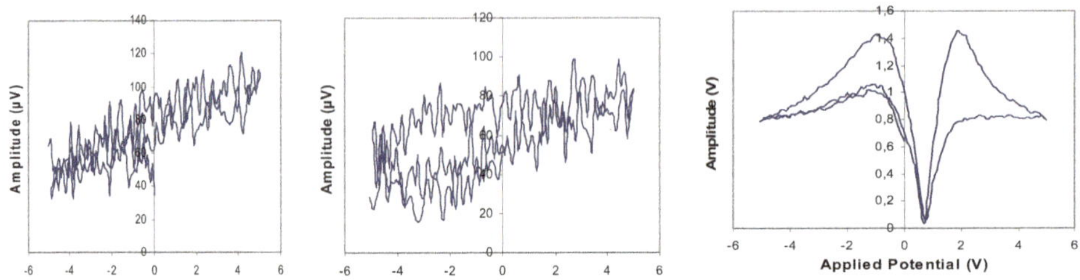

In the etched zone | At 20µm outside the etched zone | At 100µm outside the etched zone

Fig. 7. Extension of the degradations outside the etched zone: influence of the ion beam spatial extension: Gaussian beam.

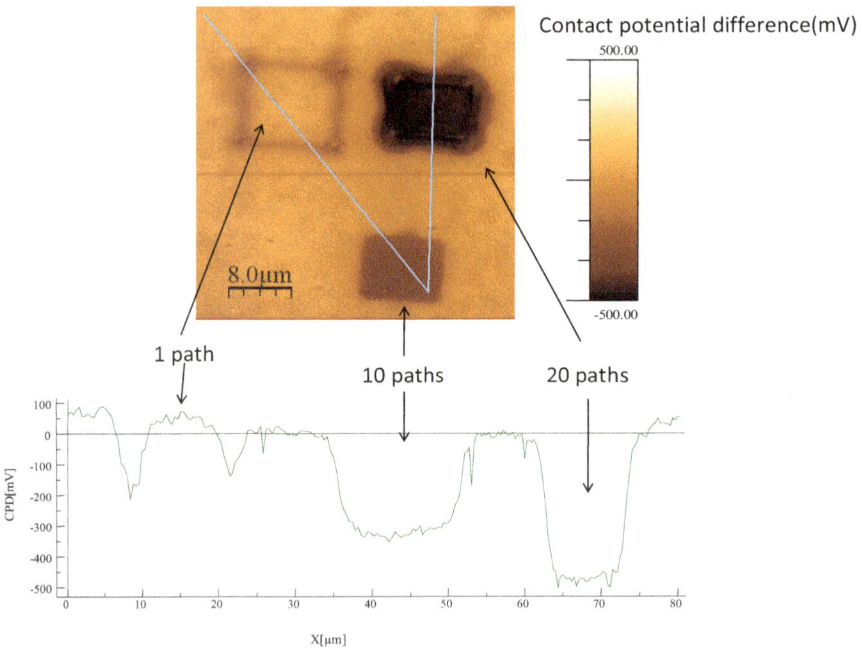

Fig. 8. KFM signal on conventional etching sample measure on areas subjected to different number of paths.

The KFM image of Figure 8 shows the evolution of the tip grounded CPD (Contact Potential Difference) response as a function of the number of paths on the conventional etching sample. The KFM and the CAFM (see later) measurements were made at the same time that means with CAFM configuration the tip is virtually connected to the ground and the bias is applied to the sample. The change of contrast is, of course, directly related to the variation of the surface properties such as the composition and / or amorphisation of the material and induced structural defects. We cannot make more definitive conclusion for the moment but it is clear that after 20 paths the etched region is completely modified. A CPD profile (here the sample is grounded) over the three etched area indicates almost no modification of CPD for 1 path and apparent decrease by 350 mV and 500 mV for 10 and 20 paths, respectively. In our conditions (room and instrument calibration), the CPD of the surface was -50mV relative to a sample grounded. To compare the KFM signals obtained on the samples (b) and (c), we perform a second experiment under the same condition (tip, instrumentation configuration and sample grounding).

The results are presented in Figure 9. The strong contrast difference in the sample (c) (between etched and not-etched areas) confirms the strong damage induced and a consequent modification of the surface potential. On the contrary, only weak damage (small contrast variation) is observed for the sample (b) (etched in the amorphous state). To point out these variations more clearly, we indicate the CPD value across the etched region (Figure 9b).

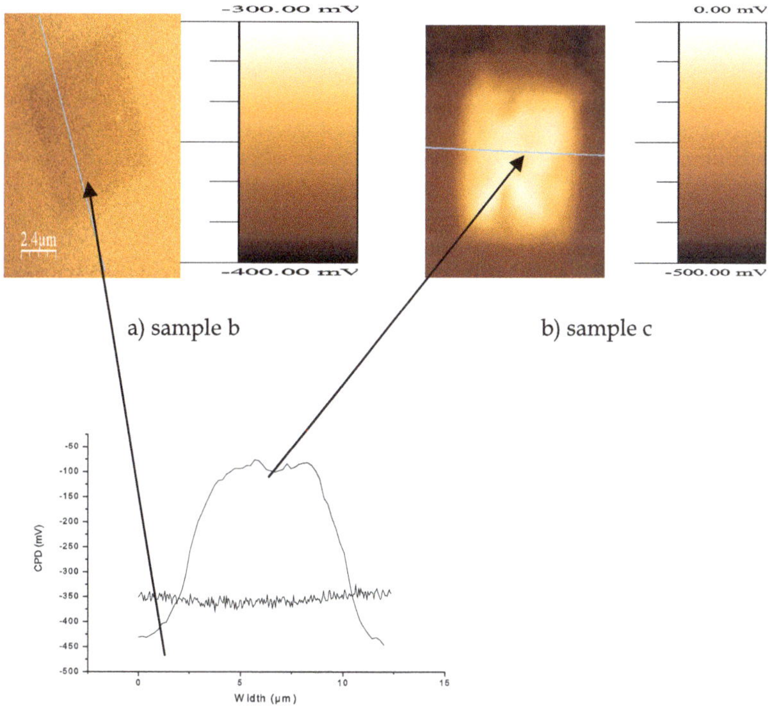

Fig. 9. a) KFM signal measured on amorphous and crystallized film after etching.
b) Cross-section of the contact potential measured sample c. The position of the scan is précised figure a.

It confirms that the potential increase between the un-etched surface and the surface of PZT etched after crystallization is very large (about 300 mV). The implanted charges are the same in both samples but for amorphous sample we can suppose that the annealing treatment induces these charges neutralization and a slight decrease (10 mV) of the CPD is only detected. The properties of the etched area are certainly different depending of the process. The difference, in terms of material is not only due the presence of charge on the sample surface but some crystal defects (dislocations), substitution between Pb and Ga [17], implantation of metallic Ga^{3+} probably were maintained on sample (c) but not in sample (b) due to the annealing treatment. Figure 4 shows the piezo-response amplitude A measured on etched zone after 1 path and 20 paths in the sample (c) (etched after crystallization). As we can observe, only weak piezoelectric activity is obtained even after one path. The poor ferroelectric properties are confirmed by the absence of the hysteresis on both samples. Therefore, the FIB etching is detrimental to functional properties of PZT. Annealing of these etched zones (in air or in O_2 atmosphere at different temperatures and times) had no effect on piezo-response. On the contrary, for PZT etched before crystallization (sample b), the piezoelectric signal (measured after crystallization) is the same as the one obtained for the sample (a) irrespective of the ion dose. Even if degradation is induced on the sample (b), the annealing treatment (crystallisation of the film) permits to eliminate the damages. These results confirm that the strong degradation occurs on the initially crystallized samples and no degradation happens if the FIB is conducted on amorphous samples.

To conclude this work it is evident that the surface potential is changed after etching, especially for the region exposed to conventional FIB, i.e. on crystallized PZT film. To see it more clearly, we shows the absolute surface potential value across the etching region (Figure 10). It implies that the potential difference between the un-etched surface and the surface after amorphous etching is very small (only about 15mv), while the difference between the un-etched surface and the surface after conventional FIB etching is very large (about 300mv).

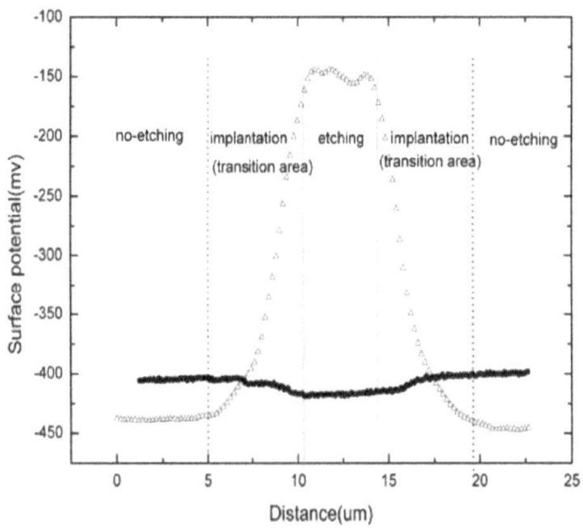

Fig. 10. Absolute film surface potential measured by KFM

So, it is evident that for nanostructures fabrication, the FIB process is very well adapted when the film is in amorphous state (probably due to the lake of grain). So, no degradation is observed and maybe the post annealing treatment (at temperature equal to the temperature formation of the perovskite phase formation) following the etching process could suppressed the eventual degradations induced is the amorphous one.

In this second part of this work, we focused the results and discussions about the fabrication of PZT nano structures, named nano islands. The typical structures obtained by SEM are shown in Figure11. The size of the island shown in Figure 11 is 50 nm.

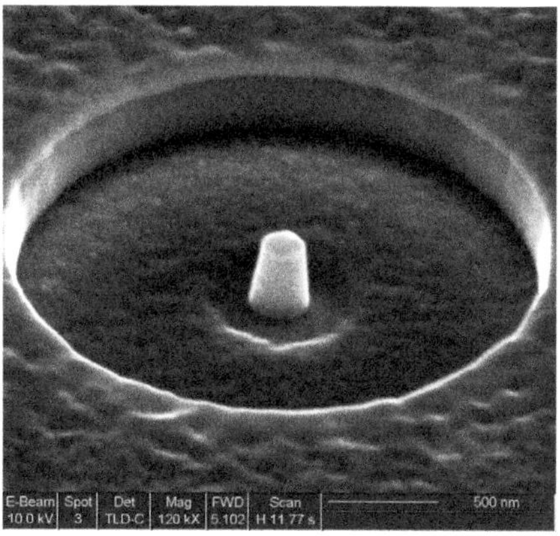

Fig. 11. PZT nano island of 50 nm in diameter

A Pt top electrode was used as a mask to protect the top of the island and, therefore, we take into account only the defects introduced into the islands walls. The etching anisotropy is probably due to the selectivity between Pt and PZT; this effect could be a limitation of the FIB process and it seems more difficult to decrease the island diameter with this technique. The comparative results of the PFM measurements on the PZT islands of 50 nm of size are shown in Figure 12. These results confirm that with conventional etching (on crystallized PZT) the piezoelectric response disappears completely and even after a post annealing treatment the signal could not be restored. On amorphous PZT etching followed by a post annealing treatment at perovskite formation temperature, the piezoelectric response is excellent.

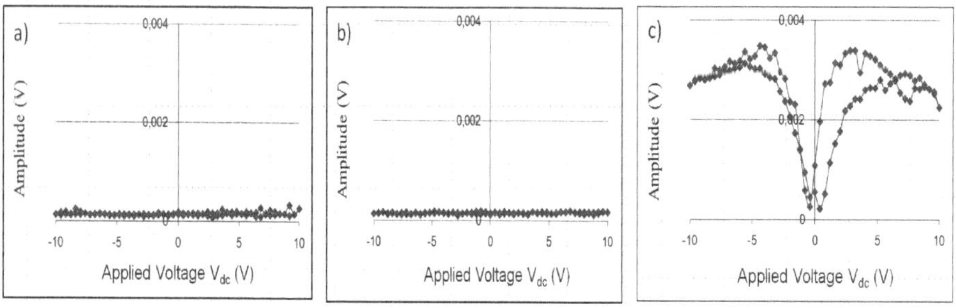

Fig. 12. PFM signal (amplitude) measured on PZT islands of 50 nm after (a) conventional etching process, (b) post-annealing at 625 °C under O_2, (c) amorphous FIB etching process.

These results contradict to those obtained by Stanishevsky *et al.* [17] where the acceleration voltage of the Ga^{3+} ions is larger (50 kV) and so it had to induce stronger degradation. In order to explain why the piezoelectric signal disappeared we have performed a local I (V) measurements with CAFM on the similar PZT island obtained with the conventional process (etching on crystallized PZT) with 250 nm diameter. The result is shown in Figure 13 for these experiments, a V_{DC} bias was applied to the substrate holder and the top surface Pt was grounded.

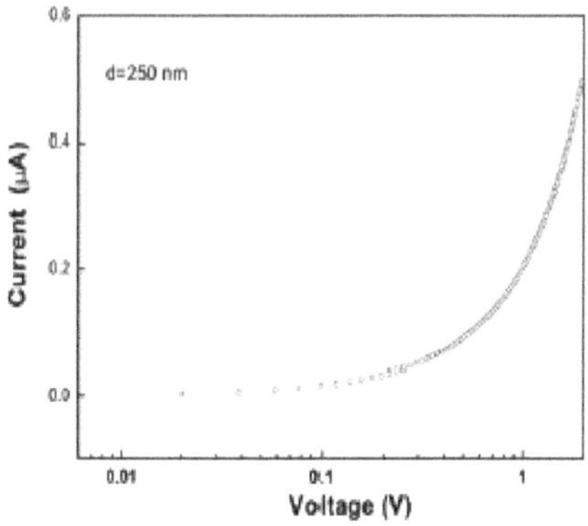

Fig. 13. I(V) measurement on conventional etching sample obtained by Conducting Atomic Force Microscopy (CAFM).

The current increased rapidly with the applied voltage indicating strong local conductivity due to the presence of an electrical conducting channel. This proves a very strong damage due to Ga^{3+} ion implantation in the PZT nano-islands and associated loss of the piezoelectric activity. The leakage current becomes more important when the island diameter decreases, it is thus can be responsible for the results on 50 nm islands. This observed conductivity increase is thus responsible for the weak electromechanical activity observed with the conventional process.

To conclude, we present a new device for BioNEMs applications. It consist of a networks of nano islands (Figure 14) fabricated by FIB. The network is constituted by nano islands of 100 nm; the dimension of the network is 600 x 600 nm^2

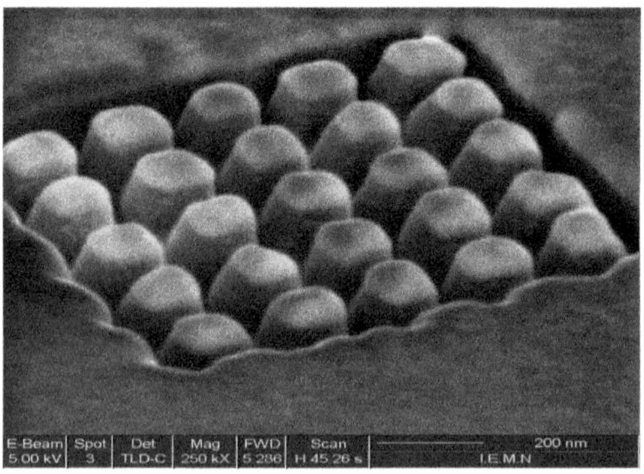

Fig. 14. Network (600 x 600nm2) of islands (100 nm) for bio application: functionalization of the surface

We measure now the PFM responses of all the nano islands.

3. Conclusion

In conclusion, in this work we have evaluated the damages induced by the $Ga^{3°}$ FIB process for the fabrication of PZT nano structures. Two processes have been developed:

- - the conventional process where the films are crystallized before etching followed by a post annealing to restore the initial piezoelectric electric properties,
- - the amorphous process where the etching process is made on amorphous films followed by a post annealing treatment at the Perovskite formation temperature.

In order to estimate the induced degradations by these two processes we have firstly etched large PZT film zones. The main conclusion is that when the PZT is crystallized in the Perovskite phase a strong degradation is observed and the piezoelectric of the film disappears even after a post annealing treatment. In amorphous state no degradation is observed. So, we have choice this process to fabricate nano piezoelectric structures.

By an optimization of the FIB Ga^{3+} beam and by using amorphous PZT films we have fabricated piezoelectric nano structures: the minimum diameter size of the nano structure is 50 nm. Even at this dimension the nano structure gives always a piezoelectric response. To measure this piezoelectric activity on very small material we have developed and optimized many experimental set up such as : PFM, KFM, CAFM; these methods are relatively news and very well adapted to measurements at nano scale but these methods are very sensitive to external parameters: electrostatic charges, contact between the tip and the material surface,... Many precautions must be taken if you want that these measurements are a true reality of the nano structures performances.

For the nano piezoelectric performances, the conclusions are clear and similar to those obtained on film: with the conventional process, the piezoelectric response disappears (even for large islands (diameter = 800 nm) and even with a post annealing treatment the properties can't be restored. The piezoelectric activity of the nano structure is maintained with the amorphous process. So, the amorphous process is used to fabricate piezoelectric nanostructures. PZT islands of 50 nm in diameter were fabricated and the piezoelectric response was unchanged in comparison with un-etched PZT films. Identical results have been obtained on PMN-PT films. We try in future work to decrease the island diameter and maybe found a diameter where the piezoelectric activity disappears as predicted by some authors (when the dimension of the island = dimension of the domains). A limit exist with the FIB process (due to the anisotropy) and we will try to find this limit and maybe the electron beam lithography will be more adapted for very narrow PZT island (< 50 nm in diameter).

4. Acknowledgement

The authors gratefully acknowledge all the persons who are engaged in this work: Prof. R. Desfeux, and Dr. J.F. Blach from UCCS. Prof. Wang, Prof. Dong and Dr. R.H. Liang from SICCAS - Shanghai.

5. References

[1] C. Ayela, T. Aleva, D. Lagrange, D. Rémiens, C. Soyer, T. Ondarcuhu, A. Greve, and L. Nicu, IEEE Sensors Journal 8, 210 (2008).

[2] J. Verd, A. Uranga, G. Abadal, J. Teva, F. Torres, F. Pérez-Murano, J. Fraxeda, J. Esteve, and N. Bardiol, Appl. Phys. Lett. 91, 013501 (2007).
[3] R. R. He, X. L. Feng, M. L. Roukes, and P. D. Yang, Nano Lett. 8, 1756 (2008).
[4] Y. Cho, K. Fujimoto, Y. Hiranaga, Y. Wagatsuma, A. Onoe, K. Terabe, and K. Kitamura, Nanotechnology 14, 637 (2003).
[5] M. Alexe, C. Harnagea, D. Hesse, U. Gösele, Appl. Phys. Lett. 75, 1793 (1999).
[6] C. S. Ganpule, A. Stanshevsky, S. Aggarwal, J. Melngailis, E. Williams, R. Ramesh, V. Joshi, and Carlos Paz de Araujo, Appl. Phys. Lett. 75, 3874 (1999).
[7] R. H. Liang, D. Rémiens, D. Deresmes, C. Soyer, D. Troadec, X.L. Dong, L.H. Yang, R. Desfeux, A. Da Costa, and J.F. Blach, J. Appl. Phys. 105, 044101 (2009).
[8] K. Saito and M. Kaise, Jpn. J. Appl. Phys. 31, 3533 (1992).
[9] M. Detalle, G. Wang, D. Rémiens, P. Ruterana, P. Roussel, and B. Dkhil, J. Cryst. Growth, 305, 137 (2007).
[10] R. Herdier, M. Detalle, D. Jenkins, C. Soyer, and D. Rémiens, Sensors & Actuators A 148, 122 (2008).
[11] S. Hong, J. Woo, H. Shin, J.U. Jeon, Y.E. Park, E. L. Colla, N. Setter, E. Kim, and K. No, J. Appl. Phys. 89, 1377 (2001).
[12] A. L. Kholkin, V. V. Shvartsman, A .Yu. Emelyanov, R. Poyato, M. L. Calzada, L. Pardo, Appl. Phys. Lett. 82, 2127 (2003).
[13] S.V. Kalinin, A. Gruverman and D.A. Bonnel, Appl. Phys. Lett. 85, 795 (2004).
[14] A. L. Kholkin, V. V. Shvartsman, D. A. Kiselev, I.K. Bdikin, Ferroelectrics 341, 3 (2006).
[15] A. Gruverman and S.V. Kalinin, J. Mater. Sci. 41, 107 (2006).
[16] R. Desfeux, C. Legrand, A. Da Costa, D. Chateigner, R. Bouregba, and G. Poullain, Surf. Sci. 600, 219 (2006).
[17] A. Ferri, A. Da Costa, R. Desfeux, M. Detalle, G.S. Wang, and D. Rémiens, Integr. Ferroelectr. 91, 80 (2007).
[18] J. F. Ziegel, The stoppingand Range of Ion in Solid (Pergamon, New York, 1985).
[19] J. F. Meng, R. S. Katiyar, and G. T. Zou, Phys. Status Solidi A 164, 851, 1997.
[20] X. Lou, X. Hu, M. Zhang, F. D. Morrison, S. A. T. Redfern, and J. F. Scott, J. Appl. Phys. 99, 044101, 2006.
[21] J. Frantti, V. Lantto, S. Nishio, and M. Kakihana, Phys. Rev. B 59, 12, 1999.
[22] J. F. Meng, R. S. Katiyar, G. T. Zou, and X. H. Wang, Phys. Status Solidi A 164, 851, 1997.
[23] K. Nomura, Y. Takeda, M. Maeda, and N. Shibata, Jpn. J. Appl. Phys., Part 1 39, 5247, 2000.
[24] C. Soyer, E. Cattan, and D. Rémiens, J. Appl. Phys. 92, 1048, 2002.
[25] R. H. Liang, D. Rémiens, C. Soyer, N. Sama, X. L. Dong, and G. S. Wang, Microelectron. Eng. 85, 670 _2008_.
[26] A. Stanishevsky, B. Nagaraj, J. Melngailis, R. Ramesh, L. Khriachtchev, and E. McDaniel, J. Appl. Phys. 92, 3275 (2002).

14

Annealing Effects on the Particle Formation and the Optical Response

Akira Ueda*, Richard R. Mu and Warren E. Collins
Fisk University
USA

1. Introduction

We present two examples for annealing effects on the implanted ions in substrates: (1) Se in SiO_2 and (2) Au in MgO. For the first system, Se in SiO_2, we analyze the absorption edges for selenium quantum dots for samples with several doses and annealed at different temperatures. From the absorption edge data, we discuss the formation of Se particles. For the second system, Au in MgO, we have seen the change in surface plasmon resonance (SPR) energy by annealing in different atmospheres from reducing to oxidizing or vice-versa. In this system, we discuss the role of substrate interacting with the implanted ions.

2. Temperature dependence and annealing effects of absorption edges for selenium quantum dots formed in silica glass

We have fabricated Se nanoparticles in silica substrates by ion implantation followed by thermal annealing up to 1000°C, and studied the Se nanoparticle formation by optical absorption spectroscopy, Rutherford backscattering spectrometry, X-ray diffraction, and transmission electron microscopy. The sample with the highest dose (1×10^{17} ions/cm^2) showed the naonparticle formation during the ion implantation, while the lower dose samples (1 and 3×10^{16} ions/cm^2) required thermal treatment to obtain nano-sized particles. The Se nanoparticles in silica were found to be amorphous. After thermal annealing, the particle sizes became larger than the exciton Bohr radius for bulk Se. Thus, the absorption edges for different doses approached the value of bulk after thermal annealing. The temperature dependent absorption spectra were also measured for this system in a temperature range from 15 to 300 K.

2.1 Overview of selenium

Selenium is a semiconductor that has been extensively studied in the past because of the applications in photovoltaic cells, rectifiers, xerography, and so on. Solid selenium is known to occur as four major allotropes: an amorphous, trigonal, and two monoclinic phases (Murphy et al., 1977). In the solid and liquid states, selenium exists in two kinds of unit structures: rings and chains. The monoclinic allotropes are composed of eight-membered

* Corresponding Author

rings in two different stacking patterns, while the trigonal selenium is composed of helical chains. The glassy selenium consists of a mixture of rings and polymeric chains. For the trigonal phase, bulk selenium has the band gap of 1.7-2.2 eV, while monoclinic phase has the band gap of 2.4-2.6 eV (Pankove, 1971). However, the values are varied in the literature and they are not yet well determined. It is also known that the temperature effect on the band gap of bulk selenium is relatively high ($[\Delta E_g/\Delta T]_P$ = -1.4×10^{-3} eV/K), compared with the other semiconductors ($-2\sim7 \times 10^{-4}$ eV/K).

Recently, quantum size effects on the optical properties on semiconductor nano-sized particles have been observed. The third-order optical susceptibility for small semiconductor crystallites is found to be very large (Hanamura, E. 1988). The nano-sized selenium quantum dots are expected to exhibit changes in electronic and vibrational characteristics due to the quantum size effects. The band gap is generally expected to be blue shifted from bulk value as the particle size decreases (Brus, L.E. ,1982).

2.2 Experimental method

Selenium ions were implanted into fused silica substrates (Corning 7940) at an energy of 330 keV with doses of 1, 3, 6, and 10 x 10^{16} ions/cm^2. The samples were annealed at 600, 800, and 1000°C for 1 h in a 5%H_2 + 95%Ar atmosphere in a tube furnace. The optical absorption spectra were measured from 185 to 2500 nm with a dual beam spectrophotometer (Hitachi U-3501) before and after annealing. Reflectance spectra were measured using a refractive index matching fluid with the incident angle of 5°. For temperature dependent measurements in the range of 15-300 K, we used a helium closed cycle type cryostat (Leybold, RGD 210) fitted with silica windows. The spectra were collected at 25 K intervals, which were monitored with a temperature controller (Lakeshore, 320 Autotuning). The errors due to the fluctuation in temperatures can be on the order of ±5 K. The depth profiles of the implanted layer were determined by Rutherford backscattering spectrometry (RBS) using 2.2 MeV He^{2+} ions. X-ray diffraction (XRD) and transmission electron microscopy (TEM) measurements were conducted to find the crystal structure and size of the Se quantum dots in silica glass.

2.3 Experimental results

The depth profiles of Se in fused silica measured by RBS are shown in Fig. 1. The shape of the ion concentration distribution is approximately Gaussian with a peak concentration located at 0.2 μm below the surface with a full width at half maximum of about 0.2 μm.

Fig.2(a) shows the transmission spectra of the as-implanted samples with four different Se doses, which were measured at RT. In the UV region, there are two absorption bands at ~250 nm and ~350 nm. The band at 350 nm can be clearly observed for 3 and 6 x 10^{16} ions/cm^2. The optical absorption edge changes for each dose. Fig. 2(b) shows the transmission spectra of the samples annealed at 1000°C in a reducing atmosphere for 1 h, which were also measured at RT. The intensity of the band at ~350 nm is no longer resolved. On the other hand, the band at ~250 nm became stronger for the samples of 3, 6, 10 x 10^{16} ions/cm^2. After annealing the samples, X-ray diffraction measurements were carried out and demonstrated that the nanoparticles were in an amorphous phase. (The X-ray diffraction spectrum is not shown here.)

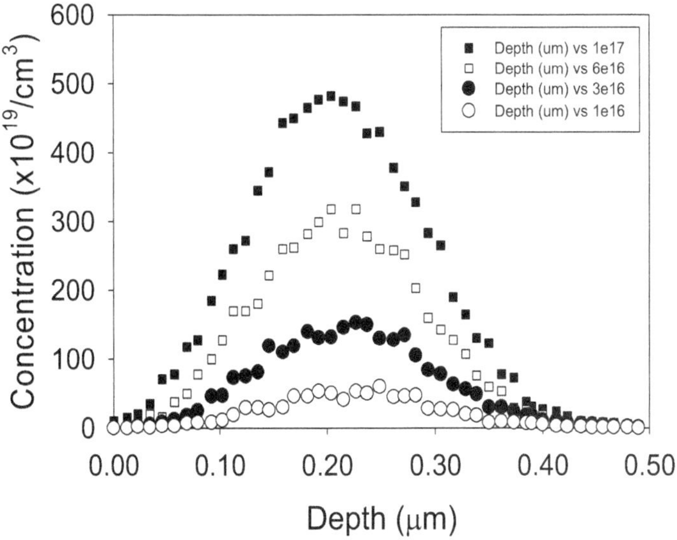

Fig. 1. The depth profile of Se ions in silica by RBS. (Ueda et al., 2007)

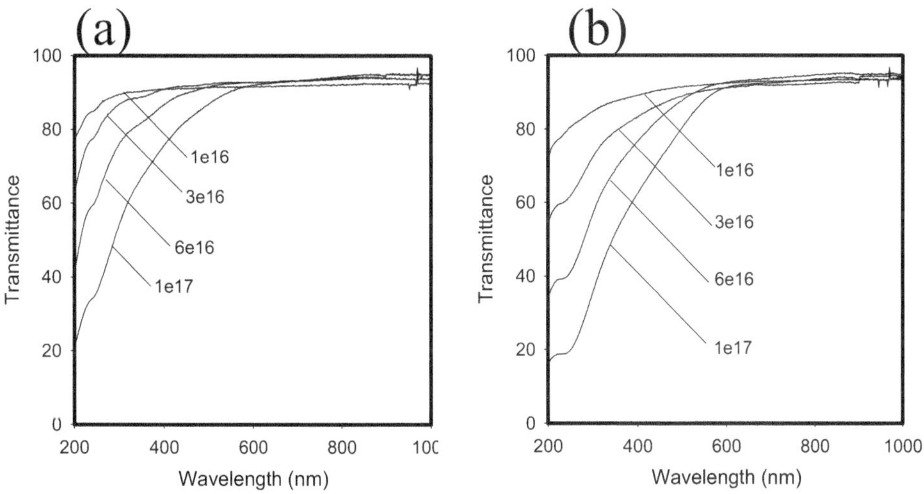

Fig. 2. Transmission spectra of Se/SiO_2 (a) as-implanted samples and (b) after annealing at 1000°C for 1 h.

Fig. 3 shows the transmission spectra of the samples with a dose of 1 x 10^{17} ions/cm^2 annealed at 600, 800, 1000°C in a reducing atmosphere for 1 h and the as-implanted sample measured at RT. According to these data, it is observed that the absorption edge is near 600 nm (~2 eV). (Ueda et al., 2007)

For the samples with a dose of 1x10^{17} ions/cm^2 annealed at 800 and 1000°C, the low-temperature absorption measurements were carried out between 15 K and 325 K, as shown in Fig. 4. We repeated these measurements three times for each sample and the statistical uncertainty can be on the order of ±0.05 eV for the absorption edge. However, the trend is clear for the temperature effects and annealing temperature dependence of absorption edge. The sample annealed at 1000°C has the lowest absorption edge at each temperature as compared to the other annealing. As the temperature approaches 0 K, the traces become flat. The coefficient of temperature effects at 300 K for every trace is on the order of $\sim 10 \times 10^{-4}$ eV/K (10.8, 12.4, 9.0×10^{-4} eV/K for as-implanted, annealed at 800°C, and 1000°C, respectively), which is similar to literature values of bulk Se (-14×10^{-4} eV/K) (Pankove, 1971).

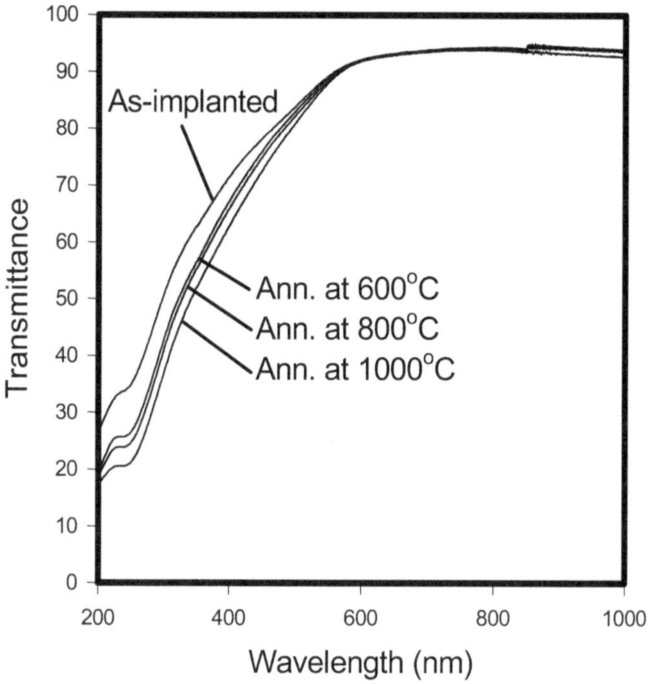

Fig. 3. Transmission spectra of the samples with a dose of 1×10^{17} ions/cm^2. (Ueda et al., 2007)

Transmission electron micrographs have been taken for the samples with a dose of 1×10^{17} ions/cm^2. Fig.5 (a), (b), and (c) show nanocrystal formation in samples annealed at 600°C, 800°C, and 1000°C, respectively. The micrographs indicate that the size of the Se nanoparticles at the peak of Se concentration distribution is about 10 nm for every sample. However, there exist smaller sized particles (≤ 3 nm) in the samples annealed at lower temperatures, which indicates that annealing of samples at higher temperature causes the increase of particle sizes, while smaller particles played a role as a material source (Ostwald Ripening).

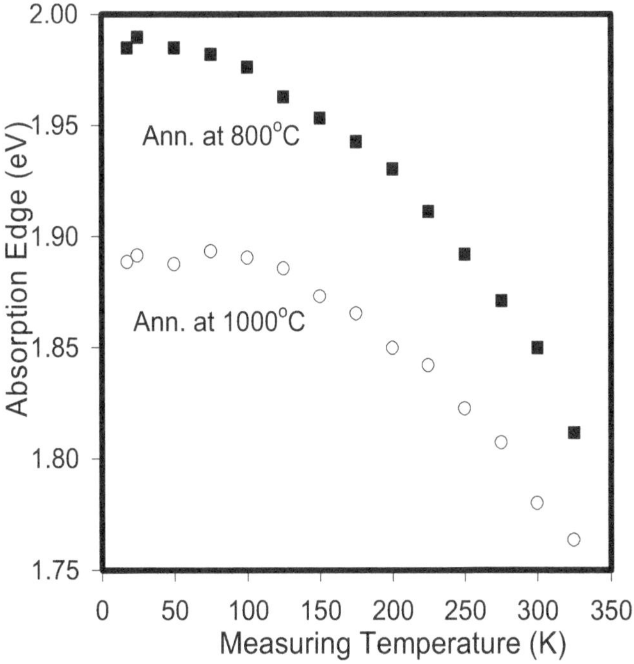

Fig. 4. Temperature dependence of absorption edge measured in the range of 15-325 K. (Ueda et al., 2007)

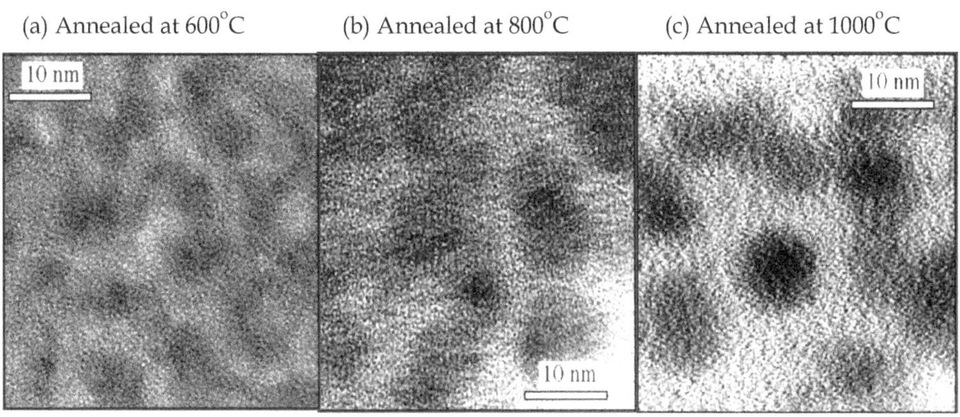

Fig. 5. High-resolution TEM micrographs of Se nanocrystals in SiO_2 annealed at (a)600°C, (b)800°C, and (c)1000°C. (Ueda et al., 2007)

2.4 Discussions

Considerable work has been published on the optical absorption in amorphous semiconductors, however its interpretation is not straightforward. As is well known, amorphous semiconductors do not have a sharp band gap unlike that for crystalline semiconductors. In amorphous materials, there exist some states smeared into the band gap of corresponding crystalline semiconductors. Most of the experimental data on amorphous semiconductors in the strong absorption region were found to be in agreement with the so-called Tauc law (Nagels et al., 1995; Tauc, 1972):

$$\alpha hv = C\left(hv - E_g\right)^2 \tag{1}$$

where hv is the photon energy, E_g. is the optical band gap, C is a constant. For the thin Se films, Nagels *et al.* (Nagels et al., 1995) used the transmission spectrum of the sample together with the reflectance spectrum for a single surface of the sample to determine the band edge. For our samples, Se is an implanted layer in the glass and not a film, however, we utilized Nagels' formalism as the first approximation for band edge determination. By inverting the following relationship among the transmittance, the reflectance, and the absorption coefficient, we can obtain the absorption coefficient α:

$$T = \frac{(1-R)^2 exp(-\alpha d)}{1 - R^2 exp(-2\alpha d)} \tag{2}$$

where T is the transmittance, R is the reflectance for a single surface, and d is the sample thickness. Since T and R are functions of wavelength, α is also a function of wavelength. Inserting α into the Tauc equation and taking the square root, we plotted $(\alpha dhv)^{1/2}$ versus hv. A linear fitting was carried out for the linear segment corresponding to the absorption edge. Then the intersection of the extrapolated line with the photon energy axis (α = 0) gives us the optical band gap E_g.

Fig. 6(a) shows the Tauc plots of the spectra of Fig. 2(a) for the as-implanted samples. The samples with doses of 6 and 10×10^{16} ions/cm^2 have relatively long linear segments, allowing for extrapolation of the linear part. We obtained E_g = 2.29 eV and 2.12 eV for 6 and 10×10^{16} ions/cm^2, respectively. For the 3×10^{16} ions/cm^2 sample, the linear segment is shorter, leading to more uncertainty in the band edge, however, an approximate value of 3.11 eV was obtained. For 1×10^{16} ions/cm^2, it is difficult to find the linear part probably because the implanted ion dose is not sufficient to form nanoparticles yet. Hence, it does not follow the Tauc Law. In this sample, the implanted ions are expected to be in the form of isolated ions or small clusters. Fig. 6(b), (c), and (d) show the Tauc plots of the spectra for the samples annealed at 600, 800, and 1000°C, respectively. In Fig. 6(d), every spectrum has a linear part corresponding to the absorption edge, indicating that these spectra follow Tauc's law. The extrapolated lines cross the photon energy axis (α = 0) around one point: For 1, 3, 6, and 10 x 10^{16} ions/cm^2, the values of E_g are 1.83, 1.70, 1.77, and 1.75 eV, respectively. In Figs. 6(b) and (c), it is apparent that the onset of absorption edge shifts to larger values for the lower ion doses (1, 3, 6, and 10 x 10^{16} ions/cm^2).

According to the Tauc analysis for the samples of 10×10^{16} ions/cm^2, the absorption edges are 1.75, 1.81, 1.82, and 2.12 eV for the samples annealed at 1000°C, 800°C, 600°C, and as-implanted, respectively. Although these absorption edges are similar to each other, the sample annealed at higher temperature has a lower energy gap. These results indicate that for the highest dose samples (1x10[17] ions/cm[2]) the sizes of nanoparticles become sufficiently large to be greater than the exciton Bohr radius of bulk Se even during the ion implantation. Since the melting and boiling points of Se are relatively low (T_m = 217°C and T_b = 685°C) and ion implantation can cause a temperature increase in the substrate, this temperature increase may lead to the formation of large nanoparticles.

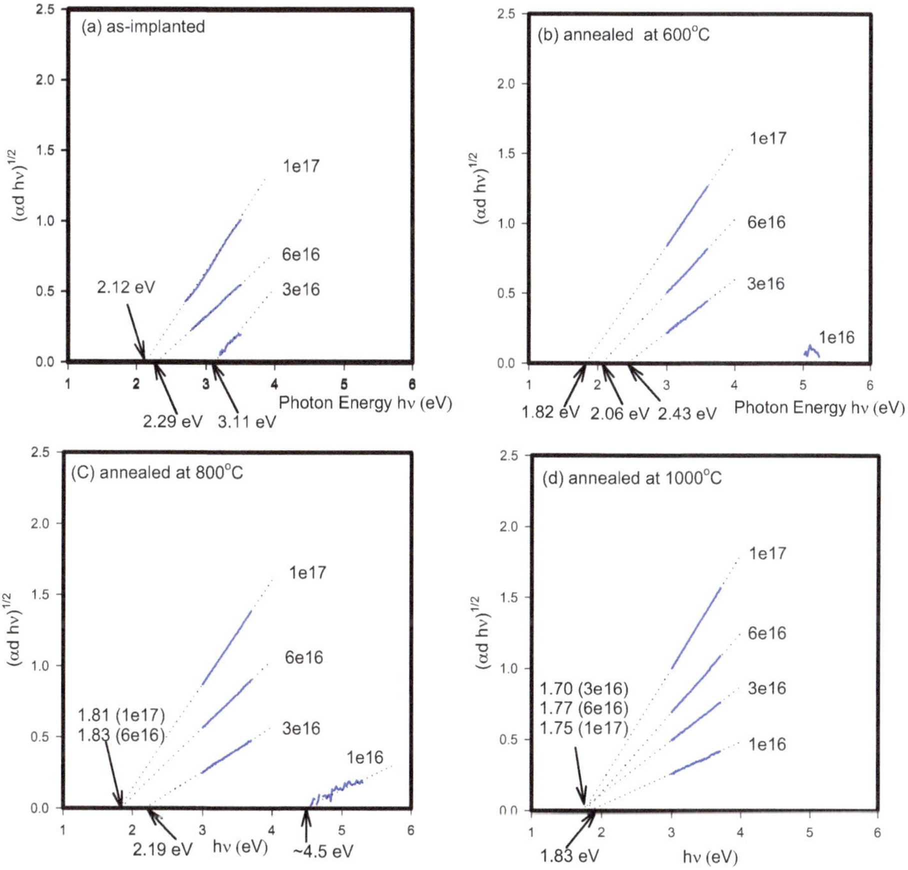

Fig. 6. Absorption edges of Se nanoparticles in silica by Tauc Analysis. (a) as-implanted, (b) annealed at 600°C, (c) annealed at 800°C, and (d) annealed at 1000°C. (Ueda et al., 2007)

Such colloid formation during ion implantation has been reported previously for Ga implanted in glass (Hole et al., 1995). For the lower dose samples (1 and 3 x 10[16] ions/cm[2]), however, according to the results shown in Figs.6(a)-(d), the temperature rise during the implantation is not sufficient to form the final "ripened" size of nanoparticles that was obtained by annealing at 1000°C.

The summary of the annealing temperature dependence on the optical absorption edges for each dose is given in Fig.7. There is no data point of as-implanted for the dose of 1 x 10[16] ions/cm^2. As we can see the trend in Fig.7, we may control the size of Se particles by annealing low dose samples (1 x 10^{16} ions/cm^2) at relatively low temperature between 200°C and 800°C for a certain time.

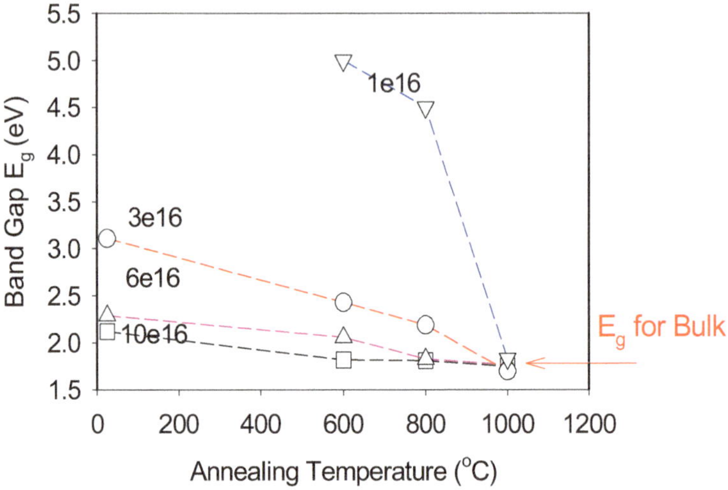

Fig. 7. The summary of Tauc Analysis; the optical absorption edges are given as functions of annealing temperature with annealing time of 1 hour. (Ueda et al., 2007)

The TEM observation is consistent with the results of the absorption edges, because the absorption edge for all samples annealed at 1000°C is similar and the size of the particles should reflect the absorption edge, as mentioned before (Brus ,1982). However, according to the estimation by Huber (Huber & Huber, 1988), who impregnated porous Vycor Glass with Se to investigate the quantum confinement effects of Se, the exciton Bohr radius in bulk Se is 6.8 Å and the confinement energy is 0.011 eV for the cylindrical confinement of 50 Å in diameter. For 10 nm sized nanoparticles, the exciton radius is too small to experience quantum confinement and the energy shift corresponds to ~0.003 eV according to Brus' theory (Brus,1982), which can be neglected for absorption edge. The value of the Bohr radius, however, may not be reliable because the values of electron effective mass for Se have not been established yet in the literature. If the exciton Bohr radius is relatively larger than 6.8 Å, the smaller nanoparticles can contribute to shifting the absorption edge.

For the lower dose samples, however, the annealing treatment strongly affects the absorption edges because the nanoparticle formation requires more time or a higher temperature to reach the final size distribution. Since the annealing time was fixed at 1 h, the annealing temperature mainly controlled the sizes of the particles in this study.

2.5 Conclusions for this section

We have fabricated Se nanoparticles in silica substrates by ion implantation followed by thermal annealing. The Se nanoparticles are found to be amorphous. The sample with the highest dose (1x10^{17} ions/cm^2) showed the naonparticle formation even during ion implantation, while the lower dose samples (1 and 3x10^{16} ions/cm^2) did not. After thermal annealing of the samples, the Se particle sizes become larger than the exciton Bohr radius for bulk Se. Thus, the absorption edges for different doses approached a common bulk value after thermal annealing. The TEM measurements for lower dose samples annealed at several temperatures will clarify the evolution of the nanoparticles. The temperature dependent spectra were also measured in a temperature range from 15 to 325 K. The value of $[\Delta E_g/\Delta T]_P$ is on the same order as the one for bulk. For the lower dose samples, the annealing treatment strongly affects the absorption edges. Since the annealing time was fixed at 1 h, the annealing temperature mainly controlled the sizes of the particles in this study. As we can see the trend in Fig.7, we may control the size of Se particles by annealing low dose samples (1 x 10^{16} ions/cm^2) at relatively low temperatures between 200°C and 800°C for a certain time.

3. Annealing effects on the surface plasmon of MgO implanted with gold

Gold ion implantation was carried out with the energy of 1.1 MeV into (100) oriented MgO single crystal. Implanted doses were 1, 3, 6, and 10 x 10^{16} ions/cm^2. The gold ion irradiation resulted in the formation of gold ion implanted layer with a thickness of 0.4 μm and defect formation. In order to form gold colloids from the as-implanted samples, we annealed the gold implanted MgO samples in three different kinds of atmospheres: (1)Ar only, (2)H_2 and Ar, and (3)O_2 and Ar. The annealing over 1200°C enhanced the gold colloid formation which shows surface plasmon resonance band of gold. The surface plasmon bands of samples annealed in three kinds of atmospheres were found to be at 535 nm (Ar only), 524 nm (H_2+Ar), and 560 nm (O_2+Ar). The band positions of surface plasmon can be reversibly changed by an additional annealing. Fe^{3+} impurities in MgO samples are utilized to monitor the degree of reduction of the samples due to annealing. From the absorption measurements of Fe^{3+} band, the diffusion constants of oxygen vacancies (F-centers) was obtained. We found that the shifts of the surface plasmon bands are correlated to the oxygen vacancy concentration and that F_n-center (aggregates of F-centers) is responsible for the shifting. According to the experimental results, we propose a model to explain the shift of surface plasmon band.

3.1 Overview of the surface plasmon resonance for small metal particles

In recent years, optical nonlinearity of small metal particles in dielectric materials has been paid much attention because of the possibility of the fast nonlinear response. The decreasing of materials to nanometer size often leads to significant changes in optical, structural, and thermodynamic properties. The appearance of the surface plasmon (SP) resonance for metal colloid reduced to the nanometer scale is one example of how the optical properties of metals change when they are reduced in size. To form nanometer sized metal particles, we used ion implantation methods (White, 1989). Metal colloid doped glasses, for example, show high optical nonlinearity, and are attractive candidates for utilization in optical devices (Fukumi, 1994; Haglund, 1992). The enhancement of the third order optical susceptibility depends on the intensity and frequency of the SP resonance of the metal colloids. According to Mie's scattering theory and the effective medium theory (Perenboom, 1981), we can predict the SP resonance frequency ω_{SP} if we know the dielectric functions of the metal colloid $\varepsilon(\omega)$ and the host material ε_m. As a first-order approximation, the following equation gives us the SP resonance frequency ω_{SP} (Perenboom, 1981; Kreibig & Vollmer, 1995):

$$\varepsilon(\omega_{SP}) + 2\varepsilon_m = 0 \quad (3)$$

We have previously reported on the SP resonance of gold nanoparticles in several substrates: Al_2O_3 (sapphire), CaF_2, and Muscovite mica (Henderson et al., 1995, 1996). MgO is also a material widely used in optics, such as windows and coatings. Although many of the alkaline earth oxides are hygroscopic and thus not widely used in optical systems, MgO is relatively insoluble, hard, and durable. Moreover, MgO is a good transparent material in the range between 300 nm and 7000 nm, and the optical absorption edge has been measured to be about 7.8 eV (160 nm) (Briggs, 1975). Transition metals, for example Fe^{3+}, are common multivalent impurities in MgO with order of 50 ppm. Absorption bands due to Fe^{3+} are found at 210 and 285 nm in MgO (Briggs, 1975). It has reported that the bands around 250 nm are related to F-center (248 nm) and F^+-center (252 nm), whose absorption bands overlap each other at room temperature (RT). F-center is defined as an oxygen vacancy with two trapped electrons, while F^+-center is defined as an oxygen vacancy with one electron. They can be created, for example, by neutron- and electron-irradiation or Mg-vapor-deposition at high temperature, which is so called additively colored crystal (Chen et al., 1969a, 1969b, 1967). The bands caused by neutron- and electron-irradiation can be annealed out at 600°C, while these bands generated by Mg-vapor-deposition could not be annealed out until the annealing temperature reaches the region of 1000-1200°C. The reason for this is as follows: in the sample, excessive Mg^{2+} ions are introduced from Mg vapor at 1800°C and then the Mg ions migrate into the MgO bulk to cause oxygen ion vacancies because of excessive Mg ions. In this case there exist no interstitial oxygen ions in the system, while neutron- or electron-irradiated samples have interstitial oxygen ions to re-combine to oxygen vacancies. The oxygen interstitials are mobile even below 600°C, while oxygen ion vacancies are immobile below 900°C (Chen et al., 1969a, 1969b, 1967). In this paper, we mainly report on how the annealing atmosphere affects to the SP resonance.

3.2 Experimental methods

MgO single crystal (15x15x0.5 mm^3) oriented (100) were obtained from two venders (Harrick Scientific and Commercial Crystal Laboratories). The major impurities are Fe^{3+} of ~100 ppm and Al^{3+} of ~40 ppm. The samples are implanted with 1.1 MeV gold $^{197}Au^+$ with doses of 1, 3, 6, and $10x10^{16}$ Au ions/cm^2 at RT. The depth profile of gold concentration was measured by Rutherford Backscattering Spectrometry (RBS), using 2.4 MeV α particles. Thermal annealing was carried out with a tube furnace at temperatures between RT and 1300°C in a reducing (10%H_2+90%Ar), an oxidizing (10%O_2+90%Ar), and an inert atmosphere (100%Ar) flow. The transmission spectra were measured between 185-3200 nm with a UV-Vis-NIR double beam spectrophotometer (Hitachi, U-3501).

3.3 Experimental results

3.3.1 Rutherford ackscattering Spectrometry (RBS) and TEM image

The RBS spectra show that the depth profile of gold is similar to the Gaussian distribution, whose peak concentration is located about 0.4 μm from the surface (The spectra are not shown here). The distribution is slightly asymmetric, which has a longer tail into the sample than surface side. The dominant gold ions are located within 1.0 μm region from the surface of (100) MgO with thickness of 0.5 mm. According to the results of RBS with the [100]-

channeling direction and the tilted angle measurements, MgO surface has not been amorphized.

Fig. 8 shows a TEM image of gold nanocrystals fabricated in MgO single crystal by ion implantation with a post annealing. The shape of the gold nanocrystals is cubic and they align along the crystal axis of host MgO [100]. This shape indicates that implanted ions interact with host materials during particle growth.

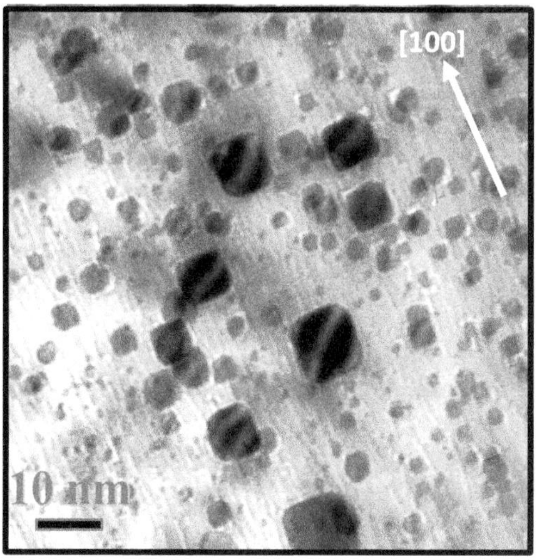

Fig. 8. TEM micrograph of MgO implanted with gold ions. (Cited from our paper, Bryant, E. M. et al., 2001)

3.3.2 Transmission of as-implanted samples

The transmission spectra of as-implanted Au/MgO samples and an unimplanted MgO sample (called "virgin MgO") are shown in Fig.9. The virgin sample (a) has high transmittance region (85% transmittance, 15% loss is mainly due to the reflection at both surfaces) down to 350 nm, and below 320 nm the transmittance suddenly drops due to Fe^{3+} ion impurities whose absorption bands are located at 285 nm and 210 nm. The spectra (b), (c), (d), and (e) are the transmission spectra for the gold implanted samples with doses of 1, 3, 6, and 10 x10^{16} ions/cm^2, respectively. Every spectrum from (b) to (e) has a broad absorption band at 576 nm, which can be attributed to F_n-center (aggregates of F-centers) due to the irradiation damage by ion implantation (Weber, 1986). As the gold dose increases, the transmittance decreases in the range between 300 nm and 1000 nm, mainly due to scattering from small gold particles. The color of the as-implanted samples is dark brownish. Below 300 nm, the Fe^{3+} band is overlapped with the F- and F^+- center bands (at 248 nm and 252 nm).

3.3.3 Transmission of annealed samples

Fig.10 shows two sets of spectra of Au/MgO samples with the dosage of 6x10^{16} ions/cm^2 for sequentially annealed in both (a) reducing atmosphere [RA] and (b) oxidizing atmosphere

[OA]. The annealing temperature and time for each spectrum are shown in Fig.10. The band at 576 nm disappeared after annealing at 500°C in both atmospheres. It was reported by Chen et al. that the band at 576 nm was completely annealed out at 600°C. (Chen et al., 1969a, 1969b, 1967). Between 200°C and 1000°C, weak broad bands appeared around 340 and 365 nm. When the annealing temperature reached 1200°C, a strong broad band

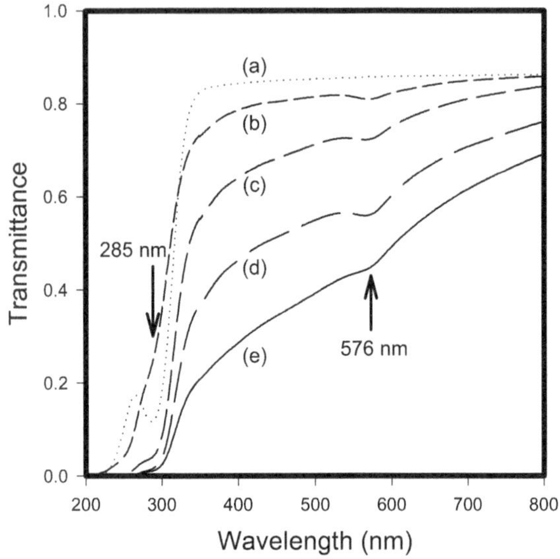

Fig. 9. Transmission spectra of (a) virgin MgO and as-implanted MgO with 1.1 MeV Au of (b) 1x10^{16}, (c) 3x10^{16}, (d) 6x10^{16}, and (e) 1x10^{17} ions/cm^2. (Ueda et al., 1998)

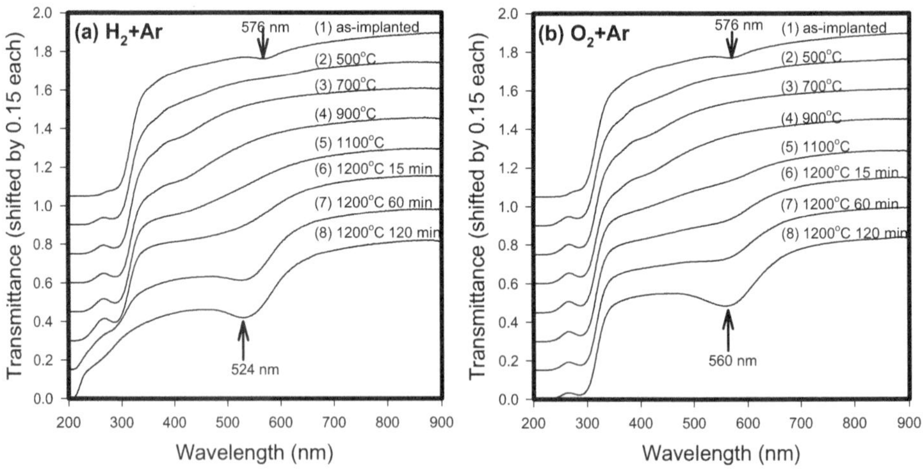

Fig. 10. Temperature development of transmission spectra of Au implanted in MgO annealed in (a)H_2+Ar and (b)O_2+Ar. In order to see easily, each spectrum has been vertically offset. (Ueda et al., 1998)

appeared at 524 nm for the samples annealed in the reducing atmosphere and at 560 nm for the oxidizing atmosphere, which are attributed to the SP resonance of gold colloids. As the annealing time increased at 1200°C, the SP bands became stronger and stronger, which indicates that at this temperature (1100 -1200°C) the implanted gold ions begin to diffuse and aggregate to form metallic colloids. In the case of the annealing in the reducing atmosphere, as shown in Fig.10(a), the Fe^{3+} band at 285 nm became weaker at high temperatures. It is coincident that oxygen vacancies become mobile at this temperature, as mentioned before (Chen et al., 1969a, 1969b, 1967).

3.3.4 Measurement of Fe^{3+} band as a monitor for creation and annihilation of F-centers

In order to investigate the correlation between the SP resonance and the reduction of the samples, we measured the transmission spectra of Fe^{3+} band at 285 nm of a virgin MgO sample that was sequentially annealed in the similar way to the previous annealing of gold implanted samples from RT to 1200°C. The Fig.11(a) shows the spectra of virgin MgO annealed in reducing atmosphere: the spectrum (a) is for the virgin MgO before annealing. Spectra (b), (c), and (d) are of the same sample after sequential annealing at 1000°C, 1100°C, and 1200°C, respectively, for 15 min each. Spectra (e) to (i) are of the same sample additionally annealed at 1200°C for 15 min for each step. During annealing between 200°C and 900°C with a step of 100°C for 15 minutes each, the spectra did not exhibit any significant changes (they are not shown in the figure.) After annealing at 1200°C for 90 minutes in total, (i) of Fig.11 (a), the surface of the sample became slightly cloudy. However, before it became cloudy, the Fe^{3+} band at 285 nm became weaker and weaker. Apparently, the Fe^{3+} ions were reduced to Fe^{2+} ions in the reducing atmosphere at high temperature, which resulted in an increased transmittance to 75% at 285 nm (Briggs, 1975). After annealing in the reducing atmosphere, we switched the atmosphere to O_2+Ar. The transmittance suddenly dropped just after annealing at 1200°C in the oxidizing atmosphere for 15 minutes, spectrum (j), as shown in Fig.11(b). We assume that Fe^{3+}ions uniformly distributed in the entire sample before annealing and that the reducing reaction occurs at the surface and proceeds to inside the sample until Fe^{3+} ions are completely converted to Fe^{2+} ions in reducing atmosphere. When annealing the samples in the oxidizing atmosphere, the conversion from Fe^{2+} to Fe^{3+} has taken place, as shown in Fig.11(b). However, this does not mean that both hydrogen and oxygen atoms diffuse into the sample. According to Griggs' experiments, the annealing MgO samples in several atmospheres showed that the gas used for annealing did not diffuse into the sample using deuterium gas (Briggs, 1975). In the reducing atmosphere at high temperature, a hydrogen molecule takes an oxygen atom at the surface with forming a water molecule, and leaves an oxygen vacancy with two electrons (F-center) behind, with giving one electron to an Fe^{3+} ion to form an Fe^{2+} ion if there exist Fe^{3+} ions nearby (Briggs, 1975; Yager & Kingery, 1981a, 1981b; Ueda et al., 2001a). In the case where Fe^{3+} ions do not exist near the surface, oxygen vacancies with electrons (F- or F^+-center) diffuse into the sample. (The diffusion of oxygen vacancies into the sample means that oxygen ions diffuse out from the sample.) At lower temperature this reaction occurs only at the surface with low rate and will not continue to react because oxygen vacancies are not mobile at low temperature (<1000°C). This is consistent with our observation that the spectra Fe^{3+} absorption decreases only at high temperature (≥1200°C). In the oxidizing atmosphere, oxygen vacancies are filled with oxygen atoms, and an oxygen takes an electron from Fe^{2+} to convert it to Fe^{3+}. Oxygen ions diffuse into the sample when the temperature is high enough to make oxygen vacancies mobile.

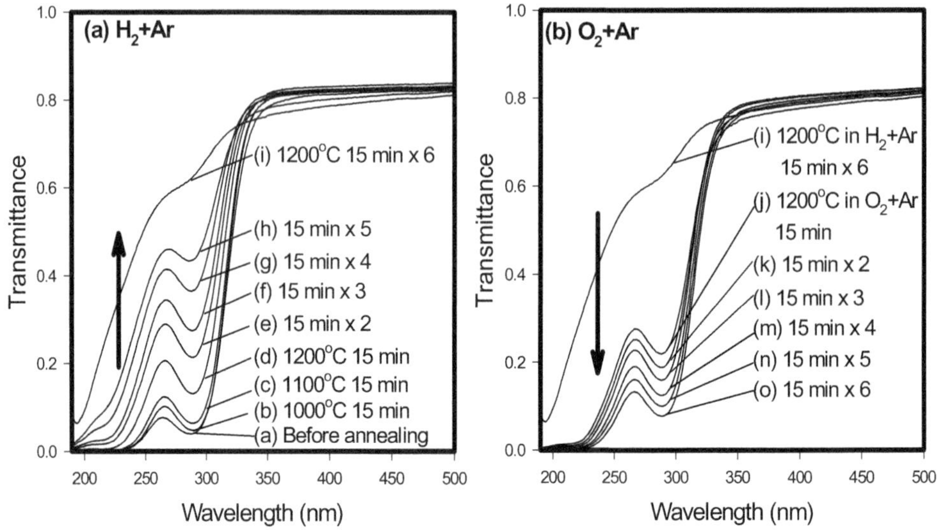

Fig. 11. Transmission spectra of virgin MgO annealed in (a) H_2+Ar and (b) O_2+Ar in the range of the Fe^{3+} absorption band. Cited from (Ueda et al., 2001)

3.3.5 Annealing temperature dependence of Fe^{3+} bands

A single virgin MgO crystal was annealed at 1200°C in O_2+Ar atmosphere for 5 hours to prepare a starting sample (the sample state O). The sample was then annealed in the reducing atmosphere for 5 h and transmission spectra were obtained at each temperature from 400°C to 1300°C with a step of 100°C to evaluate the degree of reduction as a function of annealing temperature. The sample annealed at 1300°C in the reducing atmosphere for 5 h is called as the sample state R. The sample (state R) was subsequently annealed in an oxidizing atmosphere for 5 h at each temperature from 400°C to 1300°C. This measurement is similar to the results shown in Fig.10, however, the annealing time and temperatures are different.

The area under Fe^{3+} absorption band at 285 nm was integrated for each temperature from RT (25°C) to 1300°C. In Fig.12, the annealing temperature dependence of the Fe^{3+} band is plotted for the samples annealed in both reducing and oxidizing atmospheres. Trace (1) shows the integrated band intensity as a function of annealing temperature under the reducing atmosphere, started with the sample state O. The integrated intensity of the Fe^{3+} band did not change until the annealing temperature reached 900°C. At this temperature, Fe^{3+} started to be reduced to Fe^{2+}. The reduction of Fe^{3+} then indicates the creation of oxygen vacancies, as will be presented. The band completely disappeared at 1300°C. This sample (sample state R) was used as the starting sample for annealing in the oxidizing atmosphere. Trace (2) shows that the Fe^{3+} band did not appear until the annealing temperature reached 900°C, and at 1000°C the sample started to be oxidized. The Fe^{3+} absorption band completely recovered at 1300°C.

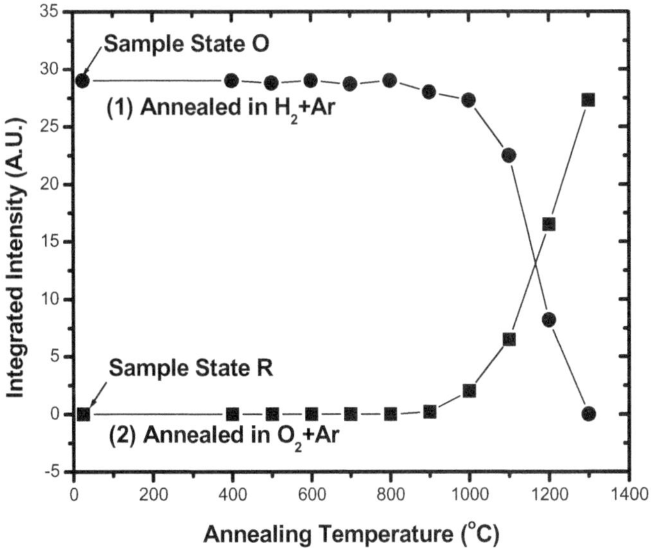

Fig. 12. Annealing temperature dependence of Fe^{3+} band in (1)H_2+Ar and (2)O_2+Ar. The starting sample for (1) was the oxidized state O, and that for (2) was the reduced state R. (Ueda et al., 2001)

3.3.6 Correlation between Fe^{3+}-band and F_n-band at 575 nm

After annealing the sample (state O) in the reducing atmosphere, an absorption band developed at 575 nm. Fig.13 shows the temperature dependence of the Fe^{3+} and 575 nm-bands. The intensity of the Fe^{3+} decreased as the annealing temperature increased, while the absorption at 575 nm increased. The band at 575 nm has been observed previously for MgO irradiated with neutrons and ions (Chen, 1969; B. Henderson & Bowen, 1971; Evans et al., 1972), and was attributed to aggregates of F-centers. Therefore, the correlation between the reduction of Fe^{3+} and the development of 575 nm-band suggests that the reduction of Fe^{3+} coincides with the creation of F-centers. Fig.13(b) shows temperature dependence of 575 nm-band.

3.3.7 Diffusion of oxygen vacancies

From the similar transmission data of virgin MgO annealed in reducing atmosphere for 5 hours at each temperature, we can estimate the diffusion behavior (Ueda et al., 2001a). The reason we took a longer annealing time of 5 hours is to establish the thermodynamic equilibrium at each temperature. The transmission spectra were converted to absorbance, and the absorption bands around 285 nm are integrated for each temperature from 800°C to 1200°C. Since below 700°C we did not see any significant changes, these data below 700°C are not useful to estimate the diffusion constants. The ratio (A/A_0) of the integrated absorbance at each temperature to that at RT can be related to the amount of the Fe^{3+} in the sample at each temperature. According to Fick's law, the diffusion constant is given by

$$\frac{d}{2}\left(1-\frac{A}{A_0}\right) = (D_0 t)^{1/2} exp\left(-\frac{E_a}{2kT}\right) \quad (4)$$

Here, d is the sample thickness (0.5 mm), and the annealing time t should be taken 5 hours (18 000 sec). Taking the logarithm of both sides and plotting these quantities versus 1/2T, we can obtain the values of E_a and D_0, as shown in Fig.14(a). A linear fitting of the plot gave values of E_a=3.08 $\pm$0.18eV and D_0 =3.63 x10^{10} μm^2 s^{-1}. The value of the activation energy is between the values for F$^+$- and F-centers (2.72 eV and 3.13 eV, respectively) that Kotomin *et al.* theoretically calculated (Kotomin et al., 1996). From these experimental values obtained, we re-plot the diffusion length as a function of annealing temperature in Fig.14(b). See ref. (Ueda et al., 2001a) for the detailed explanation.

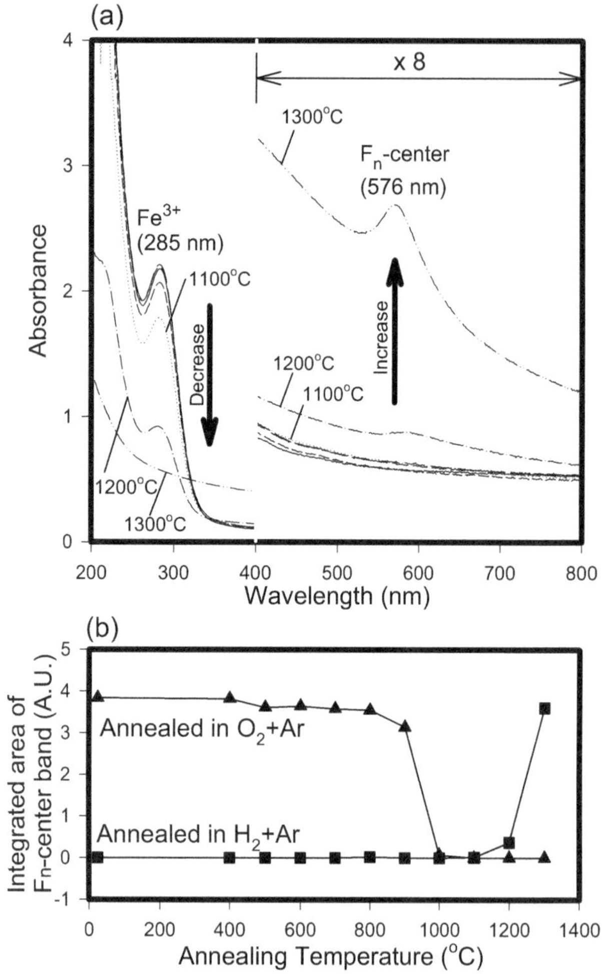

Fig. 13. (a) The temperature dependence of Fe^{3+} and 575 nm-bands during annealing in the reducing atmosphere. Fe^{3+} band decreased as annealing temperature increased, while 575 nm-band developed. (The vertical scale was magnified in the range of 400-800 nm by a factor of 8.) (b) Integrated area of 575 nm-band as a function of annealing temperature for both cases of annealing in O_2+Ar and H_2+Ar. Cited from (Ueda et al., 2001)

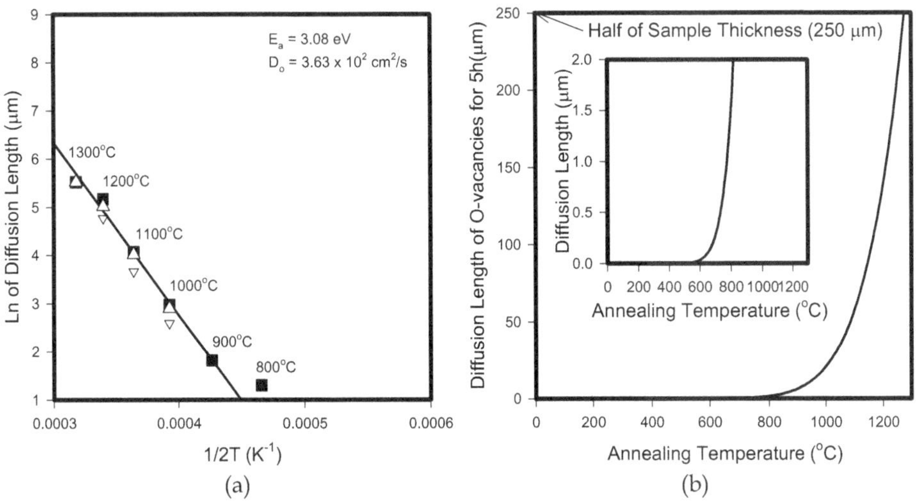

Fig. 14. (a) Linear fitting for the estimation of diffusion constants. (b) Diffusion length of oxygen vacancies as a function of annealing temperature for 5 h, using the values of the diffusion coefficient and the activation energy obtained from this study.(Ueda et al., 2001)

According to this plot, the diffusion of oxygen vacancies started around 900°C. (In the inset, the vertical scale is little magnified to see the small change.) This is consistent with the data which Chen *et al.* found that the oxygen vacancies in MgO become mobile around 1000°C (Chen et al., 1969a, 1969b, 1967). Note that this reduction process of MgO is essentially reversible, according to the data of Fe^{3+} transmission measurements, as shown later.

3.3.8 Annealing atmosphere dependence of surface plasmon band of Au in MgO

The Au implanted MgO samples showed SP bands after annealing at high temperature ($T \geq$ 1200°C), as mentioned previously. The SP band positions depend on the annealing atmospheres. To confirm the annealing atmosphere dependence of the SP band position, we sequentially annealed an implanted sample (Au/MgO with dose of $10x10^{16}$ inos/cm^2) at 1200°C in (1)Ar only, (2)O_2+Ar, (3)H_2+Ar, and (4)O_2+Ar again for 2 hours each. (1)After annealing in Ar only, SP appeared at 535 nm. (2)After annealing in O_2+Ar, SP shifted to 560 nm. (3)Annealing in H_2+Ar made SP shift to 524 nm. (4) Again, after annealing in O_2+Ar, SP moved back to 560 nm. From these results, the SP position is reversibly depending on the annealing atmosphere (Ueda et al., 1997).

To investigate this annealing atmosphere dependence of SP position, we annealed one piece of MgO implanted with gold ($1x10^{17}$ ions/cm^2) with a piece of virgin MgO as a reference at 1200°C in O_2+Ar atmosphere for 5 hours as a starting sample , whose SP position was 560 nm. These samples were annealed in the reducing atmosphere for 5 hours at each temperature from 400°C to 1300°C to find the SP position as a function of annealing temperature. In Fig.15(a), the trace (1) presents the SP position change under the reducing condition: Below 700°C, the SP position stayed at 560 nm, and SP position started to gradually change to the shorter wavelength from 800°C up to 1300°C. The final SP position was 527 nm, which was slightly longer wavelength than that stated in previous paragraph

(Ueda et al., 1997). Then this sample was annealed in the oxidizing atmosphere for 5 hours each at certain temperature from 400°C to 1300°C. In Fig.15(a), the trace (2) presents the SP position change: Below 700°C, the SP position stayed at 527 nm, and SP position started to suddenly change to the longer wavelength (~560 nm) between 800°C and 900°C. The SP position essentially stayed at about 560 nm above 900°C up to 1300°C. The final SP position was 563 nm, which is also slightly longer wavelength than that mentioned previously (Ueda et al., 1997). This sample was then annealed in the inert atmosphere (Ar only) in the same way as previous annealing. In the Fig.15(b), the trace (3) shows that the SP position essentially stayed around 560 nm, although there was a small dip at 1100°C. This sample was then first annealed at 1200°C in the reducing atmosphere for 5 hours to obtain the starting sample for annealing in the inert atmosphere, whose SP position was 523 nm as a starting SP position. In the Fig.15(b), the trace (4) shows that the SP position changed similarly to the trace (2) for annealing in O_2+Ar. From this similarity, the annealing in O_2+Ar and Ar only atmosphere are essentially thermal processes, which is the diffusion of oxygen vacancies from inside to the surfaces. On the other hand, the annealing in H_2+Ar atmosphere involves chemical reactions at the surface as well as thermal processes together. The chemical reaction mainly determines the rate of SP position change, which gradually changed from 800°C to 1300°C.

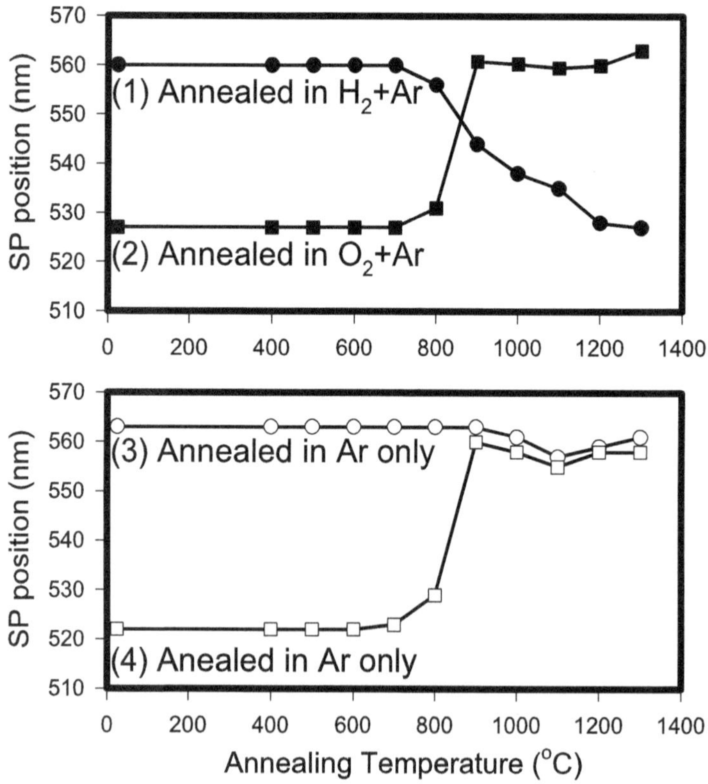

Fig. 15. Surface Plasmon (SP) position as a function of annealing temperature in different atmospheres. (Ueda et al., 1998)

3.3.9 Annealing temperature dependence of F_n-center and Fe^{3+}-bands

The temperature between 800°C and 900°C is a transition temperature for SP position shifting. This temperature, T_{OV}, is coincidently the temperature at which oxygen vacancies become mobile. Therefore, it is natural to mention that oxygen vacancies play an important role in the SP position shifting. A piece of virgin MgO was annealed with the previous annealing of the MgO implanted with Au ions for 5 hours at each temperature in order to see the roles of oxygen vacancies without the effects caused by the ion implantation. This is similar to the results shown in Fig.12, however, the annealing conditions are slightly

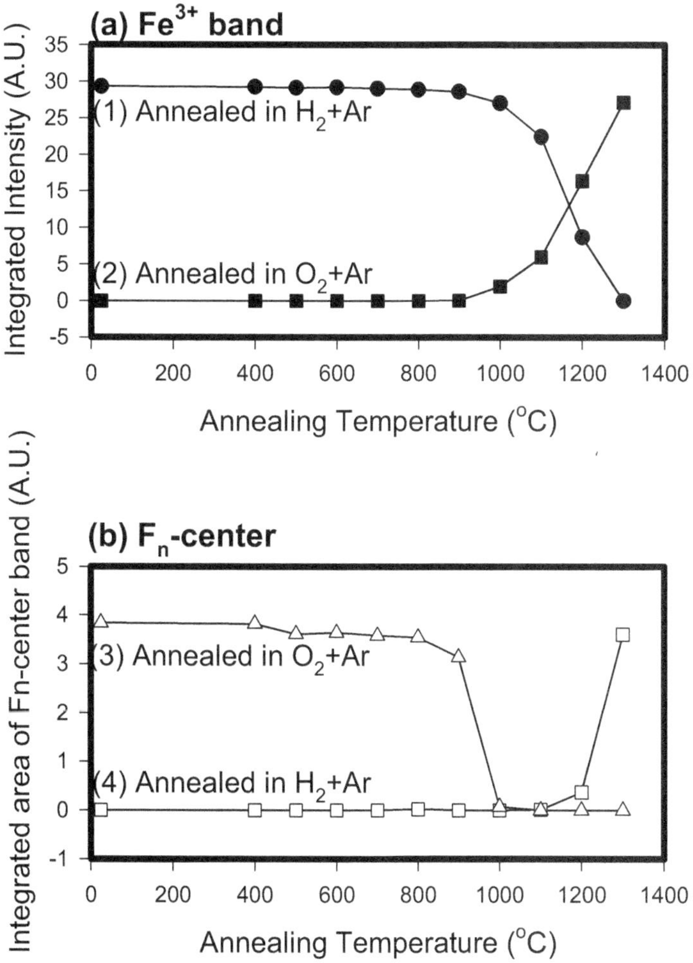

Fig. 16. Integrated intensities as a function of annealing temperature for Fe^{3+} and F_n-center bands. (Ueda et al., 1998)

different in the annealing time and temperatures. The transmission spectra were converted to absorbance, and the Fe^{3+} absorption band (285 nm) and F_n-center band (576 nm) are integrated for each temperature from RT (25°C) to 1300°C. In Fig.16(a), the annealing temperature dependence of Fe^{3+} band is plotted in both reducing and oxidizing atmosphere. The trace (1) shows the integrated band intensity as a function of annealing temperature under the reducing atmosphere, started with the sample annealed at 1300°C in the oxidizing atmosphere for 5 hours. The Fe^{3+} band did not change until the annealing temperature reached 900°C, and then at 1000°C the sample started to be reduced, which is the onset of F-center creation. The band completely disappeared at 1300°C, as shown. This final sample was used as the starting sample for the following annealing in the oxidizing atmosphere. Trace (2) shows that the Fe^{3+} band did not appear until the annealing temperature reached 900°C, and then at 1000°C the sample started to be oxidized, which is the beginning of F-center annihilation. The band completely came back at 1300°C, as shown. Trace (2) shows that oxidizing behavior was symmetrically going on with trace (1) for the reducing atmosphere. In Fig.16(b), the annealing temperature dependence of F_n -center band of the same sample is plotted. The traces (3) and (4) show the integrated F_n -center band intensity as functions of the annealing temperature under the oxidizing and reducing atmospheres, respectively. F_n-centers were gradually formed until 1200°C in the reducing atmosphere, however F_n -center were abruptly decomposed at relatively lower temperature ($\sim T_{OV}$), as shown. These traces (3) and (4) do not symmetrically behave, while the Fe^{3+} band did. It is a meaningful observation that Fn-center formation take place after enough F-centers are created, with comparison between traces (b)-(4) and (a)-(1), and that there were still F-centers in the entire sample even after F_n-centers were decomposed (1000-1200°C) because F_n--centers are aggregates of F-centers that play as the sources of F-centers during the decomposition of F_n-centers, with comparison between traces (b)-(3) and (a)-(2). Note that there is a similarity between the SP position shift and the F_n -center changes as functions of annealing temperature: Compare Fig.16(b)-(3) with Fig.15(a)-(2). There is a sudden change at T_{OV} (800-900°C) in SP position, while a sudden change occurs at 900-1000°C in F_n-center intensity. This is probably because the implanted gold layer is located at very shallow range from surface. This gold layer is about 1000 times thinner than whole sample thickness, so that before the entire sample has the final F_n -center distribution the implanted gold layer near the surface has already some degree of F_n -centers were formed.

3.4 Discussion of mechanism for surface plasmon resonance position shifts

We have seen the SP position of gold colloids can be shifted by annealing in a certain atmosphere. The SP position shifts reversibly by changing the annealing atmosphere from reducing to oxidizing or vice versa. Although the surface of the MgO sample becomes cloudy during annealing in the reducing atmosphere, this shifting is repeatable several times. Does the size or shape of gold colloid explain the SP position shifting? It is difficult to imagine that the size can be changed by annealing in an atmosphere after a gold colloid with a certain size had been formed. Since the state of as-implanted stage is not a thermodynamically equilibrium state that is metastable, the colloid formation during annealing is an irreversible process. Thus once most of the metastable gold ions formed larger colloids, they will not become smaller colloids at the temperatures where gold is supposed to be liquid or solid phase and where the vapor pressure of gold is not considerably high. The shape of gold colloids may be changed by varying the surrounding environment of a colloid to lower the sum of surface energy for all facets of the colloid surface. However, since bulk gold forms fcc crystal structure, the shape of a colloid should be reflected in the cubic symmetry of MgO and the shape could be a truncated-octahedron or a sphere-like shape. Among these shapes, the SP position may not be so changed. Thus we need to seek the other explanations of the mechanism for the SP shifting. This reversibility suggests that there are two possible mechanisms to explain the SP position shifting according to the equation (1): (i) the dielectric function of host material (MgO) near gold colloids is influenced by the annealing. (ii) the dielectric function of gold colloids is changed by the annealing.

Using these values of ω_{SP} from experimental data, Eq.(3), and the optical constants of bulk gold from literature (Palik, 1991; Lide et al., 1998), we inversely calculated the local ε_m.. These calculated values ε_m are supposed to be 3.01 (Ar only), 2.45 (H_2+Ar), and 3.90 (O_2+Ar), while the standard value of ε_m for MgO is 3.02. However, the dielectric constants ε_m of these samples measured by ellipsometry gave the following values: 2.99 (Ar only), 2.98 (H_2+Ar), and 2.99 (O_2+Ar). The reason why the calculated values ε_m of the samples vary so much and the experimental values are unchanged is as follows: If $-2\varepsilon_m$ and $\varepsilon(\omega)$ were drawn against wavelength, the crossing point satisfies the SP resonance condition according to Eq.(3). Fig.17 schematically depicts two possible situations. Since the dielectric function of gold is relatively steep compared with that of MgO host, horizontal lines present the function $-2\varepsilon_m$ and the steep function presents $\varepsilon(\omega)$. In Fig.17(a), the dielectric function $\varepsilon(\omega)$ of gold is assumed to be unchanged after annealing, and ε_m is supposed to be responsible for the SP shifting. On the other hand, in Fig.17(b) the dielectric function ε_m of MgO is assumed to be unchanged after annealing and $\varepsilon(\omega)$ is supposed to be responsible for the SP shifting. The experimental data support the case of Fig.17(b), that is, ε_m does not change so much and the dielectric function of gold colloids changes after annealing. In Fig.17(c), the model plot is given for the case where gold dielectric function's change is responsible for the shift of SP. However, there are other possibilities such as the combination of these two cases. Although the dielectric function ε_m of MgO did not change for samples annealed under different atmospheres, according to the ellipsometric measurement mentioned above, those values for ε_m were the averaged value of wide area of the sample, and the SP shift might be affected by the environment near gold particles.

Recalling the similarity between the SP shifting and the F_n-center intensity change for the case of annealing in oxidizing atmosphere, we had better compare Fig.15.(2) and Fig.16(3). Both traces suddenly change in the range of 800-1000°C. We can see the correlation between these two quantities: Just above the temperature where F_n-center intensity started to decrease, the SP position abruptly shifted to longer wavelength. On the other hand, for the case of the reducing atmosphere, SP shift gradually proceeded from 800°C and it saturated at 1200-1300°C, compared Fig.15.(1) with Fig.16(4). However, F_n-center apparently started to increase at 1200-1300°C. It is reasonable to imagine that F-centers and F_n-centers were formed near the surface due to the reduction of the MgO surface. Around the critical temperature T_{OV}, F-centers diffuse into the sample, while F_n-centers do not diffuse because F_n-center are too large to diffuse. Thus, the F-centers *evaporated* from F_n-centers near the

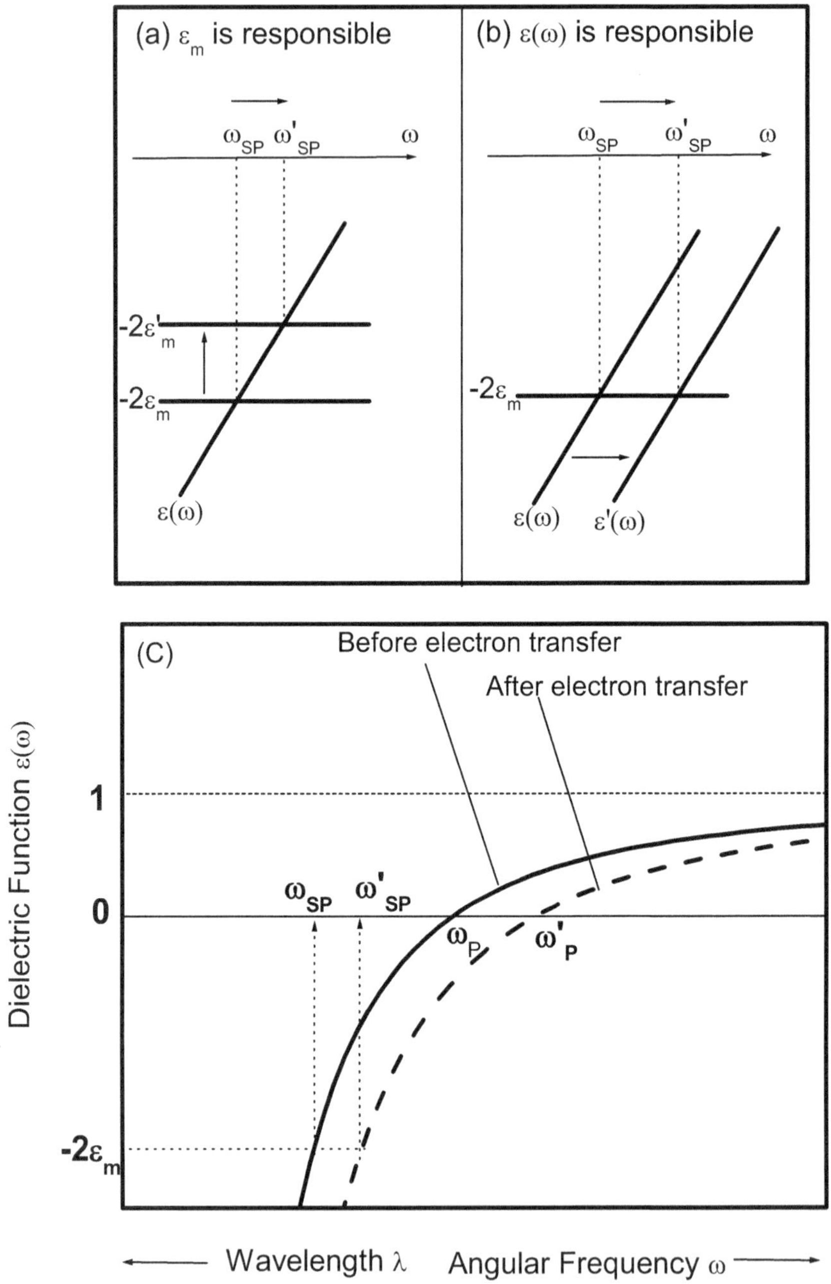

Fig. 17. Schematics for SP band shift situation. (a) the case where ε_m is responsible, (b) the case where $\varepsilon(\omega)$ is responsible, and (c) the model plot for case where $\varepsilon(\omega)$ is responsible for the shift of SP band.

surface can diffuse in, and the F-centers form their aggregates only when the concentration of F-centers reaches a threshold. However, until the F_n-centers and F-centers reach the equilibrium, the F_n-centers may be partly decomposed and then the evaporated F-centers diffuse into the sample according to the gradient of the concentration of F-centers. Thus, as we can see, at the very shallow surface near the implanted gold layer, the degree of F_n-centers gradually increase, which corresponds that SP position gradually changed until F_n-centers suddenly increased at 1200°C. However, between 800°C and 1200°C the evaporation and diffusion are in progress until the F-center concentration in the entire sample reaches the threshold to form F_n-centers around 1200°C.

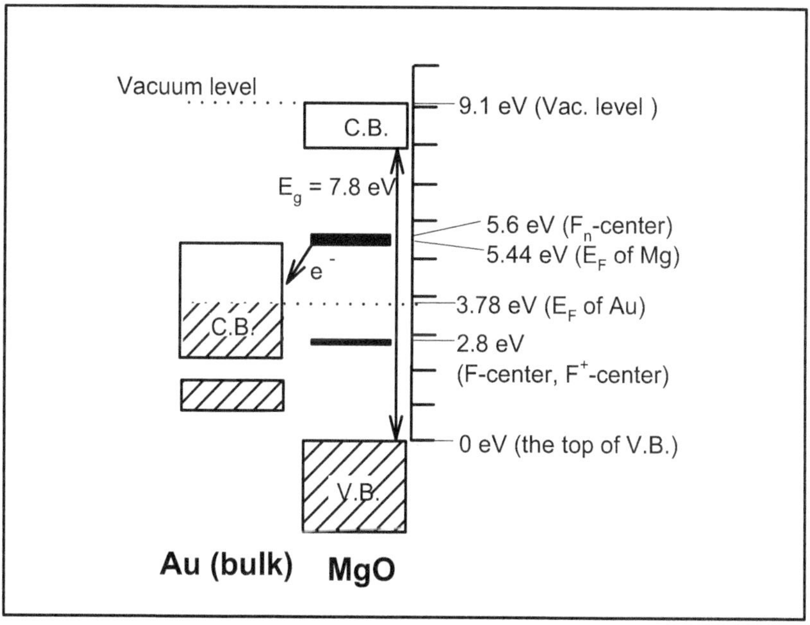

Fig. 18. Energy diagram of Au (bulk), MgO, Fn-center, F-center, and the other related energy levels. Relative positions are determined according to literature, as explained in text. (Ueda et al., 1998)

Let us consider how the dielectric function of gold colloid can change the value. F_n-center is defined as an aggregate of F-centers, which may be regarded as a metallic Mg colloid. In alkali halides irradiated with photons or electrons, metallic colloids were found as F-center aggregates (Fughes & Jain, 1979). If F_n-centers are formed close to gold colloids, electrons in the F_n-centers can transfer to the gold colloids because the Fermi level of bulk Mg metal is higher than that of bulk gold (Lide et al., 1998). The work function for bulk Mg and bulk gold are 3.66 eV and ~5.4 eV, respectively (Lide et al., 1998). The band gap of MgO is 7.8 eV and the vacuum level of MgO was found to be 9.1eV from the top of valence band by UPS (Tjeng et al., 1990). The energy levels of F- and F^+-centers are given by Chen et al. (Chen et al., 1969a, 1969b, 1967). The absorption band of F_n-center is located around 576 nm (2.15 eV) from our experimental data and literature values (Weber, 1986). Assuming that the final state of this transition is the bottom of the conduction band of MgO, we propose that the energy level for F_n-center is located about 2.2 eV below the bottom of the conduction band of MgO, which is 5.6 eV above the top of valence band. Summary of these data are shown in Fig.18. Note that the level of F_n-center is coincident with the Fermi level of bulk Mg metal, as mentioned before. The Fermi level of bulk gold is located much lower than the F_n-center level by 1.8 eV. Therefore, the electrons in F_n-center are able to transfer to gold colloids nearby. We did not consider the quantum confinement effect on electrons in gold colloids in order to estimate the energy scheme in Fig.18. The Fermi level of the gold colloids may have higher value than that of bulk gold, as shown in Fig.18.

Leaving this question at this stage, we would like to explain a simple mechanism of the dielectric function change of gold colloid due to electron transfer from F_n-centers to gold colloid, which correspond to Fig.17(c). In the free-electron gas model for metals, the dielectric function is given by the following equation (Kittel, 2005):

$$\varepsilon(\omega) = 1 - \frac{{\omega_p}^2}{\omega^2} \tag{5}$$

with the plasma frequency of gold colloid defined by, in SI units,

$${\omega_P}^2 = \frac{ne^2}{\varepsilon_0 m_e} \tag{6}$$

where e and m_e are the electron charge and mass, ε_0 is the permittivity of vacuum, and n is the electron density of a gold colloid we are considering. When some electrons in F_n-center are transferred into the gold colloid, the electron density of the gold colloid will increase so that the plasma frequency will increase. Fig.17(c) shows that the dielectric functions before and after electrons are transferred, in which the SP resonant frequencies are schematically given, according to the resonance condition: $\varepsilon(\omega_{SP}) = -2\varepsilon_m$. From this figure, we can see that the SP frequency blue-shifts after electrons are transferred. This is the case of the sample annealed in the reducing atmosphere and F_n-center plays a role of an electron injector.

Let n_0 be the electron density of a gold colloid in the oxidized MgO and n′ to be that of the same colloid in the reduced MgO. The potential difference between the Fermi level of gold and F_n-center level is about $\Delta\varphi = 1.8\, eV$. To compensate this potential difference, electrons can transfer from F_n-centers to the gold colloid. Although our gold colloid has a cubic shape, we regard it as a sphere for simplicity and suppose that the radius of the gold colloid is r. Thus, the volume of this colloid is $V_{colloid} = 4\pi r^3/3$. Let the transferred charge be ΔQ, and then the change in the number of electrons in the gold colloid will be

$$\Delta N \equiv \frac{\Delta Q}{e} \tag{7}$$

The charge transferred ΔQ is related to the potential difference $\Delta\varphi$ (volt) as follows:

$$\Delta\varphi = \frac{\Delta Q}{4\pi\varepsilon_0 r} \tag{8}$$

Eliminating ΔQ from (7) and (8), we obtain

$$\Delta N = \frac{4\pi\varepsilon_0 r\Delta\varphi}{e} \tag{9}$$

Assuming that the electron density of gold colloid in oxidized MgO is similar to the one of bulk gold; n_0=5.9x10^{28} m^{-3} (Lide et al., 1998), the total number of electrons in the colloid is given by

$$N_0 = n_0 V_{colloid} = \frac{4\pi r^3 n_0}{3} \tag{10}$$

The ratio of ΔN to the total number of electrons of an neutral Au colloid, N_0, is

$$\frac{\Delta N}{N_0} = \frac{\left(\frac{4\pi\varepsilon_0 r\Delta\varphi}{e}\right)}{\left(\frac{4\pi r^3 n_0}{3}\right)} = \frac{3\varepsilon_0\Delta\varphi}{e n_0 r^2} \approx \frac{0.005}{r^{*2}} \tag{11}$$

Here, r^* is defined as a numerical part of r in nm as follows: $r = r^* \times 1\, nm$. The dielectric function for gold before transfer is given by Eq.(5) with (6):

$$\varepsilon(\omega) = 1 - \frac{{\omega_P}^2}{\omega^2}, \qquad {\omega_P}^2 \equiv \frac{n_0 e^2}{\varepsilon_0 m_e}. \tag{12}$$

From Eq.(3), The surface plasmon position ω_{SP} satisfies

$$\varepsilon(\omega_{SP}) = 1 - \frac{{\omega_P}^2}{{\omega_{SP}}^2} = -2\varepsilon_m \tag{13}$$

After the charge transfer, the dielectric function for gold can be given by

$$\varepsilon'(\omega) = 1 - \frac{{\omega'_P}^2}{\omega^2}, \qquad {\omega'_P}^2 \equiv \frac{n' e^2}{\varepsilon_0 m_e} \tag{14}$$

After the charge transfer, the surface plasmon position ω'_{SP} satisfies

$$\varepsilon(\omega'_{SP}) = 1 - \frac{{\omega'_P}^2}{{\omega'_{SP}}^2} = -2\varepsilon_m \tag{15}$$

Here, although ε_m may have been changed, for simplicity, we assume that ε_m stays the same. By subtracting Eq.(14) from Eq.(12) each side, we get

$$\frac{{\omega'_P}^2}{{\omega'_{SP}}^2} = \frac{{\omega_P}^2}{{\omega_{SP}}^2} \quad \rightarrow \quad \frac{{\omega'_{SP}}^2}{{\omega_{SP}}^2} = \frac{{\omega'_P}^2}{{\omega_P}^2} = \frac{n'}{n_0} = \frac{n_0 + \Delta n}{n_0} = 1 + \frac{\Delta n}{n_0} \tag{16}$$

The shift for ω_{SP} is qualitatively given by

$$\frac{\omega'_{SP}}{\omega_{SP}} = \sqrt{1 + \frac{\Delta n}{n_0}} \cong 1 + \frac{1}{2}\frac{\Delta n}{n_0} = 1 + \frac{1}{2}\frac{\Delta N}{N_0} \cong 1 + \frac{0.0025}{r^{*2}}. \tag{17}$$

If we insert $\Delta N/N \approx 0.005$ for r =1 nm, then the calculated shift for ω_{SP} becomes about 1 nm in wavelength, while the actual shift for ω_{SP} is more than 30 nm in wavelength. Inversely, if we seek for the radius that causes 30 nm wavelength shift, then we obtain r =0.2 nm. Therefore, this model may not explain the actual such large shift. Since the electric charge due to free electrons in a gold particle is compensated with the positive charge of nucleus, thus the gold particle is neutral as a whole. The transferred charge, however, is an additional negative charge, so the transferred charge should be located on the surface of the gold particle. This localized surface charge may play an important role.

3.5 Sectional summary

We have found that the annealing atmosphere affects the position of SP band of Au implanted in MgO. The SP band position of the sample annealed in the reducing atmosphere is found to be about 524 nm, and that for the sample annealed in the oxidizing atmosphere is about 560 nm. These SP position shifting is reversible and repeatable by changing the annealing atmosphere from reducing to oxidizing or vice versa. We observed that F_n-centers are created during annealing in the reducing atmosphere at higher temperature (≥1200°C). Since F_n-center energy level is higher than the Fermi level of the gold colloids in MgO, F_n-centers play the role of electron injectors. Therefore, the electron density of gold colloid is changed so that the SP position shifts. However, our simple model cannot quantitatively explain the experimental results. The shifting and its reversibility may be useful for application to the development of nonlinear optical devices or sensor of gases.

4. Conclusion

We have presented two kinds of samples: First, by using samples of Se implanted in silica glass, we explained Se particle growth by Tauc analysis of optical absorption spectra on annealed samples, where we mainly concentrate on the particle growth of implanted materials. Secondly, by using MgO single crystalline substrate implanted with Au ions, we explained the interaction between implanted species and host material MgO. We focused on the surface plasmon resonance position shift due to the state of host material MgO annealed under reducing and oxidizing atmospheres.

5. Acknowledgments

We gratefully acknowledge research that was previously supported by DOE grant No. DE-F605-94ER45521. The research at ORNL was sponsored by the Division of Materials Sciences, U. S. Department of Energy, under Contract No. DE-AC05-9OR22464 with Lockheed Martin Energy Systems. Currently, the authors are partly supported by US National Science Foundation NSF-CREST-CA: HRD-0420516 and NSF-STC CLiPS, grant No. 0423914.

6. References

Briggs, A., (1975), J. Mater. Sci. 10, 737.

Bryant, E.M., Ueda, A., Mu, R.R., Wu, M.H., Meldrum, A., Henderson, D.O., (2001) Mat. Res. Soc. Symp. Proc. Vol. 635, C1.5.1-6, MRS.

Brus, L.E. (1983) J. Chem. Phys. 79, 5566.

Brus, L.E. (1984) J. Chem. Phys. 80, 4403-4409.

Chen, Y., Williams, R.T. & Sibley, W.A., (1969). Phys. Rev. 182, 960.

Chen, Y. , Kolopus, J.L. & Sibley, W.A. (1969). Phys. Rev. 186, 865

Evans, B.D., Comas, J., & Malmberg, P.R. (1972). Phys. Rev. B 6, 2453.

Fughes, A.E.,& Jain, S.C., (1979) in: *Metal Colloids in Ionic Crystals, Advances in Physics*, 28, 717.

Fukumi, K., *et al.*, (1994). J. Appl. Phys. 75, 3075.

Haglund, R.F. *et al.*, (1992). Nucl. Instr. & Method, B65, 405.

Hanamura, E. (1988). Phys. Rev. B 37, 1273.

Henderson, B. & Bowen, D.H. (1971). J. Phys. C: Solid State Physics 4, 1487.

Henderson, D.O. *et al.*, (1995). J. Vac. Sci. Technol. B 13, 1198.

Henderson, D.O. *et al.*, (1996). J. Vac. Sci. Technol. A 14, 1199.

Hole, D.E. et al. (1995). J. Non-Cryst. Solids 180, 266.

Huber, C.A. & Huber, T.E. (1988). J. Appl. Phys. 64, 6588.

Kittel, C., (2005). *Introduction to Solid State Physics*, 7th Ed., John Wiley & Sons, ISBN: 047141526X, New Jersey, USA.

Kotomin, E.A., Kuklja, M.M., Eglitis, R.I., & Popov, A.I. (1996). Materials Science and Engineering B37, 212.

Kreibig, U. & Vollmer, M. (1995). *Optical Properties of Metal Clusters*, Springer Series in Materials Science 25, ed by Toennies, J.P., Springer-Verlag, Berlin, Heidelberg.

Lide et al. (1998) CRC Handbook of Chemistry and Physics, 79th Edition, eds by Lide, D.A. et al. CRC Press, ISBN: 0849304792, Boca Raton, Florida.

Murphy, K.E. et al. (1977). J. Appl. Phys. 48, 4122.

Nagels, P.; Sleeckx, E., Callaerts, R., & Tichy, L. (1995). Solid State Communications 94, 49.

Palik, E.D.(1991) "*Handbook of Optical Constants of Solids* I & II," Ed. by E.D. Palik, Academic Press, ISBN: 0125444222, San Diego, California, USA.

Pankove, J.I. (1971). *Optical Processes in Semiconductors*, Dover, ISBN: 0486602753, New York.

Perenboom, J.A.A.J., Wyder, P. & Meier, F. (1981). Phys. Rep. 78, 173.

Sibley, W.A. & Chen, Y. (1967). Phys. Rev. 160, 712.

Tjeng, L.H., Vos, A.R. & Sawatzky, G.A. (1990) Surface Science 235, 269.

Tauc, J. (1972). Chapter 5 of *Optical Properties of Solids,* ed. by F. Abeles, North-Holland Publishing Company, ISBN: 0720402042, Amsterdam.

Ueda, A., Mu, R., Tung, Y-S., Henderson, D.O., White, C.W., Zuhr, R.A., Zhu, J.G. & Wang, P.W. (1997). Materials Forum Vols. 239-241, pp. 675-678.

Ueda, A., Mu, R., Tung, Y-S., Wu, M.H., Collins, W.E., Henderson, D.O. Henderson, White, C.W., Zuhr, R.A., Budai, J.D., Meldrum, A., Wang, P.W., Li, Xi (1998), Nuclear Instrument and Methods in Physics Research B 141 261-267.

Ueda, A., Mu, R., Tung, Y-S., Wu, M.H., Zavalin, A., Wang, P.W., & Henderson, D.O. (2001). J. Phys.: Condens. Matter 13, 5535-5544.

Ueda, A., Wu, M.H., Aga, R., Meldrum, A., White, C.W., Collins, W.E., & Mu, R. (2007), Surface and Coating Technology 201, 85.

White, C.W.*et al.*(1989). Mat. Science Reports, 4, 43.

Weber, M.J., 1986, *Handbook of Laser Science and Technology*, CRC Press, ISBN: 0-8493-3507-8, Boca Raton, Florida (1986).42-8546.

Yager, T.A. & Kingery, W.D, (1981), Journals of Materials Science 16, 483; Yager, T.A. & Kingery, W.D. (1981). Journals of Materials Science 16, 489.

15

Ion Implantation for the Fabrication of Plasmonic Nanocomposites: A Brief Review

Umapada Pal[1] and Ovidio Peña Rodríguez[2]

[1]Instituto de Física, Benemérita Universidad Autónoma de Puebla, Puebla

[2]Centro de Microanálisis de Materiales (CMAM), Universidad Autónoma de Madrid (UAM), Cantoblanco, Madrid

Instituto de Óptica, Consejo Superior de Investigaciones Científicas (IO-CSIC), C/ Serrano Madrid

[1]Mexico

[2]Spain

1. Introduction

Metal nanoparticles (NPs) have been the subject of intense research in recent years, largely due to their characteristic surface plasmon resonance (SPR), a phenomenon arising from the collective oscillation of their conduction electrons. The SPR characteristics of metal NPs have been exploited to utilize them in medical diagnostics (Allain & Vo-Dinh, 2002), immunoassay (Hirsch *et al.*, 2003a; Cui *et al.*, 2006), cancer treatments (Hirsch *et al.*, 2003b; Zhang, 2010), biological or chemical sensing (Sun & Xia, 2002; Alivisatos, 2004; Cao *et al.*, 2009), and surface enhanced Raman scattering (SERS) (Jackson & Halas, 2004). Large third-order optical susceptibility and picosecond optical response of metal NPs made them excellent candidates for non-linear optical devices (Mazzoldi *et al.*, 1996; White *et al.*, 1998; Pavesi *et al.*, 2000; Borsella *et al.*, 2001). Strong SPR-induced optical absorption of these nanostructures has also been utilized to enhance their photocatalytic activity (Tihay *et al.*, 2001) and to increase the efficiency of dye-sensitized solar cells (Catchpole & Polman, 2008; Atwater & Polman, 2010; Ding *et al.*, 2011).

Frequently these NPs are incorporated into dielectric or semiconducting matrices to fabricate dielectric-metal or semiconductor-metal nanocomposites with the aim of manipulating their functionalities, exploiting the characteristics of both the matrix and the metal NPs. Among the nanocomposites, those in which the SPR effect is prominent are called plasmonic nanocomposites (PNCs). Frequently these nanocomposites are fabricated by incorporating nanoparticles of noble metals like Au, Ag and Cu in insulating matrices like silica (a-SiO_2), quartz (c-SiO_2), and other metal oxides like zinc oxide (ZnO), alumina (Al_2O_3), and titania (TiO_2). Though, in principle, several other metal particles can be used for the fabrication of PNCs this is seldom done, due to their high proneness to oxidation or interband transitions in the usable spectral range. Consequently, in this short review we will mainly focus on these three important metals.

Two characteristics, particularly important for practical use of plasmonic nanocomposites, are: the location of the SPR in the desired spectral position and a high optical extinction. For the very small metal NPs (*i.e.* those much smaller than the optical wavelength) the extinction is dominated by absorption and the absorption coefficient (α) can be expressed as (Arnold & Borders, 1977):

$$\propto (\text{cm}^{-1}) = \frac{18\pi Q n_d^3}{\lambda} \frac{\varepsilon_2}{\left(\varepsilon_1 + 2n_d^2\right)^2 + \varepsilon_2^2}, \tag{1}$$

where Q is the volume fraction occupied by the metallic nanoparticles, n_d is the refractive index of the host matrix, ε_1 and ε_2 are the real and imaginary parts of the dielectric constant of the bulk metal, respectively, and λ is the optical wavelength. The value of α becomes maximum at the surface plasmon frequency (ω_{SPR}), when

$$\varepsilon_1 (\omega_{SPR}) + 2n_d^2 = 0. \tag{2}$$

As we see, the SPR frequency of metal nanoparticles depends on both the optical constant of bulk metal and the refractive index of the host matrix. Estimating the wavelength variation of ε_1 from the optical constants given by Johnson and Christy (Johnson & Christy, 1972) for Au, Ag, and Cu, and refractive indices of common oxide matrices we can calculate the position of the SPR band for different nanocomposites, as presented in Table 1. Though the position of SRP in different nanocomposites depends also on the size and size distribution of the metal NPs, for which we must introduce size correction terms (Hövel *et al.*, 1993) in equation (2), we can see that the tabulated SPR energies are in good agreement with the reported values for the nanocomposites fabricated by different methods (Ballesteros *et al.*, 1997; De, 1997; Pal *et al.*, 2000, 2001; Vazquez-Cuchillo *et al.*, 2001; Wang *et al.*, 2003; García-Serrano & Pal, 2003; Barreca *et al.*, 2004; Subramanian *et al.*, 2004; Zeng *et al.*, 2004; Hazra *et al.*, 2004; Okumu *et al.*, 2005; Ryasnyansky *et al.*, 2005; Serna *et al.*, 2006; Sonawane & Dongare, 2006; Mishra *et al.*, 2008; Zheng *et al.*, 2008; Kalidindi & Jagirdar, 2008; Kim *et al.*, 2008; Manoilov, 2009; Jiang *et al.*, 2009; Ma *et al.*, 2009; Sangpour *et al.*, 2010; Tohidi *et al.*, 2010; Duhan *et al.*, 2010). As the absorption coefficient depends on the volume fraction occupied by metallic NPs in the composites (eq. 1), the intensity of the SPR band increases on increasing the amount of incorporated metal NPs.

Several methods have been utilized to fabricate plasmonic nanocomposites like co-sputtering, ion-implantation, sol-gel, laser ablation, etc. (De, 1997; Pal *et al.*, 2000; García-Serrano & Pal, 2003; Barreca *et al.*, 2004; Kim *et al.*, 2008; Manoilov, 2009). However, irrespective of the fabrication method, practical application of these materials strongly depends on parameters like the size and shape of the incorporated metal NPs, volume fraction occupied by them, homogeneity of NP distribution, and their composition. In this respect, each of the utilized methods has certain constrains. Although ion-implantation is not as versatile as other techniques used for the fabrication of plasmonic nanocomposites, it still has several attractive advantages. For example, ion implantation followed by an additional thermal annealing has proven to be a very useful method to obtain large volume fractions of NPs embedded in solid matrices. Additionally, those NPs are located in a well defined depth below the surface, chosen mainly by the ion energy (Oliver *et al.*, 2002). Some other advantages of ion implantation include controllability of depth profile and concentration, high purity, and the possibility to overcome low solubility restrictions. Particularly, deep ion implantation using energies of the order of MeV produces an ion depth distribution located some micrometers underneath the surface and wide enough to be convenient for the production of optical waveguides. Ion implantation has been also successfully used to create more complex structures such as alloys (Mattei, 2002), sophisticated aggregates (Mattei *et al.*, 2003b; Pellegrini *et al.*, 2007a, 2007b), core-shell structures (Peña *et al.*, 2009) and nanorods (which are aligned, unlike those produced by other methods) (Rodríguez-Iglesias *et al.*, 2010a; b).

In this chapter, we intend to review the current status of the synthesis of plasmonic nanocomposites by means of ion-implantation. In the year 1984, Picraux (Picraux, 1984) has published a useful review on the status of ion-implantation in metals, improving the understanding on the kinetics and equilibria in metal alloy reactions, surface modification, and defect structures. Similarly, Buchal *et al.* (Buchal *et al.*, 1994) have reviewed the status of ion-implantation in optical materials in 1994, focusing mainly on the usefulness of the technique for producing linear and non-linear optical devices. However, apart from a good review on Cu implanted alumina by Stepanov and Khaibullin (Stepanov & Khaibullin, 2005) there exists no recent review summarizing the current advances made on the utilization of this technique for the fabrication and processing of plasmonic nanocomposites which have emerging applications in diverse fields.

Incorporated species	Matrix material (refractive index)	Predicted (eV)	Measured (eV)	References
Au	SiO_2 (1.48)	2.35	2.48-2.21	(Barreca *et al.*, 2004; Kim *et al.*, 2008)
Ag	SiO_2 (1.48)	3.12	3.92-3.02	(Roiz *et al.*, 2004; Ma *et al.*, 2009; Duhan *et al.*, 2010)
Cu	SiO_2 (1.48)	2.70	2.31-1.59	(De, 1997; Peña *et al.*, 2006; Tohidi *et al.*, 2010)
Au	Al_2O_3 (1.76)	2.24	2.38-1.79	(García-Serrano & Pal, 2003; Hazra *et al.*, 2004)
Ag	Al_2O_3 (1.76)	2.85	3.08-2.06	(Zeng *et al.*, 2004; Manoilov, 2009)
Cu	Al_2O_3 (1.76)	2.20	2.10-2.00	(Ballesteros *et al.*, 1997; Serna *et al.*, 2006)
Au	ZnO (2.00)	2.14	2.48-2.25	(Wang *et al.*, 2003; Mishra *et al.*, 2008)
Ag	ZnO (2.00)	2.64	2.91-2.53	(Shan *et al.*, 2007; Zheng *et al.*, 2008)
Cu	ZnO (2.00)	2.09	2.14-2.06	(Ryasnyansky *et al.*, 2005; Kalidindi & Jagirdar, 2008)
Au	TiO_2 (2.80)	1.80	2.34-1.83	(Subramanian *et al.*, 2004; Sonawane & Dongare, 2006)
Ag	TiO_2 (2.80)	2.09	2.58-2.10	(Okumu *et al.*, 2005; Jiang *et al.*, 2009)
Cu	TiO_2 (2.80)	1.82	2.15	(Sangpour *et al.*, 2010)

Table 1. SPR positions in common plasmonic nanocomposites.

2. Implantation of metal ions in dielectrics

Since long ago, ion implantation has been used to modify the optical properties of materials. Dielectrics such as silica glass (SiO_2), alumina (Al_2O_3), and quartz have been used traditionally for this purpose; but recently several other metal oxides like ZnO and TiO_2 have also been utilized. Historically, ion implantation has been used to change the optical properties of dielectric target materials like reflectivity, absorption, refractive index, luminescence, and electro-optic coefficients (Mazzoldi, 1987; Townsend, 1987). Those modifications are not only associated to the implanted species, but also to a number of other phenomena occurring during the process like radiation damage caused by the ion beam, and formation of colloidal precipitates of the implanted species with diameters in between 2 and 20 nm (Hughes, 1983). As the mechanism of formation of radiation damages and their elimination processes have been widely discussed in the literature (Agulló-López *et al.*, 1988; Pal *et al.*, 2000; Bianconi *et al.*, 2009; Rivera *et al.*, 2010), we will concentrate our discussion on the latter mechanism only.

In fact, the formation of colloidal precipitate of metals like Au and Ag has been utilized for hundreds of years to design beautiful colored glasses (Bamford, 1977). While the color of the implanted dielectrics depends mostly on the implanted species (metals) and to some extent the size of the formed colloidal particles, their intensity/contrast depends on the volume fraction occupied by the colloidal precipitates in the matrix. So, apart from the nature of implanted metal, a control over the implanted dose (amount of implanted species incorporated into the dielectric matrix) is essential to control the optical properties of the substrates (dielectrics). In contrast to the conventional melt growth process, where only about 1% of impurities (guest atoms) can be incorporated into a matrix, there is no such restriction in ion implantation, as the process occurs without thermodynamic restrictions.

For the ion implantation purpose, the ion beams of selected metals doses in the range of 10^{16}-10^{18} cm^{-2} have been utilized to expose solid target surfaces as big as a few cm^2. In fact, the process has the ability to fabricate technological grade composite wafers with uniform distribution of metal NPs at precise depths below the target surface. The range of metal NP distribution, their occupied volume fraction, size (diameter), and composition can also be controlled by suitable adjustment of implantation parameters. In the following section, we will discuss how these parameters can be controlled and their effects on the characteristics of the final product (nanocomposites).

2.1 Controlling parameters in ion implantation process

Essentially any specie can be introduced into a solid matrix by means of ion implantation, thereby changing its physical properties. For this purpose, ions of the desired element are produced in an *ionization chamber*; then they are electrostatically accelerated to the desired energy and, finally, allowed to impinge on a solid target. As typically each ion is a single atom, the amount of material incorporated or implanted into the target can be precisely calculated by integrating the ion current over the exposure time. The amount incorporated is called *dose* or *fluence*. Generally the ion currents used for implantation are small, in the range of nano- to microamperes. Ion beam energies may vary from a few keV up to hundreds of MeV, depending on the requirement, though energies in the range of a few MeV are commonly used to produce PNCs. Greater ion energy causes higher structural damage to the target. However, due to broader depth distribution, the change in chemical composition at a certain point in the target would be small.

When high energy ions penetrate into a solid, they gradually lose energy due to non-elastic collisions with the free electrons of the solid and elastic collisions with the nuclei until they become immobilized. As the second process is an elastic one, most of the energy loss of the incident ion bean occurs very close to the surface of the solid, due to the first process. The penetration depth of the ions in the solid target depends on the ion energy, nature of ionic species, and the nature and composition of the target. Typically, the penetration depth or the *range* of the ions in a solid target varies in between a few nanometers to a few micrometers, and hence all the changes occurring during the implantation process remain near the surface of the target material (Figure 1).

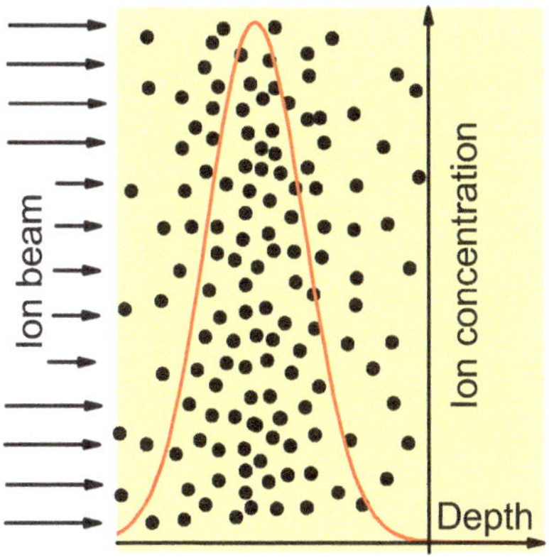

Fig. 1. Schematic presentation of the ion implantation process in a solid and the implantation profile.

Typical ion doses utilized for fabricating plasmonic nanocomposites vary in between 1.0×10^{16} cm^{-2} and 2.0×10^{17} cm^{-2}. While a greater dose frequently induces the formation of undesired nanovoids or nanocavities in the target matrix (Kooi *et al.*, 2000) or inside metal precipitates (particles) (Ren *et al.*, 2006) due to aggregation of vacancies driven by the heat produced during the high-dose ion irradiation, lower doses frequently incorporate metallic species in ionic state inside the dielectric matrices (Haug *et al.*, 2009). While higher ion energy enhances the penetration depth of the implanted species, a higher ion dose shifts the maximum of the distribution of the implanted species in the target towards lower depths, due to the diffusion to the heavily-damaged region produced just before the ions are stopped (Figure 2). Though the penetration depth of ionic species remains within a few micrometers beneath the surface of the implanted solid, it can extend to a larger depth on increasing the temperature of the solid target.

In the case of an amorphous solid target, the implanted ions get distributed randomly following a Gaussian profile inside the matrix, which can be expressed as:

$$N(x) = N_0 \exp\left[-\frac{(R_p - x)^2}{2\Delta R_p^2}\right] = \frac{\phi_0}{\sqrt{2\pi}\,\Delta R_p} \exp\left[-\frac{(R_p - x)^2}{2\Delta R_p^2}\right], \quad (3)$$

where $N(x)$ is the concentration of ions at a distance x from the surface; N_0 is the concentration of ions at $x = R_p$, ϕ_0 is the total implanted dose and R_p is the effective range of penetration by an ion, dependent of real range R_c, where R_c, the real range of penetration, i.e. distance covered by the ion in the implanted solid, and ΔR_p is the standard deviation of the R_p value. The location of R_p depends predominantly on the energy of ions and their atomic mass. The relative dispersion of implanted specie, i.e. $\Delta R_p/R_p$ depends both on the atomic mass of the implanted species and the atomic mass of the solid target.

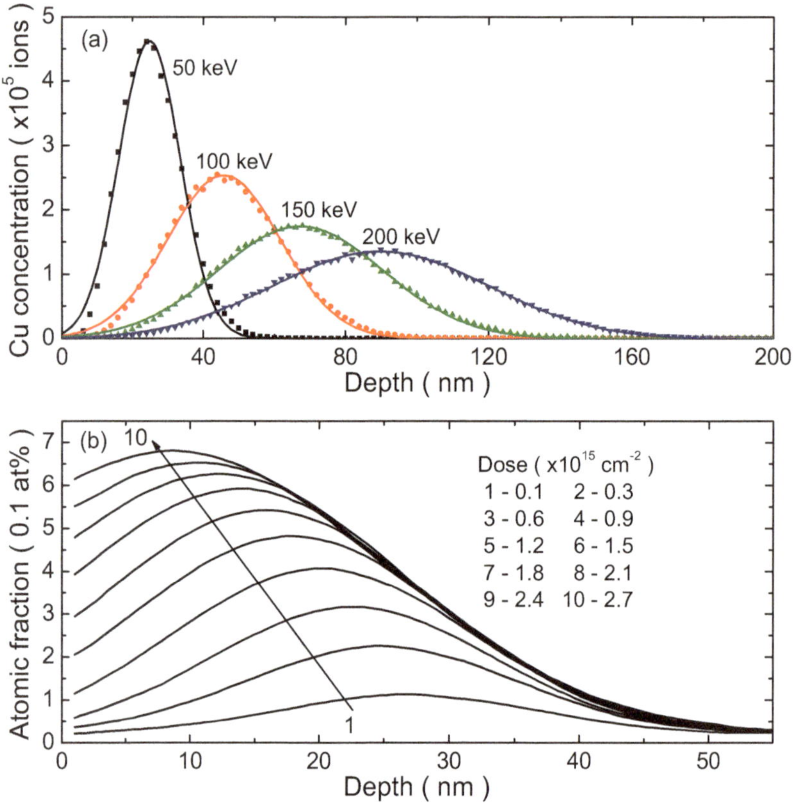

Fig. 2. a) Variation of profile and depth of implantation of Cu^+ ions into Al_2O_3 with the variation of ion energy, calculated using SRIM 2008 (Ziegler, 1985, 2008); b) Implantation depth versus dose, calculated using dynamic computer code DYNA of Cu^+ ions of 40 keV in Al_2O_3. Adapted with permission from AIP Conf. Proc. 680 (2003) 601-604 (Stepanov & Popok, 2003).

On the other hand, for the case of a crystalline solid target, the range of penetration and the implant distribution strongly depend on the orientation of the structure with respect to the direction of beam incidence. Generally the penetration depth or the implantation range is about one order higher when the ion beam is projected along a given crystallographic direction of the target like [111], [100], etc. (Burakowski & Wierzchon, 1998). As can be seen from Figure 3, the defect distribution profile remains at a shallower level with respect to the implant distribution.

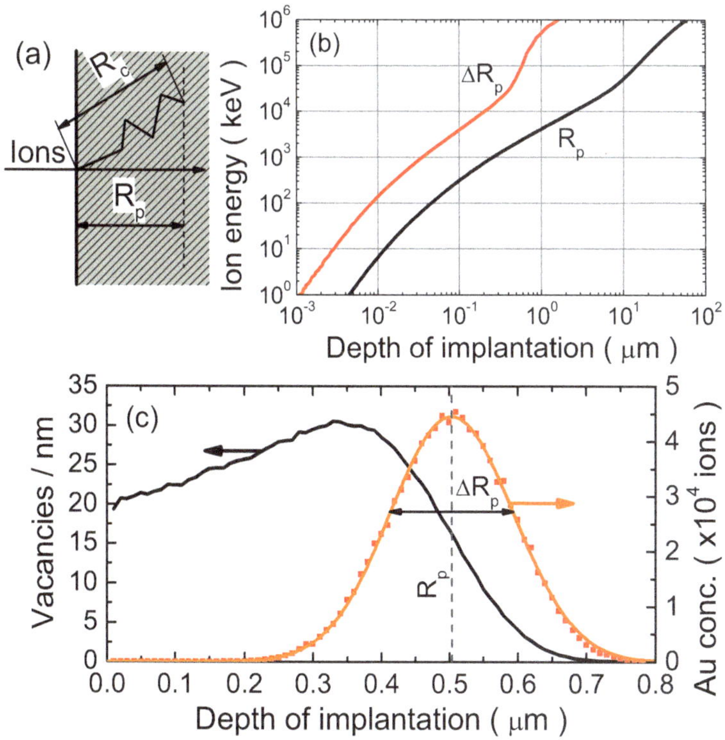

Fig. 3. a) Schematic illustration of the range of penetration; b) variation of implantation range R_p and standard deviation ΔR_p with ion energy for 2 MeV Au ions implanted in silica; c) distributions of implanted ions and defects. All simulations were performed using SRIM 2008 (Ziegler, 1985, 2008).

3. Controlled growth of noble metal nanoparticles through ion-implantation

The noble metals like Au, Ag and Cu have been incorporated into dielectric matrices to produce several plasmonic nanocomposites with varied combinations of metals and dielectrics. The main objective of such works was to utilize the intense local field, strong linear optical absorption or SPR in applications such as optical sensing, biological and chemical sensing, and fabrication of all-optical devices. The electronic transitions responsible for plasmon absorption in metal NPs are also responsible for the generation of optical nonlinearities in the same spectral range, i.e. at the SPR position. To exploit the advantage of this SPR absorption in nanocomposites, it is desirable to have a high concentration of metal NPs (i.e., high filling factor) in the matrix (Flytzanis *et al.*, 1991). Therefore, as has been mentioned earlier, apart from the adequate energy of implantation, a high dose of implanted ions is required. By the terms "high metal ion dose" ϕ_0, we mean a dose exceeding the solubility limit of the metal atoms in the target (dielectric in the present case), which generally vary in between 10^{15} - 10^{17} cm^{-2}, depending on the nature of ion species and target materials. The characteristics of the incorporated metal NPs also depend on the utilized ion dose. While there is a limiting value of implanted dose for each combination of metal ions and target material to obtain well dispersed metal NPs with controlled nucleation and growth (i.e. with definite geometrical shape and size), doses higher than this limiting value cause aggregation or coalescence of the formed NPs, strongly affecting their optical properties (Figure 4). For example, for 25 keV Ag^+ ion implantation into $LiNbO_3$, the limiting dose was found to be $\phi_0 \sim 5.0 \times 10^{15}$ cm^{-2} (Deying *et al.*, 1994), and $\sim 10^{16}$ cm^{-2} for 30 keV Ag^+ ions in solid epoxy resin (Stepanov *et al.*, 1995). In fact, the implantation range and concentration distribution of the metal NPs strongly depend on the implantation dose (Figure 2b).

Fabrication of plasmonic nanocomposites by ion implantation dates back to 1973, when Davenas and co-workers (Davenas *et al.*, 1973) pioneered this method to fabricate metal NPs of silver, calcium, etc. inside crystalline LiF and MgO matrices. Later, the method was extended for the fabrication of other metal NPs in several other matrices like sapphire (Al_2O_3), silica (SiO_2) and even polymers. For practical applications, plasmonic nanocomposites containing metal NPs of desired sizes, high filling fraction, and controlled composition are the basic requirements. While the ion implantation technique can introduce virtually any element in any matrix without solubility restrictions, the typical cluster size distributions in ion implanted nanocomposites are not narrow. On the other hand, progress has been made to fabricate metal-dielectric nanocomposites with controlled composition of metal NPs like elemental metallic clusters, bimetallic alloy clusters and core-shell type bimetallic nanoclusters. Depending on the nature of the metallic element and the matrix, adequate adjustments of implantation conditions (e.g. implantation energy, dose) and post implantation treatments (thermal annealing, laser annealing, etc.) in adequate atmosphere (e.g. oxidizing atmosphere like air or oxygen, inert atmosphere like argon and reducing atmosphere like hydrogen or a mixture of hydrogen and argon) are necessary to fabricate each type of the above mentioned architectures. In the following sections we will discuss on the fabrication of each of these nanoarchitectures, their plasmonic behaviors, and application potentials.

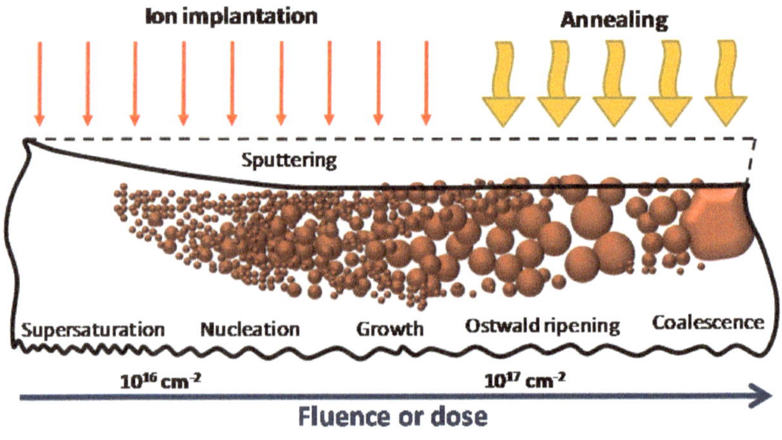

Fig. 4. Basic physical processes (left to right) involved in the formation of metal nanoparticles from an implant at different ion doses.

3.1 Formation of nanocomposites

At a certain temperature, when the solute concentrations are above their solubility limits in a matrix, there occurs spontaneous condensation of solute. Within a homogeneous matrix and in conditions which may rule out coalescence, this condensation usually leads to the formation of spherical clusters. The theory of growth of spherical clusters has been developed and used by many authors (Lifshitz & Slezov, 1959; Aaron *et al.*, 1970). According to this theory, at relatively low temperatures (in comparison with the melting temperature of bulk materials), formed clusters predominantly go through a stage of nucleation and growth. The exact analytical solution of the growth process of a spherical cluster is very difficult to obtain. For this reason, various approaches have been used to represent the real solution to a greater or lesser extent. In particular, we want to solve the field equation:

$$D\nabla^2 C = \frac{\partial C}{\partial t} \tag{4}$$

where D is the volumetric coefficient of diffusion (assumed to be independent of position) and $C = C(r,t)$ is the field of concentration in the matrix around the precipitate. Also, the independent equation of flow balance must be satisfied:

$$\left(C_p - C_I\right)\frac{dR}{dt} = D\frac{\partial C}{\partial r}\bigg|_{r=R}. \tag{5}$$

C_P is the concentration of the precipitate (constant), C_I is the solute concentration in the matrix, at the precipitate-matrix interface and R is the value of r on that interface. For the condition $(C_P - C_M) >> (C_I - C_M)$, the $R(t)$ is a slowly varying function of time and a way to simplify the field equation (4) is by using the invariant-field approximation in time (Laplace approximation), which is $\partial C/\partial t = 0$, thus being the Laplace equation $\nabla^2 C = 0$. Now, for the steady state of the cluster, we can solve the equation (4) to obtain:

$$R = \lambda (Dt)^{1/2}, \text{ where } \lambda = (-k)^{\frac{1}{2}}, k = 2\frac{C_I - C_M}{C_P - C_I}, \tag{6}$$

where C_M is the concentration of solute in the matrix far away from the precipitate. The problem is almost identical for the dissolution of clusters. In this case $R(t = 0) = R_0 > 0$, we obtain the expression:

$$R^2 = R_0^2 - kDt. \tag{7}$$

Note that k is positive during the dissolution and negative during the growth.

3.2 Fabrication of nanocomposites containing elemental noble metal clusters

Controlling the plasmonic behavior of noble metal clusters (nanoparticles), i.e. their SPR position and intensity in nanocomposites is of immense importance for their practical applications. Manipulation of the implantation variables (energy and dose) coupled with a post-implantation annealing allows the fabrication of metal nanoclusters with the desired SPR characteristics. Since the first report on ion implantation fabrication of such nanocomposites

Metal	Matrix	Ion energy (keV)	Dose (cm^{-2})	Post-implantation annealing	References
Au	SiO_2	3000	$6.0x10^{15}$	in air, 550-750, 1050 ^0C	(Mohapatra *et al.*, 2007)
Au	SiO_2	60	$2.0x10^{17}$	-	(Takeda *et al.*, 2006)
Au	SiO_2	2000	$(0.7\text{-}6.0)x10^{16}$	in air & (N_2+H_2), 300-1100 ^0C, 1h	(Oliver *et al.*, 2002)
Au	Al_2O_3	400	$6.8x10^{16}$	-	(Ohkubo & Suzuki, 1988)
Au	Al_2O_3	160	$1.0x10^{17}$	-	(Marques *et al.*, 2004)
Au	Al_2O_3	2000	$(0.2\text{-}2.0)x10^{17}$	in air, 978-1323 ^0C, 1h	(Ila *et al.*, 1997)
Au	Al_2O_3	1800	$(0.1\text{-}1.0)x10^{18}$	1000 ^0C & 1200 ^0C, 1h	(Dhara *et al.*, 2004)
Au	Al_2O_3	60	$2.0x10^{17}$	-	(Takeda *et al.*, 2006)
Au	TiO_2	260	$5.0x10^{16}$	-	(Zhu *et al.*, 2006)
Au	TiO_2	1800	$8.5x10^{16}$	480 ^0C	(Nakao *et al.*, 2000)
Au	c-TiO_2	2000	$6.0x10^{16}$	1002 ^0C	(Wang, 2004; Wang *et al.*, 2005)
Au	c-ZnO	300	$8.0x10^{13}$-$4.0x10^{16}$	-	(Kucheyev *et al.*, 2003)
Ag	SiO_2	35, 1700, 2400	$(3.0\text{-}7.0)x10^{16}$	in air, 500 ^0C, 1h	(Joseph *et al.*, 2007; Xiao *et al.*, 2007)
Ag	SiO_2	10	$5.0x10^{15}$	in Ar, 500-800 ^0C	(Arai *et al.*, 2006)
Ag	SiO_2	200	$1.0x10^{15}$-$1.0x10^{17}$	-	(Liu *et al.*, 2005)
Ag	SiO_2	2000	$(0.7\text{-}6.0)x10^{16}$	in air & (N_2+H_2), 300-1100 ^0C, 1h	(Oliver *et al.*, 2002)
Ag	Al_2O_3	25-30	$(0.2\text{-}2.0)x10^{17}$		(Steiner *et al.*, 1998)
Ag	Al_2O_3	1500	$(0.2\text{-}2.0)x10^{17}$	in air, 773 ^0C, 1h	(Ila *et al.*, 1997)
Ag	Al_2O_3	1800	$1.2x10^{17}$	in air, 1100 ^0C, 2h	(White *et al.*, 1992, 1993)
Ag	TiO_2	65	$3.0x10^{16}$-$1.0x10^{17}$	in air , 400 ^0C	(Tsuji *et al.*, 2002)
Ag	ZnO	30, 60	$(0.3\text{-}1.0)x10^{17}$	-	(Zhou *et al.*, 2011)
Ag	ZnO	20, 40	$(3.0\text{-}7.0)x10^{16}$	-	(Xiao *et al.*, 2010)
Ag	ZnO	40	$1.0x10^{16}$	-	(Mendoza-Galván *et al.*, 2006)
Cu	SiO_2	2000	$(0.7\text{-}6.0)x10^{16}$	in air & (N_2+H_2), 300-1100 ^0C, 1h	(Oliver *et al.*, 2002)
Cu	SiO_2	2000	$6.0x10^{16}$	in vacuum, 600-800 ^0C, 1h	(Pan *et al.*, 2007)
Cu	SiO_2	60	$(0.5\text{-}2.0)x10^{17}$	200 to 700 ^0C for 1 h per step with an interval of 100 ^0C	(Xu *et al.*, 2006)
Cu	SiO_2	60	$1.0x10^{17}$	-	(Bandourko *et al.*, 2005)

Metal	Matrix	Ion energy (keV)	Dose (cm^{-2})	Post-implantation annealing	References
Cu	SiO_2	2000	$(1.0\text{-}2.2)x10^{17}$	in air, 600-1050 ^0C, 1h	(Pal *et al.*, 2000)
Cu	Al_2O_3	110	$1.0x10^{17}$		(Donnet *et al.*, 1991)
Cu	Al_2O_3	160	$(0.2\text{-}1.0)x10^{17}$	in air, 673 ^0C, 1h	(Ila *et al.*, 1997)
Cu	Al_2O_3	40	$1.0x10^{17}$		(Stepanov *et al.*, 2001)
Cu	Al_2O_3	2100	$1.0x10^{17}$	in vacuum, 497-997 ^0C, 1h	(Ikeyama *et al.*, 2001)
Cu	TiO_2	20	$5.0x10^{15}$	-	(Wang *et al.*, 2006)
Cu	TiO_2	60	$1.0x10^{17}$	-	(Takeda *et al.*, 2005)
Cu	TiO_2	50	$(0.3\text{-}1.0)x10^{17}$	in air, 400 ^0C, XX	(Tsuji *et al.*, 2002)
Cu	ZnO	160	$(0.1\&1.0)x10^{17}$	-	(Kono *et al.*, 2003; Norton *et al.*, 2003; Stepanov *et al.*, 2004)

Table 2. List of noble metal ions used to incorporate in common oxide matrices with different implantation conditions and post implantation treatments.

by Davenas' group in 1973 (Davenas *et al.*, 1973), a vast effort has been devoted to their controlled fabrication using different combinations of metal particles and matrices. As can be seen from table 2, Au, Ag and Cu NPs have been incorporated into several dielectric matrices with different implantation conditions and the conditions of subsequent annealing.

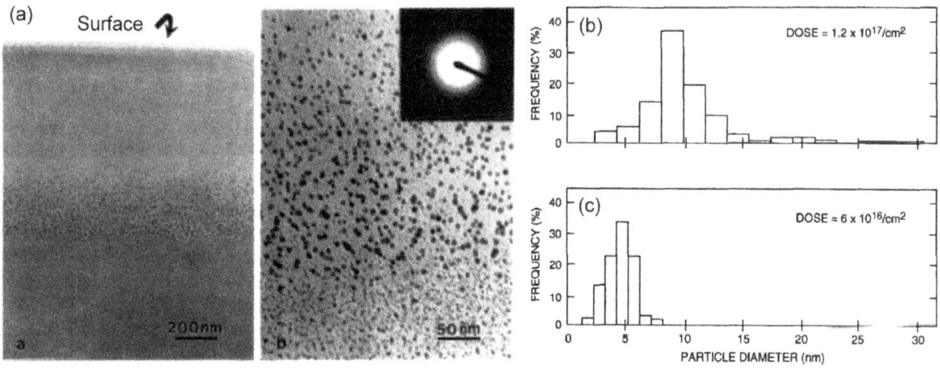

Fig. 5. Cross-sectional TEM micrographs showing colloidal Ag precipitates formed in fused silica by the implantation of Ag (1.8 MeV, $6x10^{16}$ cm^{-2}) ions at room temperature. Adapted with permission from Ann. Rev. Mater. Sci. 24 (1994) 125-158 (Buchal *et al.*, 1994).

Implantation of these metal ions of moderate energy (typically up to 200 keV) and low doses (~10^{16} cm^{-2}) generally produces metallic nanoclusters of homogeneous sizes, while a high dose and a high temperature post implantation treatment produce nanoclusters of bigger sizes with higher dispersion. However, both the size and the size dispersion of the formed nanoclusters depend on the nature of implanted metal, their solubility and diffusion behavior in the embedding matrix. While Ag nanoclusters of relatively uniform sizes and narrow size distributions can be grown in silica matrix even for a high ion dose (Figure 5) without post-implantation thermal treatment, the same is not true for the Au and Cu nanoclusters in silica matrix (Figure 6). As can be seen from Figure 6, Au nanoparticles formed with similar implantation conditions and thermal treatments have a broad size distribution.

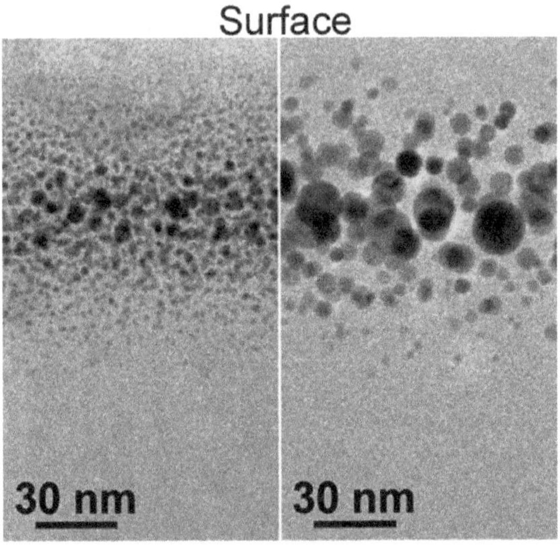

Fig. 6. Cross-sectional bright field TEM images of Au implanted (190 keV, $3x10^{16}$ cm^{-2}) silica: a) before and b) after 3h of thermal annealing at 900 ^{0}C, in air. Reproduced with permission from Nucl. Instrum. Meth. B 191 (2002) 323-332 (Mattei, 2002).

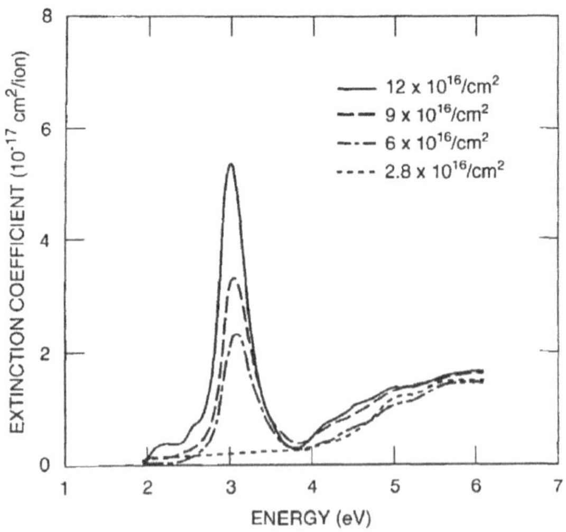

Fig. 7. Optical absorption spectra of 1.8 MeV Ag implanted fused silica at room temperature prepared with different doses. Reproduced with permission from Ann. Rev. Mater. Sci. 24 (1994) 125-158 (Buchal *et al.*, 1994).

While a very low ion dose produces metal nanoclusters of extremely small sizes (typically below 1.0 nm) that exhibit no SPR band in absorption spectra, on increasing the ion dose, the clusters grow and the intensity of the SPR extinction band increases (Figure 7). On the other hand, for a high energy of the implanted ions, the size of the formed metallic clusters can be big enough to reveal an SPR peak in the absorption spectra of the composites (Figure 8B). While a high energy implantation can place the metallic clusters at micrometer depth below the irradiated surface, a higher implantation dose can extend the cluster distribution towards the target surface (Figure 8A). Such a broadening of cluster distribution towards lower depths is due to the interaction of the incident ions with the already formed metallic clusters, which effectively reduces the projected range of the metal ions that impinge at a later stage.

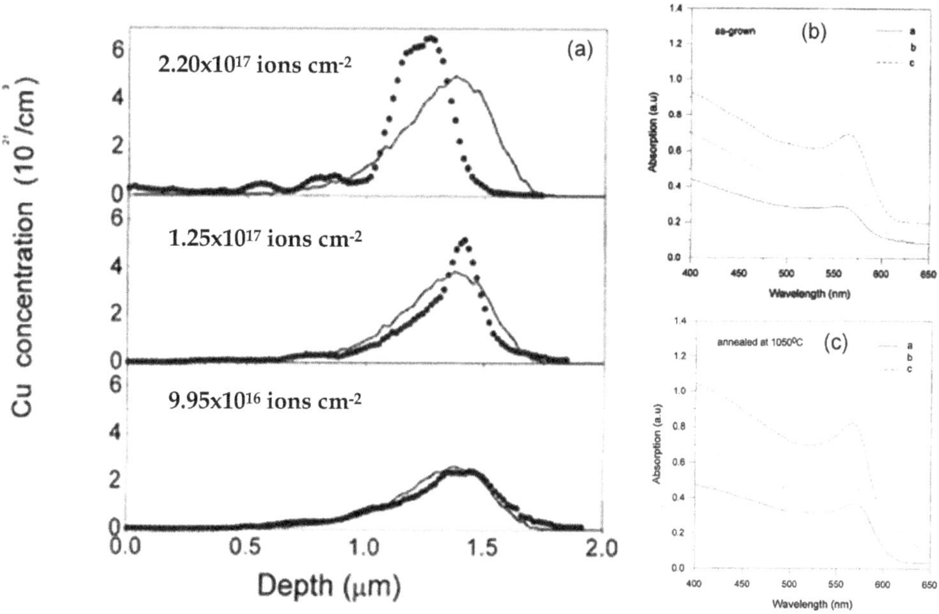

Fig. 8. a) The depth profiles of Cu concentration in silica matrix prepared with 2.0 MeV Cu ions of $9.95x10^{16}$, $1.25x10^{17}$, and $2.20x10^{17}$ ions cm^{-2} doses without post implantation treatment, estimated through RBS analysis. The continuous lines show the TRIM simulated results. Optical absorption spectra of the Cu implanted silica before (b) and after (c) thermal annealing at 1050 ^{0}C. Adapted with permission from J. Non-Cryst. Sol. 275 (2000) 65-71 (Pal *et al.*, 2000).

Overall, incorporation of monometallic NPs in dielectric matrix through ion implantation has had a good success so far. At the present time, researchers can fabricate specific noble metal clusters of desired sizes and size distributions at precise depths inside most common dielectric matrices by controlling the implantation dose, energy and the conditions of post implantation treatment. However, from the application point of view the success has been limited. Though the SPR position in a nanocomposite could be modified by varying the nature of implanted metal ions (e.g. the typical SPR position of Au nanoclusters in SiO_2 is at about 540 nm, of Ag at about 420 nm, and of Cu at about 560 nm) and its intensity could be raised by increasing the implantation dose, it is not possible to tune the position of the SPR peak for a wider spectral range using monometallic nanoclusters. The fabricated monometallic plasmonic nanocomposites are suitable for the fabrication of nonlinear optical switches, effective waveguides, optical sensors, and devices covering a limited spectral range, but the limitation of spectral tunability of SPR peak position makes them unsuitable for most biological applications, such as medical diagnosis, immunoassay, and therapeutic treatments.

Therefore, researchers felt the need of fabricating plasmonic nanocomposites containing bimetallic nanoclusters to vary the SPR position of the composites in between the SPR positions of individual monometallic metal clusters, which we will discuss in the following section.

3.3 Fabrication of nanocomposites containing alloy- and core-shell type bimetallic clusters

Sequential ion implantation, where the ions of a particular metal are first implanted at a certain depth of the matrix and then the second type of metal ions are implanted at similar depths, has been utilized frequently to fabricate bimetallic nanocomposites. Depending on the conditions of implantation, the nature of metallic species and dielectric matrix, the process can lead to the formation of i) separated monometallic nanoclusters of individual metals, ii) alloy clusters, and iii) core-shell type bimetallic clusters. Formation of each of these clusters depends on their thermodynamic stability, and hence on the process of post-implantation thermal treatments along with the annealing atmosphere. Although the composition of alloy type clusters can be varied easily by sequential ion implantation of two elements in a matrix by adjusting their energy and doses to have maximum overlap between their implant-profiles and desired local relative concentrations, post implantation thermal treatments can frequently lead to (promote) separation of a component specie from the other or oxidation, instead of alloying. Formation of alloy type clusters also depends on the miscibility of the component elements, their chemical interaction with the matrix, and the creation of radiation induced defects, which influences the nucleation and growth of the nanoclusters.

As the criterion of bulk miscibility for two elements is not very stringent in the case of nanocrystals due to their high specific surface area (surface area/volume ratio) and characteristics closer to that of a molecule than to those of a massive (bulk) system (Yasuda & Mori, 1994), alloy type nanoclusters of several metal pairs like Au-Cu, Au-Ag, Pd-Ag, and Pd-Cu, have been successfully fabricated (Gonella *et al.*, 1999; Battaglin *et al.*, 2000, 2001a) through sequential ion implantation. The SPR position in such noble metal alloy clusters remains in between those of pure elements, depending on the complex interplay between the modified free electrons and interband absorptions. As can be seen from Figure 9, the measured absorption spectra of pure Au, Ag and $Au_{0.4}Ag_{0.6}$ alloy clusters of about 3 nm size in silica matrix reported by Mattei and co-workers (Mattei, 2002) agree well with the simulated extinction spectra of the NPs of similar compositions reported by Kreibig & Vollmer (Kreibig & Vollmer, 1995).

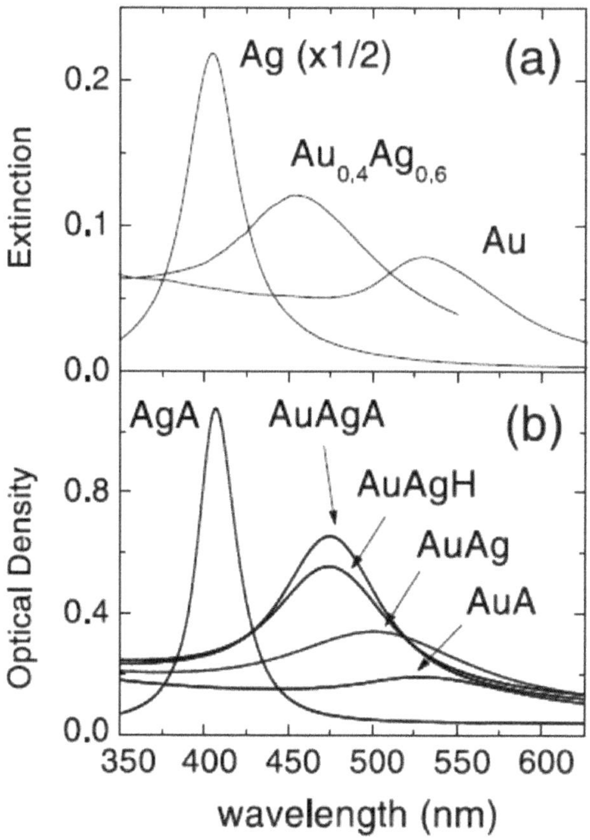

Fig. 9. Comparison between Mie-theory based calculated extinction spectra of 3 nm clusters of Au, Ag, and $Au_{0.4}Ag_{0.6}$ alloy in silica (9a) with the experimental absorption spectra of similar clusters prepared by ion implantation (9b). The notations "A" and "H" at the end of sample names indicate the samples were annealed in air and hydrogen atmosphere, respectively. Reproduced with permission from Nucl. Instrum. Meth. B 191 (2002) 323-332 (Mattei, 2002).

Using sequential implantation technique, Mazzoldi and collaborators were able to fabricate alloy nanoclusters of different compositions in fused silica glass at room temperature using 190 keV gold and 90 keV Cu ions (Gonella *et al.*, 1999). The alloy clusters of compositions close to their nominal values were formed in the silica matrix even without any post-implantation thermal treatment (Figure 10). However, Cu migrates towards the particle surface during thermal treatment and a structural reordering of the alloy occurs. The effect was more prominent when the implanted samples were annealed in air, and Cu migrated to the surface gets oxidized, causing a phase separation (Au core and Cu_2O shell). A complete separation of Cu from the alloy cluster and formation of CuO shell around the Au core could be obtained after 5 h of thermal annealing in air (Battaglin *et al.*, 2001b). The process could be reversed by annealing the sample again in a reducing atmosphere. It must be noted that for the thermal treatment of metal ion implanted composites we frequently use the value of the bulk melting point temperature, even though this quantity is substantially lower for a metallic cluster (either elemental or bimetallic), due to thermodynamic size effect (Buffat & Borel, 1976). Thermal annealing of a sequentially implanted sample helps to migrate one component elemental species towards the small clusters of the other element through diffusion, enhancing their size and improving their stoichiometry. In fact, the probability of formation of homogeneous alloy clusters in sequential implantation process strongly depends on the solubility of the second elemental specie in the nanoclusters of the first element. To increase the solubility and helping the structural rearrangement of metal ions, sometimes it is useful to perform the implantation of the second element at elevated target (substrate) temperatures.

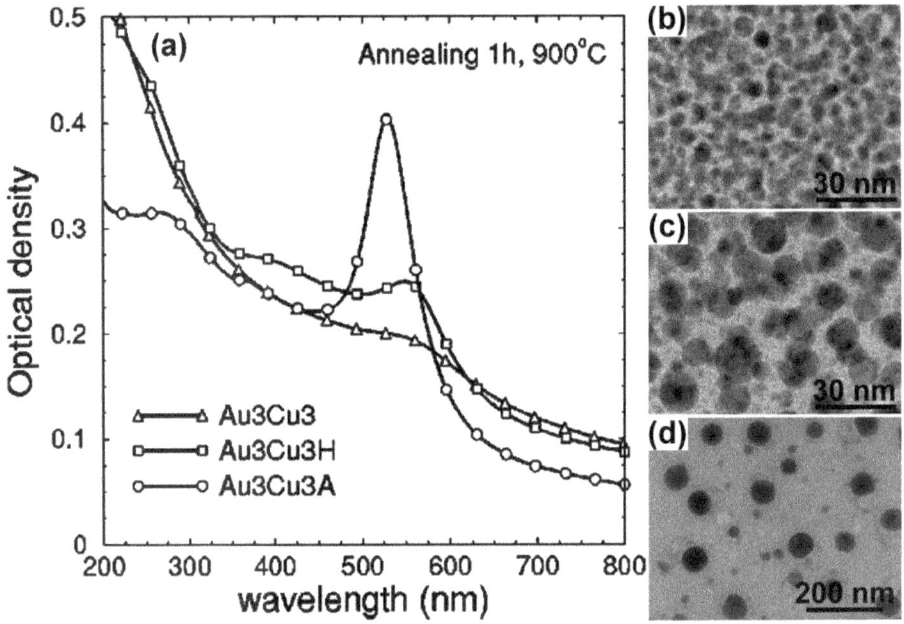

Fig. 10. (a) Optical absorption spectra of Au_3Cu_3 (as-implanted), Au_3Cu_3H, and Au_3Cu_3A (annealed at 900 ^0C for 1h in H_2-Ar and air, respectively) samples. The bright-field TEM planer views of the corresponding samples are presented as (b), (c), and (d), respectively, at right. Reproduced with permission from Appl. Phys. Lett. 75 (1999) 55-57 (Gonella *et al.*, 1999).

Utilizing the above mentioned strategy, Kluth and coworkers (Kluth *et al.*, 2006) could fabricate core-shell type Co-Au nanoclusters of about 5.3 nm diameter in SiO_2 matrix. Utilization of elevated target temperature (400 ^0C in the present case) during second metal (Co in this work) implantation helped to avoid the post-implanted thermal annealing process for the formation of bimetallic clusters. However, for the formation of Au clusters of uniform size before Co implantation, they used a thermal annealing treatment of the Au implanted sample at 1000 ^0C for one hour, in air. We can see the formation of uniform nanoparticles with Co core and Au shell (Figure 11) in their samples. As has been mentioned earlier, the formation of core-shell type bimetallic clusters depends strongly on the segregation of the component metals controlled by the minimization of surface free energy, resulting a phase separation as has been demonstrated theoretically for Co-Au (Johannessen *et al.*, 2005), Au-Pt (Esparza *et al.*, 2005), Au-Pd (Liu *et al.*, 2005, 2006), Au-Cu (Ascencio *et al.*, 2006) and many other systems.

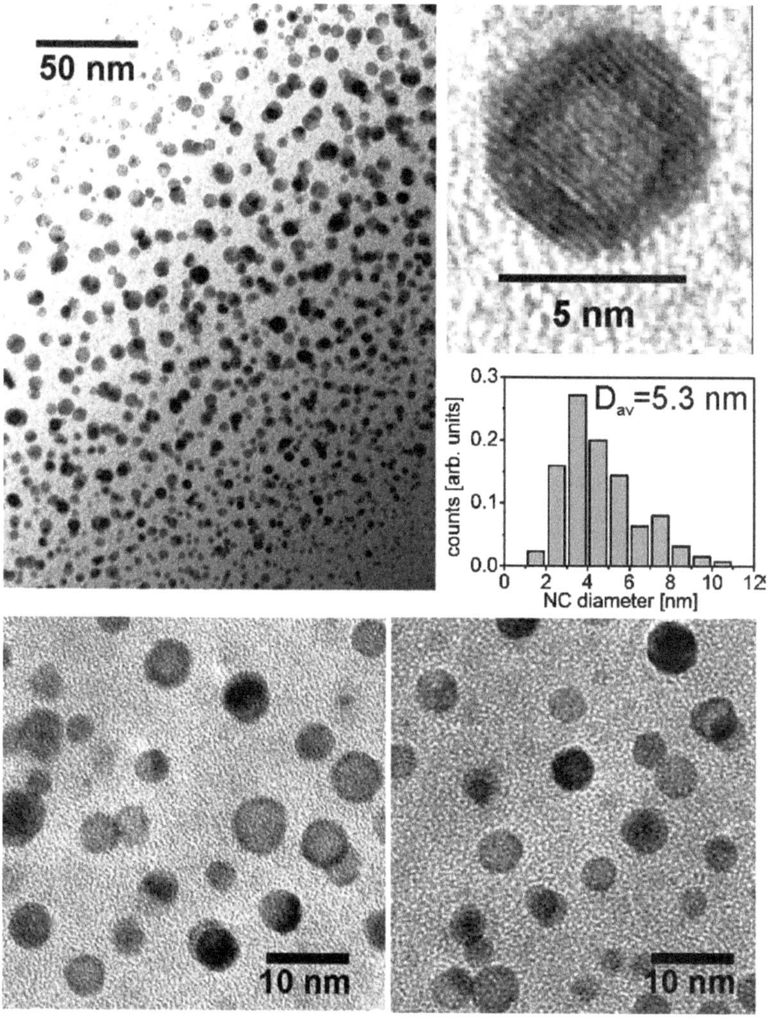

Fig. 11. Cross-sectional TEM images and size distribution of Co-Au nanoclusters as well as monometallic nanoclusters formed by sequential implantation. The high resolution TEM image of the core-shell cluster reveals a significant bending of the lattice fringes due to interfacial strain (lattice mismatch). Reproduced with permission from Appl. Phys. Lett. 89 (2006) 153118 (Kluth *et al.*, 2006).

Formation of Au-Ag core-shell nanoclusters in silica matrix through sequential implantation and subsequent thermal annealing (in air) has been demonstrated by Peña and coworkers (Peña *et al.*, 2009). Considering the miscibility and segregation behavior of the two metals, they first implanted 1.8 MeV Ag^+ ions at a 5.0×10^{16} ions cm^{-2} dose (fluence) to place most of them at about 1.0 μm depth below the target surface (Figure 12). Then, the Au^+ ions of higher energy (3.7 MeV) were implanted at a relative higher dose (6.7×10^{16} ion cm^{-2}) to place them at greater depth (about 1.1 μm) keeping a good portion of the implantation profiles overlapped. However, it has been observed that during the implantation of Au, pre-implanted Ag segregates towards the sample surface (distribution maximum moves to ~0.87 μm depth). On thermal annealing at 900 °C for 1 h in air, the co-implanted samples produced core-shell type bimetallic nanoclusters with Au core and Ag shell. The formation of both the monometallic nanoclusters (embedded in silica matrix) before thermal treatment and bimetallic nanoclusters (after thermal treatment) could be verified through the experimental optical absorption spectra in comparison with the Mie theory based simulated spectra of similar structures (Figure 13).

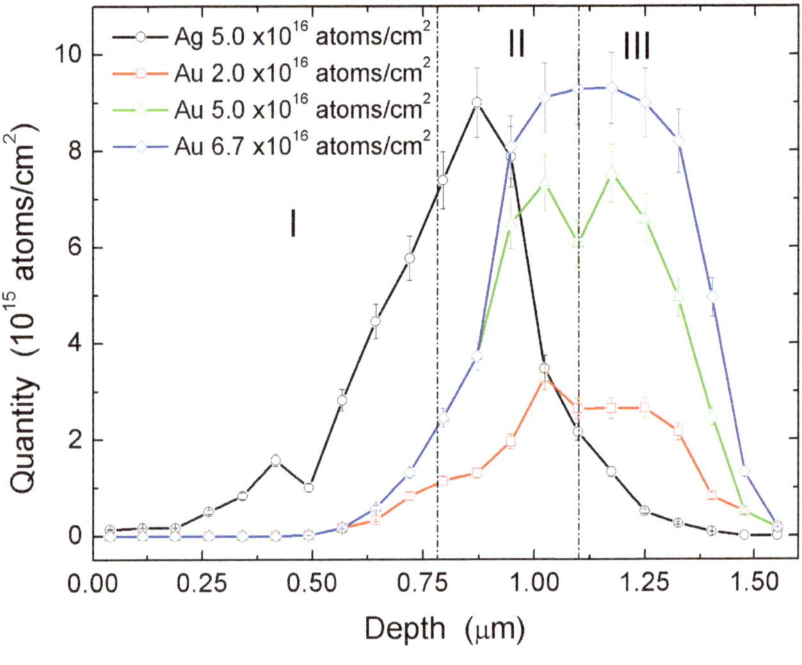

Fig. 12. Depth profiles of Au and Ag distribution for the silica samples sequentially implanted with Ag (1.8 MeV) and Au (3.5 MeV) ions, obtained from the fitting of the corresponding RBS spectra. Reproduced with permission from J. Phys. Chem. C 113 (2009) 2296-2300 (Peña *et al.*, 2009).

Though it was not possible to differentiate the core and shell regions from their HRTEM images, due to very similar lattice parameters of Au and Ag (4.078 Å for Au, JCPDS 4-0784; 4.077 Å for Ag, JCPDS 87-0720), the formation of core-shell structures could be demonstrated using high angle annular dark field (DAADF) images of the nanoclusters (Figure 14).

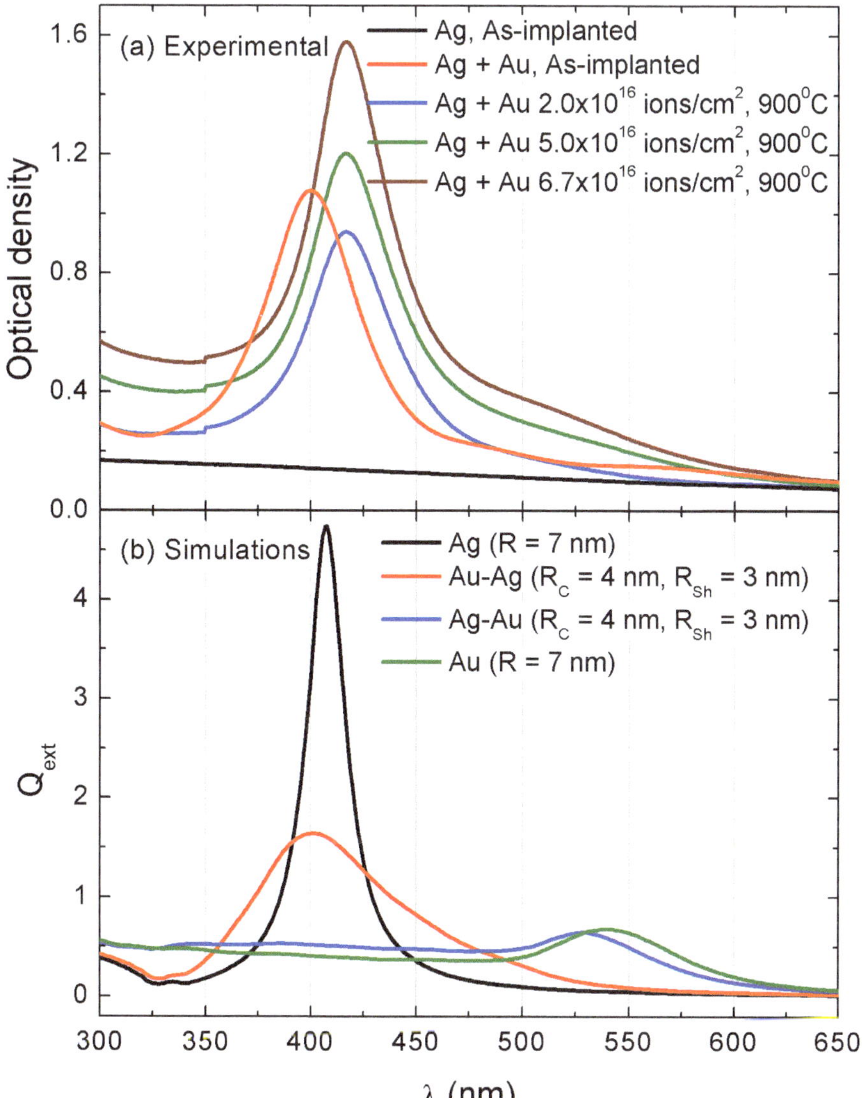

Fig. 13. Experimental optical extinction spectra for the samples implanted with $5x10^{16}$ ions cm^{-2} of Ag at 1.8 MeV and coimplanted with different fluences of Au at 3.5 MeV, before and after thermal annealing in air (a), and simulated (Mie theory) extinction spectra for some possible configurations in silica matrix (b). Au-Ag represents Au core-Ag shell, whereas Ag-Au means the opposite. Reproduced with permission from J. Phys. Chem. C 113 (2009) 2296-2300 (Peña *et al.*, 2009).

Indeed, the ion implantation process can produce both, alloy and core-shell type bimetallic nanoclusters in silica and other dielectric matrices by adequate manipulation of implantation parameters and proper post-implantation treatments. However, apart from producing monometallic and bimetallic nanoclusters of common types, ion implantation can produce several special metallic structures and aggregates, which we will discuss in the next section.

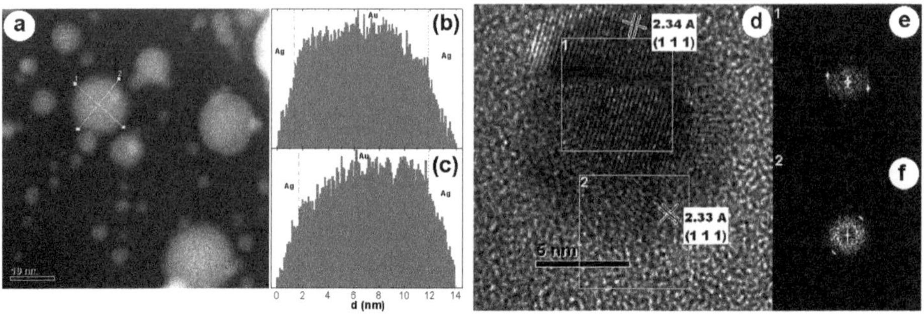

Fig. 14. Typical HAADF micrograph of the sample co-implanted with 5×10^{16} ions cm^{-2} of Ag (1.8 MeV) and 5×10^{16} ions cm^{-2} of Au (3.5 MeV) after thermal annealing at 900 °C (a) along with the intensity profiles for the selected particle (b) and (c). Typical HRTEM micrograph of a core-shell structure of about 12 nm diameter and the fast Fourier transforms (FFTs) of core (e) and shell (f) regions. Adapted with permission from J. Phys. Chem. C 113 (2009) 2296-2300 (Peña *et al.*, 2009).

3.4 Fabrication of nanocomposites containing unusual metallic nanostructures

The anisotropic plastic deformation process of amorphous materials under MeV ion irradiation has been reported by Klaumünzer and Schumacher as early as 1983 (Klaumünzer & Schumacher, 1983). The process can be understood by means of a visco-elastic model, in which the thermal expansion of the cylindrically-shaped high temperature ion track region causes shear stresses that relax by viscous flow, with the viscous strains subsequently frozen in upon rapid cooling of the ion track region (Trinkaus & Ryazanov, 1995; van Dillen *et al.*, 2005). The deformation is induced by the electron energy loss, and thus occurs mostly with ions in the MeV energy range.

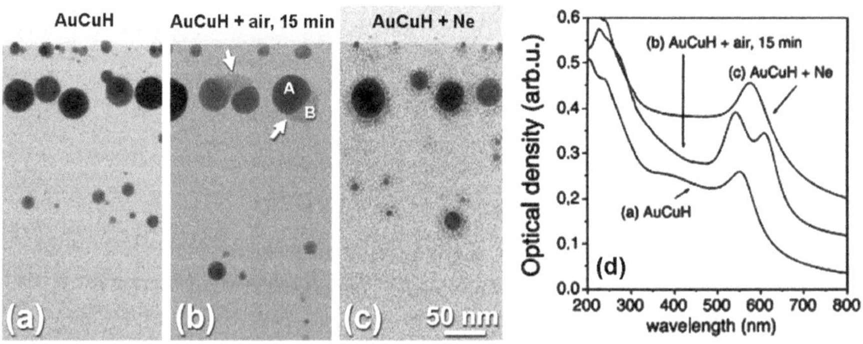

Fig. 15. Bright-field TEM cross-sectional micrographs of the AuCuH (annealed in H_2 (4.0%)-N_2 atmosphere at 900 ^{0}C, 1h) before (a), after a thermal annealing in air at 900 ^{0}C for 15 min (b) and after irradiation at room temperature with 190 keV Ne ions, at a dose of 1.0x10^{17} ions cm^{-2} (c). Adapted with permission from J. Non-Cryst. Sol. 322 (2003) 17-21 (Mattei *et al.*, 2003a).

Utilizing the above mentioned concept, ion implantation technique has been utilized for the fabrication of several unusual nanostructures either inside dielectric matrices or in other dielectric-metal combinations with distinct plasmonic behaviors. Though all of these nanostructures do not fall into the class of commonly known nanocomposites (where nanoparticles of one material are embedded/dispersed into another matrix material), as they are in combination with at least one material apart from the metal in question, we can call them as *composite nanostructures*. For example, Mattei and co-workers (Mattei *et al.*, 2003a, 2005) could disintegrate alloy clusters like Au-Cu and Au-Ag embedded in silica matrix by high energy Ne, He and Kr ions to form Au-rich satellite clusters around the initial bimetallic clusters (Figures 15 & 16).

While the optical absorption spectrum of the co-implanted (190 keV Au and 90 keV Cu, $3.0\text{x}10^{16}$ ions cm^{-2}) AuCu sample after annealing in inert atmosphere revealed SPR peak of AuCu alloy particles (550 nm, curve (a) of the Figure 15), after irradiation of 190 keV Ne ions ($1.0\text{x}10^{17}$ ions cm^{-2}) the SPR peak shifts to 575 nm, which correspond to the SPR band of the main particle along with the satellite ones (aggregate). Similar phenomenon occurred for the AuAg nanoclusters embedded in silica matrix. It has been observed that in both cases, mainly the Au segregates out of the main bimetallic cluster to form Au-rich satellite clusters of smaller sizes. The size of the satellite clusters depends on the atomic mass of the irradiating ions. Higher atomic mass of the irradiating ions produces satellite clusters of bigger sizes (Figure 16). However, the phenomenon of disintegration of the preformed noble metal clusters in dielectric matrix occurs when the atomic mass of the irradiating ions is lower than the atomic mass of the embedded metal nanoclusters. The process was tentatively related to the nuclear component of the energy released to the system during irradiation with different ions.

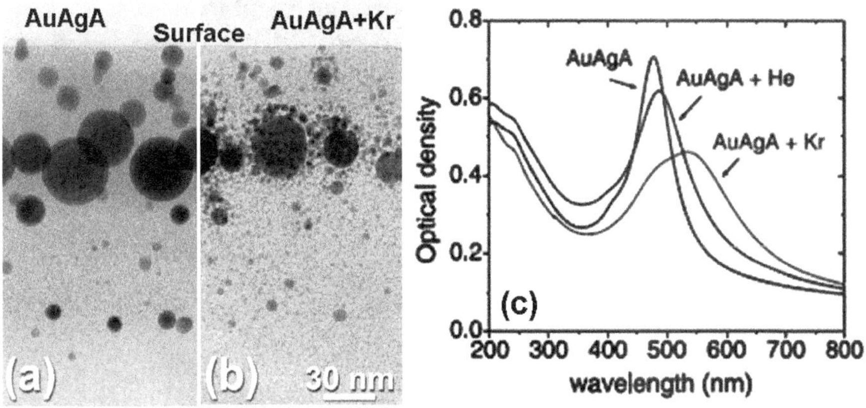

Fig. 16. Bright-field TEM cross-sectional micrographs of the sample AuAgA (annealed in air atmosphere at 800 ^{0}C, 1h) before (a) and after irradiation at room temperature with 380 keV Kr ions at a dose of $2.0\text{x}10^{16}$ ions cm^{-2} (b) (Mattei et al., 2003a). Adapted with permission from J. Non-Cryst. Sol. 322 (2003) 17-21 (Mattei *et al.*, 2003a).

On the other hand, irradiation by high energy ions of higher atomic mass (with respect to the implanted metals in a dielectric matrix) generates very different phenomena. Generally, when an embedded metallic nanocluster is irradiated by heavier ions of higher energy (in comparison to the energy of implantation of the metal clusters), the preformed nanocluster can grow in bigger size or get deformed. For the low dose/fluence of irradiation, the preformed metallic nanoclusters grow in size maintaining their initial shape. However, an increased dose of irradiation generates deformation of the nanoclusters. In fact, several researchers have reported this deformation process for the noble metal nanoclusters embedded in amorphous dielectric matrices. While D'Orléans and co-workers (D'Orléans *et al.*, 2003) could observe this effect for the preformed Co nanoclusters embedded in silica matrix upon irradiating them by 200 MeV ^{127}I ions, Awazu and co-workers (Awazu *et al.*, 2009) could observe this effect for the pre-implanted gold and silver nanoclusters embedded in silica matrix (Figure 17).

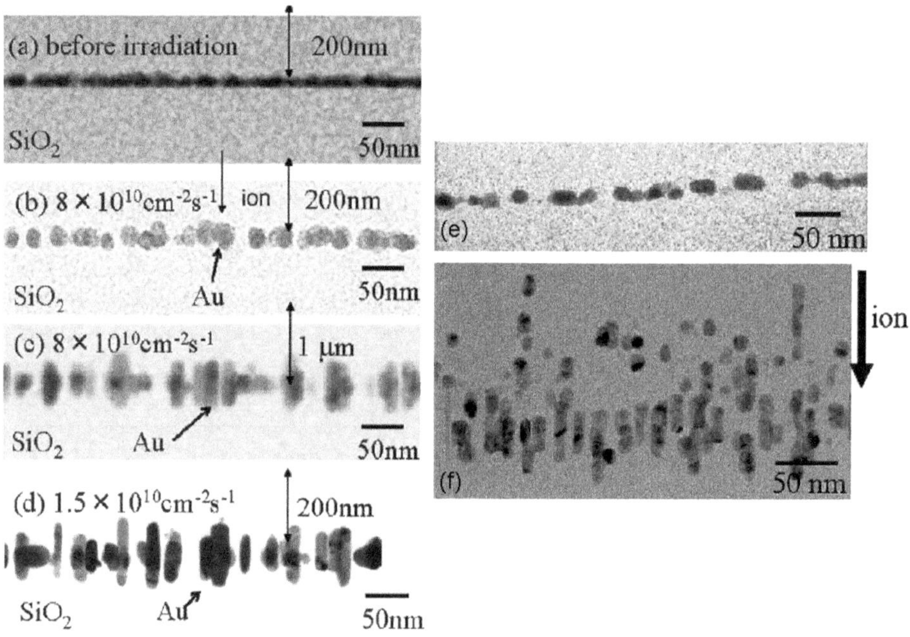

Fig. 17. Cross-sectional TEM image of Au (left) and Ag (right) nanoparticles embedded in SiO_2 before and after irradiation with 110 MeV Br^{10+} ions at different doses. The Ag nanoparticles were irradiated with 1.0×10^{14} ions cm^{-2} dose. Adapted with permission from Nucl. Instrum. Meth. B 267 (2009) 941-943 (Awazu *et al.*, 2009).

Ag and Au nanoclusters could also be deformed on irradiating them by high energy Si ions to form elongated or rod shaped aligned nanoclusters in silica matrix (Rodríguez-Iglesias *et al.*, 2010a; b). As we can see from Figure 18, the elongated nanoclusters are well aligned, and generate SPR absorption bands corresponding to longitudinal and transversal vibrational modes of prolate nanoclusters on oblique incidence. Indeed, the metallic clusters get deformed upon irradiation of high energy heavy ions along the direction of incident ions; i.e. they get elongated in this direction. On the contrary, the deformation or expansion of nonmetallic or dielectric clusters occurs in the perpendicular direction of the irradiating ion beams.

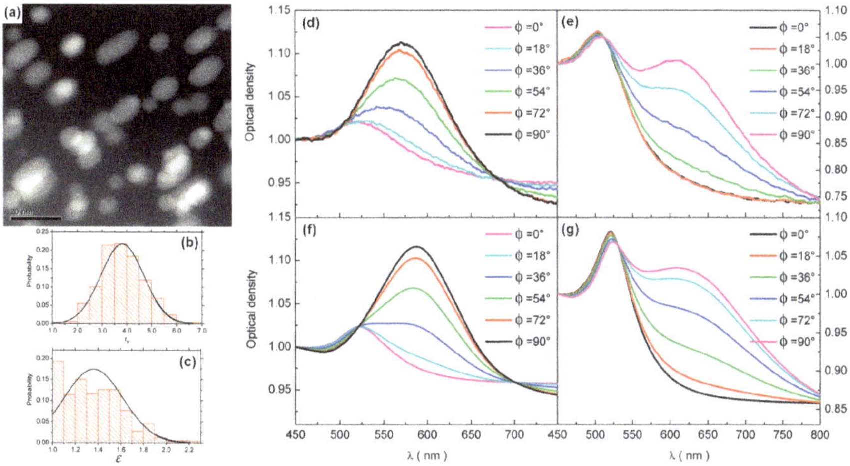

Fig. 18. Experimental optical extinction spectra of elongated Au nanoclusters (a)-(c), measured for the light incident at (d) 90°, and (e) 29° and polarization angles varying between 0° (p-polarized) and 90° (s-polarized). Corresponding simulated spectra (Mie theory) are presented in (f), and (g), respectively. All the spectra were normalized to 1 at the 450 nm wavelength for better visualization. Adapted with permission from J. Phys. Chem. C 114 (2010) 746-751 (Rodríguez-Iglesias *et al.*, 2010b).

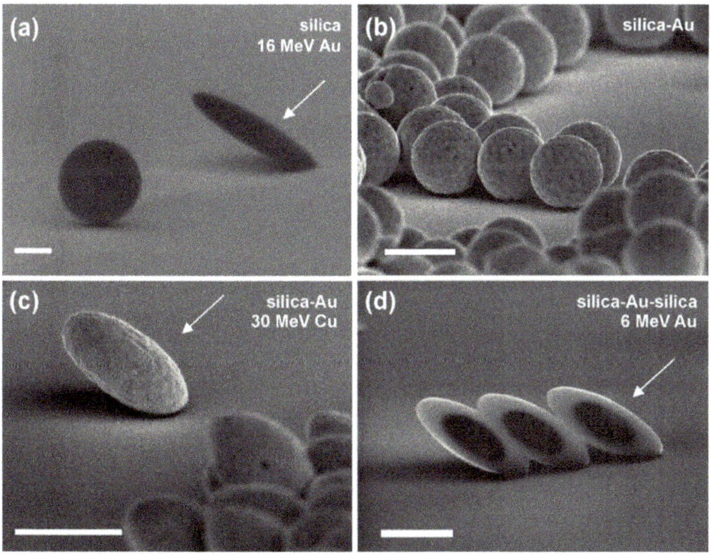

Fig. 19. Scanning electron microscopy images of (a) silica colloid (1.0 μm diameter) before (left) and after (right) 16 MeV Au irradiation (1.1×10^{15} ions cm^{-2}); (b) silica-Au core-shell colloid (d_{SiO2} = 438 nm, t_{Au} = 46 nm); (c) same as (b), after 30 MeV Cu irradiation (5.0×10^{14} ions cm^{-2}); (d) silica-Au-silica core-shell-shell colloid (d_{SiO2} = 300 nm, t_{Au} = 25 nm, t_{SiO2} = 133 nm) after 6 MeV Au irradiation (6.5×10^{14} ions cm^{-2}). Scale bars represent 500 nm. Reproduced with permission from Nucl. Instrum. Meth. B 242 (2006) 523-529 (Penninkhof *et al.*, 2006).

Utilizing this deformation behavior of metal and dielectric nanoparticles, Pennikhof and co-workers (Penninkhof *et al.*, 2006) could produce colloidal prolate core-shell type Au-SiO_2 and reverse nanoparticles as big as 1 μm in size (Figure 19) by irradiating high energy Au or Cu ions. As the ion irradiation induced deformation is more prominent for a dielectric like SiO_2 than a metal like Au, the extent of deformation of the SiO_2-core Au-shell particles is not as drastic as of Au-core SiO_2-shell particles. In fact, the presence of a thin layer (shell) of metal imposes a mechanical constrain on the deforming silica core, reducing the net deformation compared to that of pure silica (Penninkhof *et al.*, 2006).

Optical characteristics of all the above plasmonic structures, which we defined as 'special structures' are quite different from that of conventional nanocomposites where the shape of incorporated/embedded noble metal nanoclusters remains in its most stable or spherical form. As the technological application of plasmon-based nanocomposites remains on the ability of manipulating their optical behavior, which is possible mainly by controlling the size, shape and other geometrical parameters, rather than just changing their composition. Ion implantation technique has demonstrated its great potential for their fabrication and manipulations.

4. Prospects of ion-implantation for the fabrication of plasmonic nanocomposites

Significant advances have been made in the last few decades on the fabrication of noble metal plasmonic nanocomposites through ion implantation. By controlling the implantation parameters and post implantation annealing, noble metal nanocomposites with embedded elemental, alloy and core-shell type NPs could be produced with moderate control over the NP size, size distribution, composition and location in the matrix. The significant advantage of ion implantation technique is that the treated surface is an integral part of the sample and consequently does not suffer from possible adhesion problems associated with coatings. Therefore, the implanted structures remain as the integral parts of an optical device. The moderate heating associated with the process virtually eliminates any risks of distortion or oxidation effects. On the other hand, ion implantation produces no macroscopic change in the workpiece. From the point of view of applications, though several plasmonic nanocomposites with embedded noble metal clusters of tailored compositions and even of core-shell geometries could be fabricated extending their SPR peak position to a broader spectral range (in comparison with monometallic nanocomposites), tuning the relative thickness of the core and shell layer has not been sufficiently demonstrated. As has been shown recently (Peña-Rodríguez & Pal, 2010, 2011a; b; Peña-Rodríguez *et al.*, 2011a; b), application potential of plasmonic nanostructures depends on their enhanced SPR extinction, along with a precise control over its peak position. For this purpose, it is possible that in the near future we will need to fabricate multi-layer metallic nanoclusters with precise control over the thickness of each layer. Though the modern ion implanters are capable of introducing virtually any element in a solid matrix with a wide variation of energy and implantation dose, there has been no attempt to fabricate nanocomposites with trimetallic nanoclusters incorporated into dielectric matrix. On the other hand, fewer efforts have been devoted to control the shell-thickness in a core-shell type nanocluster or the fabrication of multilayer metallic nanoclusters. However, very recently this technique has been utilized for the fabrication of unusual composite nanostructures with distinct plasmonic behaviors, which are highly attractive for several applications from optical sensing to biological treatments. Considering the current interest on plasmonic nanostructures and nanocomposites and the demonstrated potential of available commercial ion implanters, utilization of this technique for fabricating new exotic nanostructures in the near future is very much expected.

5. Acknowledgments

The authors extend their apology to the several other authors who have contributed significantly in this field, but were not cited due to limited space of this review. The work was supported by CONACyT, Mexico (Grant # 151767) and VIEP-BUAP (Grant # VIEP/EXC/2011). OPR thanks CONACyT, Mexico for extending a postdoctoral fellowship.

6. References

Aaron, H. B., Fainstein, D. & Kotler, G. R. (1970). Diffusion-limited phase transformations: A comparison and critical evaluation of the mathematical approximations. *J. Appl. Phys.* Vol.41, No.11, pp.4404-4410, ISSN 0021-8979.

Agulló-López, F., Catlow, C. R. A. & Townsend, P. D. (1988). *Point defects in materials.* San Diego, CA, USA: Academic Press., ISBN 0120445107.

Alivisatos, P. (2004). The use of nanocrystals in biological detection. *Nat. Biotechnol.* Vol.22, No.1, pp.47-52, ISSN 1087-0156.

Allain, L. R. & Vo-Dinh, T. (2002). Surface-enhanced Raman scattering detection of the breast cancer susceptibility gene BRCA1 using a silver-coated microarray platform. *Anal. Chim. Acta* Vol.469, No.1, pp.149-154, ISSN 0003-2670.

Arai, N., Tsuji, H., Ueno, K., Matsumoto, T., Gotoh, N., Aadachi, K., Kotaki, H., Gotoh, Y. & Ishikawa, J. (2006). Formation of silver nanoparticles aligned near the bottom of SiO_2 film on silicon substrate by negative-ion implantation and post-annealing. *Nucl. Instrum. Meth. B* Vol.242, No.1-2, pp.217-220, ISSN 0168-583X.

Arnold, G. W. & Borders, J. A. (1977). Aggregation and migration of ion-implanted silver in lithia-alumina-silica glass. *J. Appl. Phys.* Vol.48, No.4, pp.1488-1496, ISSN 0021-8979.

Ascencio, J. A., Liu, H. B., Pal, U., Medina, A. & Wang, Z. L. (2006). Transmission electron microscopy and theoretical analysis of AuCu nanoparticles: Atomic distribution and dynamic behavior. *Microsc. Res. Tech.* Vol.69, No.7, pp.522-530, ISSN 1097-0029.

Atwater, H. A. & Polman, A. (2010). Plasmonics for improved photovoltaic devices. *Nat. Mater.* Vol.9, No.3, pp.205-213, ISSN 1476-1122.

Awazu, K., Wang, X., Fujimaki, M., Tominaga, J., Fujii, S., Aiba, H., Ohki, Y. & Komatsubara, T. (2009). Mechanism of elongation of gold or silver nanoparticles in silica by irradiation with swift heavy ions. *Nucl. Instrum. Meth. B* Vol.267, No.6, pp.941-943, ISSN 0168-583X.

Ballesteros, J. M., Serna, R., Solís, J., Afonso, C. N., Petford-Long, A. K., Osborne, D. H. & Haglund, R. F. (1997). Pulsed laser deposition of Cu:Al_2O_3 nanocrystal thin films with high third-order optical susceptibility. *Appl. Phys. Lett.* Vol.71, No.17, pp.2445, ISSN 0003-6951.

Bamford, C. R. (1977). *Colour generation and control in glass.* 1. ed. Amsterdam: Elsevier Science Ltd., ISBN 0444416145.

Bandourko, V., Umeda, N., Plaksin, O. & Kishimoto, N. (2005). Heavy-ion-induced luminescence of amorphous SiO_2 during nanoparticle formation. *Nucl. Instrum. Meth. B* Vol.230, No.1-4, pp.471-475, ISSN 0168-583X.

Barreca, D., Gasparotto, A., Tondello, E., Bruno, G. & Losurdo, M. (2004). Influence of process parameters on the morphology of Au/SiO_2 nanocomposites synthesized by radio-frequency sputtering. *J. Appl. Phys.* Vol.96, No.3, pp.1655-1665, ISSN 0021-8979.

Battaglin, G., Catalano, M., Cattaruzza, E., D'Acapito, F., De Julian Fernandez, C., De Marchi, G., Gonella, F., Mattei, G., Maurizio, C., Mazzoldi, P., Miotello, A. & Sada, C. (2001a). Influence of annealing atmosphere on metal and metal alloy nanoclusters produced by ion implantation in silica. *Nucl. Instrum. Meth. B* Vol.178, No.1-4, pp.176-179, ISSN 0168-583X.

Battaglin, G., Cattaruzza, E., Gonella, F., Mattei, G., Mazzoldi, P., Sada, C. & Zhang, X. (2000). Formation of metal-alloy nanoclusters in silica by ion implantation and annealing in selected atmosphere. *Nucl. Instrum. Meth. B* Vol.166-167, No.0, pp.857-863, ISSN 0168-583X.

Battaglin, G., Cattaruzza, E., de Julian Fernandez, C., De Marchi, G., Gonella, F., Mattei, G., Maurizio, C., Mazzoldi, P., Miotello, A., Sada, C. & D'Acapito, F. (2001b). Influence of post-implantation thermal and laser annealing on the stability of metal-alloy nanoclusters in silica. *Nucl. Instrum. Meth. B* Vol.175-177, pp.410-416, ISSN 0168-583X.

Bianconi, M., Bentini, G. G., Chiarini, M., De Nicola, P., Montanari, G. B., Nubile, A. & Sugliani, S. (2009). Defect engineering and micromachining of Lithium Niobate by ion implantation. *Nucl. Instrum. Meth. B* Vol.267, No.17, pp.2839-2845.

Borsella, E., Garcia, M. A., Mattei, G., Maurizio, C., Mazzoldi, P., Cattaruzza, E., Gonella, F., Battaglin, G., Quaranta, A. & D'Acapito, F. (2001). Synthesis of GaN quantum dots by ion implantation in dielectrics. *J. Appl. Phys.* Vol.90, No.9, pp.4467-4473, ISSN 1089-7550.

Buchal, C., Withrow, S. P., White, C. W. & Poker, D. B. (1994). Ion implantation of optical materials. *Ann. Rev. Mater. Sci.* Vol.24, pp.125-158, ISSN 0084-6600.

Buffat, P. & Borel, J.-P. (1976). Size effect on the melting temperature of gold particles. *Phys. Rev. A* Vol.13, No.6, pp.2287-2298, ISSN 1094-1622.

Burakowski, T. & Wierzchon, T. (1998). Implantation Techniques (Ion Implantation). *Surface Engineering of Metals: Principles, Equipment, Technologies*. CRC Press. (Materials Science & Technology)., ISBN 9780849382253.

Cao, M., Wang, M. & Gu, N. (2009). Optimized surface plasmon resonance sensitivity of gold nanoboxes for sensing applications. *J. Phys. Chem. C* Vol.113, No.4, pp.1217-1221.

Catchpole, K. R. & Polman, A. (2008). Plasmonic solar cells. *Opt. Express* Vol.16, No.26, pp.21793-21800, ISSN 1094-4087.

Cui, Y., Ren, B., Yao, J.-L., Gu, R.-A. & Tian, Z.-Q. (2006). Synthesis of Ag_{core}-Au_{shell} bimetallic nanoparticles for immunoassay based on surface-enhanced Raman spectroscopy. *J. Phys. Chem. B* Vol.110, No.9, pp.4002-4006, ISSN 1520-5207.

D'Orléans, C., Stoquert, J. P., Estournès, C., Cerruti, C., Grob, J. J., Guille, J. L., Haas, F., Muller, D. & Richard-Plouet, M. (2003). Anisotropy of Co nanoparticles induced by swift heavy ions. *Phys. Rev. B* Vol.67, No.22, pp.220101, ISSN 1550-235X.

Davenas, J., Perez, A., Thevenard, P. & Dupuy, C. H. S. (1973). Correlation between absorption bands and implanted alkali ions in LiF. *Phys. Stat. Sol. A* Vol.19, No.2, pp.679-686, ISSN 1521-396X.

De, G. (1997). Sol-gel synthesis of metal nanoclusters-silica composite films. *J. Sol-Gel Sci. Technol.* Vol.11, No.3, pp.289-298, ISSN 1573-4846.

Deying, S., Saito, Y. & Suganomata, S. (1994). Optical properties of $LiNbO_3$ implanted with Ag^+ ions. *Jpn. J. Appl. Phys.* Vol.33, pp.L966-L969, ISSN 0021-4922.

Dhara, S., Sundaravel, B., Ravindran, T. R., Nair, K. G. M., David, C., Panigrahi, B. K., Magudapathy, P. & Chen, K. H. (2004). "Spillout" effect in gold nanoclusters embedded in c-Al_2O_3(0 0 0 1) matrix. *Chem. Phys. Lett.* Vol.399, No.4-6, pp.354-358, ISSN 0009-2614.

van Dillen, T., Polman, A., Onck, P. R. & van der Giessen, E. (2005). Anisotropic plastic deformation by viscous flow in ion tracks. *Phys. Rev. B* Vol.71, No.2, pp.024103.

Ding, I., Zhu, J., Cai, W., Moon, S., Cai, N., Wang, P., Zakeeruddin, S. M., Grätzel, M., Brongersma, M. L., Cui, Y. & McGehee, M. D. (2011). Plasmonic dye-sensitized solar cells. *Adv. Ener. Mater.* Vol.1, No.1, pp.52-57, ISSN 1614-6840.

Donnet, C., Marest, G., Moncoffre, N., Tousset, J., Rahioui, A., Esnouf, C. & Brunel, M. (1991). Copper, iron and zirconium implantation into polycrystalline α-Al_2O_3. *Nucl. Instrum. Meth. B* Vol.59-60, No.2, pp.1205-1210, ISSN 0168-583X.

Duhan, S., Devi, S. & Srivastava, M. (2010). Characterization of nanocrystalline Ag/SiO_2 nanocomposites and synthesis by wet chemical method. *Ind. J. Pure & Appl. Phys.* Vol.48, No.4, pp.271-275, ISSN 0019-5596.

Esparza, R., Ascencio, J. A., Rosas, G., Sánchez Ramírez, J. F., Pal, U. & Perez, R. (2005). Structure, stability and catalytic activity of chemically synthesized Pt, Au, and AuPt nanoparticles. *J. Nanosci. Nanotechnol.* Vol.5, No.4, pp.641-647, ISSN 1533-4899.

Flytzanis, C., Hache, F., Klein, M. C., Ricard, D. & Roussignol, P. (1991). Nonlinear optics in composite materials. In: Wolf, E. (Ed.) *Progress in Optics.* pp 321-411. Elsevier Science Pub Co., ISBN 0444889515.

García-Serrano, J. & Pal, U. (2003). Synthesis and characterization of Au nanoparticles in Al_2O_3 matrix. *Int. J. Hydrog. Energy* Vol.28, No.6, pp.637-640, ISSN 0360-3199.

Gonella, F., Mattei, G., Mazzoldi, P., Sada, C., Battaglin, G. & Cattaruzza, E. (1999). Au-Cu alloy nanoclusters in silica formed by ion implantation and annealing in reducing or oxidizing atmosphere. *Appl. Phys. Lett.* Vol.75, No.1, pp.55-57, ISSN 0003-6951.

Haug, J., Dubiel, M., Kruth, H. & Hofmeister, H. (2009). Structural characterization of bimetallic Ag-Au nanoparticles in glass. *J. Phys.: Conf. Ser.* Vol.190, No.1, pp.012124, ISSN 1742-6596.

Hazra, S., Gibaud, A. & Sella, C. (2004). Tunable absorption of Au-Al_2O_3 nanocermet thin films and its morphology. *Appl. Phys. Lett.* Vol.85, No.3, pp.395-397, ISSN 0003-6951.

Hirsch, L. R., Jackson, J. B., Lee, A., Halas, N. J. & West, J. L. (2003a). A whole blood immunoassay using gold nanoshells. *Anal. Chem.* Vol.75, No.10, pp.2377-2381, ISSN 1520-6882.

Hirsch, L. R., Stafford, R. J., Bankson, J. A., Sershen, S. R., Rivera, B., Price, R. E., Hazle, J. D., Halas, N. J. & West, J. L. (2003b). Nanoshell-mediated near-infrared thermal therapy of tumors under magnetic resonance guidance. *Proc. Natl. Acad. Sci. U.S.A.* Vol.100, No.23, pp.13549-13554, ISSN 0027-8424.

Hövel, H., Fritz, S., Hilger, A., Kreibig, U. & Vollmer, M. (1993). Width of cluster plasmon resonances: Bulk dielectric functions and chemical interface damping. *Phys. Rev. B* Vol.48, No.24, pp.18178.

Hughes, A. E. (1983). Colloid formation in irradiated insulators. *Rad. Eff. Def. Solids* Vol.74, No.1-4, pp.57-76, ISSN 0033-7579.

Ikeyama, M., Nakao, S., Tazawa, M., Kadono, K. & Kamada, K. (2001). Optical property changes of silica glass and sapphire induced by Cu and O implantation. *Nucl. Instrum. Meth. B* Vol.175-177, pp.652-657, ISSN 0168-583X.

Ila, D., Williams, E. K., Sarkisov, S., Poker, D. B. & Hensley, D. K. (1997). Change in the optical properties of sapphire induced by ion implantation. *Mat. Res. Soc. Proc.* Vol.504, pp.381-386, ISSN 0272-9172.

Jackson, J. B. & Halas, N. J. (2004). Surface-enhanced Raman scattering on tunable plasmonic nanoparticle substrates. *Proc. Natl. Acad. Sci. U.S.A.* Vol.101, No.52, pp.17930-17935, ISSN 0027-8424.

Jiang, T., Miao, L., Tanemura, S., Tanemura, M., Xu, G. & Wang, R. P. (2009). Tunable Ag surface-plasmon-resonance wavelength and its application on the photochromic behavior of TiO_2-Ag films. *Superlattices Microstruct.* Vol.46, No.1-2, pp.159-165, ISSN 0749-6036.

Johannessen, B., Kluth, P., Glover, C. J., de M. Azevedo, G., Llewellyn, D. J., Foran, G. J. & Ridgway, M. C. (2005). Structural characterization of Cu nanocrystals formed in SiO_2 by high-energy ion-beam synthesis. *J. Appl. Phys.* Vol.98, No.2, pp.024307, ISSN 0021-8979.

Johnson, P. B. & Christy, R. W. (1972). Optical constants of the noble metals. *Phys. Rev. B* Vol.6, No.12, pp.4370-4379, ISSN 1550-235X.

Joseph, B., Sandeep, C. S. S., Sekhar, B. R., Mahapatra, D. P. & Philip, R. (2007). Nonlinear optical properties of MeV and keV ion beam synthesized Ag nanoclusters. *Nucl. Instrum. Meth. B* Vol.265, No.2, pp.631-636, ISSN 0168-583X.

Kalidindi, S. B. & Jagirdar, B. R. (2008). Synthesis of Cu@ZnO core-shell nanocomposite through digestive ripening of Cu and Zn nanoparticles. *J. Phys. Chem. C* Vol.112, No.11, pp.4042-4048, ISSN 1932-7447.

Kim, D.-G., Koyama, E., Tokuhisa, H., Koshizaki, N. & Kim, Y. D. (2008). Tunable aggregation of Au nanoparticles in Au/SiO_2 composite film and its photo-absorbance. *Appl. Phys. A* Vol.92, pp.263-266, ISSN 1432-0630.

Klaumünzer, S. & Schumacher, G. (1983). Dramatic growth of glassy $Pd_{80}Si_{20}$ during heavy-ion irradiation. *Phys. Rev. Lett.* Vol.51, No.21, pp.1987-1990, ISSN 1079-7114.

Kluth, P., Hoy, B., Johannessen, B., Dunn, S. G., Foran, G. J. & Ridgway, M. C. (2006). Co-Au core-shell nanocrystals formed by sequential ion implantation into SiO_2. *Appl. Phys. Lett.* Vol.89, No.15, pp.153118, ISSN 0003-6951.

Kono, K., Arora, S. K. & Kishimoto, N. (2003). Modification in optical properties of negative Cu ion implanted ZnO. *Nucl. Instrum. Meth. B* Vol.206, pp.291-294, ISSN 0168-583X.

Kooi, B. J., van Veen, A., De Hosson, J. T. M., Schut, H., Fedorov, A. V. & Labohm, F. (2000). Rectangular nanovoids in helium-implanted and thermally annealed MgO(100). *Appl. Phys. Lett.* Vol.76, No.9, pp.1110-1112, ISSN 0003-6951.

Kreibig, U. & Vollmer, M. (1995). *Optical properties of metal clusters*. 1. ed. Berlin, Heidelberg: Springer., ISBN 3540578366.

Kucheyev, S. O., Williams, J. S., Jagadish, C., Zou, J., Evans, C., Nelson, A. J. & Hamza, A. V. (2003). Ion-beam-produced structural defects in ZnO. *Phys. Rev. B* Vol.67, No.9, pp.094115, ISSN 1550-235X.

Lifshitz, M. & Slezov, V. V. (1959). Kinetics of diffusive decomposition of supersaturated solid solutions. *Soviet Phys. JETP* Vol.35, pp.331-339, ISSN 0038-5646.

Liu, H. B., Pal, U., Medina, A., Maldonado, C. & Ascencio, J. A. (2005). Structural incoherency and structure reversal in bimetallic Au-Pd nanoclusters. *Phys. Rev. B* Vol.71, No.7, pp.075403, ISSN 1550-235X.

Liu, H. B., Pal, U., Perez, R. & Ascencio, J. A. (2006). Structural transformation of Au-Pd bimetallic nanoclusters on thermal heating and cooling: a dynamic analysis. *J. Phys. Chem. B* Vol.110, No.11, pp.5191-5195, ISSN 1520-6106.

Ma, Z. Y., Dosev, D. & Kennedy, I. M. (2009). A microemulsion preparation of nanoparticles of europium in silica with luminescence enhancement using silver. *Nanotechnol.* Vol.20, No.8, pp.085608, ISSN 1361-6528.

Manoilov, E. G. (2009). Optical and photoluminescent properties of Ag/Al_2O_3 nanocomposite films obtained by pulsed laser deposition. *Semicond. Phys. Quant. Electron. & Opt.* Vol.12, No.3, pp.298-301, ISSN 1560-8034.

Marques, C., Alves, E., da Silva, R. C., Silva, M. R. & Stepanov, A. L. (2004). Optical changes induced by high fluence implantation of Au ions on sapphire. *Nucl. Instrum. Meth. B* Vol.218, pp.139-144, ISSN 0168-583X.

Mattei, G. (2002). Alloy nanoclusters in dielectric matrix. *Nucl. Instrum. Meth. B* Vol.191, No.1-4, pp.323-332, ISSN 0168-583X.

Mattei, G., Battaglin, G., Bello, V., De Marchi, G., Maurizio, C., Mazzoldi, P., Parolin, M. & Sada, C. (2003a). De-alloying behaviour of metal nanoclusters in SiO_2 upon irradiation and thermal treatments. *J. Non-Cryst. Sol.* Vol.322, No.1-3, pp.17-21, ISSN 0022-3093.

Mattei, G., Bello, V., Mazzoldi, P., Pellegrini, G., Sada, C., Maurizio, C. & Battaglin, G. (2005). Modification of composition and structure of bimetallic nanocluster in silica by ion beam irradiation. *Nucl. Instrum. Meth. B* Vol.240, No.1-2, pp.128-132, ISSN 0168-583X.

Mattei, G., Marchi, G. D., Maurizio, C., Mazzoldi, P., Sada, C., Bello, V. & Battaglin, G. (2003b). Chemical- or radiation-assisted selective dealloying in bimetallic nanoclusters. *Phys. Rev. Lett.* Vol.90, No.8, pp.085502, ISSN 1079-7114.

Mazzoldi, P. (1987). *Ion beam modifications of insulators*. New York: Elsevier Science Ltd., ISBN 044442816X.

Mazzoldi, P., Arnold, G. W., Battaglin, G., Gonella, F. & Haglund, R. F. (1996). Metal nanocluster formation by ion implantation in silicate glasses: Nonlinear optical applications. *J. Nonlinear Opt. Phys. Mater.* Vol.5, No.2, pp.285-330, ISSN 0218-8635.

Mendoza-Galván, A., Trejo-Cruz, C., Lee, J., Bhattacharyya, D., Metson, J., Evans, P. J. & Pal, U. (2006). Effect of metal-ion doping on the optical properties of nanocrystalline ZnO thin films. *J. Appl. Phys.* Vol.99, No.1, pp.014306, ISSN 0021-8979.

Mishra, Y. K., Mohapatra, S., Singhal, R., Avasthi, D. K., Agarwal, D. C. & Ogale, S. B. (2008). Au-ZnO: A tunable localized surface plasmonic nanocomposite. *Appl. Phys. Lett.* Vol.92, No.4, pp.043107, ISSN 0003-6951.

Mohapatra, S., Ghatak, J., Joseph, B., Lenka, H. P., Kuiri, P. K. & Mahapatra, D. P. (2007). Anomalous diffusion of Au in mega-electron-volt Au implanted $SiO_2/Si(100)$. *J. Appl. Phys.* Vol.101, No.6, pp.063542, ISSN 0021-8979.

Nakao, S., Nonami, T., Jin, P., Miyagawa, Y. & Miyagawa, S. (2000). High-energy metal ion implantation into titanium dioxide films. *Surf. Coat. Technol.* Vol.128-129, pp.446-449, ISSN 0257-8972.

Norton, D. P., Overberg, M. E., Pearton, S. J., Pruessner, K., Budai, J. D., Boatner, L. A., Chisholm, M. F., Lee, J. S., Khim, Z. G., Park, Y. D. & Wilson, R. G. (2003). Ferromagnetism in cobalt-implanted ZnO. *Applied Physics Letters* Vol.83, No.26, pp.5488-5490, ISSN 0003-6951.

Ohkubo, M. & Suzuki, N. (1988). Morphology of small gold crystals formed inside sapphire by ion implantation. *Phil. Mag. Lett.* Vol.57, No.5, pp.261-265, ISSN 1362-3036.

Okumu, J., Dahmen, C., Sprafke, A. N., Luysberg, M., von Plessen, G. & Wuttig, M. (2005). Photochromic silver nanoparticles fabricated by sputter deposition. *J. Appl. Phys.* Vol.97, No.9, pp.094305, ISSN 0021-8979.

Oliver, A., Cheang-Wong, J. C., Roiz, J., Rodríguez-Fernández, L., Hernández, J. M., Crespo-Sosa, A. & Muñoz, E. (2002). Metallic nanoparticle formation in ion-implanted silica after thermal annealing in reducing or oxidizing atmospheres. *Nucl. Instrum. Meth. B* Vol.191, No.1-4, pp.333-336, ISSN 0168-583X.

Pal, U., Aguila-Almanza, E., Vázquez-Cuchillo, O., Koshizaki, N., Sasaki, T. & Terauchi, S. (2001). Preparation of Au/ZnO nanocomposites by radio frequency co-sputtering. *Sol. Energy Mater. Sol. Cells* Vol.70, No.3, pp.363-368, ISSN 0927-0248.

Pal, U., Bautista-Hernández, A., Rodríguez-Fernández, L. & Cheang-Wong, J. . (2000). Effect of thermal annealing on the optical properties of high-energy Cu-implanted silica glass. *J. Non-Cryst. Sol.* Vol.275, No.1-2, pp.65-71, ISSN 0022-3093.

Pan, J., Wang, H., Takeda, Y., Umeda, N., Kono, K., Amekura, H. & Kishimoto, N. (2007). Effect of indentation and annealing on 2 MeV Cu ion-implanted SiO_2. *Nucl. Instrum. Meth. B* Vol.257, No.1-2, pp.585-588, ISSN 0168-583X.

Pavesi, L., Dal Negro, L., Mazzoleni, C., Franzo, G. & Priolo, F. (2000). Optical gain in silicon nanocrystals. *Nature* Vol.408, No.6811, pp.440-444, ISSN 0028-0836.

Pellegrini, G., Bello, V., Mattei, G. & Mazzoldi, P. (2007a). Local-field enhancement and plasmon tuning in bimetallic nanoplanets. *Opt. Express* Vol.15, No.16, pp.10097-10102, ISSN 1094-4087.

Pellegrini, G., Mattei, G., Bello, V. & Mazzoldi, P. (2007b). Interacting metal nanoparticles: Optical properties from nanoparticle dimers to core-satellite systems. *Mater. Sci. Eng. C* Vol.27, No.5-8, pp.1347-1350, ISSN 0928-4931.

Penninkhof, J. J., van Dillen, T., Roorda, S., Graf, C., van Blaaderen, A., Vredenberg, A. M. & Polman, A. (2006). Anisotropic deformation of metallo-dielectric core-shell colloids under MeV ion irradiation. *Nucl. Instrum. Meth. B* Vol.242, No.1-2, pp.523-529, ISSN 0168-583X.

Peña, O., Pal, U., Rodíguez-Fernández, L., Silva-Pereyra, H. G., Rodríguez-Iglesias, V., Cheang-Wong, J. C., Arenas-Alatorre, J. & Oliver, A. (2009). Formation of Au-Ag core-shell nanostructures in silica matrix by sequential ion implantation. *J. Phys. Chem. C* Vol.113, No.6, pp.2296-2300, ISSN 1932-7447.

Peña, O., Rodríguez-Fernández, L., Cheang-Wong, J. C., Santiago, P., Crespo-Sosa, A., Muñoz, E. & Oliver, A. (2006). Characterization of nanocluster formation in Cu-implanted silica: Influence of the annealing atmosphere and the ion fluence. *J. Non-Cryst. Sol.* Vol.352, No.4, pp.349-354, ISSN 0022-3093.

Peña-Rodríguez, O. & Pal, U. (2010). Geometrical tunability of linear optical response of silica-gold double concentric nanoshells. *J. Phys. Chem. C* Vol.114, No.10, pp.4414-4417, ISSN 1932-7447.

Peña-Rodríguez, O. & Pal, U. (2011a). Au@Ag core-shell nanoparticles: Efficient all-plasmonic Fano-resonance generators. *Nanoscale* Vol.3, No.9, pp.3609-3612, ISSN 2040-3364.

Peña-Rodríguez, O. & Pal, U. (2011b). Enhanced plasmonic behavior of bimetallic (Ag-Au) multilayered spheres. *Nanoscale Res. Lett.* Vol.6, No.1, pp.279, ISSN 1556-276X.

Peña-Rodríguez, O., Pal, U., Campoy-Quiles, M., Rodríguez-Fernández, L., Garriga, M. & Alonso, M. I. (2011a). Enhanced Fano resonance in asymmetrical Au:Ag heterodimers. *J. Phys. Chem. C* Vol.115, No.14, pp.6410-6414, ISSN 1932-7447.

Peña-Rodríguez, O., Pal, U., Rodríguez-Iglesias, V., Rodríguez-Fernández, L. & Oliver, A. (2011b). Configuring Au and Ag nanorods for sensing applications. *J. Opt. Soc. Am. B* Vol.28, No.4, pp.714-720, ISSN 1520-8540.

Picraux, S. T. (1984). Ion implantation in metals. *Ann. Rev. Mater. Sci.* Vol.14, pp.335-372, ISSN 0084-6600.

Ren, F., Jiang, C. Z., Wang, Y. H., Wang, Q. Q. & Wang, J. B. (2006). The problem of core/shell nanoclusters formation during ion implantation. *Nucl. Instrum. Meth. B* Vol.245, No.2, pp.427-430, ISSN 0168-583X.

Rivera, A., Crespillo, M. L., Olivares, J., García, G. & Agulló-López, F. (2010). Effect of defect accumulation on ion-beam damage morphology by electronic excitation in lithium niobate: A Monte Carlo approach. *Nucl. Instrum. Meth. B* Vol.268, No.13, pp.2249-2256, ISSN 0168-583X.

Rodríguez-Iglesias, V., Peña-Rodríguez, O., Silva-Pereyra, H. G., Rodríguez-Fernández, L., Cheang-Wong, J. C., Crespo-Sosa, A., Reyes-Esqueda, J. A. & Oliver, A. (2010a). Tuning the aspect ratio of silver nanospheroids embedded in silica. *Opt. Lett.* Vol.35, No.5, pp.703-705, ISSN 1539-4794.

Rodríguez-Iglesias, V., Peña-Rodríguez, O., Silva-Pereyra, H. G., Rodríguez-Fernández, L., Kellermann, G., Cheang-Wong, J. C., Crespo-Sosa, A. & Oliver, A. (2010b). Elongated gold nanoparticles obtained by ion implantation in silica: Characterization and T-matrix simulations. *J. Phys. Chem. C* Vol.114, No.2, pp.746-751, ISSN 1932-7447.

Roiz, J., Oliver, A., Munoz, E., Rodriguez-Fernandez, L., Hernandez, J. M. & Cheang-Wong, J. C. (2004). Modification of the optical properties of Ag-implanted silica by annealing in two different atmospheres. *J. Appl. Phys.* Vol.95, No.4, pp.1783-1791, ISSN 1089-7550.

Ryasnyansky, A., Palpant, B., Debrus, S., Ganeev, R., Stepanov, A., Can, N., Buchal, C. & Uysal, S. (2005). Nonlinear optical absorption of ZnO doped with copper nanoparticles in the picosecond and nanosecond pulse laser field. *Appl. Opt.* Vol.44, No.14, pp.2839-2845, ISSN 1539-4522.

Sangpour, P., Hashemi, F. & Moshfegh, A. Z. (2010). Photoenhanced degradation of methylene blue on cosputtered M:TiO_2 (M = Au, Ag, Cu) nanocomposite systems: A comparative study. *J. Phys. Chem. C* Vol.114, No.33, pp.13955-13961, ISSN 1932-7447.

Serna, R., Suárez-García, A., Afonso, C. N. & Babonneau, D. (2006). Optical evidence for reactive processes when embedding Cu nanoparticles in Al_2O_3 by pulsed laser deposition. *Nanotechnol.* Vol.17, No.18, pp.4588-4593, ISSN 1361-6528.

Shan, G., Xu, L., Wang, G. & Liu, Y. (2007). Enhanced raman scattering of ZnO quantum dots on silver colloids. *J. Phys. Chem. C* Vol.111, No.8, pp.3290-3293, ISSN 1932-7447.

Sonawane, R. S. & Dongare, M. K. (2006). Sol-gel synthesis of Au/TiO_2 thin films for photocatalytic degradation of phenol in sunlight. *J. Mol. Catal. A: Chem.* Vol.243, No.1, pp.68-76, ISSN 1381-1169.

Steiner, G., Pham, M. T., Kuhne, C. & Salzer, R. (1998). Surface plasmon resonance within ion implanted silver clusters. *Fresenius J. Anal. Chem.* Vol.362, No.1, pp.9-14, ISSN 1432-1130.

Stepanov, A. & Popok, V. (2003). Synthesis of Cu nanoparticles in Al_2O_3 by ion implantation and subsequent laser annealing.(Duggan, J. L., Morgan, I. L., & Hall, M., Eds.) *AIP Conf. Proc.* Vol.680, No.1, pp.601-604, ISSN 0094-243X.

Stepanov, A. ., Kreibig, U., Hole, D. ., Khaibullin, R. ., Khaibullin, I. . & Popok, V. . (2001). Laser annealing of sapphire with implanted copper nanoparticles. *Nucl. Instrum. Meth. B* Vol.178, No.1-4, pp.120-125, ISSN 0168-583X.

Stepanov, A. L. & Khaibullin, I. B. (2005). Fabrication of metal nanoparticles in sapphire by low-energy ion implantation. *Rev. Adv. Mater. Sci.* Vol.9, No.2, pp.109-129, ISSN 1605-8127.

Stepanov, A. L., Abdullin, S. N., Khaibullin, R. I., Valeev, V. F., Osin, Y. N., Bazarov, V. V. & Khaibullin, I. B. (1995). Ion synthesis of colloidal silver nanoclusters in the organic substrate. *MRS Online Proceedings Library* Vol.392, pp.267-272, ISSN 1946-4274.

Stepanov, A. L., Khaibullin, R. I., Can, N., Ganeev, R. A., Ryasnyansky, A. I., Buchal, C. & Uysal, S. (2004). Application of ion implantation for synthesis of copper nanoparticles in a zinc oxide matrix for obtaining new nonlinear optical materials. *Tech. Phys. Lett.* Vol.30, No.10, pp.846-849, ISSN 1090-6533.

Subramanian, V., Wolf, E. E. & Kamat, P. V. (2004). Catalysis with TiO_2/gold nanocomposites. Effect of metal particle size on the fermi level equilibration. *J. Am. Chem. Soc.* Vol.126, No.15, pp.4943-4950, ISSN 0002-7863.

Sun, Y. & Xia, Y. (2002). Increased sensitivity of surface plasmon resonance of gold nanoshells compared to that of gold solid colloids in response to environmental changes. *Anal. Chem.* Vol.74, No.20, pp.5297-5305, ISSN 1520-6882.

Takeda, Y., Plaksin, O. A., Kono, K. & Kishimoto, N. (2005). Nonlinear optical properties of Cu nanoparticles in various insulators fabricated by negative ion implantation. *Surf. Coat. Technol.* Vol.196, No.1-3, pp.30-33, ISSN 0257-8972.

Takeda, Y., Plaksin, O. A., Wang, H., Kono, K., Umeda, N. & Kishimoto, N. (2006). Surface plasmon resonance of Au nanoparticles fabricated by negative ion implantation and grid structure toward plasmonic applications. *Opt. Rev.* Vol.13, No.4, pp.231-234, ISSN 1349-9432.

Tihay, F., Pourroy, G., Richard-Plouet, M., Roger, A. C. & Kiennemann, A. (2001). Effect of Fischer-Tropsch synthesis on the microstructure of Fe-Co-based metal/spinel composite materials. *Appl. Catal. A* Vol.206, No.1, pp.29-42, ISSN 0926-860X.

Tohidi, S. H., Grigoryan, G., Sarkeziyan, V. & Ziaie, F. (2010). Effect of concentration and thermal treatment on the properties of sol-gel derived CuO/SiO_2 nanostructure. *Iran. J. Chem. Chem. Eng.* Vol.29, No.2, pp.27-35, ISSN 1021-9986.

Townsend, P. D. (1987). Optical effects of ion implantation. *Rep. Prog. Phys.* Vol.50, No.5, pp.501-558, ISSN 1361-6633.

Trinkaus, H. & Ryazanov, A. I. (1995). Viscoelastic model for the plastic flow of amorphous solids under energetic ion bombardment. *Phys. Rev. Lett.* Vol.74, No.25, pp.5072-5075, ISSN 1079-7114.

Tsuji, H., Sagimori, T., Kurita, K., Gotoh, Y. & Ishikawa, J. (2002). Surface modification of TiO_2 (rutile) by metal negative ion implantation for improving catalytic properties. *Surf. Coat. Technol.* Vol.158-159, No.0, pp.208-213, ISSN 0257-8972.

Vazquez-Cuchillo, O., Bautista-Hernandez, A., Pal, U. & Meza-Montes, L. (2001). Study of the optical absorption of Cu clusters in the Cu/ZnO system. *Mod. Phys. Lett. B* Vol.15, No.17-19, pp.625-629, ISSN 0217-9849.

Wang, C. M. (2004). Microstructure of precipitated Au nanoclusters in TiO_2. *J. Appl. Phys.* Vol.95, No.12, pp.8185, ISSN 0021-8979.

Wang, C. M., Zhang, Y., Shutthanandan, V., Baer, D. R., Weber, W. J., Thomas, L. E., Thevuthasan, S. & Duscher, G. (2005). Self-assembling of nanocavities in TiO_2 dispersed with Au nanoclusters. *Phys. Rev. B* Vol.72, No.24, pp.245421, ISSN 1550-235X.

Wang, D.-Y., Lin, H.-C. & Yen, C.-C. (2006). Influence of metal plasma ion implantation on photo-sensitivity of anatase TiO_2 thin films. *Thin Solid Films* Vol.515, No.3, pp.1047-1052, ISSN 0040-6090.

Wang, X.-H., Shi, J., Dai, S. & Yang, Y. (2003). A sol-gel method to prepare pure and gold colloid doped ZnO films. *Thin Solid Films* Vol.429, No.1-2, pp.102-107, ISSN 0040-6090.

White, C. W., Budai, J. D., Withrow, S. P., Zhu, J. G., Sonder, E., Zuhr, R. A., Meldrum, A., Hembree, J., Henderson, D. O. & Prawer, S. (1998). Encapsulated semiconductor nanocrystals formed in insulators by ion beam synthesis. *Nucl. Instrum. Meth. B* Vol.141, No.1-4, pp.228-240, ISSN 0168-583X.

White, C. W., Thomas, D. K., Hensley, D. K., Zuhr, R. A., McCallum, J. C., Pogany, A., Haglund, R. F., Magruder, R. H. & Yang, L. (1993). Colloidal Au and Ag precipitates formed in Al_2O_3 by ion implantation and annealing. *Nanostr. Mater.* Vol.3, No.1-6, pp.447-457, ISSN 0965-9773.

White, C. W., Thomas, D. K., Zuhr, R. A., McCallum, J. C., Pogany, A., Haglund, R. F., Magruder, R. H. & Yang, L. (1992). Modification of the optical properties of Al_2O_3 by ion implantation. *Mat. Res. Soc. Proc.* Vol.268, pp.331-336, ISSN 0272-9172.

Xiao, X. H., Ren, F., Wang, J. B., Liu, C. & Jiang, C. Z. (2007). Formation of aligned silver nanoparticles by ion implantation. *Mater. Lett.* Vol.61, No.22, pp.4435-4437, ISSN 0167-577X.

Xiao, X. H., Ren, F., Zhou, X. D., Peng, T. C., Wu, W., Peng, X. N., Yu, X. F. & Jiang, C. Z. (2010). Surface plasmon-enhanced light emission using silver nanoparticles embedded in ZnO. *Appl. Phys. Lett.* Vol.97, No.7, pp.071909, ISSN 0003-6951.

Xu, J.-X., Ren, F., Fu, D.-J. & Jiang, C.-Z. (2006). Effect of thermal annealing on the optical properties of low-energy Cu-implanted silica glass. *Physica B* Vol.373, No.2, pp.341-345, ISSN 0921-4526.

Yasuda, H. & Mori, H. (1994). Cluster-size dependence of alloying behavior in gold clusters. *Z. Phys. D* Vol.31, No.1-2, pp.131-134, ISSN 1434-6079.

Zeng, H., Qiu, J., Jiang, X., Zhu, C. & Gan, F. (2004). Effect of Al_2O_3 on the precipitation of Ag nanoparticles in silicate glasses. *J. Cryst. Growth* Vol.262, No.1-4, pp.255-258, ISSN 0022-0248.

Zhang, J. Z. (2010). Biomedical applications of shape-controlled plasmonic nanostructures: a case study of hollow gold nanospheres for photothermal ablation therapy of cancer. *J. Phys. Chem. Lett.* Vol.1, No.4, pp.686-695, ISSN 1948-7185.

Zheng, Y., Chen, C., Zhan, Y., Lin, X., Zheng, Q., Wei, K. & Zhu, J. (2008). Photocatalytic activity of Ag/ZnO heterostructure nanocatalyst: Correlation between structure and property. *J. Phys. Chem. C* Vol.112, No.29, pp.10773-10777, ISSN 1932-7447.

Zhou, X. D., Xiao, X. H., Xu, J. X., Cai, G. X., Ren, F. & Jiang, C. Z. (2011). Mechanism of the enhancement and quenching of ZnO photoluminescence by ZnO-Ag coupling. *Euro Phys. Lett.* Vol.93, No.5, pp.6, ISSN 1286-4854.

Zhu, S., Sun, K., Wang, L. M., Ewing, R. C. & Fromknecht, R. (2006). TEM characterization of Au nano-particles in TiO_2 single crystals by ion implantation. *Nucl. Instrum. Meth. B* Vol.242, No.1-2, pp.152-156, ISSN 0168-583X.

Ziegler, J. *SRIM - The Stopping and Range of Ions in Matter: http://www.srim.org/.* [online] (2008). Available from: http://www.srim.org/. [Accessed 2009-09-16].

Ziegler, J. F. (1985). *The stopping and range of ions in solids.* Pergamon Pr., ISBN 008021603X.

16

Surface Modification by Ion Implantation to Improve the Oxidation Resistance of Materials for High Temperature Technology

Hans-Eberhard Zschau and Michael Schütze
Dechema e. V., Karl-Winnacker-Institut, Frankfurt am Main
Germany

1. Introduction

The intermetallic Titaniumaluminides are expected to have a high potential as material in high temperature technology. Their high specific strength at temperatures above 700°C offers the possibility for manufacturing components of aerospace and automotive industries. With a specific weight of 50% of that of the widely used Ni-based superalloys TiAl is very suitable for fast rotating parts like turbine blades in aircraft engines and land based power stations, exhaust valves or turbocharger rotors. Thus lower mechanical stresses and a reduced fuel consumption are expected. In contrast to these benefits TiAl shows insufficient oxidation resistance at temperatures above 750°C (Rahmel et al., 1995). To reach higher service temperatures a protective alumina scale would be needed. A surface modification avoids any detrimental influence on the excellent mechanical properties of the material. By using the "halogen effect" a dense protective alumina scale was formed after doping the metal surface with small amounts of chlorine (Kumagai et al., 1996; Schütze & Hald, 1997; Donchev et al., 2003; Schumacher et al., 1999a; Schumacher et al., 1999b; Hornauer et al., 1999). The halogen effect can be explained by a thermodynamic model assuming the preferred formation and transport of volatile Al-halides through pores and microcracks within the metal/oxide interface and their conversion into alumina, forming a protective oxide scale on the surface (Donchev et al., 2003). However the alumina scale fails during thermocyclic loading of Cl-implanted TiAl-samples. Based on this model comprehensive calculations have been performed for fluorine and TiAl predicting a positive effect. The results of thermodynamical calculations have to be transformed into F-concentrations. The beam line ion implantation is chosen as a method of F-application because of its accuracy and reproducibility. By varying the implantation parameters optimal conditions have to be determined to meet the region of F-amounts necessary for a positive F-effect. The implanted F-depth profiles can be verified by the non-destructive ion beam analysis method PIGE (Proton Induced Gamma-ray Emission). The alumina scale formed via the F-effect is adherent even under thermocyclic conditions. However technical use is only possible if the fluorine effect can be stabilized for a time of at least 1000 hours. Hence after the alumina scale formation the stability of oxidation protection depends on the behaviour of the implanted fluorine vs.

oxidation time. This work is dedicated to the role of fundamental material parameters related to the stability of the fluorine effect at TiAl.

Ni-base superalloys with Al-contents of less than 10 wt.% are widely used in high temperature technology due to their superior mechanical properties. In contrast to this their oxidation resistance may be insufficient at temperatures above 950°C. Oxidation of these Ni-base alloys does not form a pure protective continuous alumina scale on the surface, but rather a complex layer structure of spinel phases and a significant amount of internal oxidation is formed (Litz et al., 1989). However the formation of a dense continuous alumina scale without significant internal oxidation would theoretically be possible, if a „critical" Al-concentration N_c is realized. In this work a new concept for the formation of a protective alumina scale on the surface is presented. The change from a non-protective discontinuous internal to a continuous dense protective external alumina scale can be achieved by an "artificial" increase of the Al-activity on the surface. This can be realized via the halogen effect at Ni-based alloys. Thermodynamical calculations show the existence of a region for a positive fluorine effect for the Ni-based alloys, e. g. IN738 and IN939 at temperatures between 900-1200°C. These results have to be transformed into fluorine concentrations by a screening using fluorine ion implantation.

2. Materials and methods

Cast γ-TiAl (Ti-50 at.% Al) and the technical γ-Met alloy (46.6 at.% Al) manufactured by powder metallurgy were prepared as coupons of size 8*8*1 mm^3 and polished with SiC paper down to 4000 and 1200 grit resp. Microstructural investigations showed minor amounts of the α_2-Ti_3Al phase (lamellar structure) within the γ-TiAl phase. For the F-implantation fluences between $1x10^{16}$ and $4x10^{17}$ F cm^{-2} were used. The ion energy of 20 keV corresponds to a mean projected range of 35 nm for F-ions in TiAl. The coupons of Ni-based alloy IN738 with a size of 10*10*1 mm^3 were polished down to 4000 grit. Fluorine ions with fluences between $1x10^{16}$ and $4x10^{17}$ F cm^{-2} were chosen for the implantation. With an energy of 38 keV a projected mean range of 35 nm in IN738 has been achieved.

All beam line implantations were carried out at the 60 kV implanter of the Institute of Nuclear Physics (IKF) of the Johann Wolfgang Goethe - University in Frankfurt/Main using CF_4 in the gas source. The fluorine concentration depth profiles were determined by using the non-destructive PIGE-technique (Tesmer & Nastasi, 1995). The PIGE measurements were performed at the 2.5 MV Van de Graaff accelerator of the IKF using the nuclear reaction $^{19}F(p, \alpha\gamma)^{16}O$ at resonance energies of E_p=340 keV and 484 keV, resp., and detecting the high energetic γ-rays (5-6 MeV) with a 10 inch NaI-detector. The information depth of the PIGE depth profiling is within 1.4 – 1.5 μm, whereas the depth resolution near the surface is about 10 nm. The non-destructive RBS (Rutherford Backscattering Spectrometry) using He-ions of 2 and 3.53 MeV enables the determination of O, Al and Ti-depth profiles simultaneously. For the TGA (Thermo-Gravimetric Analysis) mesurements both sides and all edges of the specimens were implanted.

The TiAl-samples were oxidized isothermally and thermocyclically (cycles of 1 h and 24 h) in a furnace at 900°C and 1000°C in lab air.

The Ni-alloy samples were oxidized isothermally between 12 h and 1000 h at 1050°C. Finally all samples were inspected by metallographic methods and SEM.

3. Surface modification of TiAl by fluorine ion implantation

3.1 Principles of the halogen effect for TiAl

γ-TiAl with an Al-content between 44 and 50 at.% has a very promising prospect in high temperature application (fig. 1). In contrast to the good mechanical properties at temperatures above 750°C the TiAl-surface is covered with a porous titania scale on top and a porous scale of fast growing mixed titania and alumina underneath (fig. 2). This scale does not hinder the further inward diffusion of oxygen and nitrogen leading to the continuous formation of this mixed oxide scale. Already after 12 h at 900°C the scale thickness exceeds 10 µm. Finally, these processes will lead to a rapid destruction of the alloy (fig. 2). The insufficient oxidation resistance of TiAl can be improved by using the halogen effect. The principles of the halogen effect at TiAl are illustrated in fig. 3 for fluorine.

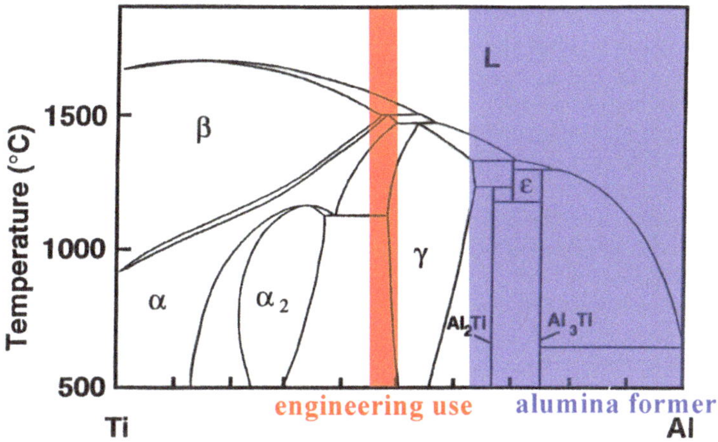

Fig. 1. Phase diagram of TiAl (Calphad).

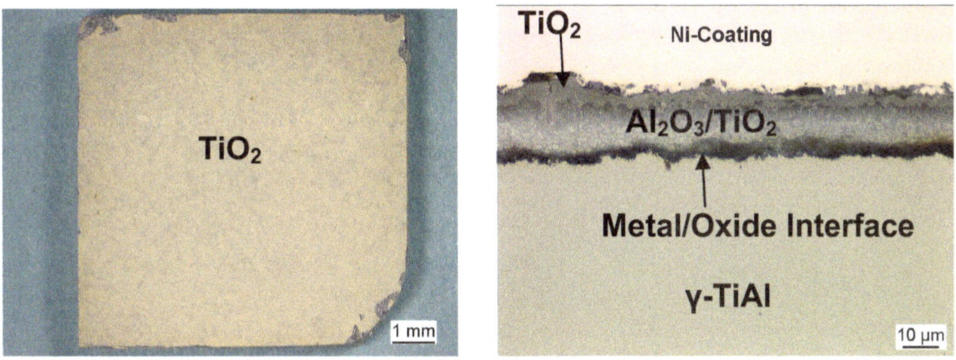

Fig. 2. Untreated γ-TiAl after oxidation (12h/900°C/air). Left image: A brittle white TiO_2-layer is visible on top. Right image: Metallographic cross-section. A TiO_2-top scale and underlying mixed Al_2O_3/TiO_2-scale covers the surface. The Ni-coating comes from the metallographic preparation.

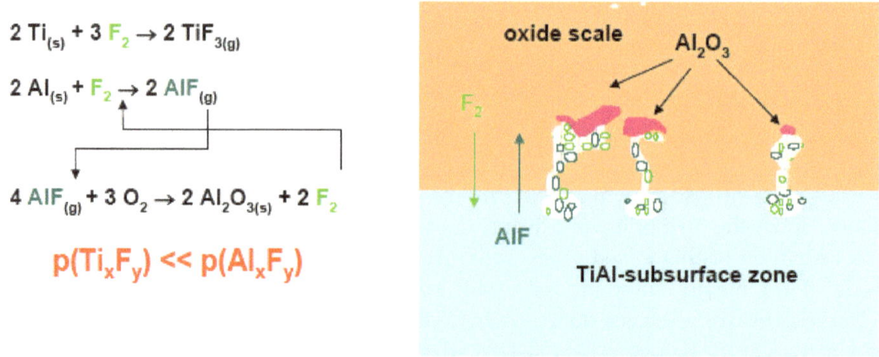

Fig. 3. Principle reactions of the F-effect at TiAl (Donchev et al., 2003).

A preferred formation of Al-fluorides occurs, if their partial pressure is significantly higher than that of the Ti-fluorides formed. Thus, the Al-fluorides migrate to the surface where they disintegrate to F_2 and Al, which is oxidized to alumina due to the increasing oxygen partial pressure. The free gaseous fluorine is able to return into the subsurface zone to form Al-fluorides again. This cycle process leads to the formation of a thin protective Al_2O_3 - scale. Thermodynamic calculations for TiAl with Al-amounts between 40-50 at.% using the software FactSage (FactSage) show the existence of a region with significantly higher partial pressures of Al-fluorides compared to Ti-fluorides as depicted in fig. 4 (Masset & Schütze, 2008). In this case a AlF-partial pressure of at least 10^{-8} bar is required to form and to transport a significant amount of Al to the surface (Fergus, 2002). By increasing the total F-partial pressure to $p(F)_{min}$ at first the partial pressure of AlF meets this condition and the window for the F-effect is open. If the total F-partial pressure exceeds the value of $p(F)_{max}$ the partial pressures of competing Ti-fluorides become significant. Thus the window for the F-effect is closed. The existence of a window for the F-effect at TiAl was found for temperatures between 900 and 1200°C defining a corridor for the F-effect for this temperature range (Donchev et al., 2003).

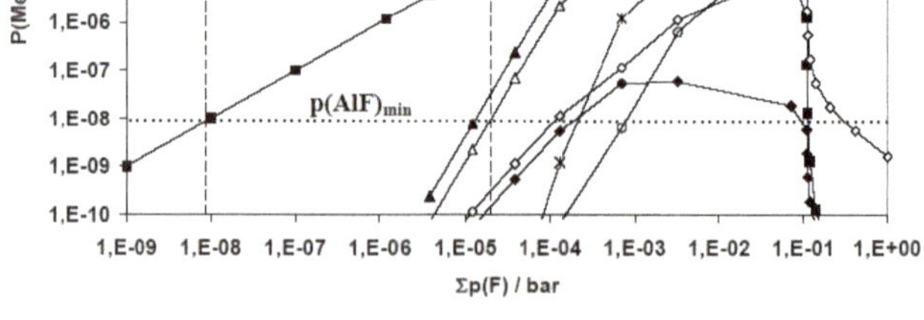

Fig. 4. Partial pressure of several metal fluorides vs. F-total partial pressure for Ti60Al40 at 900°C (calculated with FactSage). The region for a positive F-effect is located between the total F-partial pressures of $p(F)_{min}$ and $p(F)_{max}$.

3.2 Determination of the optimal F-concentration for the F-effect in TiAl

3.2.1 F-implantation and non-destructive analysis of the implantation profiles

The corridor for the F-effect found in 3.1 is expressed in terms of fluorine partial pressures. However, for practical applications the fluorine concentration in TiAl is needed. Thus this corridor has to be transformed into a corridor for F-concentrations (in at.%). However there is no equation between F-pressures and F-concentrations to allow a direct calculation. Therefore a screening has to be performed to study the oxide scale structure after oxidation at 900°C with respect to different fluorine amounts within the near-surface region. The beam line ion implantation was chosen because of its accuracy and reproducibility. Based on TRIM-calculations (Ziegler et al., 1995) a F-energy of 20 keV was chosen corresponding to a projected range of about 35 nm (fig. 5).

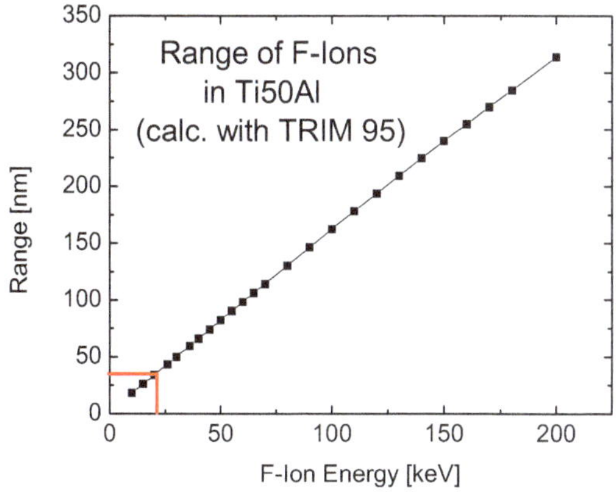

Fig. 5. The projected range for F-ions in TiAl calculated by using TRIM.

The calculation of the F-implantation profiles by using the Monte Carlo-software T-DYN (Biersack, 1999) allows to design the ion implantation prior to the experiment. The calculated

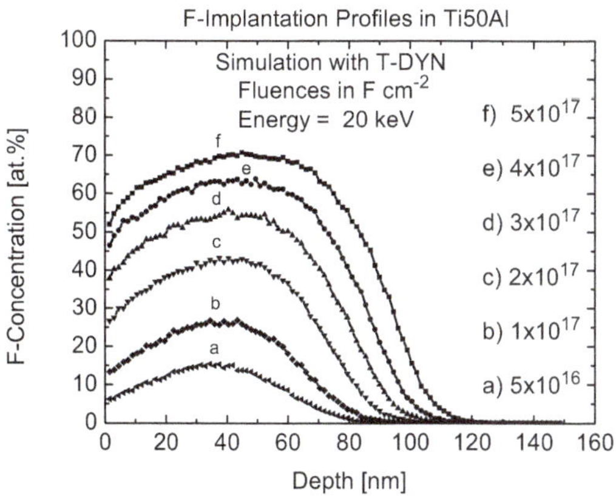

Fig. 6. F-implantation profiles in γ-TiAl calculated with Monte Carlo simulation software T-DYN.

F-depth profiles between the fluences of 5 x 10^{16} F cm^{-2} and 5 x 10^{17} F cm^{-2} /20 keV are illustrated in fig. 6 revealing the corresponding maximal F-concentrations between 15 and 70 at.%. Due to a sputtering yield of 0.60 the saturation effects occur remarkably above a fluence of 4 x 10^{17} F cm^{-2}. The non-destructive PIGE-technique was used to determine the fluorine concentration depth profiles within the first 1.5 µm of the ion implanted samples before and after oxidation. By using the resonance depth profiling the implantation profiles can be verified as depicted in fig. 7. A CaF_2 - single crystal is used as a standard for the quantification of fluorine in at.%.

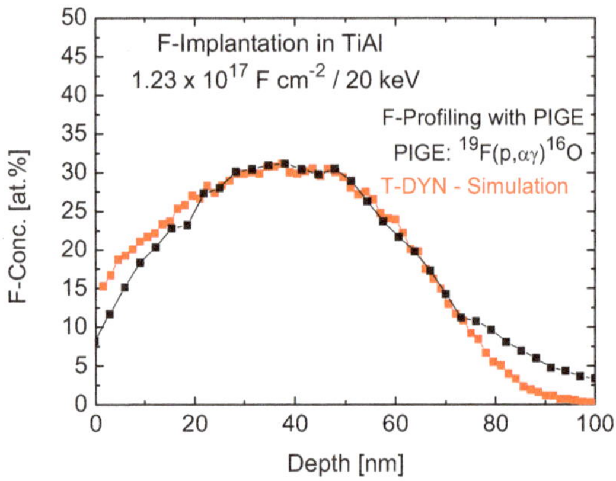

Fig. 7. F-implantation profiles in γ-TiAl calculated with Monte Carlo simulation software T-DYN and measured by using PIGE.

The F-maximum is shifted into a depth of 35-45 nm in order to obtain a lower loss of fluorine during the heating process. The small deviations at the surface may be caused by sputtering whereas the differences at the end of the ion range may be influenced by secondary ion collisions.

3.2.2 Screening of the implantation parameters

The screening consists of the following steps:

1. Calculation of the F-depth profiles by using Monte Carlo-software T-DYN.
2. F-ion implantation by using the implantation parameters (fluence, ion energy) determined in item 1.
3. Verifying of the F-depth profiles by using PIGE.
4. Oxidation of the implanted samples (24-120h/900°C/air).
5. Measurement of F-depth profiles by using PIGE.
6. Metallographic cross-section preparation.
7. Metallographic inspection of the oxide scale by using SEM/EDX.
8. Definition of the next set of implantation parameters and switch to item 1.

As a criterion for the screening the stability of the F-effect for an oxidation of at least 120h/900°C is required.

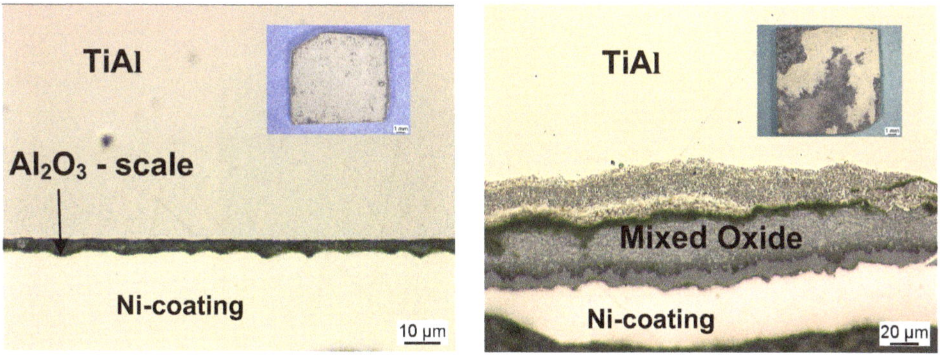

Fig. 8. Cross-section of implanted γ-TiAl (5 x 10^{16} F cm^{-2} /20 keV). The small pictures show the surface. Left image: After oxidation (12h/900°C/air) the surface is covered with a grey thin alumina scale. Right image: After oxidation (120h/900°C/air) the surface is covered with a grey alumina scale, but a mixed oxide scale with a TiO_2 – scale on top grows partly on the surface. The Ni-coating is from the metallographic preparation.

A fluence of 10^{16} F cm^{-2} /20 keV gives a F-concentration which is too low for the F-effect. Fig. 8 illustrates the initial formation of a thin alumina scale of a TiAl-sample implanted with a fluence of 5 x 10^{16} F cm^{-2} /20 keV. After short oxidation (12h/900°C/air) the implanted surface is covered completely with an alumina scale. However after an oxidation time of 120h the surface is covered partly by a fast growing mixed oxide scale with TiO_2 on top. After implantation of 10^{17} F cm^{-2} /20 keV the alumina scale formed remains stable after oxidation of 120h/900°C/air (fig. 9). To obtain a higher F-concentration a fluence of 2 x 10^{17} F cm^{-2} /20 keV shows the best results even after long term oxidation (fig. 10, 11). The samples implanted with fluences of 3 x 10^{17} F cm^{-2} /20 keV and more show no sufficient alumina scale growth.

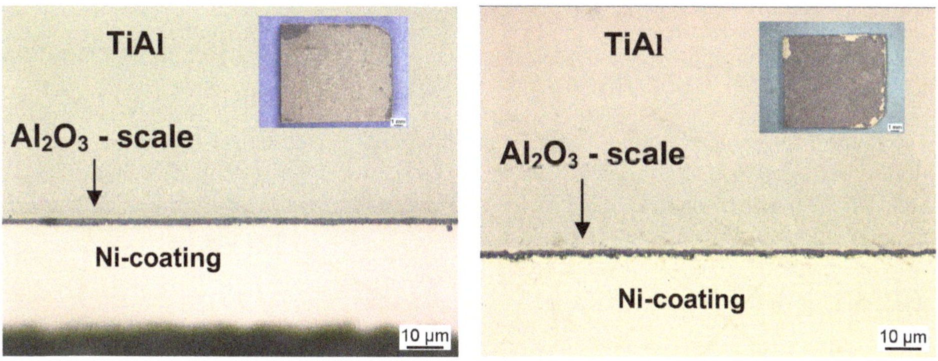

Fig. 9. Cross-section of implanted γ-TiAl (10^{17} F cm^{-2} /20 keV). The small pictures show the surface. Left image: After oxidation (12h/900°C/air) the surface is covered with a grey thin alumina scale. Right image: After oxidation (120h/900°C/air) the surface is covered with a grey alumina scale, only minor amounts of a mixed oxide scale have been formed at the non-implanted edges with a TiO_2 – scale on top.

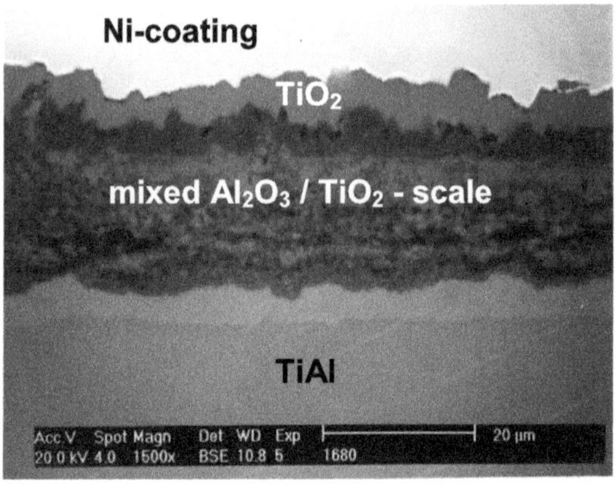

Fig. 10. Non-protective mixed oxide scale on an untreated γ-TiAl- sample after oxidation (12h/900°C/air).

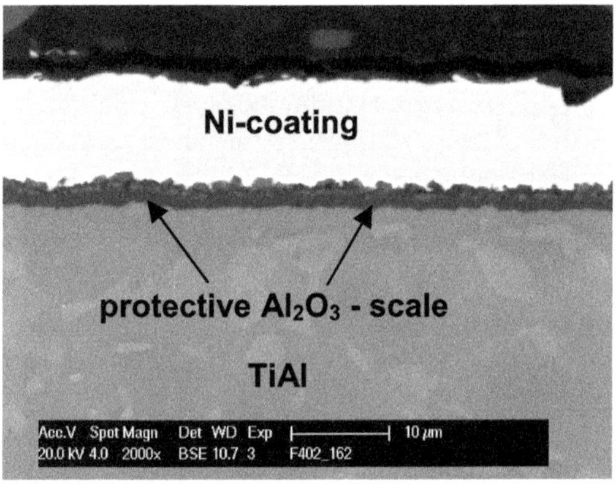

Fig. 11. Protective alumina scale on a F-implanted γ-TiAl-sample (2×10^{17} F cm^{-2} /20 keV) after oxidation (1000h/900°C/air).

As a result of the screening an optimal set of implantation parameters (2×10^{17} F cm^{-2} / 20 keV) has been obtained showing the formation of a pure and dense protective alumina scale on the TiAl-surface (Zschau et al., 2002). This fluence corresponds to a maximal initial F-amount of about 44 at.% as known from the calculation with T-DYN (fig. 6).

3.3 Behaviour of the implanted fluorine during the oxidation of TiAl

3.3.1 Fluorine profile and alumina scale formation

The non-destructive measurement of the F-depth profiles of implanted and oxidized TiAl-samples by using PIGE reveals a typical shape which can be divided into 3 characteristic regions as shown in fig. 12 (Zschau et al., 2002). Starting with a negligible F-amount on the surface and within the initially formed alumina scale the F-concentration increases to a maximum value C_F^{max} located at the metal/oxide interface. The position of the metal/oxide interface can be verified by RBS-measurements revealing the depth profiles of O, Al and Ti (fig. 13). Within the metal the F-concentration decreases. The shape of the F-profile reveals a distinct F-enrichment in the metal/oxide-interface supporting the suggested mechanism. The alumina scale acts as a diffusion barrier and prevents the F-loss via the surface. On the other hand, the fluorine diffuses into the metal. Thus, the behaviour of the F-reservoir at the metal/oxide-interface during long term oxidation determines the stability of the F-effect. Therefore a stable oxidation protection of TiAl is connected to the presence of a F-reservoir at the interface between the formed Al_2O_3-scale and the metal. Additionally the fluorine stimulates the repeated formation of the protective alumina scale once it has been damaged locally.

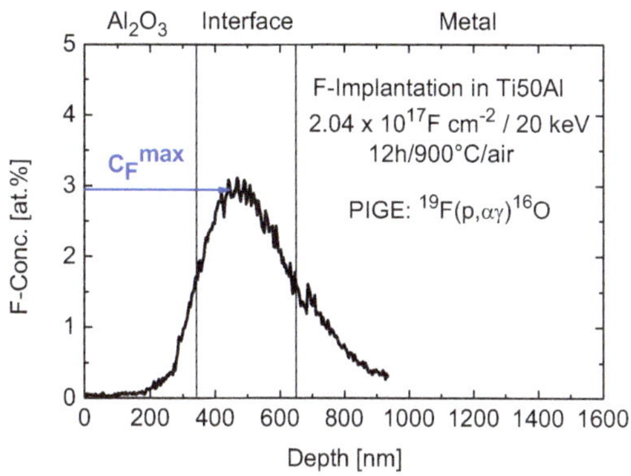

Fig. 12. F-depth profile of a γ-TiAl sample after oxidation (12h/900°C/air) obtained by PIGE.

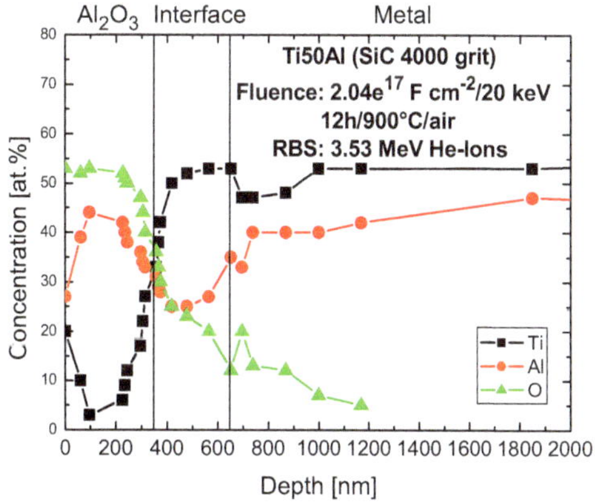

Fig. 13. Depth profiles of O, Al and Ti of a γ-TiAl sample after oxidation (12h/ 900°C/air) obtained by RBS.

3.3.2 Dependence of the F-profiles during heating

The fluorine profiles implanted into TiAl undergo a rapid alteration during the heating process. Starting as a typical implantation profile determined by the fluence and the energy of the F-ions, the maximum F-concentration C_F^{max} decreases rapidly after reaching a temperature of 500°C. By using a fluence of 2 x 10^{17} F cm^{-2} /20 keV the F-maxima drop to a value of about 5 at.% after oxidation of 1h/air as depicted in fig. 14. The pronounced F-loss is caused by the evaporation of gaseous Ti-fluorides, especially TiF_4 (Zschau et al., 2005). For temperatures up to 700°C fluorine is present within the near-surface region as a result of the transport to the surface. If the fluorine could escape also at higher temperatures, the F-effect would not be able to work. However at 800°C the oxide growth changes from a mixed Al_2O_3/TiO_2 - type to pure Al_2O_3 as was proven by RBS-measurements. The alumina scale formed inhibits the outward diffusion of fluorine drastically as illustrated by the F-profiles at 900 and 1000°C in fig. 14.

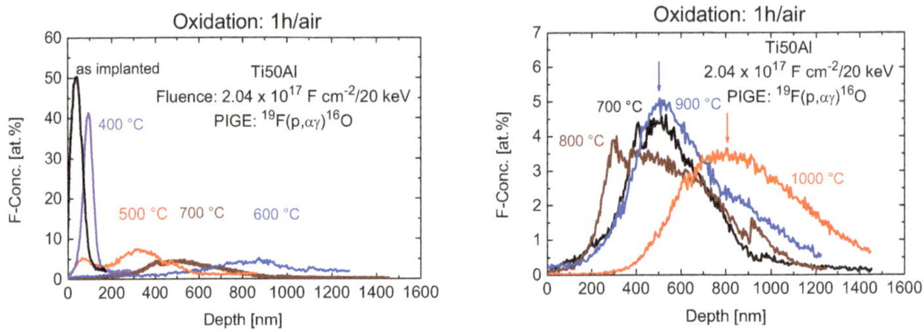

Fig. 14. F-depth profiles of implanted γ-TiAl- samples (2 x 10^{17} F cm^{-2} /20 keV) and after oxidation (1h/air) at temperatures of 400-1000°C obtained by PIGE. The arrows denote the position of the metal/oxide-interface.

The development of the fluorine maxima vs. temperature for oxidation (1h/air) is summarized in fig. 15. After the high F-loss between 400-500°C the F-maxima decrease slowly reaching values of about 5 at.%. The integral fluorine concentration (dose) shown in fig. 16 drops to a value between 1.2 and 1.4 x 10^{17} F cm^{-2}. A stable oxidation protection is based on the further behaviour of the F-maxima C_F^{max}. The following sections are focused on the formulation of basic parameters and their influence for the long term stability of the F-effect at TiAl .

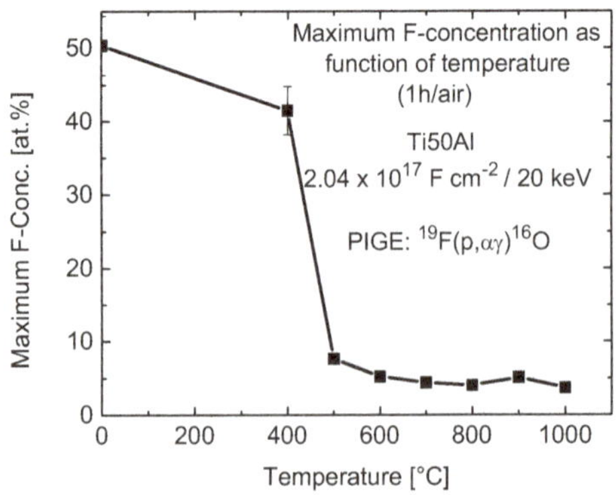

Fig. 15. Behaviour of the F-maxima C_F^{max} of implanted γ-TiAl (2 x 10^{17} F cm^{-2} /20 keV) during oxidation (1h/air) obtained by PIGE.

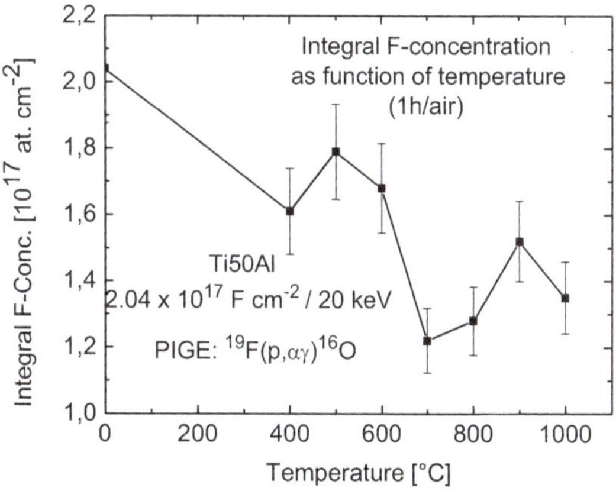

Fig. 16. Behaviour of the F-dose of implanted γ-TiAl (2×10^{17} F cm^{-2} / 20 keV) during oxidation (1h/air) obtained by PIGE.

3.3.3 Dependence of the F-concentration during long-term oxidation of TiAl

From a technical point of view the parts manufactured from TiAl have to withstand oxidation for service times of at least 1000 hours under cyclic conditions. Thus the time behaviour of the F-reservoir during long-term oxidation has been investigated by using PIGE after isothermal and cyclic oxidation at 900°C and 1000°C. The time dependence of the F-maximum C_F^{max} is chosen to characterize the stability of the F-effect. The formation of a protective alumina scale is confirmed by metallographic preparation and SEM studies. For longer oxidation times the F-maximum drops to 1-2 at.% after 500h/900°C (see fig. 17). By considering the F-profiles after isothermal oxidation up to 500h/900°C/air (fig. 17) a nearly stable F-concentration can be expected for oxidation times > 500 h.

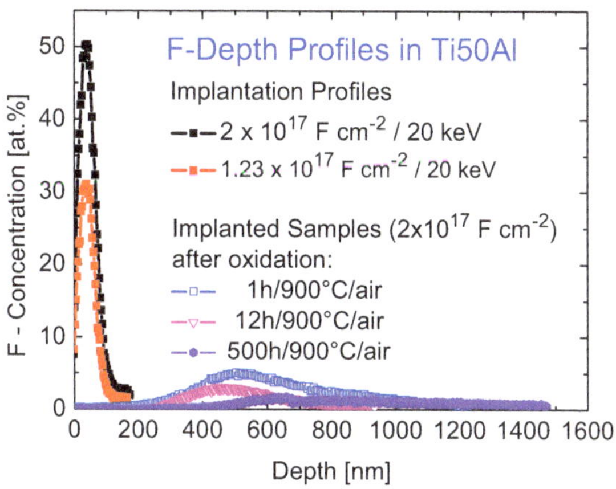

Fig. 17. Time behaviour of the F-depth profiles of implanted TiAl and after isothermal oxidation (900 °C/air).

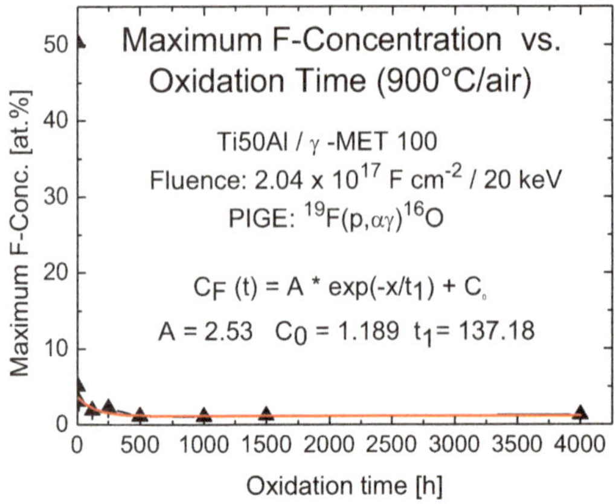

Fig. 18. An exponential decay function (1) with a constant term c_0 fits the time behaviour of C_F^{max} (2 x 10^{17} F cm^{-2} /20 keV) during isothermal oxidation (4000 h/900 °C/air).

Treatment/Oxidation	**Parameters** A	C_0	t_1
F-Implant. (900 °C/isoth.)	2.53	1.19	137.18
	6.70	*3.33*	*13.65*
F-Implant. (900 °C/cycl.)	3.18	1.06	65.77
	5.25	*2.06*	*37.70*
F-Implant. (1000 °C/isoth.)	4.18	0.27	97.27
	5.41	*1.37*	*59.03*
F-Implant. (1000 °C/cycl.)	3.69	0.77	80.02
	4.80	*1.23*	*76.34*

Table 1. Parameters of the function (1) describing the C_F^{max} – dependence after single F-implantation (2 x 10^{17} F cm^{-2} / 20 keV) and oxidation of TiAl. The parameters for double implantation are in italics (see section 3.3.4).

In the case of isothermal oxidation at 900°C first results have been published in (Zschau & Schütze, 2008). After extending the oxidation time to 4000 h (fig. 18) a nearly constant time behaviour of C_F^{max} is obtained as depicted in tab. 1. The SEM inspection confirms a dense protective alumina scale on the sample surface revealing that the F-effect is still working. The time dependence of C_F^{max} after passing the heating phase can be expressed by an exponential decay function starting at t=12 hours of the following type (1), where t denotes the time, t_1 a time constant, A the F-amount after the alumina scale has been established and c_0 the asymptotic F-concentration:

$$c_F^{\max}(t) = c_0 + A \exp\left(-\frac{t}{t_1}\right) \quad (1)$$

Equation 1 suggests that after formation of a dense alumina scale C_F^{max} decreases only slightly approaching a constant value of c_0. The fit parameters summarized in tab. 1 allow the description of the F-reservoir essential for the oxidation protection reaching a constant value of about 1 at.%. By increasing the oxidation temperature to 1000 °C the kinetics of C_F^{max} corresponds also to eq. 1. The parameters in tab. 1 indicate that a fluorine content of 0.27 at.% stabilizes the F effect even at 1000 °C at least within 1000 h. Cyclic oxidation (1 h-cycles/900 °C/air) was performed up to an oxidation time of 2592 h. Before starting the cyclic process a preoxidation was performed to establish a dense alumina scale on the surface. The time behaviour above 12 h of the F-maximum in fig. 19 has been fitted by eq. 1, whereas the parameters in tab. 1 reveal the presence of about 1 at.% fluorine at the metal/oxide interface. The remaining dose shows a constant value of about 7.5 x 10^{16} F cm^{-2} as can be seen in fig. 20.

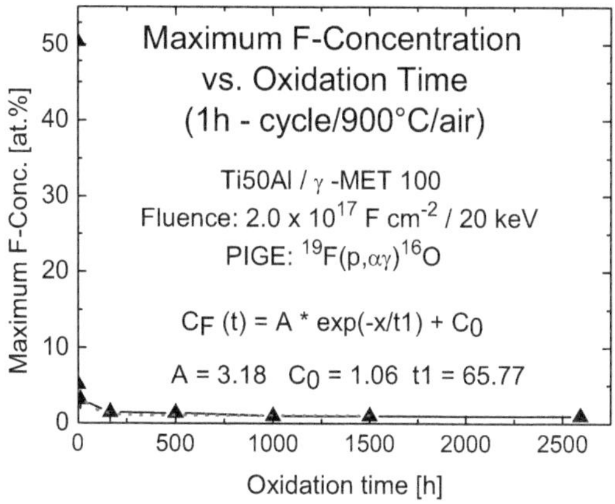

Fig. 19. Exponential decay function with a constant term fits the time behaviour of C_F^{max} (2 x 10^{17} F cm^{-2} /20 keV) for cyclic oxidation (2592h/900°C/air).

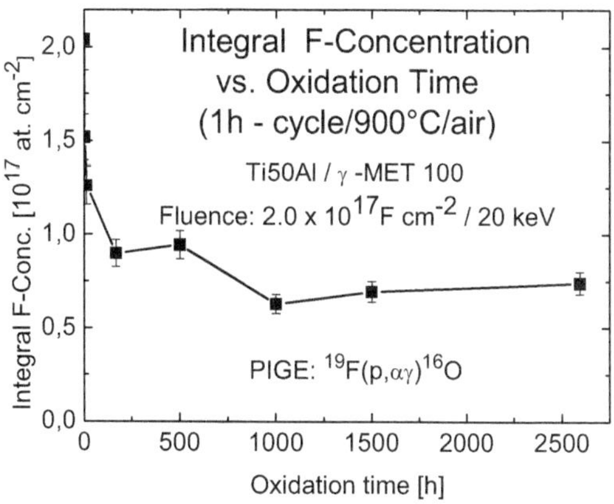

Fig. 20. The fluorine dose remains nearly constant for long oxidation times at TiAl (2 x 10^{17} F cm^{-2} /20 keV) and cyclic oxidation (2592h/900°C/air).

The results for cyclic loading at 1000 h/1000 °C indicate the stability of the F-content also under these extreme conditions as summarized in tab. 1. The F-maximum reaches values of about 0.7 at.%.

In agreement with the metallographic inspections in fig. 21 it was demonstrated that an excellent oxidation protection can be obtained already after single fluorine implantation.

The mass gain obtained by TGA-measurements depicted in fig. 22 shows the existence of an alumina kinetics i. e. a slow growing alumina scale also for cyclic oxidation (24 h - cycle/1000 h/900 °C/air). The excellent oxidation protection of a TiAl-alloy via the F-effect has been demonstrated by the oxidation of a turbocharger rotor with and without fluorine treatment for 1000 h at 1050°C in air (fig. 23). While severely corroded blades are observed for the unprotected rotor, the F-treated rotor is intact and is covered with a protective alumina scale (Donchev et al., 2006).

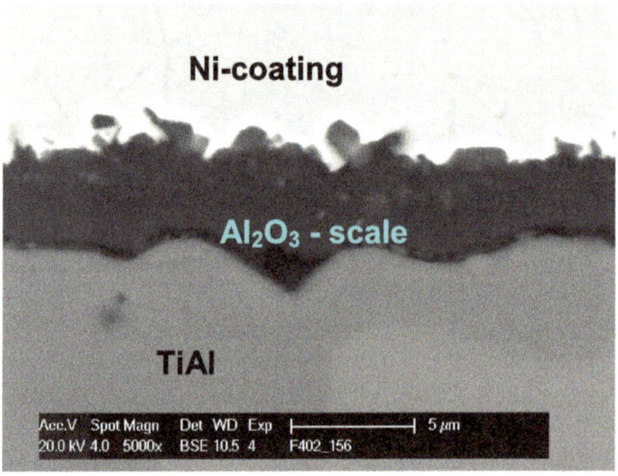

Fig. 21. A 2 μm thick protective alumina scale on F-implanted Gamma-Met (2 x 10^{17} F cm^{-2} /20 keV) after oxidation (1000h/1000°C/air).

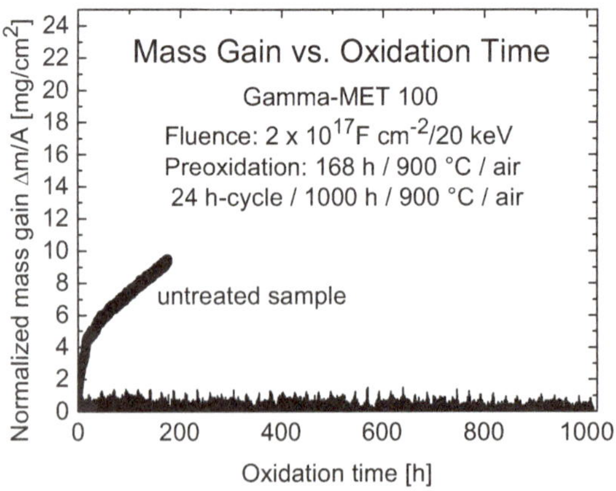

Fig. 22. Mass gain vs. oxidation time of F-implanted Gamma-Met (2 x 10^{17} F cm^{-2} /20 keV) during cyclic oxidation (1000h/900°C/air).

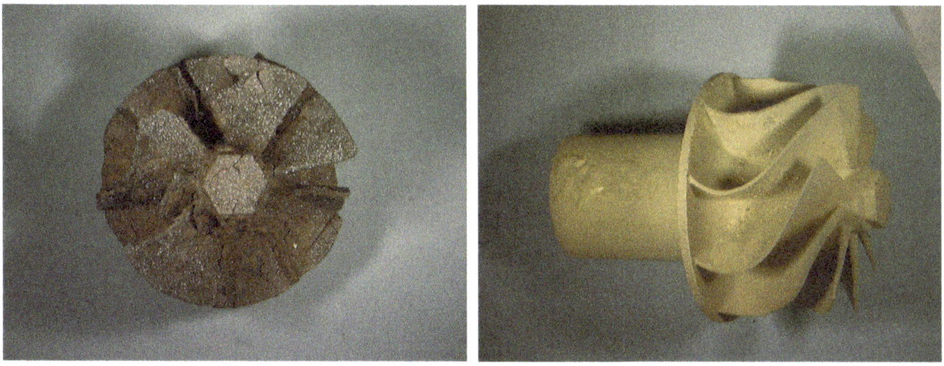

Fig. 23. Turbocharger rotor manufactured from a γ-TiAl-alloy after oxidation for 1000 h at 1050°C in air. Left image: Untreated rotor. Right image: Fluorine-treated rotor.

3.3.4 Increase of the F-reservoir by double F-implantation of TiAl

Despite the good results obtained for single step implanted samples in the present study the formation of an increased F-reservoir should be performed. The basic idea is that the near-surface implanted F - after having undergone a high F-loss during heating - has to stimulate the formation of the alumina scale, whereas the deeper implanted fluorine acts as a reservoir for longer oxidation time scales.

To optimize the set of implantation parameters numerous Monte Carlo - simulations were performed by using the software T-DYN. The influence of the implantation parameters in the case of double implantation is illustrated in fig. 24. From the results two implantation strategies have been suggested. The first implantation is done with the higher energy (55 keV) followed by the second implantation with a lower energy (20 keV). In these cases the implantation profiles show a near-surface peak with a maximum located in a depth of 30-40 nm, followed by a lower second peak in a depth of about 100 - 140 nm. The second strategy

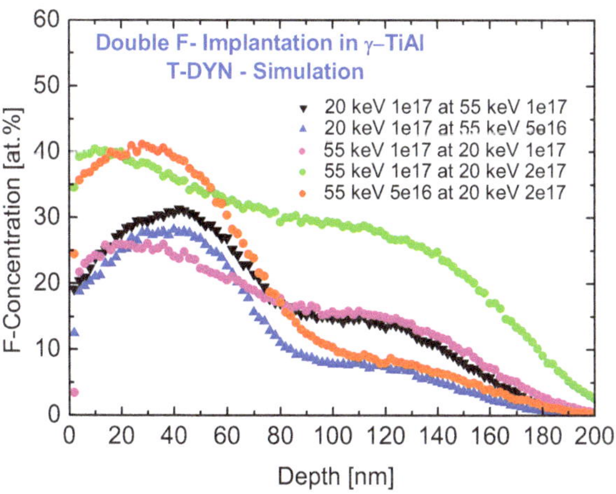

Fig. 24. F-depth profiles for double implantation into TiAl calculated by using T-DYN.

works in the reverse way: The high energy implantation is followed after a low energy implantation. These profiles show a more flat behaviour, so that the difference between the two maxima decreases. This can be explained by the implantation process: During the high energy implantation a part of the ions collides with the near-surface fluorine from the former implantation step leading to a smoother depth profile.

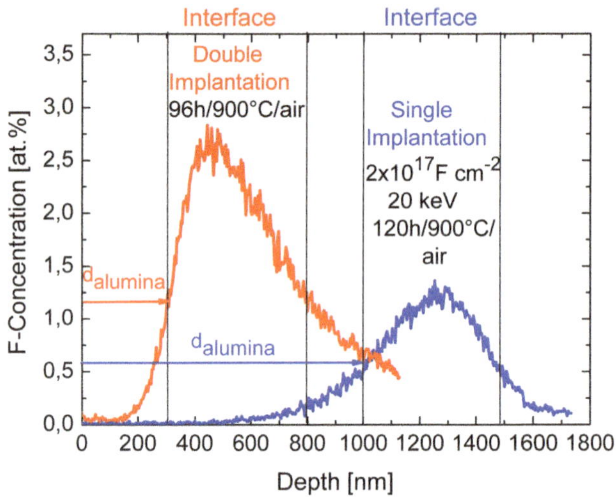

Fig. 25. F-depth profiles after oxidation of single implanted (120h/900°C) and double implanted (96h/900°C) TiAl-samples (see text).

In principle, by using more implantation steps a nearly rectangular implantation profile can be obtained. To choose a suitable set of implantation parameters one has to take into account that the near-surface implanted F is reduced during heating. In contrast to this the deeper implanted fluorine should be less affected by the fluorine loss. Therefore the F-reservoir requires a lower F-maximum compared to the near-surface level. Keeping this in mind the first strategy was chosen with 10^{17} F cm^{-2} / 55 keV as first and 10^{17} F cm^{-2} / 20 keV as the second step. The difference between single and double implantation is illustrated by the obtained F-depth profiles in fig. 25 after comparable oxidation times. Both F-maxima are located at the metal/oxide interface. While in the single-step case the F-maximum is in a depth of about 1250 nm, for the two-step sample the F-maximum occurs distinctly closer to the surface. The underlying area - this corresponds to the remaining F-dose - after double implantation is twice as high as that of the single implantation case. The near-surface region without fluorine corresponds to the alumina scale. Once formed the alumina scale acts as a diffusion barrier for fluorine and prevents further F-loss from the surface. From fig. 25 it can be concluded that the alumina growth after two-step implantation is slower compared to the single-step case. This is in agreement with the micrograph showing the cross-section in fig. 26. The alumina scale is characterized by a thickness of < 1 µm in most cases. In the buckling structures F-enriched regions are visible which serve as main F-reservoirs. Because of the 1 mm diameter of the proton beam spot of PIGE the F-depth profiles in fig. 25 represent an area averaged information. The time development of C_F^{max} for isothermal and cyclic oxidation at 900°C/1000 h reveals a significantly increased F-reservoir at the metal/oxide interface. The fit-parameters according to equ. (1) for double implantation are summarized in tab. 1.

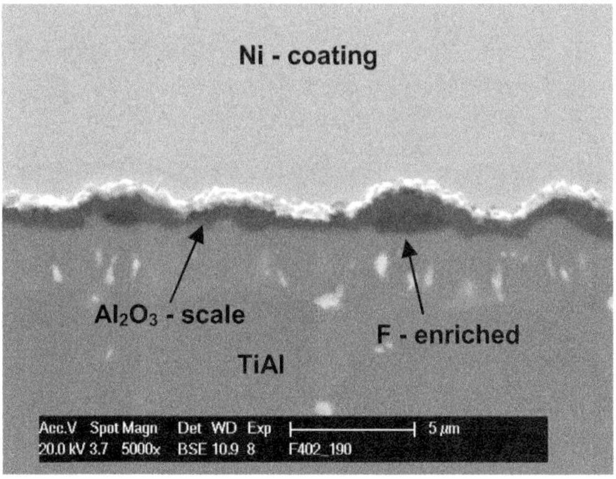

Fig. 26. Alumina scale with F-reservoirs on the surface of a Gamma-Met 100-sample after F-double implantation as described in the text and oxidation (1000h/900°C/air/isothermal).

3.3.5 Diffusion coefficient of fluorine in TiAl

After the growth of the alumina scale has been started, the remaining fluorine must be localized within the metal/oxide interface to act as a reservoir for slow alumina growth during long term oxidation protection of TiAl. Once formed, the alumina scale acts as a diffusion barrier for the remaining F into the alloy preventing a further F-loss to the surface. On the other hand, the time behaviour of the F-profiles during oxidation at 900°C, depicted in fig. 27, suggests a slow diffusion process into the metal. A basic parameter for the stability of this F-reservoir is the diffusion coefficient of F in TiAl at 900°C. There were no data available up to now. Based on the F-depth profiles after oxidation of 12, 120, 250 and 500h at 900°C illustrated in fig. 27 a first determination has been performed (Zschau et al., 2007; Zschau & Schütze, 2008) by using the diffusion equation

$$\frac{\partial c}{\partial t} = D\frac{\partial^2 c}{\partial x^2} \tag{2}$$

where c is the fluorine concentration in at./cm^3, t is the oxidation time, x the distance from the metal/oxide interface into the metal and D the diffusion coefficient in cm^2/s. The analytical solution is given by the formula (3) (Philibert, 1990)

$$c(x,t) = \frac{Q}{2\sqrt{Dt\pi}} \exp\left(-\frac{x^2}{4Dt} \right) \tag{3}$$

with Q denoting the number of fluorine atoms in at./cm^2 remaining at the metal/oxide interface after oxidation of 12 h/900°C. By getting the logarithmic form of equation (3) the diffusion coefficient D can be calculated as follows

$$\ln c = \ln\left(\frac{Q}{2\sqrt{Dt\pi}}\right) - \frac{x^2}{4Dt} \tag{4}$$

which is a linear function with respect to x^2. By supposing that after 2 hours of oxidation the initial alumina scale has been formed and hinders the F-loss outwards, the diffusion of F into the metal is the dominating process. After the transformation of the depth profiles the fit of the parameter D leads to the results summarized in tab. 2.

Oxidation time	**Diffusion Coefficient [10^{-15} cm^2/s]**
120 h	2.55 +- 0.84
250 h	0.72 +- 0.18
500 h	1.41 +- 0.51

Table 2. Diffusion coefficients of fluorine in TiAl at 900°C calculated from the fluorine depth profiles of the implanted TiAl-samples. The mean value is D = 1.56 x 10^{-15} cm^2s^{-1}.

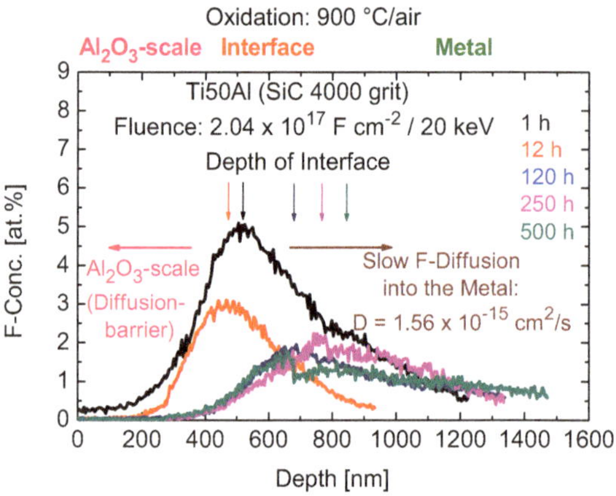

Fig. 27. Time behaviour of the F-depth profiles of implanted (2 x 10^{17} F cm^{-2} / 20 keV) TiAl after isothermal oxidation (1, 12, 120, 250 and 500 h at 900 °C in air).

The F-profiles have been determined by using PIGE without removing the Al_2O_3 – scale, which may explain the scatter within one order of magnitude of the diffusion coefficient. The influence of the chemical bonding of fluorine cannot be taken into account, because the PIGE-technique is sensitive only to the nucleus ^{19}F – independently of its chemical environment. Additionally an irregular behaviour of the metal/oxide interface may influence the scatter. The stability of the F-effect is determined by the main parameters as illustrated in fig. 27: The Al_2O_3 – scale as a diffusion barrier inhibits the F-loss to the surface, whereas the low diffusion coefficient of F in TiAl of D = 1.56 x 10^{-15} cm^2s^{-1} limits the F-diffusion into the metal. The knowledge of the diffusion coefficient allows the calculation of the F-maximum C_F^{mac} vs. oxidation time at 900°C (fig. 28). By setting Q = 1.26 x 10^{17} F-at. cm^{-2} as the number of F-atoms at the metal/oxide interface after 12 h of isothermal oxidation (fig. 29) C_F^{mac} shows a similar behaviour compared to the experimental values.

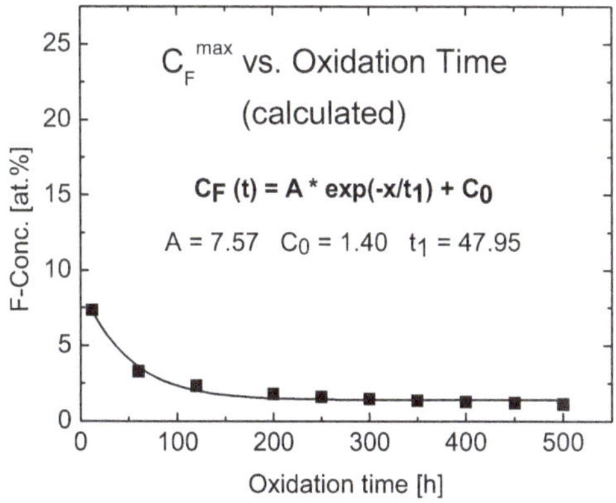

Fig. 28. Modelling of the F-maximum C_F^{max} at the metal/oxide-interface during oxidation at 900°C.

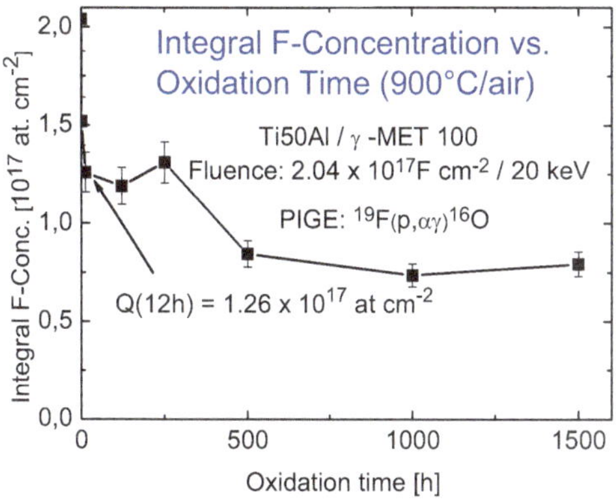

Fig. 29. Behaviour of the F-dose of implanted γ-TiAl (2 x 10^{17} F cm^{-2} /20 keV) during isothermal oxidation (1500h/900°C/air) obtained by PIGE.

3.4 Surface modification of Ni-base superalloys by fluorine ion implantation

3.4.1 Principles of the halogen effect for Ni-base superalloys

Most technical Ni-base alloys are so-called borderline materials from an oxidation point of view, i. e. their Al-contents of 2-5 wt.% are not sufficient to form a continuous dense and protective aluminium oxide layer on the surface. Unprotected Ni-base borderline alloys form a fast growing mixed oxide scale consisting of oxides and spinels from Ni, Cr, Al and other alloying elements with high metal consumption rates and limited barrier effect against ingress of species from the operating environment. Underneath the fast growing oxide scale internal oxidation takes place leading to discrete aluminium oxide particles which do not form a continuous diffusion barrier (Litz et al., 1989; Nickel et al., 1999; Pint et al., 2006). Additionally nitrides are formed at the metal/oxide interface due to the inward diffusing nitrogen. As one possibility to protect these alloys against oxidation at temperatures above

1000°C Al-rich coatings were applied. Today this method has become a state-of-the-art process for the manufacturing of components for high temperature technology like turbine blades for gas fired power stations and aircraft engines. Typical borderline alloys are IN738 and IN939 whose compositions are given in tab. 3.

Alloy	Ni	Co	Cr	Al	W	Mo	C	Ti	Ta	Others
IN738	61.13	9.0	16.00	3.30	1.70	2.60	0,17	3.50	1.70	0.90 Nb
IN939	48.35	20	22	1.4	2		0.15	3.8	1.3	1.0 Nb

Table 3. Composition of the Ni-base alloys IN738 und IN939 (in wt.%).

Wagner`s oxidation theory (Kofstad, 1988) predicts the formation of a dense alumina scale, if the Al-activity is sufficiently high. Based on this a detailed calculation of the minimum (or critical) Al-concentration N_c for several Ni-base alloys has been performed (Zschau et al., 2009a). Three groups of alloys were considered according to their Al-content: Chromia formers (group 1), alumina formers (group 2) and bond coats (group 3). The results summarized in fig. 30 reveal a strong dependence of N_c on temperature, but only a weak dependence on the chemical composition of the alloy. By considerung the alloys of group 1 the formation of a protective external Al_2O_3 - scale is possible above the borderline (i. e. the calculated N_c - curve), whereas beneath the borderline internal oxidation of Al takes place. The basic idea is to shift the borderline between alumina and chromia formers to lower Al-concentrations which allows the formation of a dense alumina scale at lower Al-contents of the bulk alloys. This can be achieved by increasing the outward mobility of Al on the surface by using the halogen effect. The halogen effect was applied already succesfully for TiAl-alloys as shown in the previous sections.

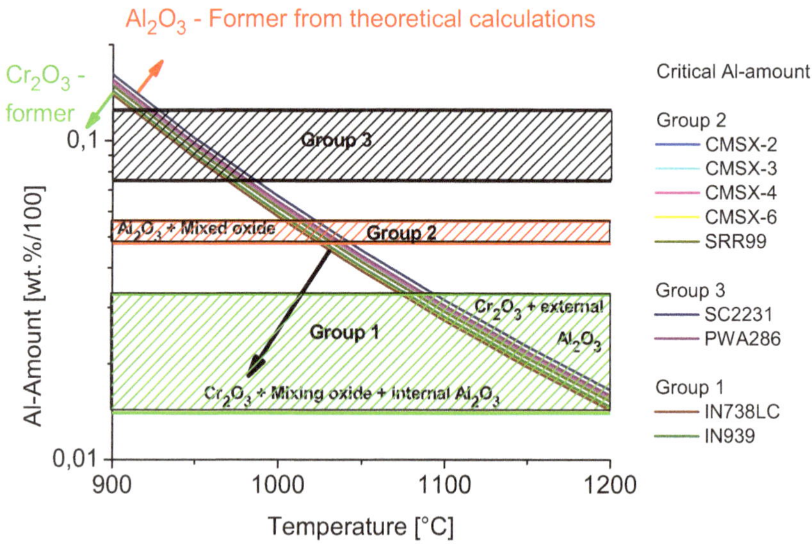

Fig. 30. Calculated minimal Al-concentration N_c required for the formation of a continuous alumina scale (external oxidation) for several Ni-based alloys (Zschau et al., 2009a).

Following the thermodynamic considerations for TiAl the ability of the selective formation of Al-fluorides as the basic condition has to be studied. The fig. 31 shows possible reactions at the metal/oxide interface. Thermodynamical calculations were carried out for the alloys IN738 and IN939 by using the software FactSage. All calculations assume thermodynamic equilibrium and an infinite system. Because no data bases for Ni-base alloys and fluorine or chlorine were available up to now, new data bases had to be developed. The calculations were performed for F and Cl and both alloys at temperatures between 900-1200°C. The results for Fluorine and IN738 at 1000°C are summarized in fig. 32, where the minimum F partial pressure $p(AlF)_{min}$ for a significant formation of AlF is denoted. By increasing the total F-partial pressure to $p(F)_{min}$ the partial pressure of gaseous AlF meets the value of $p(AlF)_{min}$ which opens the window for the F-effect. The effect works until the partial pressures of gaseous fluorides of other alloying elements (Ni, Cr, Co, Ti...) reach the value of the minimal partial pressure $p(AlF)_{min}$ too defining the partial pressure $p(F)_{max}$ and leading to the formation of a mixed oxide scale. Hence a positive F-effect is possible within the corridor between $p(F)_{min}$ and $p(F)_{max}$. The results of the calculations are summarized in fig. 33 showing the behaviour of this corridor within the temperature range of 900-1200°C. From fig. 33 it can be concluded, that the F-effect is possible for IN738 within this temperature range 900-1200°C. The calculation for IN939 shows similar results. For Cl a positive effect is predicted too (Zschau et al., 2009b). However due to the good results obtained for F at TiAl the investigations have been continued with fluorine.

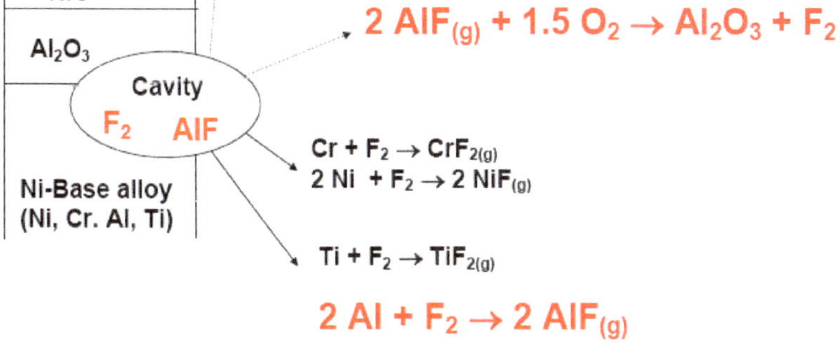

Fig. 31. Selected possible reactions in the F-doped subsurface zone of the Ni-base alloy IN738 during the initial oxidation stage.

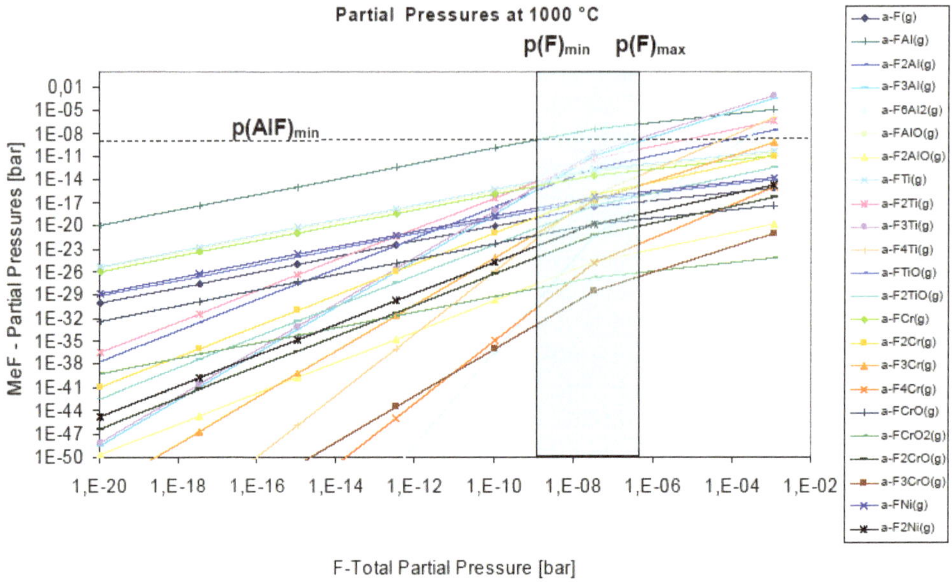

Fig. 32. Partial pressures of several metal fluorides vs. total F-partial pressure at 1000°C for the Ni-base alloy IN738 (calculated with FactSage).

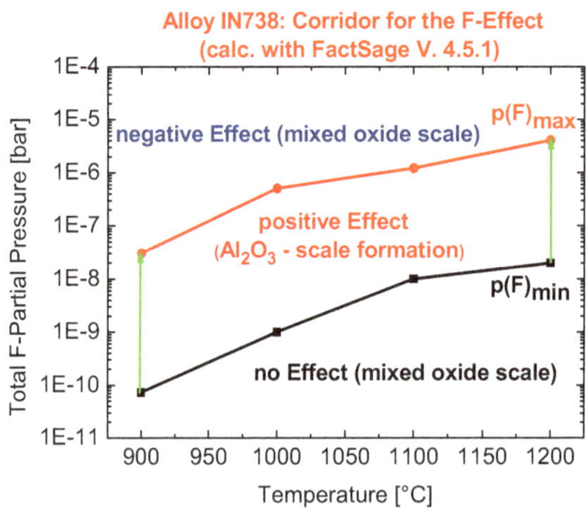

Fig. 33. Corridor of a positive fluorine effect for the alloy IN738 between 900°C and 1200°C (calculated with FactSage).

3.4.2 Screening of the implantation parameters for Ni-base superalloys

As already discussed for TiAl the calculated optimal fluorine partial pressures for the positive fluorine effect cannot be transformed into fluorine concentrations by straightforward calculations. Therefore a screening of the fluorine amount near the surface is necessary by using beam line ion implantation. The implantation profiles for fluences between 10^{16} and 4 x 10^{17} F cm^{-2} / 38 keV calculated by using the Monte Carlo software T-DYN are shown fig. 34. The energy of 38 keV meets the projected range of about 35 nm as was selected in the TiAl-case. The calculated F-profiles cover the range of fluorine amounts between 2 and 44 at%. For fluences above 4 x 10^{17} F cm^{-2} a saturation effect was found, hence higher fluences were not considered. Following these results implantations were carried out on one side of the samples with fluences between 10^{16} and 4 x 10^{17} F cm^{-2} at a constant ion energy of 38 keV. The F-profiles obtained with PIGE (Zschau et al., 2009c) reveal the good agreement between calculated and implanted F-depth profiles (fig. 35). The minor deviations at the end of the ion range may be caused by secondary ion collisions and the sputter process can explain the small differences near the surface.

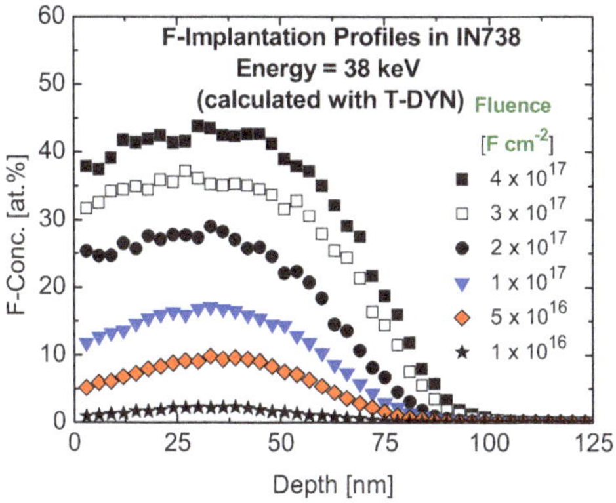

Fig. 34. F-implantation profiles in the alloy IN738 calculated by using the Monte Carlo program T-DYN.

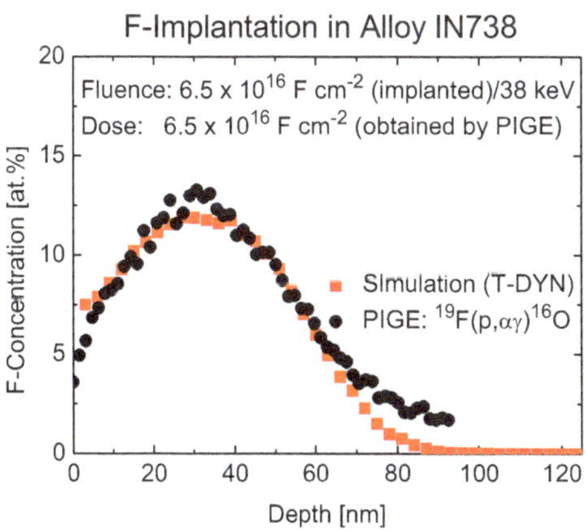

Fig. 35. Comparison of the implanted (obtained by using PIGE) and calculated F-depth profile (T-DYN) in the Ni-base alloy IN738.

After oxidation at 1050°C/air between 24 h and 168 h and metallographic preparation the SEM-micrographs reveal a change of the oxidation mechanism. The non-implanted side

(24h/1050°C/air) shows an outer porous Cr_2O_3 – scale and an inwards growing Al_2O_3 – scale (fig. 36a). The porous chromia rich layer as well as the inward growing rather discontinuous alumina scale do not protect the metal against the inward diffusion of oxygen and nitrogen. Hence also needle-like Ti-nitrides are observed up to a depth of 30-40 μm within the metal. In contrast to this the implanted side (1 x 10^{17} F cm^{-2} / 38 keV) after oxidation of 60h/1050°C/air reveals a completely different behaviour (fig. 36b): An outer Ni-Al-spinel scale covers a continuous alumina scale of 1-2 μm thickness. No internal nitrides were found. Minor inclusions of alumina are visible showing the tendency for a change from internal to external oxidation.

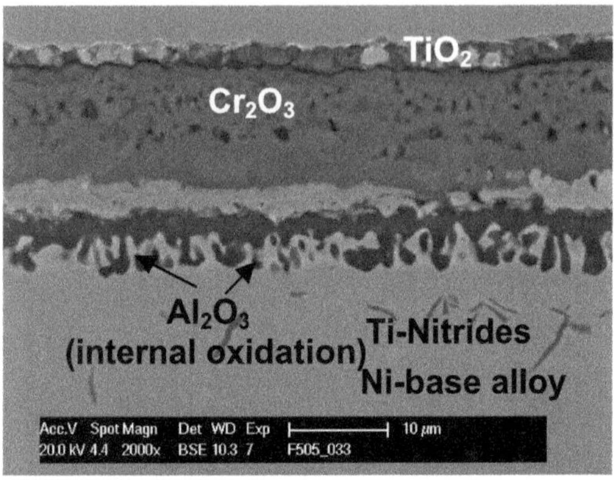

Fig. 36. a) The cross-section of the untreated Ni-base alloy IN738 after oxidation (24h/1050°C/air) shows a non-protective internal alumina scale.

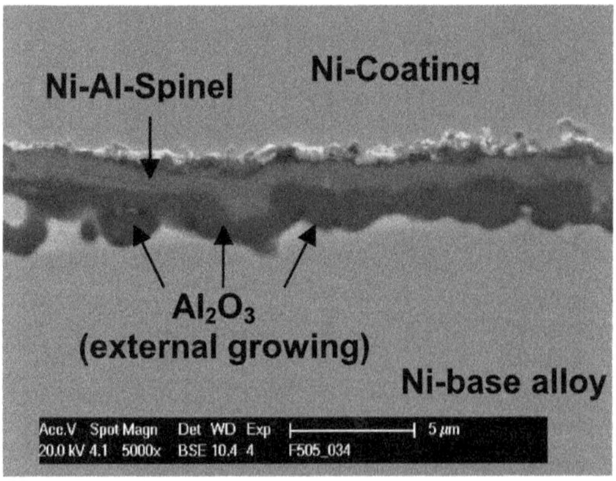

Fig. 36. b) The cross-section of the F-implanted (10^{17} F cm^{-2}/ 38 keV) Ni-base alloy IN738 after oxidation (60h/1050°C/air) reveals a protective alumina scale.

As a result of the screening the F-effect was found to be working between 5 x 10^{16} and 4 x 10^{17} F cm^{-2}. Optimum implantation parameters were determined to be between 5 x 10^{16} and 1 x 10^{17} F cm^{-2}. These fluences correspond to maximal F-contents of 10-18 at.%, as can be seen from fig. 34. At distinctly higher fluences the external alumina partly tends to the formation of an internal alumina scale.

3.4.3 Long time stability of the F-effect at Ni-base superalloys

The TGA-measurements illustrated in fig. 37 show the mass gain during long term oxidation. The TGA measurements were carried out at 1050°C for 1000 hours at samples of IN738 implanted with fluences of 5×10^{16}, 1×10^{17} and 2×10^{17} F cm^{-2} / 38 keV. Within the first 50-100 hours all samples reveal a distinct mass gain. The non-implanted sample is characterized by following a nearly linear mass loss within the whole oxidation time. The sample implanted with 2×10^{17} F cm^{-2} shows a parabolic mass gain but it did not reach slow growth kinetics. After implantation with a fluence of 1×10^{17} F cm^{-2} the parabolic mass gain reaches a slow growth rate after 400 hours, but after 600 hours a slow mass loss is obtained. After implanting with a fluence of 5×10^{16} F cm^{-2} the parabolic behaviour after 200 hours can be characterized by a kinetic constant of $k_p = 2.51 \times 10^{-9}$ g^2 cm^{-4} h^{-1}. After 900 hours a small mass loss occurs. Generally the results of the TGA curves can be influenced by several processes as formation of transient oxides, internal oxidation and nitridation, scale spallation and loss of gaseous Cr-species. The high mass gain during the first 100 h may be due to the formation of metastable transient oxides. The mass loss is most probably caused by the spallation of an outer chromia or spinel scale and/or the evaporation of Cr-based gaseous species. In order to reveal at least the final scale structure metallographic cross-sections were prepared after the TGA-measurements.

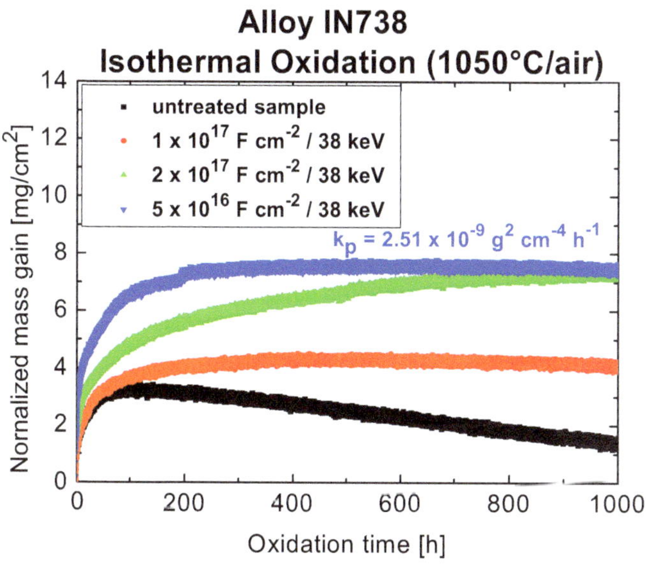

Fig. 37. Isothermal TGA-Measurements at 1050°C/air of F-implanted Ni-base alloy IN738.

The metallographic cross-section of the untreated sample (fig. 38, left micrograph) shows a thin scale (2-3 µm) of inward growing and non-protective alumina with a thin (1-2 µm) spinel layer on top whereas the outer part of the originally formed thick oxide scale had spalled completely. Underneath the alumina scale a large zone of about 30-40 µm thickness enriched with nitrides was formed.

A 1-2 µm thin alumina scale covered with a spinel (1-3 µm) phase was found on the surface of the sample implanted with 5×10^{16} F cm^{-2} (fig. 38, right micrograph). The absence of Al_2O_3

- precipitates indicates a change from internal oxidation to the formation of a continuous protective external Al_2O_3 - scale. EDX-analysis proofs the composition of this scale as pure alumina, whereas the spinel scale mainly contains Al and Ni.

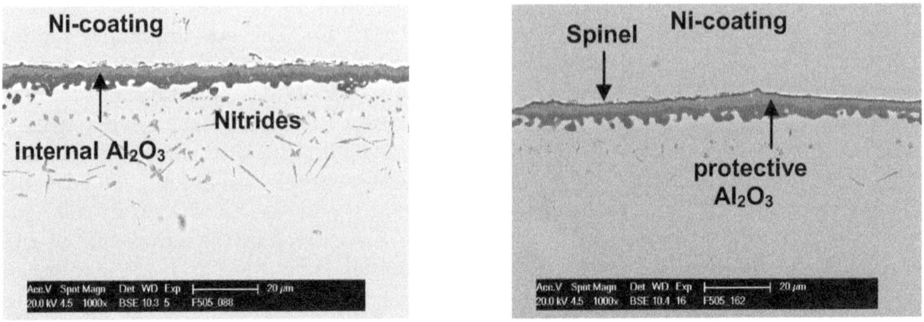

Fig. 38. Oxide scale of an untreated (left micrograph) and an implanted (5×10^{16} F cm^{-2} / 38 keV) sample (right micrograph) of IN738 after oxidation (TGA/1000h/1050°C/air).

Finally the region for a positive F-effect for the Ni-base alloy IN738 in terms of F-partial pressures, F-fluences and F-concentrations can be illustrated in fig. 39. In practice there are no sharp boundaries for fluences and concentrations. This is illustrated by the light grey regions.

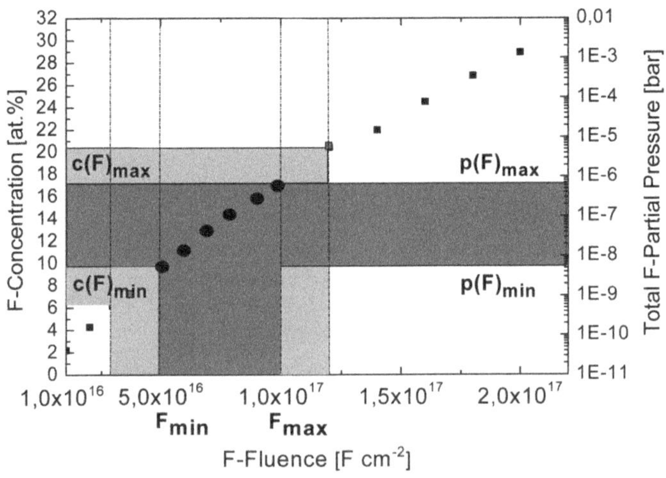

Fig. 39. Relation between F-partial pressures, F-fluences and F-concentrations for a positive F-effect at the Ni-base alloy IN738.

4. Conclusions

The ion implantation of fluorine combined with the non-destructive Ion Beam Analysis have been applied to investigate the halogen effect for fluorine at γ-TiAl and Ni-base alloys. By suitable surface doping of γ-TiAl (44-50 at.% Al) with fluorine the high temperature oxidation resistance above 750°C has been improved dramatically by stimulating the formation of a dense protective alumina scale on the metal surface. The optimal F-concentrations have been defined. Recently these results have been used to apply other methods of F-application (gas phase, liquid phase, plasma immersion ion implantation) for the treatment of technical components with a complex geometry. Based on the results of the beam line ion implantation the optimal parameters can be found also for each of these process routes. The ion beam analysis, especially PIGE, plays an important role in these investigations. The fluorine effect at TiAl-alloys is close to industrial application in high temperature technology.

The Ni-base superalloys with 2-5 wt.% Al-content form a non-protective complex oxide scale during oxidation at temperatures above 900°C. After surface modification by fluorine ion implantation the oxidation mechanism can be changed dramatically to the growth of a dense protective alumina scale. Optimal F-concentrations for the F-effect for 1st and 2nd generation Ni-base superalloys have been determined. Recently basic work is in progress to apply the halogen effect to the 3rd and 4th generation superalloys and to improve the scale adherence. Additional studies focuse on the mechanism of the initial oxidation in the presence of the fluorine effect.

5. Acknowledgment

The work consisting of several projects was supported to a large extent by the Deutsche Forschungsgemeinschaft (DFG) under contracts SCHU 729/12, SCHU 729/15-1, SCHU 729/18-1 and SCH 729/18-2 which is gratefully acknowledged.

The authors thank Prof. K. Bethge, Dr. H. Baumann, Prof. R. Dörner, Dr. L. Schmidt, Dr. K.E. Stiebing and S. Neve from the Institute of Nuclear Physics of the Goethe-University Frankfurt for their support by using the implanter and the Van de Graaff accelerator.

6. References

Biersack, J. (1999). TRIM-DYNAMIC applied to marker broadening and SIMS dept profiling. *Nucl. Instr. & Meth. Phys.. Res. B*, 153, (1999), pp. 398-409.

Calphad, www.calphad.com.

Donchev, A.; Gleeson, B. & Schütze, M. (2003). Thermodynamic considerations of the beneficial effect of halogens on the oxidation resistance of TiAl-based alloys. *Intermetallics*, 11, (2003), pp. 387-398.

Donchev, A.; Schütze, M.; Heller, J. & Scheydecker, M. (2006). Oxidationsschutz für Titan-aluminidbauteile durch Oberflächenbehandlung mit Fluor. *Proceedings Oberflächentage 2006.*

FactSage Version 4.5.1 Fa. GTT, Herzogenrath.

Fergus, J. W. (2002). Review of the effect of alloy composition on the growth rates of scales formed during oxidation of gamma titanium aluminide alloys. *Materials Science and Engineering*, A338, (2002), pp. 108-125.

Hornauer, U,; Richter, E.; Wieser, E.; Möller, W.; Schumacher, G.; Lang, C. & Schütze, M. (1999). Improvement of the High Temperature Oxidation Resistance of Ti50Al via Ion-Implantation. *Nucl. Instr. & Meth. in Phys. Res. B*, 148, (1999), pp. 858-862.

Kofstad, P. (1988). *High Temperature Oxidation of Metals*, Elsevier, New York, 1988.

Kumagai, M.; Shibue, K.; Kim, M.-S. & Yonemitsu. M. (1996). Influence of chlorine on the oxidation behavior of TiAl-Mn intermetallic compound. *Intermetallics*, 4, (1996), pp. 557-566.

Litz., J.; Rahmel, A.; Schorr, M. & Weiss, J. (1989). Scale Formation on the Ni-Base Superalloys IN 939 and IN 738 LC. *Oxidation of Metals*, 32, (1989), pp. 167–184.

Masset, P.J. & Schütze, M. (2008). Thermodynamic Assesments of the Alloy Concentration Limits for the Halogen Effect of TiAl Alloys. *Adv. Engin. Mat.*, 10, 7, (2008), pp. 666- 674.

Nickel, H.; Clemens, D.; Quadakkers, W.J. & Singheiser, L. (1999). Development of NiCrAlY Alloys for Corrosion-Resistant Coatings and Thermal Barrier Coatings of Gas Turbine Components. *Transactions of the ASME, Journal of Pressure Vessel Technology*, 121, (1999), pp. 384-387.

Philibert, J. (1990). *Diffusion et transport de matiere dans les solides*, Les Ulis, 1990.

Pint, B. A.; DiStefano, J. R. & I. G. Wright (2006). Oxidation resistance: One barrier to moving beyond Ni-base superalloys. *Materials Science and Engineering: A*, 415, Issues 1-2, (15 January 2006), pp. 255-263.

Rahmel, A.; Quadakkers, W. J. & Schütze, M. (1995). Fundamentals of TiAl Oxidation – A Critical Review. *Materials and Corrossion*, 46, (1995), pp. 271-285.

Schütze, M. & Hald, M. (1997). Improvement of the oxidation resistance of TiAl alloys by using the chlorine effect. *Mat. Sci. Eng.*, A239-240, (1997), pp. 847-858.

Schumacher, G.; Dettenwanger, F.; Schütze, M.; Hornauer, U.; Richter, E.; Wieser, E. & Möller, W. (1999a). Microalloying Effects in the Oxidation of TiAl Materials. *Intermetallics*, 7, (1999), pp. 1113-1120.

Schumacher, G.; Lang, C.; Schütze, M.; Hornauer, U.; Richter, E.; Wieser, E. & Möller, W. (1999b). Improvement of the Oxidation Resistance of gamma titanium aluminides by microalloying with chlorine using ion implantation. *Maerials and Corrosion*, 50, (1999), 162-165.

Tesmer, J. R. & Nastasi, M. (1995.), *Handbook of Modern Ion Beam Materials Analysis*, Materials Research Society, Pittsburgh 1995, PA, USA.

Ziegler, J.; Biersack, J. & Littmark, U. (1995). *The stopping and range of ions in solids, Version 95*, 1995, Pergamon Press, New York.

Zschau, H.-E.; Gauthier, V.; Schütze,, M.; Baumann, H. & Bethge, K. (2002). Investigation of the Fluorine Microalloying Effect in the Oxidation of TiAl at 900°C in Air. *Proc. International Symposium Turbomat*, June 17-19, 2002, Bonn, ed. W. Kaysser, (2002), pp. 210-214.

Zschau, H.-E.; Schütze, M.; Baumann, H. & Bethge, K. (2005). Application of Ion Beam Analysis for the Control of the Improvement of the Oxidation Resistance of TiAl at 900°C in Air by Fluorine Ion Implantation and HF-Treatment. *Nucl. Instr. & Meth. in Phys. Res. B*, 240, (2005), pp. 137-141.

Zschau, H.-E.; Schütze, M.; Baumann. H. & Bethge, K. (2007). Surface modification of titanium aluminides with fluorine to improve their application for high temperature service conditions. *Nucl. Instr. & Meth. in Phys. Res. B*, 257, (2007), pp. 383-387.

Zschau, H.-E. & Schütze, M. (2008). Modelling of the long time stability of the fluorine effect in TiAl oxidation. *Materials and Corrosion*, 59, 7, (2008), pp. 619-623.

Zschau; H.-E.; Renusch, D.; Masset, P. & and Schütze, M. (2009a). A new concept of oxidation protection of Ni-base alloys by using the halogen effect. *Materials at High Temperatures*, 26, No. 1, (2009), pp. 85-89.

Zschau, H.-E.; Masset, P. J.; Renusch, D. & Schütze, M. (2009b). Thermodynamical Considerations for applying the Halogen Effect to Ni-Base Superalloys. *Proc. TMS Spring Meeting*, San Francisco, 15.-19.2.2009, pp. 149-156.

Zschau, H.-E.; Renusch, D.; Masset, P. & Schütze, M. (2009c). The halogen effect for Ni-base alloys – a new method for increasing the oxidation protection at high temperatures. *Nucl. Instr. & Meth. in Phys. Res. B*, 267, Issue 8-9, (2009), pp. 1662-1665.

17

Ion Implantation-Induced Layer Splitting of Semiconductors

U. Dadwal[1], M. Reiche[2] and R. Singh[1]
[1]*Department of Physics, Indian Institute of Technology Delhi, Hauz Khas, New Delhi*
[2]*Max Planck Institute of Microstructure Physics, Halle*
[1]*India*
[2]*Germany*

1. Introduction

Ion implantation is considered to be a precise technology for the dopant introduction not only in semiconductors but also in metals and insulators (Plummer et al., 2000; Sze, 1998). The other way of introducing dopants is the plasma treatment (Sze et al., 1998). However, ion implantation in semiconductors is extensively used for various applications in the field of optoelectronics, micro/nano-electro-mechanical systems, power electronics etc (Plummer et al., 2000; Sze, 1998). In particular, ion implantation in silicon (Si) using light ions such as hydrogen (H) has gained significant attention since last two decades (Bruel, 1995; Tong & Gösele, 1999; Lee et al., 2004). This is because hydrogen in semiconductors, whether introduced intentionally or unintentionally, is known for its role in the defects and dopant passivation (Pankove & Johnson, 1991; Chabal et al., 1999). Moreover, as H in semiconductors is chemically reactive, it influences the transport mechanisms of the charge carriers and ultimately the device performance (Chabal et al., 1999; Höchbauer et al., 2002). The H-implantation in Si started a new era in the electronic industry when it was first time coupled with the direct wafer bonding technique (Bruel, 1995). This coupled process is now called as ion-cut process, which was commercially employed by SOITEC, France by the name of Smart-Cut™ process. This process came into existence by the end of twentieth century when first time the engineered substrates such as silicon-on-insulator (SOI) were realized (www.soitec.com). Since then SOI technology really helped in boosting the device performance because of its numerous advantages leading to high speed microprocessors, low power consumption, low leakage current, low parasitic and latch-up effects etc (Bruel, 1995; Tong & Gösele, 1999; Christiansen et al., 2006). After that, the ion-cut process was studied and successfully extended to germanium (Ge) for the fabrication of germanium-on-insulator (GeOI) substrates (Deguet et al., 2006). Since then Si and Ge processing technologies have grown upto high maturity, which are now applied on industrial scale in a regular basis. But, ion-cut process lacks its maturity when it is applied to other semiconductors due to various technical as well as commercial reasons (Radu et al., 2003; Cioccio et al., 1996; Moutanabbir et al., 2010). Nevertheless, this process is experimentally demonstrated on a small scale in silicon carbide (SiC), indium phosphide (InP) and gallium nitride (GaN) (Cioccio et al., 1996; Singh et al., 2010). In the past couple of years the extensive studies have been carried out in understanding the ion-cut process in these and some other technological important semiconductors, such as (gallium arsenide) GaAs, silicon germanium (SiGe), and aluminium nitride (AlN) (Radu et al., 2003; Singh et al., 2005, 2010). These studies have shown that ion-cut process in these semiconductors still holds promise for potential applications in devices.

1.1 Basics of H-implantation-induced Blistering in Semiconductors

Ion-cut process consists of H-implantation in semiconductors and direct wafer bonding techniques (Bruel, 1995; Christiansen et al., 2006). The implanted H ions induce damage within the semiconductor, which during implantation or after post-implantation annealing results in deformations of the implanted surface (Tong & Gösele, 1999; Hayashi et al., 2008). If these deformations are in the form of surface blisters, then it is referred as surface blistering and if they are in the form of regions from where the implanted surface is completely detached, then it is called surface exfoliation (Fig. 1) (Kucheyev et al., 2001). However, there is no difference between the H-implantation parameters that lead to blistering/exfoliation or layer splitting/layer transfer using direct wafer bonding.

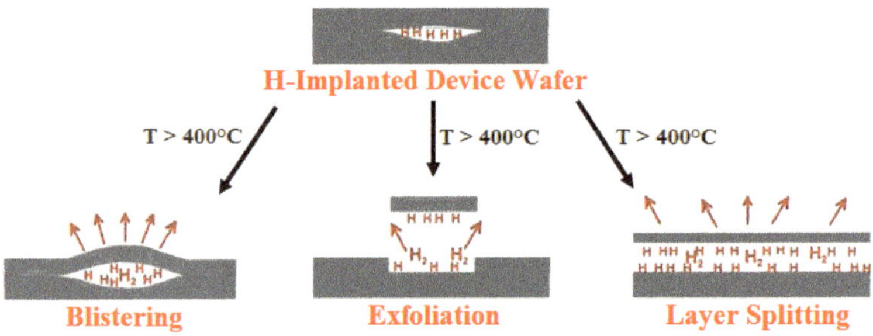

Fig. 1. Schematic illustration of the H-induced surface blistering, surface exfoliation and layer splitting process after post-implantation annealing at elevated temperature (T).

In some cases large area exfoliation, where the implanted surface is completely removed over a large regions, also occurs (Dadwal et al., 2010). The detailed understanding of the hydrogen implantation-induced surface blistering mechanisms is very essential from the prospect of successful application for ion-cut process. H-implantation induces various types of defects within the damage region, which upon annealing agglomerates to extended defects like nanocracks/microcracks (Chabal et al., 1999; Höchbauer et al., 2002). These extended defects act as efficient trapping centres for the implanted H. The implanted H gets segregate into these defects in molecular form, which thereby creates internal pressure and hence takes them to an overpressurized state. When this internal pressure reaches to the fracture limit of the implanted material, it ultimately lifts the implanted surface in the form of surface blistering/exfoliation (Kucheyev et al., 2001; Di et al., 2008; Padilla et al., 2010). However, the detail mechanisms of the surface blistering are very complex and which even in the highly matured SOI technology are still not fully understood (Moras et al., 2010).

1.2 Layer transfer of semiconductors

Hydrogen implantation-induced surface blistering/exfoliation can be extended to the transfer of a thin layer of the implanted semiconductor (donor wafer) onto an appropriate foreign substrate (handle wafer) using ion-cut process (Höchbauer et al., 2002; Singh et al., 2010). In this process the implanted wafer is bonded to a handle wafer after following the standard wafer cleaning procedure (Tong & Gösele, 1999). Then these bonded wafers are subjected to thermal annealing at elevated temperatures for the sufficient duration. The thermal treatment induces defects migration and agglomeration, which leads to layer cracking parallel to the bonded interface. Hence, it gives transfer of implanted layer from the parent substrate onto the handle substrate (Bruel, 1995; Tong & Gösele, 1999; Moutanabbir et al., 2010). However, the layer splitting is only possible if bonding energy of the bonded interface between the handle wafer and the implanted layer is much stronger than the implanted layer and the donor wafer. Moreover, one can transfer multiple layers from the single donor wafer by tuning the energy of the implanted H ions. This feature of ion-cut process makes it very attractive in the case where donor wafer is highly expensive and is mostly available in small sizes such as in the case of III-V nitride semiconductors, like GaN and AlN (Kucheyev et al., 2001; Moutanabbir et al., 2010).

1.3 Brief chapter description

This chapter is basically an attempt to shed light and give an overview about the implantation-induced layer splitting of the semiconductors using ion-cut process. The layer splitting is discussed in some of the important compound and conventional semiconductors for the realization of current and new generation technology. This chapter is divided into ten sections including introduction and reference section. Each section is further divided into different subsections. The extensive studies of the layer splitting mechanisms in Si makes ion-cut process highly matured in it and, that is why highly explored in the layer transfer applications (Tong & Gösele, 1999). Therefore, the implantation-induced ion-cut process in Si is not discussed in this chapter. The focus is given to some other technologically important semiconductors such as Ge, SiGe, GaAs, InP, GaN, AlN and other important semiconductors for the understanding of the H-induced layer splitting mechanisms.

2. Ge and SiGe alloys

Ge and its alloy with Si, which is SiGe, are two important materials which have promising applications in the integrated circuits (ICs) technology because of the higher carrier mobility and optimum forbidden bandgap (Singh et al., 2005; Huang et al., 2001). Specifically in Ge, the hole mobility is about 1800 $cm^2V^{-1}sec^{-1}$. Whereas in SiGe, depending upon the concentration of Ge in Si, mobility enhancement of the charge carriers to the values up to 40000 $cm^2V^{-1}sec^{-1}$ could be achieved in the modulation doped structure of the strained Si (sSi) on silicon germanium-on-insulator (SGOI) (Huang et al., 2001). The process of H-implantation-induced layer splitting was used to transfer thin layers of Ge and SiGe without investigating its dependence on the implantation temperatures. Therefore in the following subsections we have discussed temperature dependent behaviour of the H-induced layer splitting process in these materials.

2.1 H-implantation in Ge

H-implantation in Ge results in ion induced defects which are quite similar to that in H implanted Si. These defects are vacancies, interstitials and their complexes with the implanted H (Akatsu et al., 2005; David et al., 2009). When H is implanted in Ge, it results in surface blistering/exfoliation depending upon the implantation and post-implantation annealing parameters (Akatsu et al., 2005; Dadwal et al., 2009). For instance, Fig. 2 shows 100 keV H ion implanted Ge samples at different implantation temperatures. For the implantation at liquid nitrogen (LN_2), the surface exfoliation is observed in the form of craters of different size (Fig. 2a). The room temperature (RT) H-implantation displayed large area exfoliation after post implantation annealing at 500 °C for 30 min (Fig. 2b). This large area exfoliation is extended upto few 100 square micrometer over the implanted surface. However, H-implantation at higher temperature of 300 °C results surface exfoliation in the as-implanted state (Fig. 2c) (Dadwal et al., 2009).

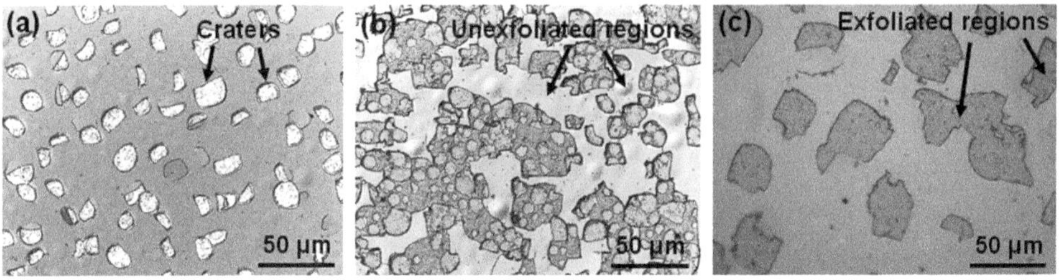

Fig. 2. Nomarski optical images of the 100 keV H implanted Ge at (a) LN_2, (b) RT after post-implantation annealing at 500 °C for 30 min, and (c) 300 °C in the as-implanted state (Dadwal et al., 2009).

The increase of implantation temperature shows pronounced diffusion of the H-induced defects and their re-arrangements within the damage band, which leads to the formation of extended defects like microcracks (Zahler et al., 2007). The role of implanted H is to passivate the internal damage lattice, which results in molecular hydrogen within these extended defects during implantation or after post-implantation annealing. These extended defects filled with H_2 molecules are responsible for the surface craters or exfoliation depending upon the implantation temperature (Dadwal et al., 2009). Since diffusion of the implanted species is also a time dependent process, that is why post-implantation annealing at temperature lower than 500 °C for time less than 30 min does not produced any surface deformation. However, one may obtain the H-induced surface blistering at an annealing temperature lower than 500 °C but the annealing should be performed for duration longer than 30 min. This blistering study of the H implanted Ge was successfully used for the fabrication of GeOI substrates where bulk Ge acted as donor substrate and the transferred layer thickness was about few thousands of angstroms (Deguet et al., 2006; David et al., 2009). The root mean square roughness (RMS) of the transferred layer was below 0.2 nm and Z range value was around 2 nm on a 5 × 5 μm^2 scan area after the chemical mechanical polishing (CMP) step.

2.2 H-implantation in SiGe

The introduction of Ge in the Si lattice results in new applications of the SiGe alloy in the fabrication of strained SiGe (sSiGe) based high speed devices and modern complementary-metal-oxide-semiconductor (CMOS) technology (Singh et al., 2005; Huang et al., 2001). When Ge adds into the Si lattice it gets strained. Depending upon the molar composition of

Ge, gate length of the sSiGe based CMOS devices and bandgap in base region of the SiGe heterojunction bipolar transistors (HBTs) varies accordingly which gives its importance in the said applications. The typical value of Ge composition in the SiGe alloy ranges between 15–30% of Si. However, with the increase in Ge concentration in Si lattice, quality of the transferred sSi layer starts degrading due to the over concentration of Ge. This allows agglomeration or precipitation of Ge during the wafer bonding or in the post-implantation annealing step (Huang et al., 2001). Hence, in order to prepare high speed electronic circuits and optoelectronics, the wider range of Ge concentration is preferable. For that, understanding of the implantation-induced blistering parameters is essential in order to normalize the layer splitting process at different Ge concentrations. In particular, if we consider H-implantation in $Si_{0.70}Ge_{0.30}$, it does not results surface blistering/exfoliation in the as-implanted state for the implantation at LN_2 and RT (Dadwal et al., manuscript under preparation). However, their post-implantation annealing at higher temperature of 400 °C for 1 hr showed surface blistering/exfoliation depending upon the implantation temperature (Fig. 3).

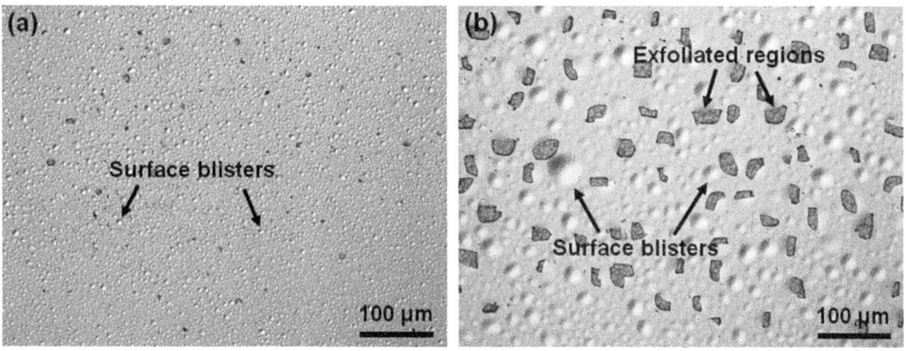

Fig. 3. Nomarski optical images of the 100 keV H implanted $Si_{70}Ge_{30}$ for a fluence of 1×10^{17} cm^{-2} at (a) RT and (b) LN_2 after post-implantation annealing at 400 °C for 1 hr (Dadwal et al., manuscript under preparation).

With the increase of post-implantation annealing temperature to 500 °C, the RT implanted SiGe samples also undergo surface exfoliation for the same annealing time of 1 hr. However in the LN_2 implanted SiGe samples, surface exfoliation dominate on a large scale after post-implantation annealing at 500 °C for 1 hr (Dadwal et al., manuscript under preparation). In comparison to H-implantation at RT, H-implantation at LN_2 undergoes effective segregation of the H-induced extended defects within the narrow damage band and is responsible for the surface exfoliation after post-implantation annealing. Such type of H-induced extended defects in the form of platelets and microcracks are very recently been reported in blistering study of the H-implantation in $Si_{0.70}Ge_{0.30}$ (Singh et al., 2011). These extended defects are not only oriented along (001) plane but also along {111} planes, and are capable of deforming top implanted layers in the form of surface blisters and craters (Fig.4).

This implantation temperature dependence of the blistering study can provide a meaningful information in the prospective of high Ge contained SiGe layer transfer of good quality. Specifically, the surface blistering/exfoliation at the lower implantation and annealing temperature seems to be more appropriate in the fabrication of sSGOI substrates with high

Ge concentration. This is because of some issues related to the thermal and lattice mismatch between the donor and handle substrate. H implanted $Si_{0.85}Ge_{0.15}$ has shown high quality layer splitting in the fabrication of strained silicon-germanium-on-insulator (sSGOI) substrates (Huang et al., 2001). This was done by bonding of the H implanted $Si_{0.85}Ge_{0.15}$ epitaxial layers with the oxidized Si wafer followed by annealing at 500–600 °C to induce layer splitting. The final RMS roughness of the transferred $Si_{0.85}Ge_{0.15}$ layer was less than 1 nm (Huang et al., 2001).

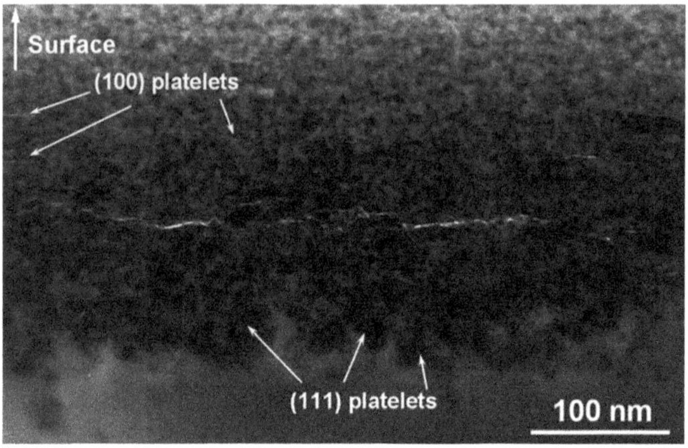

Fig. 4. Cross-sectional transmission electron microscopy (XTEM) image of the 120 keV H^+ ion implanted $Si_{0.70}Ge_{0.30}$ with a fluence of 1 × 10^{17} cm^{-2} after post-implantation annealing at 500 °C for 5 min (Singh at al., 2011). doi:10.1088/0268-1242/26/12/125001

3. GaAs

GaAs is the first binary compound semiconductor which was used in the fabrication of electronic devices after conventional semiconductors such as Si. The combination of GaAs with the low lattice matched Ge provides fruitful platform for the integration of the GeOI substrates with the III-V materials. Moreover, being a direct bandgap material, it is an alternative mean of revolutionizing GaAs technology when it combines with ion-cut process (Radu et al., 2003). If GaAs is implanted by H ions in the temperature range of about 100–250 °C, the surface blistering results depending upon implantation fluence in the range of about (0.3–2) × 10^{17} cm^{-2}. The hydrogen induced surface blistering is due to overlapping of the microcracks oriented along {111} and {100} planes (Fig. 5). It results in the formation of larger cracks of size more than 1 μm, giving layer cracking near to the peak region of the damage concentration. However, surface blistering can be probable even at a higher H fluence. In general, the H-induced surface blistering is not effective for layer splitting applications because of its much temperature sensitive nature due to the beam heating effects. In addition to this, helium (He) implantation in GaAs provides control surface blistering and even large surface area exfoliation only after post-implantation annealing for a fluence of 5 × 10^{16} cm^{-2} (Radu et al., 2003). This controlled surface blistering has laid basis for the He-induced GaAs layer splitting applications. Hence in GaAs, in comparison to He-implantation, ion fluence is higher for H-implantation in achieving the surface blistering/exfoliation.

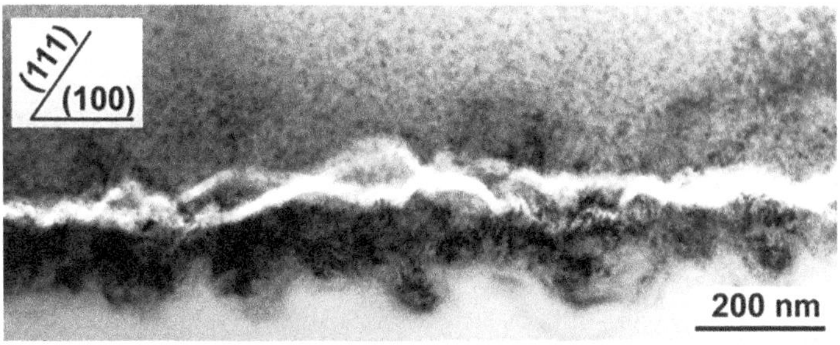

Fig. 5. XTEM image of the 130 keV hydrogen implanted (100) GaAs wafer at 100 °C with a fluence of 3.5×10^{16} cm^{-2} (Radu et al., 2003). Reprinted with permission from Radu et al., (2003), American Institute of Physics

Moreover, H and He co-implantation method is also used for the layer splitting of GaAs. The GaAs co-implantation with H_2^+ ions of energy 160 keV and fluence 3×10^{16} cm^{-2}, and with He ions of energy 105 keV and fluence 5×10^{15} cm^{-2} showed controlled surface blistering only after post-implantation annealing. This is due to the formation of relatively narrow damage band decorated with bubbles, platelets like defects over the whole damage band (Radu et al., 2003). The surface blistering in H and He co-implanted GaAs is observed only for the implantation at RT.

3.1 Role of implantation temperature

In general, the implantation temperature is very critical for the ultimate surface deformations in the form of surface blistering/exfoliation. There are many reasons responsible for this temperature dependent blistering behaviour of the implanted semiconductors (Kucheyev et al., 2001; Lee et al., 2004). The prominent facts are that, blistering is a temperature and time dependent process where implanted species should get released from the passivated defect sites for their segregation and the formation of extended defects like platelets. The platelets formation is necessary for the surface blistering/exfoliation (Moutanabbir et al., 2010). In the case of GaAs, if implantation temperature goes beyond the range of about 100–160 °C, the surface blistering does not occur due to the out diffusion of H from the damage region. Hence it does not allow the agglomeration of H for the formation of extended defects like platelets, which are the precursors for the surface blistering. Similarly if the implantation temperature is lesser than 100 °C, then implanted H does not get sufficient migration to undergo proper interactions with the implantation-induced defects, which further hinder the surface blistering/exfoliation. Hence the implanted ions should possess a sufficient minimum energy called as activation energy for the surface blistering or ion-cut process.

3.2 GaAs layer splitting

The layer splitting of GaAs using ion implantation and direct wafer bonding technique can be done by choosing an appropriate handle substrate. In comparison to H-implantation, the He implanted GaAs wafer is used in the layer transfer process onto Si substrate (Radu et al., 2003). However, the layer transfer occurred partially. But, H and He co-implantation provides complete and uniform transferred layer of the implanted GaAs onto Si substrate using intermediate Si-on-glass (SOG) layer (Fig. 6).

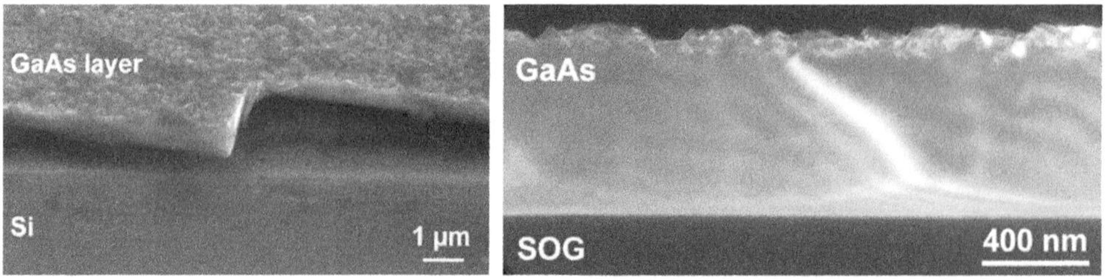

Fig. 6. Transferred GaAs layers onto Si-on-glass/Si substrate with (a) scanning electron microscopy image of the He implanted GaAs at RT after post-implantation annealing at 200 °C for 14.5 hr and (b) XTEM image of the bonding interface between the transferred GaAs layer and SOG layer using H and He co-implantation at RT followed by post-implantation annealing at 225 °C for 14 hr (Radu et al., 2003). Reprinted with permission from Radu et al., (2003), American Institute of Physics

The transferred GaAs layer obtained by the co-implantation is little bit more rough in comparison to the partially transferred GaAs layer obtained by He-implantation alone (Radu et al., 2003). This is due to the reason that surface morphology of the transferred GaAs layer depends upon the implantation-induced extended defects, like microcracks orientation and there overlapping within the damage region (Fig. 7).

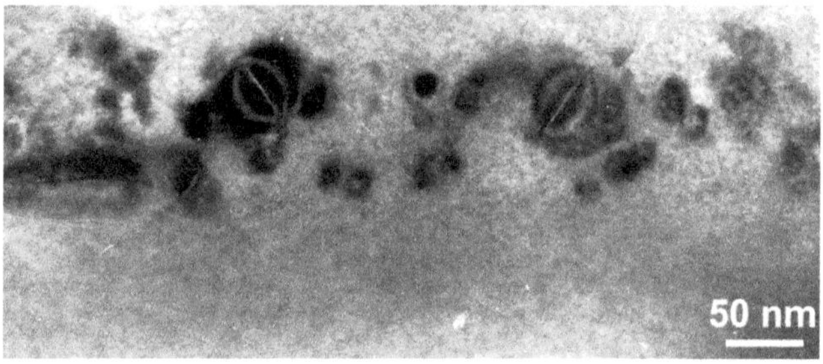

Fig. 7. XTEM image of the platelets and microcracks in the as-implanted (100) GaAs by He (105 keV, 5 × 10^{15} cm^{-2}) followed by H_2^+ (160 keV, 3 × 10^{16} cm^{-2}) implantation at RT (Radu et al., 2003). Reprinted with permission from Radu et al., (2003), American Institute of Physics

During co-implantation, H-implantation induces platelets formation oriented along (100) and (111) planes, which gives zig-zag overlapping of the microcracks in the as-implanted state (Fig. 7). Hence it results relatively high surface roughness to the transferred GaAs layer after post-implantation annealing in comparison to only He implanted GaAs, where platelets are mostly oriented along (100) planes that are parallel to the implanted surface.

4. InP

The implantation-induced layer splitting of InP using hydrogen has strong dependence on the implantation temperature as well as the post-implantation annealing parameters (Hayashi et al., 2004; Chen et al., 2008). The problem becomes more severe if the temperature of the sample holder rises during implantation. The rise in the sample holder temperature during implantation can activate the implanted hydrogen and it starts diffusing out of the damage region. This eventually results in the absence of blistering/exfoliation in

InP even after post-implantation annealing. It was reported by the Tong *et al* that in the case of InP the surface blistering or layer exfoliation occurred after post-implantation annealing only if the implantation temperature was between 150–250 °C (Tong et al., 1999). The Hayashi *et al* have shown that hydrogen implantation-induced out-of-plane tensile strain within the damage region also influences the blistering/exfoliation behaviour in InP (Hayashi et al., 2004). This out-of-plane tensile strain gives rise to in-plane compressive stress within the damage region which is dependent upon the implantation temperature. Moreover, H-implantation at higher temperature produced much less out-of-plane tensile strain in comparison to the low temperature implantation. The origin of this induced strain within the damage region is related to the implantation-induced distortion of the crystal lattice in the form of point defects and there clusters (Chen & Di et al., 2008). It was reported by Singh *et al* that instead of hydrogen, He-implantation can also be used for achieving layer splitting in InP (Singh et al., 2006). This is because in InP the diffusivity of the implanted He ions is not very sensitive to the implantation temperature. In this report four-inch diameter InP wafers were used and the implantation temperature was kept at either RT or –15 °C. This resulted in controlled surface blistering/exfoliation only after the post-implantation annealing in the temperature range of 225 to 400 °C. Moreover, in this work, 100 keV He implantation was done with a fluence of 5×10^{16} cm^{-2}. This resulted in the formation of implantation-induced extended defects. Such defects are in the form of microcracks located within the damage band after post-implantation annealing at 250 °C (Fig. 8). These microcracks after post-implantation annealing at 250 °C for sufficient long time of about 30 min undergo agglomeration and are responsible for the surface blistering/exfoliation (Singh et al., 2006).

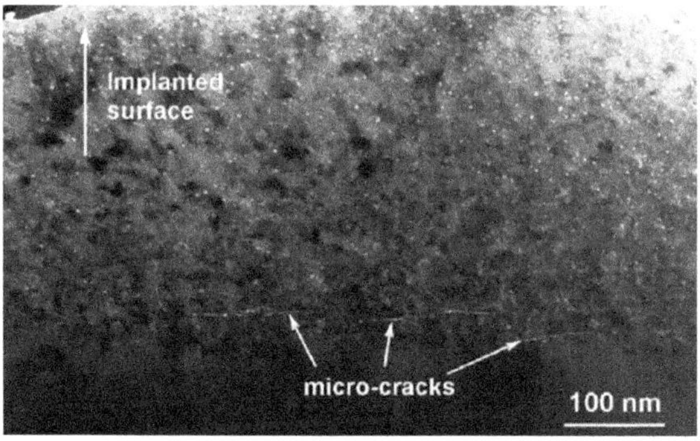

Fig. 8. XTEM image of the damage band of 100 keV He implanted InP with a fluence of 5×10^{16} cm^{-2} and annealed at 250 °C for 25 min (Singh et al., 2006). Reprinted with permission from Singh et al., (2006), Elsevier

He ion implantation in combination with direct wafer bonding was also used for the transfer of InP layer onto Si. It results in the formation of InP-on-insulator engineered substrate. For instance, He implanted InP wafer with spin-on-glass (SOG) layer was used in the wafer bonding with oxidized Si wafer. The post-implantation annealing of this bonded pair at relatively low temperature of 200 °C for sufficiently long time of 20 hr results in the transfer of thin InP layers onto the Si substrate (Singh et al., 2010). This provides an attractive platform for the fabrication of single crystalline layers of device quality which has potential to reduce the cost of fabricated devices based on this InP-on-insulator technology. The other way of layer transfer of InP can be done by using lattice matched InGaAs. The trick involves in this technique is to keep away the InP from ion implantation and actually perform the H-implantation in InGaAs using InP/InGaAs heterostructure (Chen & Jing et al., 2008; Dadwal et al., 2011). The energy of the implanted H ions is chosen in such a manner that it creates maximum damage only within the InGaAs. The advantage of using InGaAs is to achieve the control blistering only after post-implantation annealing at the elevated temperature due to the weak dependence of H diffusivity in it. Therefore, the H-implanted InP/InGaAs heterostructure can be used potentially for the transfer of thin InP layers on the in-expensive substrate like Si using direct wafer bonding technique and subsequent annealing. This allows the fabrication of high quality InP layers which are almost free from any growth related defects, and hence it results in the decrease of carrier scattering with enhanced device performance (Chen & Jing et al., 2008; Dadwal et al., 2011).

5. GaN

The nitride technology is quite dominant in the field of optoelectronics, high frequency and high power devices (Padilla et al., 2010). In the main list it includes blue, deep ultra violet (UV) high brightness light emitting diodes (HBLEDs), laser diodes (LDs), UV detectors and high electron mobility transistors (HEMTs). This has significant impact on the modern high definition (HD) display and solid state lighting. However, the main barrier in the commercialization of these nitride devices is in terms of their cost. The reasons of this are lack of availability of the high quality materials and various technical issues related to the manufacturing of their devices. In particular, GaN has drawn lot of interest in the today nitride technology from the prospect of ion-cut process in the III-V nitrides. Inspite of highly expensive free standing fs-GaN substrate, using ion-cut process, one can re-use the same substrate in multiple times which ultimately reduces the cost of fabricated devices. With this advantage, the applications of ion cut process in GaN are mostly explored by the H-implantation but on a small scale (Moutanabbir et al., 2010). Hence in order to fully employ this ion-cut process in GaN, understanding of the layer splitting mechanisms is very essential, which can be done by studying the implantation-induced surface blistering/exfoliation (Moutanabbir et al., 2010). When GaN is implanted by H ions with suitable implantation parameters followed by annealing at particular parameters, it results surface blistering in the as-implanted state as well as after post-implantation annealing (Fig. 9) (Dadwal et al., manuscript under preparation).

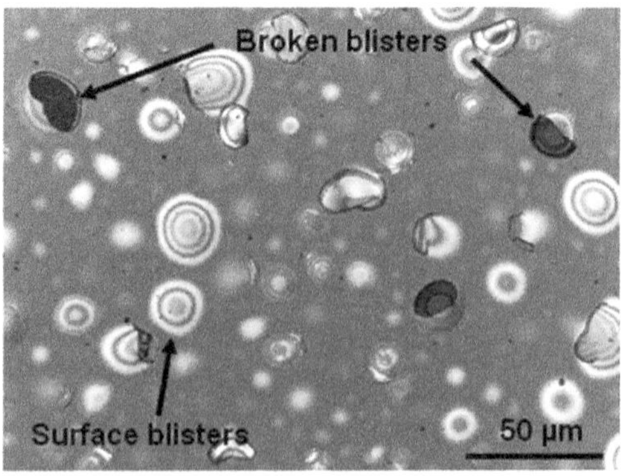

Fig. 9. Nomarski optical image of the 100 keV H^+ ion implanted GaN with a fluence of 2.5×10^{17} cm^{-2} at RT after post-implantation annealing at 500 °C for 30 min (Dadwal et al., manuscript under preparation).

H-implantation in GaN usually requires a higher fluence for the surface blistering/layer splitting to occur after post-implantation annealing at higher temperature. The critical value of a fluence in the 100 keV H implanted GaN is about 2.5×10^{17} cm^{-2} (Tauzin et al., 2005). This minimum required fluence is order of magnitude higher than the conventional H implanted semiconductors like Si and GaAs, respectively. There are many reasons which are responsible for this higher H-implantation fluence in GaN, such as very efficient dynamic annealing of the implantation-induced defects, interaction and trapping of the H with the induced defects and some particular properties of the GaN (Kucheyev et al., 2001). Moreover, due to the efficient dynamic annealing nature of the induced defects, they undergo defects annihilation even during the implantation if the implantation temperature increases to a higher value. Hence, higher temperature implantation can result in easier agglomeration of the induced defects, which shows surface blistering/exfoliation in the as-implanted state. This surface blistering/exfoliation is due to the formation of the H-induced defects within the damage region. The damage region investigations have shown that these induced defects could be of the type of point defects V_{Ga}, point and H defects complexes $V_{Ga}V_N$, V_{Ga}-H_n for n≤4 and eventually planer defects like nanovoids (Moutanabbir et al., 2010) (Fig. 10a). After post-implantation annealing or H-implantation at higher temperature results in the agglomeration of these defects to microcracks lying along different planes within the damage band (Fig. 10b). These H-induced microcracks in an overpressurized state undergo overlapping and combination to each other, and hence resulted in surface blistering/exfoliation. However, the way with which H interacts with the lattice atoms during the implantation and after post-implantation annealing is a subject of further study.

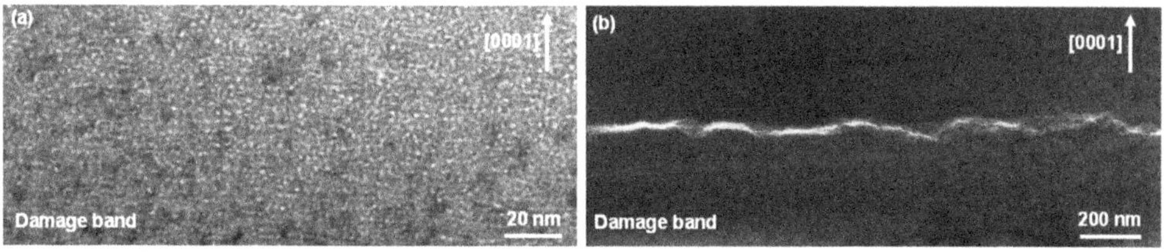

Fig. 10. High magnification XTEM images of the damage band in the 100 keV H^+ ion implanted GaN with a fluence of 2.5×10^{17} cm^{-2} at (a) RT and (b) 300 °C in the as-implanted state (Dadwal et al., manuscript under preparation).

The study of H-implantation-induced blistering can be extendable to the layer transfer work also using direct wafer bonding. The epitaxial grown thin film of GaN on sapphire has shown a layer transfer onto the foreign sapphire substrate (Tauzin et al., 2005). However in the case of free standing fs-GaN substrate, only partial area layer transfer is achievable so far (Moutanabbir et al., 2010). One of a main obstacle of this is the implantation-induced in-plane compressive stress within the damage band, which leads to bowing of the whole GaN wafer away from the implanted surface. For example, in the case of two inch H implanted fs-GaN wafer, the RT H-implantation results in increase of the wafer bow from its original value of 1–2 µm to higher value of more than 20 µm, which makes difficulty in its bonding with other wafer. One of the suggested method in order to minimize this wafer bowing problem is by performing H-implantation in the back side of the already H implanted GaN wafer. This can results in decrease of wafer bow of the H implanted GaN wafer with effective bow value is sufficiently lower for performing direct wafer bonding (Moutanabbir et al., 2010). Hence for the success of ion-cut process in GaN, understanding of the implantation-induced stress behaviour is very important. In this context the latest studies have shown that, implantation-induced stress is strongly depends upon the implantation temperature (Padilla et al., 2010). The analysis showed that there exists biaxial stress in the damage region for the implantation at different temperatures. This means that the out-of-plane tensile strain induces in-plane compressive stress in the damage region. The measured compressive stress was about -1.0 GPa in the LN_2 and about -0.5 GPa in the RT implanted GaN samples after post-implantation annealing at 500 °C for 30 min (Dadwal et al., manuscript under preparation). The H-implantation at low temperature have very less dynamic annealing effects, which results in large implantation-induced damage and hence large induced stress in comparison to the RT implanted GaN. Therefore, the RT implantation could be an effective way for the GaN layer splitting, but only partial success has achieved. This ion-cut process using fs-GaN wafer has unique technological importance because once this process is optimized completely, it will enable various types of highly efficient and low cost nitride devices.

6. AlN

The nitride technology has also attracted great attention towards AlN next to GaN due to its higher direct bandgap and thermal conductivity, and along with some other special physical properties (Dadwal et al., 2010). Hence, it makes AlN as a perfect material for the fabrication of high power lasers and highly efficient UV optoelectronic devices. Another advantage of AlN is its high melting point which is almost double than Si. This makes it as an excellent candidate for the device applications which need to work at high temperatures. In addition, its radiation hardness aspect makes it suitable material for space and military applications. However, the AlN technology is still in their infancy stages and it needs a greater look-up to study this material not only in the technological but material point of view also (Dadwal et al., 2010). But the fs-AlN wafers are very expensive and are mostly available upto two inch of diameter. Hence working on these lines, the recent studies are concentrated more towards the material aspects of AlN epitaxial layers and the possibilities of combining it with ion-cut process (Dadwal et al., 2010). Once the ion-cut process gets its maturity in AlN, one can think about the reduction of cost of this AlN technology. Even the high quality epitaxial layers of AlN in combination with ion-cut process have a potential for the fabrication of low cost nitride devices. That is why, first part of the ion-cut process which is ion implantation-induced damage study is mostly considered in the recent studies (Dadwal et al., 2010; Singh et al., 2010). It mostly includes the influence of H-implantation and post-implantation annealing parameters on the induced damage. These investigations show that depending upon the implantation and annealing parameters, H-implantation in thin film AlN results in the formation of surface blistering/exfoliation either in the as-implanted state or after post-implantation annealing at higher temperature (Fig. 11).

In addition to the surface blistering/exfoliation in H implanted AlN, large area exfoliation is also observed after post-implantation annealing for the implantation at higher temperature of 300 °C (Fig. 11b). It means that the exfoliated area extends over hundreds of square micrometers. This is a special type of the surface deformation in the H implanted AlN and is not observed in its other family of materials like GaN. Current investigations have shown that such type of deformation in the H implanted AlN is probably due to the nature of

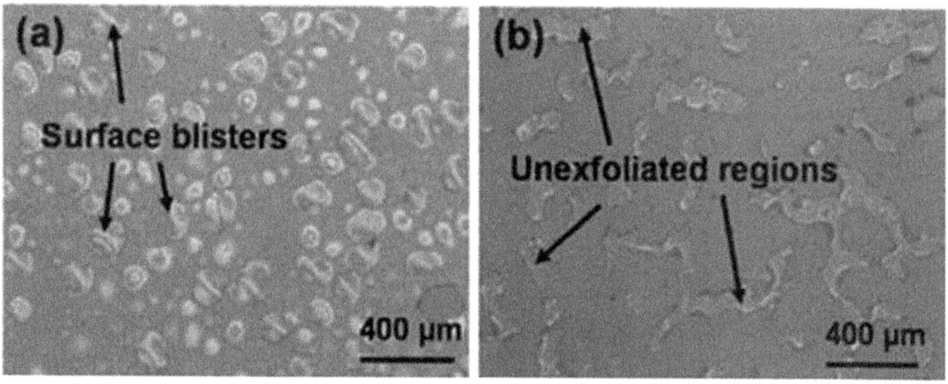

Fig. 11. Nomarski optical microscopy images of the 100 keV H^+ ion implanted AlN with a fluence of 1×10^{17} cm^{-2} at (a) RT and (b) 300 °C after post-implantation annealing at 800 °C for 1hr (Dadwal et al., 2010).

extended defects like nano/microcracks distributions with in the narrow damage band (Fig. 12). The damage band is decorated with the H filled nanovoids which serves as precursors for the surface blistering/exfoliation (Dadwal et al., 2010). These nanovoids after post-implantation annealing agglomerate to nanocracks. This agglomeration happens parallel to the implanted surface and is clearly shown in the Fig. 12b. These nanocracks finally converted to the microcracks after post-implantation annealing for the sufficient long time and which are responsible for the surface blistering/exfoliation. However, the further investigations are underway in order to understand the large area exfoliation phenomenon in detail. But if we consider these surface deformations from the ion-cut point of view, then after direct wafer bonding there is no any practical difference between the surface exfoliation and large area exfoliation. This is because of the fact that the host wafer acts as a stiffener and it allow the implantation-induced surface deformations to migrate more towards the lateral manner and parallel to the bonded interface during the annealing of the bonded wafers at elevated temperature.

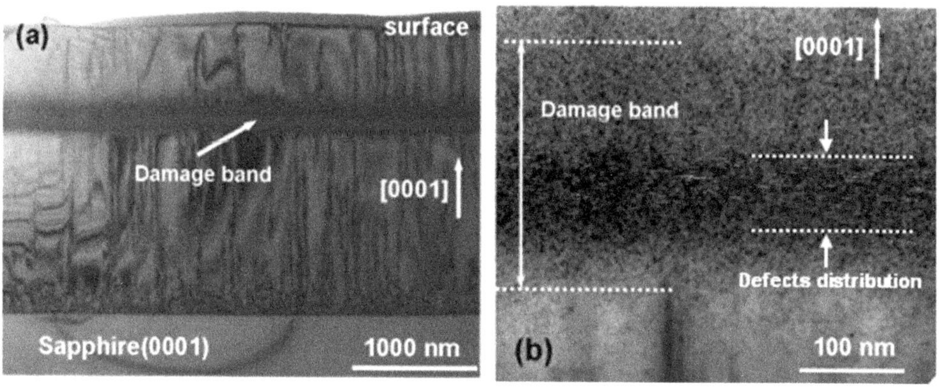

Fig. 12. (a) XTEM images of the damage band in the 100 keV H^+ ion implanted AlN at 300 °C with a fluence of 1×10^{17} cm^{-2} and (b) High magnification TEM image of the same damage band showing the narrow distribution of the extended defects (Dadwal et al., 2010).

The value of a critical fluence for the surface blistering/exfoliation in H implanted AlN after post-implantation annealing is 1×10^{17} cm^{-2} (Dadwal et al., 2010). This value of the critical fluence is less than the H implanted GaN but higher than the conventional semiconductors like Si. Since AlN and GaN has a same chemical structure, so one would expect the same reasons could be also applicable for the high implantation fluence in the H implanted AlN. The other aspect of AlN is the higher surface roughness of its thin film and of fs-AlN wafer. The typical value of a root mean square surface roughness is about 25 nm, which is a very higher and is not acceptable for the direct wafer bonding. However, unlike Si chemical mechanical polishing (CMP), the surface planarization of AlN is difficult and its available technology is different and expensive than Si. In the device point of view, AlN technology is still lacking behinds the other nitride materials. But the device quality fs-AlN wafer in combination with ion-cut process have definite possibilities to fabricate high quality AlN thin layers, which can provides fruitful platform for the low cost AlN technology.

7. Other semiconductors

Ion-cut process has also been extended to some other technological important semiconductors, like ZnO, GaSb and CdZnTe. In most of these semiconductors H-implantation is explored in the initial investigations of the implantation-induced surface blistering/exfoliation, which is then used in the layer transfer process (Hobart et al., 1999). In this section a brief overview of the ion-cut process in these semiconductors is given.

7.1 ZnO

H-implantation in ZnO is considered to be another valuable mean for the use of this material in the high temperature operating devices just like AlN. However, H-implantation in ZnO is mainly restricted onto the study of implantation-induced surface blistering/exfoliation. This is because of ion-cut process in ZnO is not completely investigated, especially related to wafer bonding part. In addition to this, the higher H-implantation fluence requirement in ZnO makes large damage to the crystal structure, which still needs greater investigations to apply this material for the layer transfer purpose. Like GaN and AlN, H-implantation in ZnO also results surface blistering/exfoliation depending upon the H fluence and annealing parameters. But, the minimum fluence for the surface blistering/exfoliation to occur after post-implantation annealing is about 2.5×10^{17} cm^{-2} H_2^+ ions (Singh et al., 2010). This minimum fluence requirement is even larger than nitride semiconductors like GaN and is could be due to the efficient dynamic annealing of the induced defects. The implantation of H ions less than this critical fluence does not show any surface blistering even after post-implantation annealing at higher temperature. However, H-implantation with a fluence of 2.8×10^{17} cm^{-2} results surface exfoliation in the as-implanted state (Singh et al., 2010). The investigation of H-implantation-induced damage shows the formation of H-induced microcracks within the damage band in the as-implanted state, which upon thermal annealing at higher temperature result in the formation of surface blistering/exfoliation (Fig. 13). The thickness of the exfoliated ZnO layer is about 450 nm, which is close to the projected range 400 nm of the 100 keV hydrogen ions in ZnO. This projected range is calculated by the Stopping and Ranges of Ions in Matter (SRIM2003) simulation program.

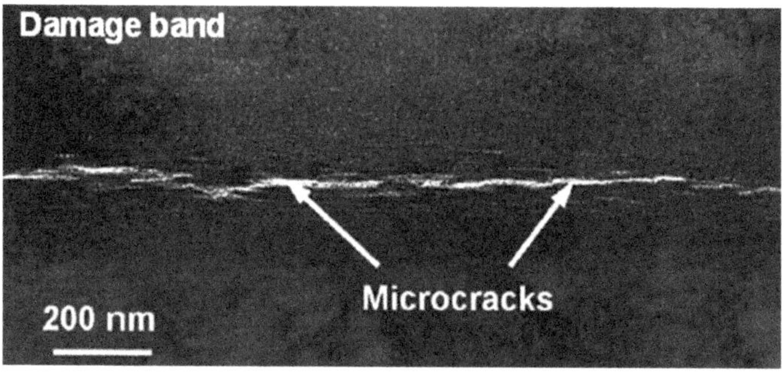

Fig. 13. XTEM image of the damage band in the 100 keV H_2^+ ion implanted ZnO with a fluence of 2.5×10^{17} cm^{-2} in the as-implanted state (Singh et al., 2010). doi:10.1088/0268-1242/22/11/003

The higher implantation fluence for the surface blistering in ZnO is directly related to the higher implantation time requirement to fulfil the required fluence. This means that implanted H ions needs sufficient time to interact with the host lattice atoms and form defect sites, which then acts as source for the extended defects like microcracks formation during the higher temperature treatment. The layer splitting of ZnO is still not achieved. Therefore it needs thorough study of the surface blistering/exfoliation and its wafer bonding.

7.2 GaSb

GaSb epitaxial layers can be transferred using H ion implantation and direct wafer bonding technique. The H-implantation in GaSb with a fluence of 5~8 × 10^{16} cm^{-2} at RT results surface blistering after post-implantation annealing at 60–600 °C (Hobart et al., 1999; Zheng et al., 2000). The thermal annealing treatment of the H implanted GaSb is studied for time 1 to 80 min under both isothermal and isochronal conditions. In both annealing conditions there is an increase of the lateral size of surface blisters/exfoliated regions (Zheng et al., 2000). The blistering kinetics study showed activation energy value each for the higher and lower annealing temperature regime. That is 0.30 eV in the annealing temperature range of 90–230 °C and 1.8 eV in the annealing temperature lower than 90 °C (Zheng et al., 2000). The higher temperature activation energy is related to diffusion of the implanted H in the damage lattice. Whereas lower temperature activation energy may be governed by the trapping-detrapping mechanisms of the induced defects similar to H implanted Si. The onset time for the surface blistering at higher annealing temperature is much shorter than the annealing at the lower temperature. These parameters of the blistering study are applied in the thin layer transfer of n-type GaSb onto GaAs wafer by using borosilicate glass (BSG) as an intermediate layer. The H implanted GaSb wafer is than bonded with GaAs at RT followed by annealing at 100 °C for 2 hr. The final layer splitting is then achieved by annealing of the bonded wafers at 160 °C for 30 min (Zheng et al., 2000).

7.3 CdZnTe

H-implantation-induced surface blistering/exfoliation study is also extended to ternary semiconductors like CdZnTe (Miclaus et al., 2005). Again, the main motive of these studies is to find out the possibilities of the layer transfer of CdZeTe depending upon its constituents molar concentration. For example, H-implantation in $Cd_{0.96}Ze_{0.04}Te$ is studied for the implantation energy in the range of 40–200 keV with an ion fluence of 5×10^{17} cm^{-2} (Miclaus et al., 2005). The co–implantation work using boron (B) (5×10^{15} cm^{-2}, 147 keV) and H (5×10^{16} cm^{-2}, 40 keV) is also used for the implantation at 77 or 273 K (Miclaus et al., 2005). The H implanted samples at low temperature of 77 K results in surface blisters after post-implantation annealing at 150–300 °C. However, implantation at 253 K does not result surface blistering even after post-implantation annealing at higher temperature. This pronounced blistering phenomenon at lower implantation temperature is could be due to the less efficient dynamic annealing of the induced defects. In addition, co-implantation of H with boron (B) at lower temperature results in surface blistering with an advantage that blistering occurs at lesser annealing time for an annealing temperature which depends upon both implanted ions fluence. The increase in H fluence leads to the decrease of implantation-induced blistering time. This means that B-implantation promotes the damage formation and H-induced internal pressure within the damage region for the surface blistering (Miclaus et al., 2005).

8. Conclusions and future perspective

We have presented an overview of the ion implantation-induced surface blistering/exfoliation and their use for the layer splitting of the different technological important compound semiconductors. In most of these semiconductors H-implantation-induced blistering/exfoliation was investigated for the potential application in ion-cut process. In the case of Si, Ge and SiGe alloys, both controlled blistering as well as layer splitting/transfer has been successfully demonstrated in the earlier studies. However, in the case of III-V semiconductors like GaN and AlN, they are reportedly very resilient to H-implantation-induced damage due to the dynamic annealing effects. The critical H fluence for the surface blistering in GaN is 2.5×10^{17} cm^{-2}, whereas in AlN it is 1×10^{17} cm^{-2}. However blistering kinetics shows that, the blistering occurs at lower post-implantation annealing temperature in the case of H implanted GaN in comparison to the H implanted AlN. This fact is very important in the case of ion-cutting process because of the heterogeneous wafer bonding where low temperature annealing is preferable. This blistering study has its main significance in the layer splitting of these nitride semiconductors for the establishment of low cost nitride technology. The free standing wafers of GaN and AlN are very expensive and are mostly available in the small sizes. But using these free standing wafers one can transfer its multiple layers onto the foreign inexpensive substrate. As a result these free standing wafers can be utilized for multiple times which ultimately reduce the cost of fabricated nitride devices. The H-implantation in others compound semiconductors like GaAs and InP for the layer splitting purpose is more or less not appropriate due to its high diffusivity in the damage region. This makes the process very sensitive to the implantation temperature and if implantation parameters go beyond the critically defined window parameters, then no blistering/exfoliation will occur at all. Hence during H-implantation, the temperature at which implantation is perform should be properly checked in order to achieve the controlled surface blistering only after post-implantation annealing. However, He-implantation is an appropriate alternative in order to achieve the controlled surface blistering and layer splitting of GaAs and InP. This is because the He is less mobile within the implantation-induced damage, which results its less sensitivity towards the implantation temperature. Hence using He-implantation and direct wafer bonding technique, GaAs layer has been successfully transferred onto the foreign substrate Si. Similarly using He-implantation, InP layer transfer is also successfully demonstrated onto the Si substrate. In addition to this, the layer transfer of nitride semiconductors like GaN and AlN is not fully achieved. Only the epitaxial layers of GaN has been successfully transferred onto the sapphire substrate using H-implantation and direct wafer bonding technique. However, till now the layer transfer from the free standing GaN substrate is only partially attained. The minimum critical fluence for the controlled surface blistering in the H implanted GaAs and InP is about two times lesser than the H implanted GaN, AlN and ZnO. Moreover, the fundamental mechanisms behind the H ion implantation-induced surface blistering/exfoliation in these compound semiconductors are not fully understood, even though in some of the compound semiconductors the layer transfer is fully achieved such as GaAs and InP. In fact, the H-induced layer splitting mechanisms is not completely understood even in the conventional semiconductors also like Si and Ge. Hence, further investigations are required in these technological important semiconductors in order to fully explore the advantages of the ion-cut process.

9. Acknowledgment

This work is financially supported by the Max Planck Society (MPS), Germany under the cooperative scheme of Max Planck India Partner Group. U. Dadwal is thankful to the MPS, Germany for the research fellowship.

10. References

Akatsu, T.; Bourdelle, K. K.; Richtarch, C.; Faure, B. & Letertre, F. (2005). Study of Extended-Defects Formation in Ge and Si Ion Implantation. *Applied Physics Letters,* Vol.86, No.18, pp. 181910-181910-3, ISSN 0003-6951

Bruel, M. (1995). Silicon on Insulator Material Technology. *Electronics Letters,* Vol.31, No.14, pp. 1201-1202, ISSN 0013-5194

Chabal, Y. J.; Weldon, M. K.; Caudano, Y.; Stefanov, B. B. & Raghavachari, K. (1999). Spectroscopic Studies of H-Decorated Interstitials and Vacancies in Thin-Film Silicon Exfoliation. *Physics B,* Vol.273-274, pp. 152-163, ISSN 0921-4526

Chen, P.; Di, Z.; Nastasi, M.; Bruno, E.; Grimaldi, M. G.; Theodore, N. D. & Lau, S. S. (2008). Effect of Hydrogen Implantation Temperature on InP Surface Blistering. *Applied Physics Letters,* Vol.92, No.20, pp. 202107-202107-3, ISSN 0003-6951

Chen, P.; Jing, Y.; Lau, S. S.; Xu, D.; Mawsk, L.; Alford, T. L.; Paulson, C. & Kuech, T. F. (2008). High Crystalline-Quality III-V Layer Transfer onto Si Substrate. *Applied Physics Letters,* Vol.92, No.9, pp. 092107-092107-3, ISSN 0003-6951

Christiansen, S. H.; Singh, R. & Gösele, U. (2006). Wafer Direct Bonding: From Advanced Substrate Engineering to Future Applications in Micro/Nanoelectronics, *Proceedings of IEEE,* ISSN 0018-9219, December 2006

Cioccio, L. D.; Tiec, Y. L.; Letertre, F.; Jaussaud, C. & Bruel, M. (1996). Silicon Carbide on Insulator Formation by The Smart-Cut® Process. *Electronics Letters,* Vol.32, No.12, pp. 1144-1145, ISSN 0013-5194

Dadwal, U.; Scholz, R.; Goesele, U. & Singh, R. (2009). Effect of Implantation Temperature on The Blistering/Exfoliation Behavior of Hydrogen Implanted Si and Ge, *Proceedings of the XVth IWPSD*, ISBN 978-93-80043-55-5, New Delhi, December 15-19, 2009

Dadwal, U.; Scholz, R.; Kumar, P.; Kanjilal, D.; Christiansen, S.; Gösele, U. & Singh, R. (2010). Hydrogen Implantation-Induced Large Area Exfoliation in AlN Epitaxial Layers. *Physics Status Solidi A,* Vol.207, No.1, pp. 29-32, ISSN 1862-6300

Dadwal, U.; Kumar, A.; Scholz, R.; Reiche, M.; Kumar, P.; Boehm, G.; Amann, M. C. & Singh, R. (2011). Blistering Study of H-Implanted InGaAs for Potential Heterointegration Applications. *Semiconductor Science and Technology,* Vol.26, No.8, pp. 085032-085032-5, ISSN 0268-1242

David, M. L.; Pizzagalli, L.; Pailloux, F. & Barbot J. F. (2009). Atomic Scale Structure of (001) Hydrogen-Induced Platelets in Germanium. *Physics Review Letters,* Vol.102, No.15, pp. 155504-155504-4, ISSN 0031-9007

Deguet, C.; Sanchez, L.; Akatsu, T.; Allibert, F.; Dechamp, J.; Madeira, F.; Mazen, F.; Tauzin, A.; Loup, V.; Richtarch, C.; Mercier, D.; Signamarcheix, T.; Letertre, F.; Depuydt, B. & Kernevez, N. (2006). Fabrication and Characterisation of 200 mm Germanium-on-Insulator (GeOI) Substrates Made From Bulk Germanium. *Electronics Letters,* Vol.42, No.7, pp. 415-417, ISSN 0013-5194

Hayashi, S.; Bruno, D. & Goorsky, M. S. (2004). Temperature Dependence of Hydrogen-Induced Exfoliation of InP. *Applied Physics Letters,* Vol.85, No.2, pp. 236-238, ISSN 0003-6951

Hayashi, S.; Sandhu, R.; Bruno, D.; Wojtowicz, M.; & Goorsky, M. S. (2005). Effect of Implantation Heating on Exfoliation of InP. *Sensors and Materials,* Vol.17, No.6, pp. 335-341, ISSN 0914-4935

Hobart, K. D. & Kub, F. J. (1999). Transfer of GaSb Thin Film to Insulating Substrate Via Seperation by Hydrogen Implantation. *Electronics Letters,* Vol.35, No.8, pp. 675-676 ISSN 0013-5194

Höchbauer, T.; Misra, A.; Nastasi, M. & Mayer, J. W. (2002). Physical Mechanisms behind the Ion-Cut in Hydrogen Implanted Silicon. *Journal of Applied Physics,* Vol.92, No.5, pp. 2335-2342, ISSN 0021-8979

Huang, L. J.; Chu, J. O.; Canaperi, D. F.; D'Emic, C. P.; Anderson, R. M.; Koester, S. J. & Wong, H. S. P. (2001). SiGe-on-Insulator Prepared by Wafer Bonding and Layer Transfer for High-Performance Field-Effect Transistors. *Applied Physics letters,* Vol.78, No.9, pp. 1267-1269, ISSN 0003-6951

Kucheyev, S. O.; Williams, J. S. & Pearton, S. J. (2001). Ion Implantation into GaN. *Materials Science and Engineering,* Vol.33, No.2-3, pp. 51-107, ISSN 0927-796X

Lee, J. K.; Nastasi, M.; Theodore, N. D.; Smalley, A.; Alford, T. L.; Mayer, J. W.; Cai, M. & Lau, S. S. (2004). Effects of Hydrogen Implantation Temperature on Ion-Cut of Silicon. *Journal of Applied Physics,* Vol.96, No.1, pp. 280-288, ISSN 0021-8979

Miclaus, C.; Malouf, G.; Johnson, S. M. & Goorsky, M. S. (2005). Exfoliation and Blistering of $Cd_{0.96}Zn_{0.04}Te$ Substrates by Ion Implantation. *Journal of Electronic Materials,* Vol.34, No.6, pp. 859-863, ISSN 0361-5235

Moras, G.; Ciacchi, L. C.; Elsässer, C.; Gumbsch, P. & Vita, A. D. (2010). Atomically Smooth Stress-Corrosion Cleavage of a Hydrogen-Implanted Crystal. *Physical Review letters,* Vol.105, No.7, pp. 075502-075502-4, ISSN 0031-9007

Moutanabbir, O. & Gösele, U. (2010). Heterogeneous Integration of Compound Semiconductors. *Annual Review of Materials Research,* Vol.40, pp. 469-500, ISSN 1531-7331

Padilla, E.; Jackson, M. & Goorsky, M. S. (2010). The Role of The Nucleation Annealing Temperature Annealing on The Exfoliation of Hydrogen-Implanted GaN. *Electrochemical Society Transactions,* Vol.33, No.4, pp. 263-270, ISSN 1938-5862

Pankove, J. I. & Johnson, N. M. (1991). *Hydrogen in Semiconductors,* Academic Press, ISBN 0127521348, Boston

Plummer, J. D.; Michael, D. D. & Griffin, B. P. (2000). *Silicon VLSI Technology,* Prentice Hall, ISBN 0130850373, Englewood Cliffs

Radu, I.; Szafraniak, I.; Scholz, R.; Alexe, M. & Gösele, U. (2003). GaAs on Si Heterostructures Obtained by He and/or H Implantation and Direct Wafer Bonding. *Journal of Applied Physics,* Vol.94, No.12, pp. 7820-7825, ISSN 0021-8979

Singh, R.; Radu, I.; Reiche, M.; Scholz, R.; Webb, D.; Gösele, U. & Christiansen, S. H. (2005). Investigation of Hydrogen Implantation-Induced Blistering in SiGe. *Materials Science and Engineering B,* Vol.124-125, pp. 162-165, ISSN 0921-5107

Singh, R.; Radu, I.; Scholz, R.; Himcinschi, C.; Gösele, U. & Christiansen, S. H. (2006). Investigation of Helium Implantation Induced Blistering in InP. *Journal of Luminescence,* Vol.121, pp. 379-382, ISSN 0022-2313

Singh, R.; Christiansen, S. H.; Moutanabbir, O. & Gösele, U. (2010). The phenomenology of Ion Implantation-Induced Blistering and Thin-Layer Splitting in Compound Semiconductors. *Journal of Electronic Materials,* Vol.39, No.10, pp. 2177-2189, ISSN 0361-5235

Singh, R.; Scholz, R.; Christiansen, S.; Mantl, S. & Reiche, M. (2011). Investigation of a Hydrogen Implantation-Induced Blistering Phenomenon in $Si_{0.70}Ge_{0.30}$. *Semiconductor Science and Technology,* Vol.26, No.12, pp. 125001-125001-5, ISSN 0268-1242

Sze, S. M. (1988). *VLSI Technology,* Second Edition, McGraw-Hill, ISBN 0070582912, New York

Tauzin, A.; Akatsu, T.; Rabarot, M.; Dechamp, J.; Zussy, M.; Moriceau, H.; Michaud, J. F.; Charvet, A. M.; Cioccio, L. D.; Fournel, F.; Garrione, J.; Faure, B.; Letertre, F. & Kernevez, N. (2005). Transfers of 2-Inch GaN Films onto Sapphire Substrates Using Smart Cut™ Technology. *Electronics Letters,* Vol.41, No.11, pp. 668-670, ISSN 0013-5194

Tong, Q. Y. & Gösele, U. (1999). *Semiconductor Wafer Bonding: Science and Technology,* Wiley, ISBN 0471574813, New York

Tong, Q. Y.; Chao, Y. L.; Huang, L. J. & Gösele, U. (1999). Low Temperature InP Layer Transfer. *Electronic Letters,* Vol.35, No.4, pp. 341-342, ISSN 0013-5194

Zahler, J. M.; Morral, A. F. I.; Griggs, M. J.; Atwater, H. A. & Chabal, Y. J. (2007). Role of Hydrogen in Hydrogen-Induced Layer Exfoliation of Germanium. *Physical Review B,* Vol.75, pp. 035309-035309-10, ISSN 1098-0121

Zheng, Y.; Moran, P. D.; Guan, Z. F.; Lau, S. S.; Hansen, D. M.; Kuech, T. F.; Haynes, T. E.; Hoechbauer, T. & Nastasi, M. (2000). Transfer of n-Type GaSb onto GaAs Substrate by Hydrogen Implantation and Wafer Bonding. *Journal of Electronic Materials*, Vol.29, No.7, pp. 916-920, ISSN 0361-5235

www.soitec.com

Permissions

The contributors of this book come from diverse backgrounds, making this book a truly international effort. This book will bring forth new frontiers with its revolutionizing research information and detailed analysis of the nascent developments around the world.

We would like to thank Prof Mark Goorsky, for lending his expertise to make the book truly unique. He has played a crucial role in the development of this book. Without his invaluable contribution this book wouldn't have been possible. He has made vital efforts to compile up to date information on the varied aspects of this subject to make this book a valuable addition to the collection of many professionals and students.

This book was conceptualized with the vision of imparting up-to-date information and advanced data in this field. To ensure the same, a matchless editorial board was set up. Every individual on the board went through rigorous rounds of assessment to prove their worth. After which they invested a large part of their time researching and compiling the most relevant data for our readers. Conferences and sessions were held from time to time between the editorial board and the contributing authors to present the data in the most comprehensible form. The editorial team has worked tirelessly to provide valuable and valid information to help people across the globe.

Every chapter published in this book has been scrutinized by our experts. Their significance has been extensively debated. The topics covered herein carry significant findings which will fuel the growth of the discipline. They may even be implemented as practical applications or may be referred to as a beginning point for another development. Chapters in this book were first published by InTech; hereby published with permission under the Creative Commons Attribution License or equivalent.

The editorial board has been involved in producing this book since its inception. They have spent rigorous hours researching and exploring the diverse topics which have resulted in the successful publishing of this book. They have passed on their knowledge of decades through this book. To expedite this challenging task, the publisher supported the team at every step. A small team of assistant editors was also appointed to further simplify the editing procedure and attain best results for the readers.

Our editorial team has been hand-picked from every corner of the world. Their multi-ethnicity adds dynamic inputs to the discussions which result in innovative outcomes. These outcomes are then further discussed with the researchers and contributors who give their valuable feedback and opinion regarding the same. The feedback is then collaborated with the researches and they are edited in a comprehensive manner to aid the understanding of the subject.

Apart from the editorial board, the designing team has also invested a significant amount of their time in understanding the subject and creating the most relevant covers. They scrutinized every image to scout for the most suitable representation of the subject and create an appropriate cover for the book.

The publishing team has been involved in this book since its early stages. They were actively engaged in every process, be it collecting the data, connecting with the contributors or procuring relevant information. The team has been an ardent support to the editorial, designing and production team. Their endless efforts to recruit the best for this project, has resulted in the accomplishment of this book. They are a veteran in the field of academics and their pool of knowledge is as vast as their experience in printing. Their expertise and guidance has proved useful at every step. Their uncompromising quality standards have made this book an exceptional effort. Their encouragement from time to time has been an inspiration for everyone.

The publisher and the editorial board hope that this book will prove to be a valuable piece of knowledge for researchers, students, practitioners and scholars across the globe.

List of Contributors

Sébastien Pezzagna and Jan Meijer
RUBION, Ruhr-Universität Bochum, Germany

R. C. Fadanelli, M. Behar and J. F. Dias
Universidade Federal do Rio Grande do Sul, Brazil

Yuancun Nie
State Nuclear Power Technology R&D Center, Beijing, China
State Key Lab of Nuclear Physics and Technology (Peking University), Beijing, China

Yuanrong Lu, Xueqing Yan and Jiaer Chen
State Key Lab of Nuclear Physics and Technology (Peking University), Beijing, China

Denis Shamiryan and Dmitriy V. Likhachev
Globalfoundries, Dresden, Germany

Caroline Moulet and Mark S. Goorsky
University of California, Los Angeles, USA

Anna Wolska
Institute of Physics, Polish Academy of Sciences, Poland

Stefania Maria Serena Privitera
Consiglio Nazionale delle Ricerche (CNR), Istituto di Microelettronica e Microsistemi (IMM), Catania, Italy

Irina Antonova
A.V. Rzhanov Institute of Semiconductor Physics SB RAS, Russia

Abdul Majid
Department of Physics, University of Gujrat, Pakistan

Valery Mikoushkin
Ioffe Institute, the Russian Academy of Sciences, Russia

Katharina Lorenz
Instituto Tecnológico e Nuclear, Instituto Superior Técnico, Universidade Técnica de Lisboa, Sacavém & Centro de Física Nuclear da Universidade de Lisboa, Lisboa, Portugal

Elke Wendler
Institut für Festkörperphysik, Friedrich-Schiller-Universität Jena, Jena, Germany

Ovidio Peña-Rodríguez and José Olivares
Centro de Microanálisis de Materiales (CMAM), Universidad Autónoma de Madrid (UAM), Cantoblanco, Madrid, Spain
Instituto de Óptica, Consejo Superior de Investigaciones Científicas (IO-CSIC), Madrid, Spain

Fernando Agulló-López
Centro de Microanálisis de Materiales (CMAM), Universidad Autónoma de Madrid (UAM), Cantoblanco, Madrid, Spain

Mercedes Carrascosa and Ángel García-Cabañes
Departamento de Física de Materiales C-IV, Universidad Autónoma de Madrid, Madrid, Spain

Antonio Rivera
Instituto de Fusión Nuclear, Universidad Politécnica de Madrid, Madrid, Spain

D. Rémiens and J. Costecalde
Institut d'Electronique de Microélectronique et Nanotechnologies (IEMN), France

D. Deresmes and D. Troadec
Université de Valenciennes et du Hainaut Cambrésis (UVHC), France

Akira Ueda, Richard R. Mu and Warren E. Collins
Fisk University, USA

Umapada Pal
Instituto de Física, Benemérita Universidad Autónoma de Puebla, Puebla, Mexico

Ovidio Peña Rodríguez
Centro de Microanálisis de Materiales (CMAM), Universidad Autónoma de Madrid (UAM), Cantoblanco, Madrid, Spain
Instituto de Óptica, Consejo Superior de Investigaciones Científicas (IO-CSIC), C/ Serrano Madrid, Spain

Hans-Eberhard Zschau and Michael Schütze
Dechema e. V., Karl-Winnacker-Institut, Frankfurt am Main, Germany

U. Dadwal and R. Singh
Department of Physics, Indian Institute of Technology Delhi, Hauz Khas, New Delhi, India

M. Reiche
Max Planck Institute of Microstructure Physics, Halle, Germany